Precision Manufacturing

Precision Manufacturing

David Dornfeld
University of California at Berkeley
Mechanical Engineering
6195 Etcheverry Hall
Berkeley, California 94720

Dae-Eun Lee
Lawrence Berkeley National Laboratory
Mechanical Engineering
1 Cyclotron Road
Berkeley, California 94720

David Dornfeld
University of California at Berkeley
Mechanical Engineering
6195 Etcheverry Hall
Berkeley, California 94720

Dae-Eun Lee
Lawrence Berkeley National Laboratory
Mechanical Engineering
1 Cyclotron Road
Berkeley, California 94720

Library of Congress Control Number: 2007935803

ISBN 978-0-387-32467-8 e-ISBN 978-0-387-68208-2

Printed on acid-free paper.

9 8 7 6 5 4 3 2 1

springer.com

To manufacturing engineers

TABLE OF CONTENTS

PREFACE & ACKNOWLEDGEMENTS

This book is the result of course notes developed for a graduate course on precision manufacturing at Berkeley and taught for the past decade or so. The course was developed to meet the growing need of mechanical engineers, and others, to understand the design and process issues associated with precision machine tools and the fabrication of precision components. It also tried to introduce some fundamental understanding of metrology and the techniques of measurement. These are big topics and we don't claim to cover all to sufficient depth in this book. But, the foundations laid here can be built upon for additional study.

The book is designed to compliment a typical 15 week semester course aimed at upper division and graduate level engineering students. The text is extensively referenced so that original sources can be consulted for more details if desired. At Berkeley, some industrial visits to local precision manufacturing firms and several laboratory exercises were included in the semester. These are not outlined in the book.

A substantial coverage of some of the historical events driving the development of manufacturing and machine tools is presented in Chapter 1. This is both fascinating as history as well as offers insight into why certain machines and processes are the way they are and the context in which they were developed. References to a number of excellent in depth histories of precision machines and instruments are given for more investigation.

This book is not so much an original creative product but the compilation, with insight, order, and some original material added of course, of a large amount of existing material and expertise. The course notes and, ultimately, this book benefited from a great many contributors, experts in precision engineering worldwide and a special few who either lectured at Berkeley in the class, wrote books covering portions of the topics of interest here, or wrote technical papers published in journals and conference proceedings covering important aspects of precision manufacturing. The actual references are given in the text but a few have had a significant impact on the notes and should be mentioned.

A great number of excellent precision engineers have contributed to the book by their lectures, publications and conversations with the authors. Specifically, Professor Pat McKeown of Cranfield University, Professor Ichiro Inasaki of Keio University and Professor Hans-Kurt Tönshoff of the University of Hannover, and the late Professor Juri Tlusty, all spent time visiting Mechanical Engineering at Berkeley as Springer Professors. Their lectures and colloquia as part of their visits offered excellent material to set the tone for and contribute to several sections of this book. Two contributors, Mr. Jim Bryan, retired Chief Metrologist, and Dr. Ken Blaedel, retired precision engineer, both from Lawrence Livermore National Laboratory, lectured to the class, and, through contact at technical meetings and seminal publications in precision engineering, provided much material for the book. The chapter on thermal errors, for example, is based on Dr. Blaedel's short course notes on thermal errors and augmented with Mr. Bryan's technical publications.

Professor Alex Slocum of MIT wrote a first excellent book on Precision Machine Design which was used for the class for several years and serves as a competent reference. It is exceptionally detailed and covers the fundamentals of design of machines, fixtures, tooling and related elements. It does not cover manufacturing processes and, hence, this book was necessary. Similarly, Professor Bharat Bhushan of Ohio State University edited a handbook on nanotechnology that includes an impressive volume of information and detail on most aspects of nano-scale device design, processes and fundamentals. This book references both Slocum and Bhushan generously.

A group of very talented researchers and engineers in precision engineering and machine tool design, control and application contributed to a seven volume review of the state of the art of machine tool technology in the late 1970's. They represented national laboratories, industry and academia from around the world. Many of the same names listed above appear as authors of major sections of that study. Sponsored by the US Air Force and officially coordinated and published in 1980 by Lawrence Livermore National Laboratory (and, sadly, promptly forgotten by most folks it was designed to benefit in academia and industry), it provided a treasure trove of information on topics ranging from machine control to error budgets to sensor technology to business practices in the industry. You will see extensive reference to this amazing study in this book in several chapters.

One of the strings that binds the precision manufacturing community together is the International Academy for Production Engineering (CIRP). Many of the contributors mentioned above are Fellows of this Academy and numerous references to CIRP publications and presentations at the annual meetings provide critical material for this book. CIRP has played a leading role in precision manufacturing research and development for almost 60 years and the CIRP publications, presentations, and conversations of industry, academic, and national laboratory and institute participants in CIRP activities were a treasure trove of basic research and industrial application information.

Finally, many past student researchers, postdoctoral researchers, and visiting scholars to the Laboratory for Manufacturing and Sustainability in Mechanical Engineering at Berkeley have contributed to this book in various ways. You will see specific reference to their work throughout the chapters on sensors, process planning, precision machining processes, precision manufacturing applications and sustainable process design. And, in the laboratory, several engineering interns contributed more practically to the production of the book. A number of undergraduate students helped with preparation of images and figures over the years and one, Ms. Maddie Cousens, lead the major effort to secure permissions for material used from other sources.

The assistance of all of these individuals and contributions are gratefully appreciated by the authors.

David Dornfeld
Dae-Eun Lee

Berkeley, California
August 2007

1 INTRODUCTION TO PRECISION MANUFACTURING

1.1 Precision engineering

Precision manufacturing is a subset of a much larger domain commonly referred to as "precision engineering." A number of definitions of precision engineering exist, depending upon whom you ask and when. Definitions range from the general esoteric to the specific. Evans[1] summarizes a few better known definitions under the general characterization of "doing the best possible work" as follows:

> "...work at the forefront of current technology"
> - from Poulter[1]
>
> "Precision engineering is shooting after the next decimal place."
> - from Liebers[1] and
>
> "...the most important and essential feature of fine mechanisms is small dimensions of working parts and that distinguishing precision instruments - accurate functioning and accurate dimension."
> - from Trylinski[1].

Another useful definition is that found within the Constitution of the American Society of Precision Engineering (ASPE).

> "Precision Engineering is a discipline encompassing the design, development, and measurement of and for high-accuracy components. By extension, the field also includes the design of systems in which high dimensional accuracy is a central concern, as well as the design of machine tools and measuring machines to accomplish the necessary manufacture and measurement."[2]

Most definitions of precision engineering refer to increasingly small and accurate parts characterized by dimensions with very small tolerances. How small? Figure 1.1, from McKeown[3], shows the relative scale of various things found in nature ranging in size from a city on the kilometer scale to atomic diameters in the Angström range. McKeown suggests that the distinction between normal and precision machining maybe the difference between 1 part in 10^4 and 1 part in 10^5. Taniguchi lends additional distinction to the difference as we will see next.

1.2 Precision manufacturing

The classic measure of progress in precision manufacturing over time is the Taniguchi diagram, Figure 1.2, which has appeared in a number of studies. This figure, from Taniguchi[4] illustrates the "attainable" machining accuracies or preciseness over time. Here is the interesting link between precision engineering and precision manufacturing. Whereas precision engineers may focus on the size of an artifact on a workpiece (and its role in the successful operation of a precision instrument, perhaps), the precision manufacturing community concerns itself with the creation of the artifact. Of course, precision machines are needed to create precision components. Taniguchi addresses these issues by casting the improvements in machining accuracies in different classes of processes, "normal," "precision," "high-precision," and "ultra-high precision." These cover the range of processes from turning and milling machines (normal) to ion beam machining (ultra-high precision.) They distinctions are, in fact, the best tolerance on a feature that can be obtained by the process.

Table I, from Taniguchi[4], relates the specific machining accuracies with physical products from a number different domains of industry.

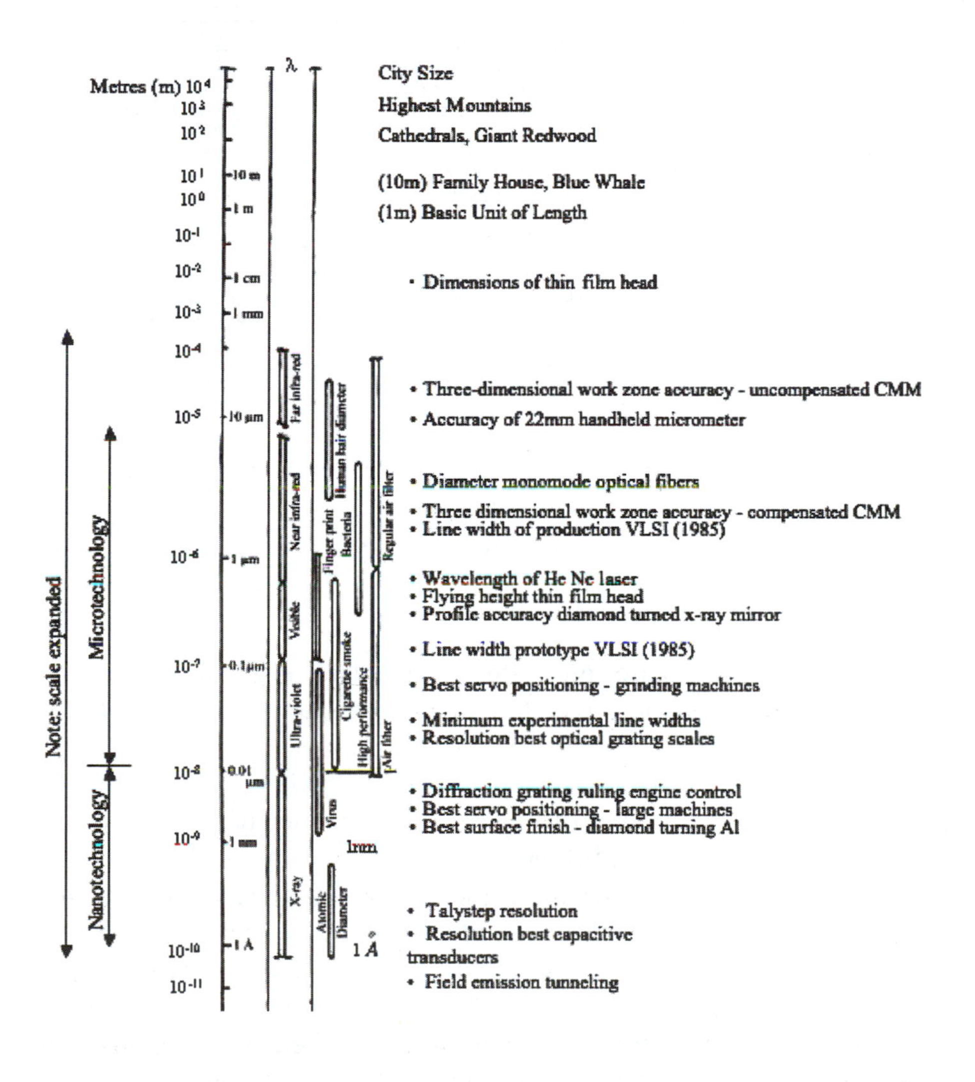

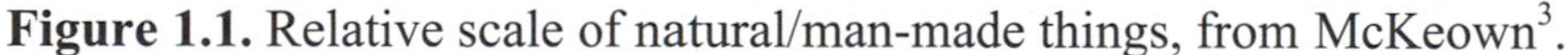
Figure 1.1. Relative scale of natural/man-made things, from McKeown[3].

Table 1.1. Tolerances on components for a range of mechanical and optical products, after Taniguchi[4].

	Tolerance band	Mechanical	Optical
Normal machining	200 μm	Normal domestic appliances and automotive parts, etc.	Camera, telescope and binocular bodies
	50 μm	General purpose mechanical parts for typewriters, engines, etc.	Camera shutters Lens holders for cameras and microscopes
Precision machining	5 μm	Mechanical watch parts Machine tool bearings Gears Ballscrews Rotary compressor parts	Lenses Prisms Optical fiber and connectors (multi-mode)
	0.5 μm	Ball and roller bearings Precision drawn wire Hydraulic servo valves Aerostatic bearings Ink-jet nozzles Aerodynamic gyro bearings	Precision lenses Optical scales IC exposure masks (photo, X-ray) Laser polygon mirrors X-ray mirrors Elastic deflection mirrors Monomode optical fibre and connectors
Ultra-precision machining	0.05 μm	Gauge blocks Diamond indentor tip radius Microtome cutter edge radius Ultra-precision X-Y tables	Optical flats Precision Fresnel lenses Optical diffraction gratings Optical video disks
	0.005 μm		Ultra-precision diffraction gratings

The products are referenced to the state of technology in the mid '80's but illustrate the differences in product tolerance requirements.

In the lower corner of Figure 1.2 is a somewhat controversial image of a distribution of dimensions measured on an artifact. This could represent the diameters of a cylindrical feature on a part produced on a lathe for several hundred samples. It shows a mean of the distribution, and it's deviation from a specified, or desired, dimension, as well as some variance on the distribution. It is controversial in that there are precision engineers who will argue with the concept of "random" behavior of a precision mechanism. They cite the fact that there are very few truly "random" effects in nature and the ones that exist rarely have anything to do with the behavior of mechanisms. What this shows is, by their reasoning, poor mechanical design (i. e. problems that can be solved but are not for lack of effort or intelligence) or poor metrology or measurement of the performance.

A classic quote is attributed to Jim Bryan[5], retired chief metrologist from the Lawrence Livermore National Lab to the effect that

"random results are the consequence of random procedures."

Dr. Robert Donaldson[6] also cited by Evans (and also retired from LLNL) elaborates but is similarly blunt:

"A basic finding of our experience in dealing with machining accuracy is that machine tools are deterministic. By this we mean that machine tool errors obey cause-and-effect relationships, and do not vary randomly for no reason. Further, the causes are not esoteric and uncontrollable, but can be explained in terms of familiar engineering principles."

We will not enter into the fray but will use this interesting representation of the degree to which a machine behaves the way we wish it to (that is the deviation from a mean- or systematic error and the variance of the behavior or scattering or "random" errors) shown in Figure 1.2a as our road map to our success as precision manufacturers. If, as a result of our improved understanding of the errors affecting the performance of machines (distortion due to static load or vibration, for example) we are able to make the actual mean dimension closer to the desired and reduce the variation in the distribution, Figure 1.2b, we will consider ourselves to be on the right track. As we will see, there are a number of ways, some more "deterministic" than others, to accomplish this.

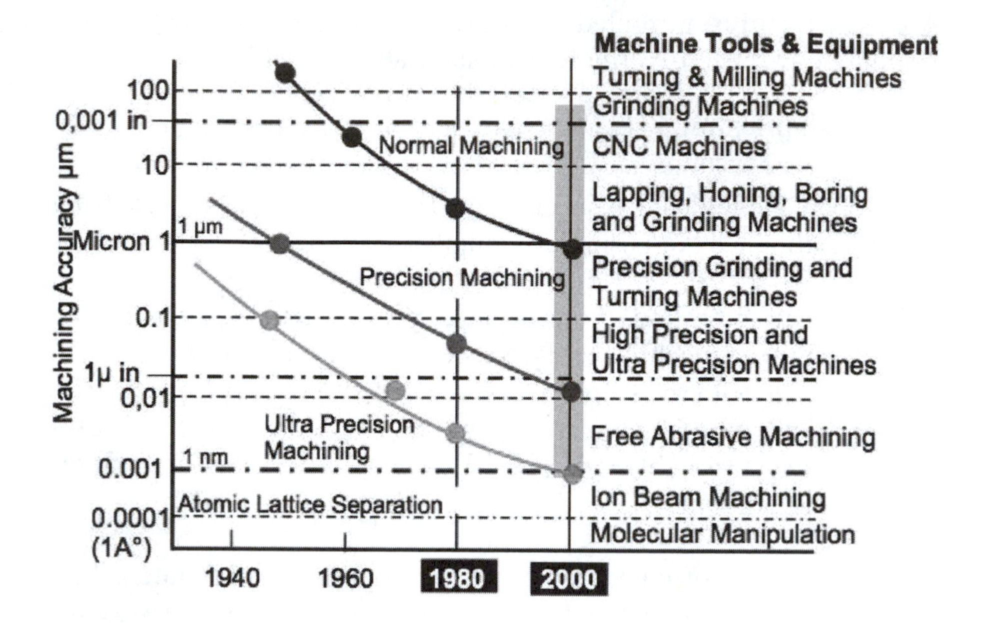

a. Progress in machining accuracy

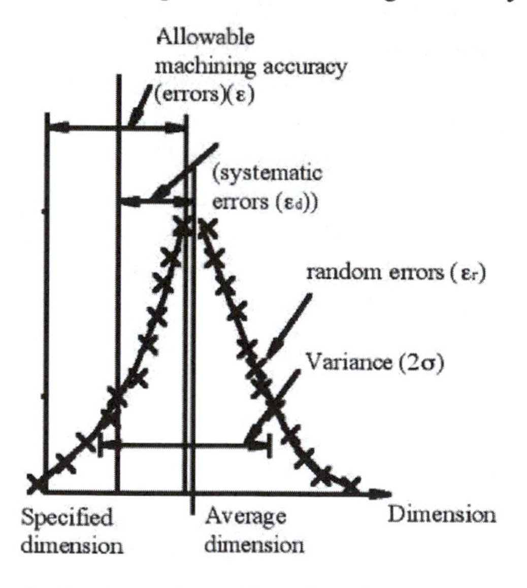

b. Systematic vs "random" errors

Figure 1.2. Taniguchi Chart, from Taniguchi[4].

1.3 Competitive drivers of precision manufacturing

The trends illustrated in Taniguchi's chart in Figure 1.2a are not occurring in a vacuum. There are and have always been driving forces behind the pursuit of ever smaller dimensions. Basically, they fall into several categories. There is an increasing technical complexity of products being manufactured today. This complexity is evident in both the number of parts or components comprising a device as well as the sophistication of device performance or operation. Excellent examples range from the ubiquitous personal computer with multi-gigabyte hard drives to VCR's and handheld cell phones. On the large end of the scale are space shuttles and large passenger aircraft- both comprised of sophisticated propulsion systems, aerodynamic structures and control systems unheard of a few decades ago. And all requiring components with complex shapes and exacting tolerances (reference the read head on a hard disk or the turbine blade in a jet engine). At the same time, the consumer is demanding that these highly complex and sophisticated devices be built to high standards of quality so they may operate with uncompromised reliability. But they must be affordable as well and usually are manufactured in relatively small lot sizes to reflect the product diversity offered to the consumer. The pressures of miniaturization and increasing complexity (but while maintaining high productivity- and, hence, low cost) are seen most in the manufacture of semiconductor devices. This product area and its best application in the microprocessor, clearly provides the most impressive drive to smaller feature sizes, larger wafers, higher process yield and process throughput). In fact, this segment alone is enough to justify our interest in precision manufacturing as we will see later in the text.

The critical elements of manufacturing that need to be maintained from a precision manufacturing point of view are summarized by McKeown[3] as follows:

- To eliminate 'fitting' and promote **assembly** especially in **automatic assembly** (emphasis in the original)
- To improve **interchangeability** of components

- To improve **quality control** through **higher machine accuracy capabilities** and thence reduce scrap, rework, and conventional inspection
- To achieve longer **wear/fatigue life** of components
- To achieve greater "**miniaturization** and **packing densities**"

and, on the general push for improvement in precision engineering and manufacturing, he adds

- To achieve further advances in **technology** and **science**.

The response of manufacturing to the small lot size, diversity and low cost issue has been to introduce "agile" or reconfigurable systems that can respond effectively to these rapid changes without necessitating a massive re-design of the manufacturing system. We learned this from the Japanese manufacturers over the last three decades. They were trying to insure a global market share in the face of fluctuating currency values for the production of high tech products. They were fortunate to be able to build on a strong academic and industrial infrastructure of precision engineering (which is a legitimate discipline of study in many leading Japanese universities) to insure that the improvements in the manufacturing system did not sacrifice the required process capability to produce precision components for these complex products.

Unfortunately, product complexity and production process sophistication can interact to create many opportunities for defects- which can frustrate attempts at increased reliability. Ayres[7] has written an excellent paper on the interaction of complexity, reliability and design and the manufacturing implications. One of the observations made by Ayres relates to this potential for defects in a complex product, and he proposes a simple model to illustrate the point. If a final product includes components of *n* distinct part types, each of which passes through *m* unit operations, the total number of actual operations is, basically, n x m. At each intersection between a component and a process there is the potential for an error (or in Ayres words "a hypothetical inspector makes a yes/no decision"- yes meaning the operation was correctly done; no meaning it was not).

He proceeds to analyze the probability that an incorrect part passes inspection and moves on to the next operation following relatively standard methodology. The bottom line is, the potential for defect introduction is impressive. Ayres cites Meister[8] who gives the example of a large US automaker who, in the course of producing an auto, provides some 3 billion opportunities for human error per day in the assembly operations (remember n x m) alone. With an optimistic assumption of 1 serious error per million operations this manufacturer could expect to see about 3000 serious undetected production flaws per day- or about 1 in 3 cars. This data from Meister is circa 1982, and surely the situation has improved a lot, but the potential for problems is sobering. As a contrast, defect rates for Japanese manufacturers were traditionally two orders of magnitude better. The development of measures of product quality and process capability, C_p and C_{pk} for example, are a result of efforts to track product quality. We will discuss these later.

We will not focus here on the organizational issues of manufacturing of precision products. But, we must keep in mind the importance of quality of the final product. The critical elements of McKeown above address the reduction of uncertainty at the interfaces between processes and products. In fact, that is the primary objective of the precision manufacturing engineer- reduction of uncertainty. With this, the determinism championed by Donaldson and Bryan will be more economically realized.

1.4 Historical developments in manufacturing

1.4.1 Background

Manufacturing processes (so-called unit processes) date historically back to some 4000 BC where hammering of metals for jewelry or simple tools were practiced. For reference, Kalpakjian[9] summarizes the developments of these processes over time. Gradually, over the first five or six millennia, the use of tools of higher sophistication, often as part of distinct unit processes (a casting process, or tube rolling, for example as part of deformation-based processes) grows. The progress in the development of these manufacturing methods derives from the increase in sophistication of the following elements (after Boothroyd[11]) necessary in any unit process:

- source of energy or relative motion
- means to secure the work
- means to secure and orient the tool
- control of the source of energy and means above

The "means to secure the work" is, of course, the fixture or jig employed, The "means to secure and orient the tool" is the tool holder or machine frame securing the tooling. Orientation is clearly important in machining operations where critical changes in tool geometry can have a large influence on surface finish and part dimensions as well as ease of chip control. Precision manufacturing is the art and science of combining these three elements in such a way as to achieve McKeown's objectives. Control is important as it allows the smooth application of the other elements to create and **maintain the necessary environment to create** the artifacts on the component in the face of variation in material, process, tooling or work geometry conditions.

These elements are comprised of sub-elements. For example, motion is often the vector sum of a primary motion (like spindle speed in a lathe) and a secondary motion (feed in the case of a lathe). Further, as a result of the application of the tool to the work material, there is a distribution of forces which can be resolved along principal directions of tool motion or any other arbitrary direction. Figure 1.3 shows the orientations of the motions and forces for a typical turning operation. The integration of forces over distance and time consumes energy and power. Expressed as a ratio to material removed (material removal rate, MRR, in machining) or deformed (in forming) it yields a specific measure of energy or power consumption which can be a basis of comparison of ease of manufacturing with different materials. On top of the orientation of the tool is the specific tool geometry. And, finally, we can calculate production time from the primary and secondary motions causing the tool to pass through the working volume of the component.

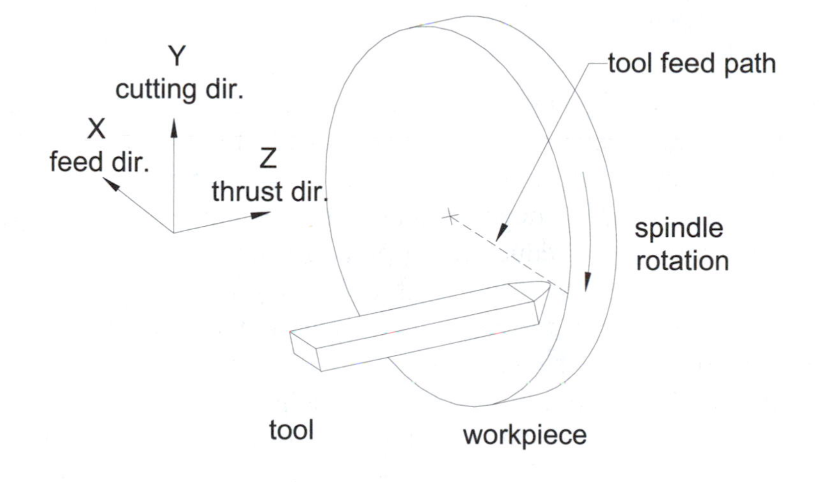

Figure 1.3. Tool orientation with respect to the workpiece in a typical turning operation.

Historically, developments in all of these elemental areas and sub areas have been at the basis of the improvement in achievable accuracy previewed by Taniguchi. With time, tradeoffs have been observed in these process parameters which has encouraged improvement in machines, tools, energy sources and control. These tradeoffs include, for a machining process:

- speed--------------------------------temperature/tool life
- tool material----------------------------toughness
- feed rate---------------------surface finish
- power/MRR-----------accuracy
- MRR-------------mass
- precision----cost

Higher primary speeds reduce machining/production time (and time is money!) but generate higher temperatures which reduce tool life. At some point, increasing speed is counterproductive as the process much be stopped frequently to change tools — thus increasing production time. This topic was extensively studied by F. W. Taylor in the early 1900's and led to, among others, the development of his famous tool life equation,

$$VT^n = C \tag{1.1}$$

where V is the cutting speed, T is the tool life, and n and C are empirically-obtained constants[11]. One solution to the tool deterioration with speed/temperature is the creation of tools with higher hot hardness. This prompted the development of brazed carbide, then disposable insert carbide (to reduce resharpening time), then ceramics and coated ceramic tools. Now diamond tools and diamond (or diamond-like) coated tools are available. The reduction in tool change frequency is offset by the reduced (generally) impact toughness of these tool materials requiring altered machining conditions and/or stiffer machines. Increased feed rate reduces machining time but also creates a poorer surface finish — hence the rule of thumb that roughing cuts are taken at maximum MRR and the finish cut (last cut taken) is at a feedrate to yield the desired surface finish. Material removal rates are desired to be as large as possible to minimize manufacturing time but small parts may not be able to withstand the accompanying forces at high MRR — hence limiting the machining conditions. Finally, there is a long-standing belief that minimizing manufacturing cost and obtaining high precision are contradictory. If we stay with the old concept that machining speed is necessarily low for precision components (recall the feedrate discussion above) this belief is hard to dismiss. However, the intelligent application of precision manufacturing principles has offered ample proof that precision components can be made competitively. The best example is the hard disk on the computer on which these notes are being prepared! As seen in Figure 1.1, dimensions on a hard disk are on the order of 0.1 micron. This for just one component of a unit that sells for a few hundred dollars and is now capable of storing several orders of magnitude more information than a few years ago at much lower cost. If that doesn't convince you, check the ratio of price to performance characteristics of computer chips, cameras, etc. over the last few years.

1.4.2 Key drivers

The historical developments of manufacturing in general, and precision manufacturing in particular, did not occur in a vacuum! There are always drivers for these and these drivers generally fall into the categories of military needs (we need a weapon that shoots further than our neighbor), artistic desires (we need a more highly polished jewel or more intricate clock than our neighbor), or business (we need to go some place to get something our neighbor can't get). The first engineers were military engineers responsible for designing and building "engines" of war (some of which we will see in a bit). Whether or not we agree with these drivers they are there and have, for better or worse, made Taniguchi's chart representative of the pace of development in precision manufacturing.

These drivers are best understood in terms of the "connections" that lead from one development to another. The concept of connections has been made popular recently by a series of public television programs by that name by J. Burke produced by the BBC. For any one interested in interesting "cross sections" through the time history of technical developments the accompanying book is highly recommended[12]. Here are presented some connections tailored to precision manufacturing developments. As engineers, we note that we generally are called on to create the technology in response to needs defined by others. The needs defined by others are the drivers referred to in this discussion. We will look at four "connections."

The first falls under the logo of *he who has the most energy generally wins*.

Consider the start of the industrial revolution in England in the early 1700's—

- industry needed water pumps to aid in mining
- mining was important to get coal
- coal was needed to fire boilers
- boilers were necessary to make steam

- steam was necessary to power engines (like pumps!)

Improvements in efficiency of pumps was dependent on the goodness of fit between the piston and cylinder of the pump as well as the steam engine — and this was dependent on the quality of the machining process creating the cylindrical feature.

The second falls under the logo of *he who shoots furthest generally wins.*

Consider the development of military hardware through out time—

- the military needed machinery to make cylinders
- the cylinders were needed as part of cannons
- the cannons were needed to shoot projectiles far and accurately
- projectiles were needed to hit one's enemy

And, of course, if you could hit your enemy and he could not hit you — due either to range, power or accuracy — that generally helped you in battle. After that, it is a "management" problem!

The third falls under the logo of *he who sells the most generally wins.*

Consider the business climate and the need to venture abroad for trade—

- businessmen needed accurate timepieces (especially those that could function at sea)
- time pieces were needed to aid in navigation
- navigation was needed to sail for trade (and return!)
- trade was necessary to sell and buy exotic goods
- selling and buying was the basis for making money

Let's look a bit more closely at this one. By the 1700's all the major improvements in time pieces had been made except for the "longitude problem" — how to accurately track one's position east-west from a point (the Royal Observatory in Greenwich to be exact).

Sir Isaac Newton said in 1714 "...for determining the Longitude at sea, there have been several projects, true in theory, but difficult to execute...one is by watch...but by reason of the motion of the ship,...variation of heat and cold...such a watch hath not yet been made..."[13] Latitude (one's distance above or below the equator) could be found from observation of the Sun at noon or the Pole star at night but longitude was still a problem. The solution to this was the development, by John Harrison, starting in the 1730's, of a marine clock that could accommodate the swaying of a ship at sea (which perturbs simple pendulum clocks) as well as the problem of great extremes of temperature (and the accommodating growth and shrinkage of sensitive parts of the mechanism) and the humidity. He did it applying principles of precision engineering and invented the caged roller race, the predecessor to the roller bearing, and the bimetallic strip in the process[13].

Finally, we combine the above three under the logo of *the ship was the plane of the past centuries.* Consider the following—

- steam engines power ships
- ships sail further with better navigation
- carry goods/guns to sell/shoot

- machines build bigger engines
- to sail ships further, faster and more accurately
- with better navigation to sell better products

OR

- machines build bigger engines
- to sail ships further, faster and more accurately
- with better navigation to shoot better guns

And so it goes!

We now turn to some interesting examples of the above drivers focused on the development of weapons and machine tools.

1.4.3 Historical examples

What follows here are a series of "plates" showing early manufacturing processes and machines. The plates are from historical literature, the most famous of which is by Diderot[14], showing scenes of machine usage or development.

Figure 1.4a. Typical 15th century castle with battlements and defenses designed for "pre-black powder" warfare.

As background on the "state of the art" consider Figure 1.4. Here is a castle of the 15th century, Figure 1.4a, which found itself vulnerable to forces using a "new" weapon, a bombard, Figure 1.4b, which could be fired up to ten times a day. The rate here is determined by the difficulty of preparing the bombard for firing. The power and distance of the bombard lobbing heavy objects at a castle designed for, at most, arrows and battering rams won the day. Once the technology of cannons became more developed, and cannons

Figure 1.4b. Wrought iron bombard used in 1400-1500 time period for shooting a 90 cm diameter projectile.

were more commonplace, the folks inside the castle took advantage of them as well, Figure 1.5. Notice that the crenulations seen on the castle wall in Figure 1.4 are gone. The occupants discovered that many casualties were caused by flying pieces of stone walls after a bombard hit. The design, developed for protection of archers, was changed to reflect the new technology of battle. Figure 1.5 points out the truth to the second logo, *he who shoots furthest generally wins,* as, clearly, the party with the most accurate weapons and longest range will dominate the engagement.

Figure 1.5. Beaumaris castle built from 1295 on and continuously remodeled; round towers were thought to be more resistant to cannon shot and would deflect all but straight on hits; crenulations are gone; provisions for cannon in the castle were made.

How these tubular products (cannons, and later cylinders for steam engines) were made over the years is the story of the development of modern manufacturing processes. Figure 1.6 illustrates early casting techniques. The mold was actually buried in the ground in the foreground of the figure where the workers are pouring molten metal into the sprue. Smaller castings are in conventional mold boxes on the left-hand side of the image. The worker on the far right is cleaning a recently cast product removed from the mold. The "tools of the trade" are shown in the lower half of the figure. Some of the products from the foundry are shown in Figure 1.7. Although these illustrate the production of cast iron pipe for water supply, the basic shape, and the cores (seen at the top of the figure) used to create the internal features of the pipe are similar to those used for cannons and other cylindrical products.

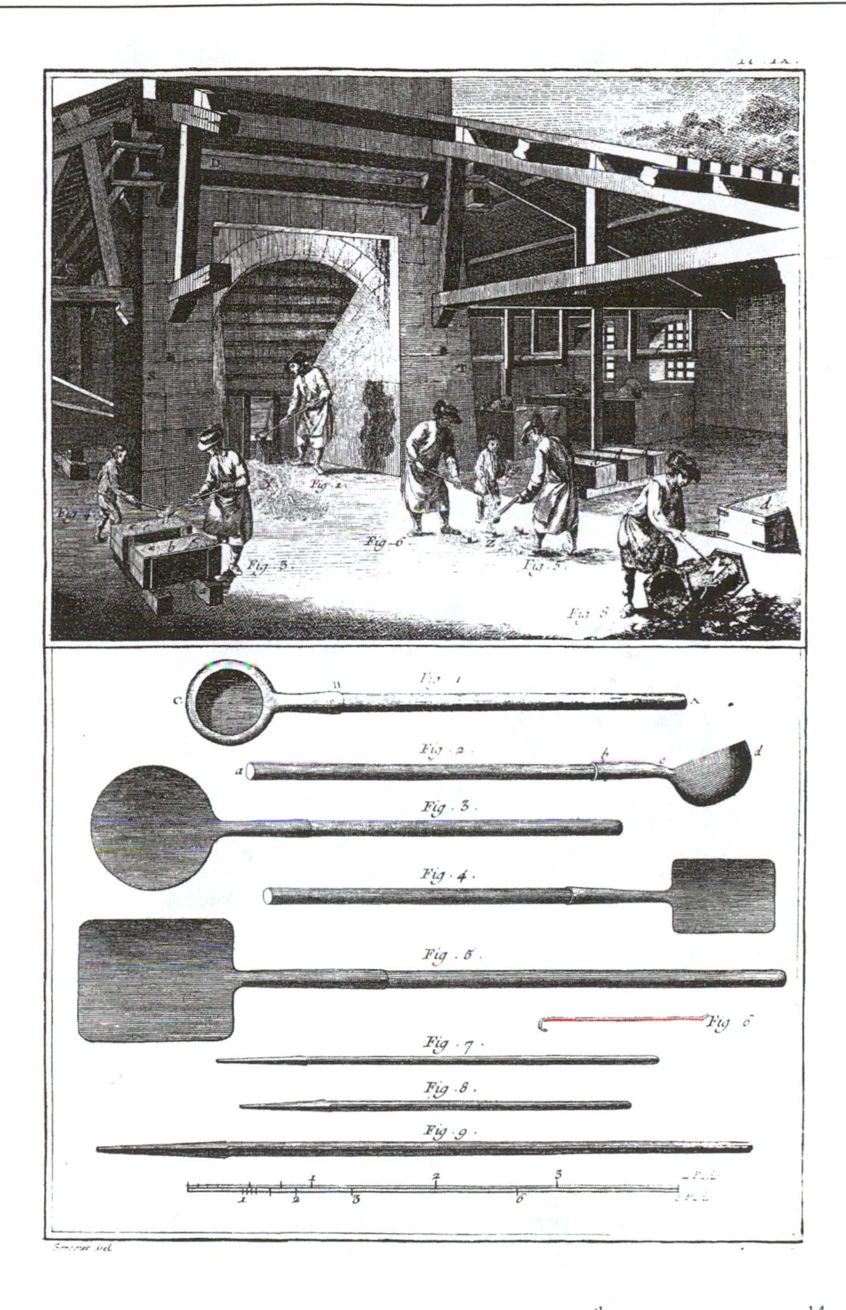

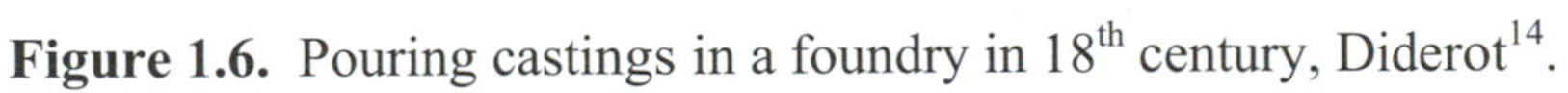

Figure 1.6. Pouring castings in a foundry in 18th century, Diderot[14].

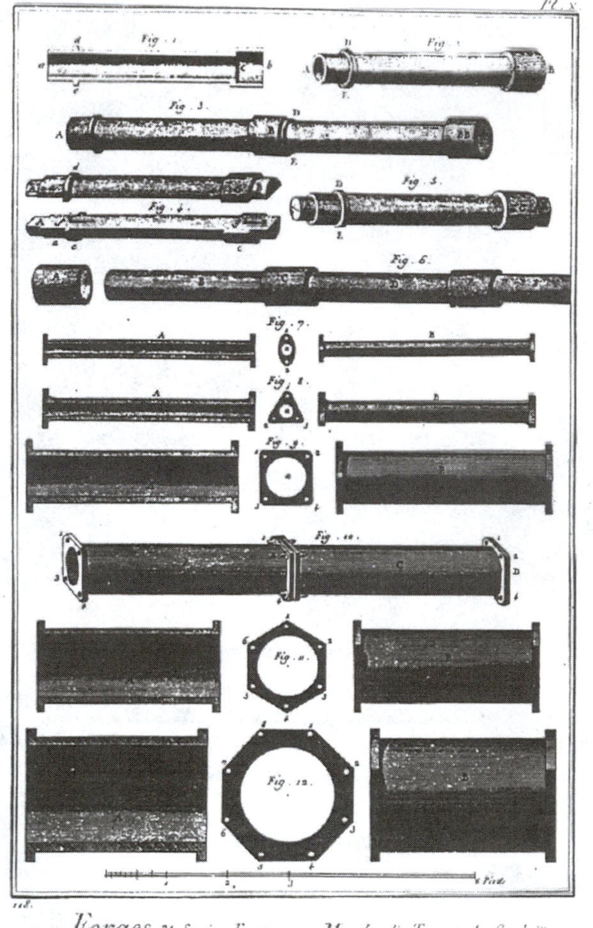

Figure 1.7. Cast products and cores, from Diderot[14].

Some fastenings and components were forged from metal by repeated hammer blows on heated metal by skilled workers who could create the complex surfaces, often non-planar, by wielding the powered hammer. In Figures 1.8 and 1.9[14], the general layout of the forge shop is seen as well as the water powered hammer. The workpiece was transferred from the hearth (furnace) for forging until it cooled enough to reduce the malleability below the power of the hammer. At this time if was re-heated for continued forging. When regarding these images keep in mind the earlier discussion of the elements of manufacturing including the need to secure and orient the tool and how this might have been accomplished here. Further, it is interesting to speculate what tolerances on the dimensions were achievable with this process.

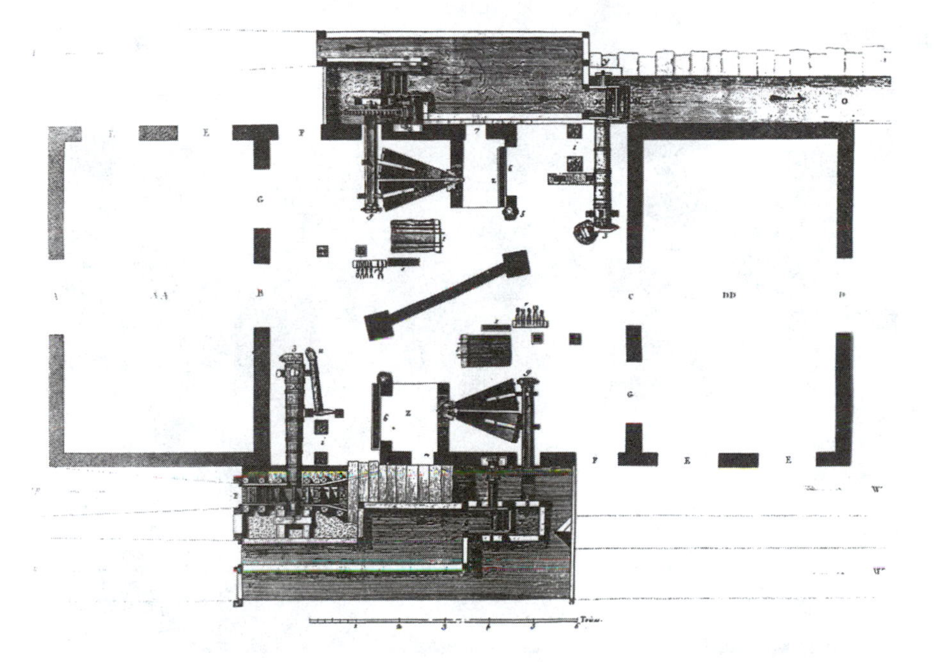

Figure 1.8. General layout of forge with two hearths, from Diderot[14].

The completion of cast or forged parts was often accomplished by machining. This would create the final surfaces and necessary finishes and dimensions. At the time of Diderot, the major machining processes were turning (single point tool) and boring (multipoint tools). Figure 1.10, from Roe[15], shows a treadle-driven screw cutting lathe from the mid-16th century. A "master screw form" seen on the left-hand side of the machine is used to guide the tool as it is fed along the workpiece. The work is turned by a falling weight rotating the "tailstock" to which the part is attached. This process allowed the duplication of screw elements (as used in the feed mechanism here) for other machines. The earliest type of lathe is the "pole lathe", shown in Figure 1.11. Here it is shown for creating wood ornamentation for furniture. The pole lathe operates on the power of a bow overhead "springing and un-springing" with the action of the worker's foot on the pedal below the lathe. The work is driven (both clockwise and counterclockwise depending on whether the bow is being extended or released!) by the cord wrapped around

Figure 1.9. Image of forger working in a forge shop with a water wheel powered hammer in the early 18th century, from Diderot[14].

it. The feed motion is provided by the worker moving the tool along the axis of the work manually on a tool support. The template over the lathe provides a guide for the turner to periodically check the dimensions of the part to the master. Both the lathes in these two figures have wooden frames.

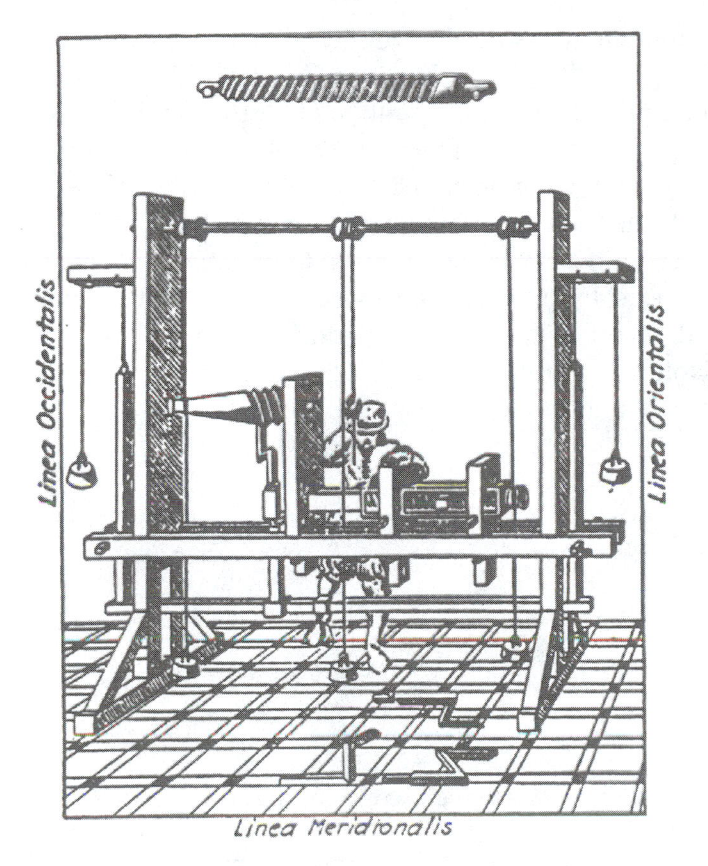

Figure 1.10. French screw-cutting lathe, prior to 1569, from Roe[15].

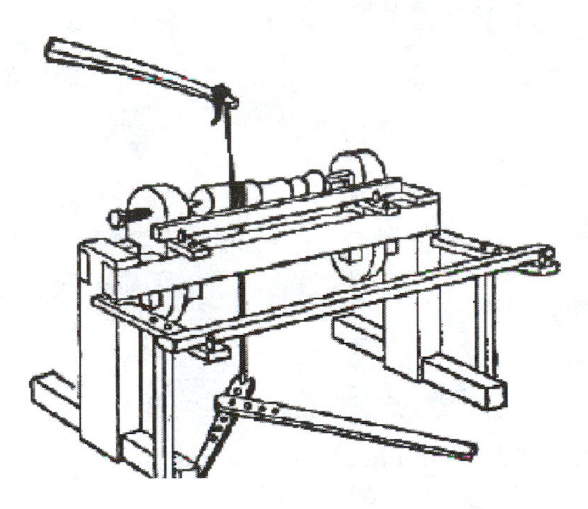

Figure 1.11. Pole lathe for wood turning.

Improvements on the wooden structure as well as in the part rotation and tool feed are shown in Figure 1.12[15] showing a French screw cutting lathe form the 1740's. Here the handle causing part rotation is geared to the feed mechanism providing better pitch control. And the metal components are less susceptible to distortion (or moisture!) than wooden components. Figures 1.13 and 1.14 illustrate workshops with early lathes at work. Generally the "bow" action has been replaced by a wheel and someone to turn it (Figure 1.14) although the worker in the foreground of Figure 1.13 is using a treadle lathe. Note the method of "securing and orienting the tool" used by the turner in Figure 1.14.

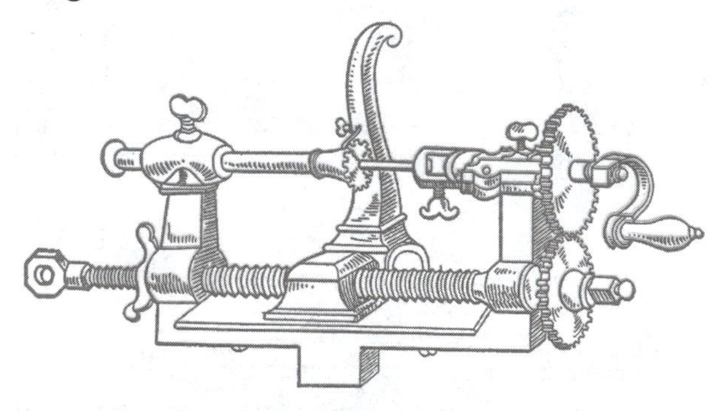

Figure 1.12. French screw cutting lathe, from Roe[15].

Figure 1.13a. Tool making shop for a manufacturer of vises, from Diderot[14].

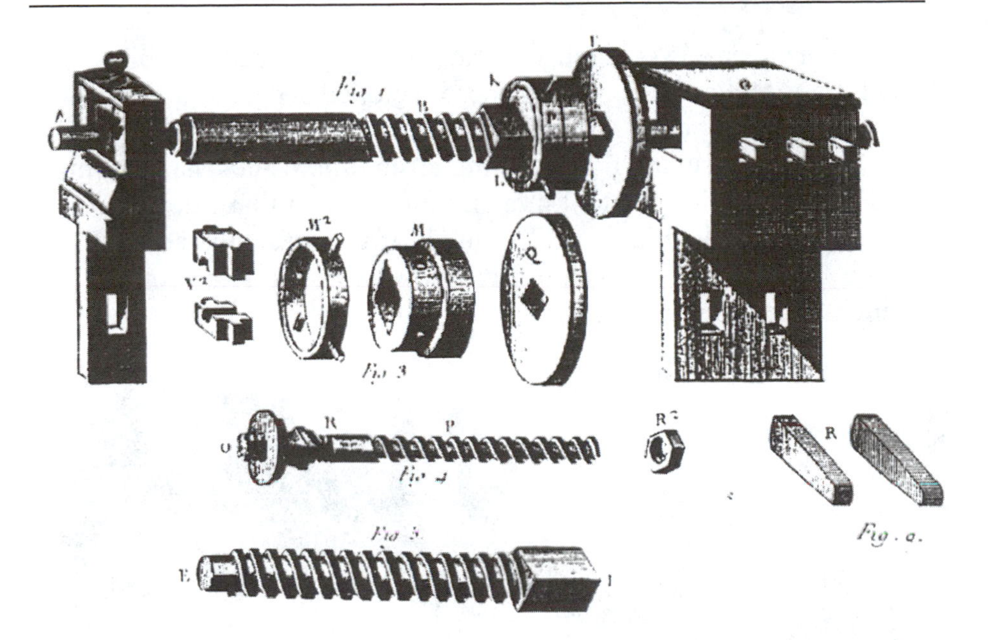

Figure 1.13b. Close up of lathe and products, from Diderot[14]

Figure 1.14. Lathes and turners working in a machine shop, from Diderot[14].

Maudslay is credited with advancing the "controllability" of machine tools in the sense that he developed lathes and other machine tools with features that insure higher accuracy, Figure 1.15[12]. The lathe shown in the figure, dating around the 1800s, has a turning lead screw carrying the tool in a sliding tool rest that rides on accurately planed triangular bars. The pitch of the screw thread could be varied by changing the gearing on the end of the lathe (much like French lathe in Figure 1.12.)

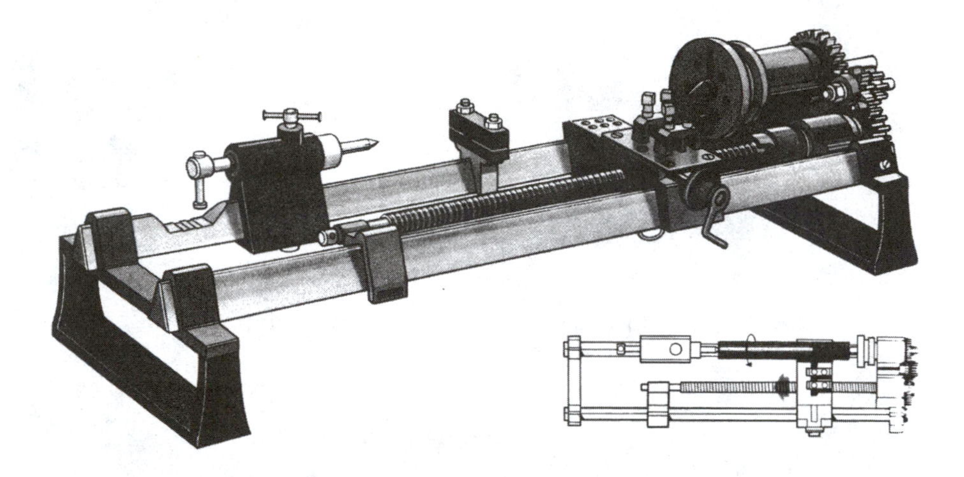

Figure 1.15. Maudslay's lathe, circa 1800, from Burke[12].

One of the metal cutting processes related to turning is reaming. Reaming was important for the finishing of the interior cylindrical cavities of workpieces like cannon bores or steam drive and pump cylinders. The early techniques for reaming these cylinders did not effectively meet the requirement to secure and orient the tool. A schematic of a cannon reaming process from the mid-1700's is seen in Figure 1.16[14]. The figure shows a cannon suspended over a reaming tool on a long shaft. The shaft is turned by horses and the cannon body is lowered over the reamer head. There is no structural linkage between the reamer/shaft and the support suspending the cannon. Hence, there is not real guidance of the tool with respect to the cannon bore. The machinist relies on the shape of the existing cast cylinder bore and the reamer to guide the tool as it advances through the bore.

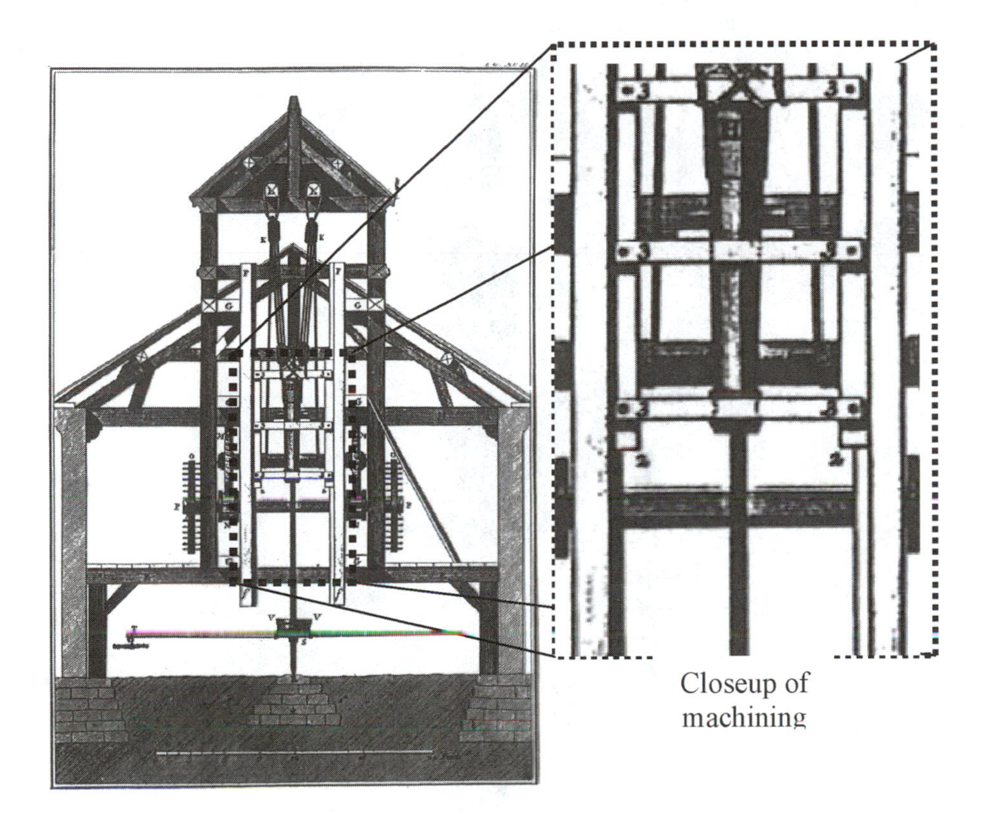

Figure 1.16. Mechanism for reaming cannon bores, from Diderot[14].

Figure 1.17, also from Diderot[14], shows the reaming tool used. Multiple tool elements are imbedded in a wooden mandrel that is attached to the shaft. The conical head of the tool allows the tool to follow the initial bore. The extended length of the tool provides some control over minute changes in direction as the tool diameter inside the "finished" bore will guide the general direction of the reamer. Clearly, this results in a reasonably circular bore but not particularly cylindrical in shape. In fact, James Watt is known to have remarked that, with respect to the cylindricity of his steam cylinders and the fit of the piston in the cylinder in the early 1700's, he could not get a fit tighter than the thickness of a schilling (about a nickel!).

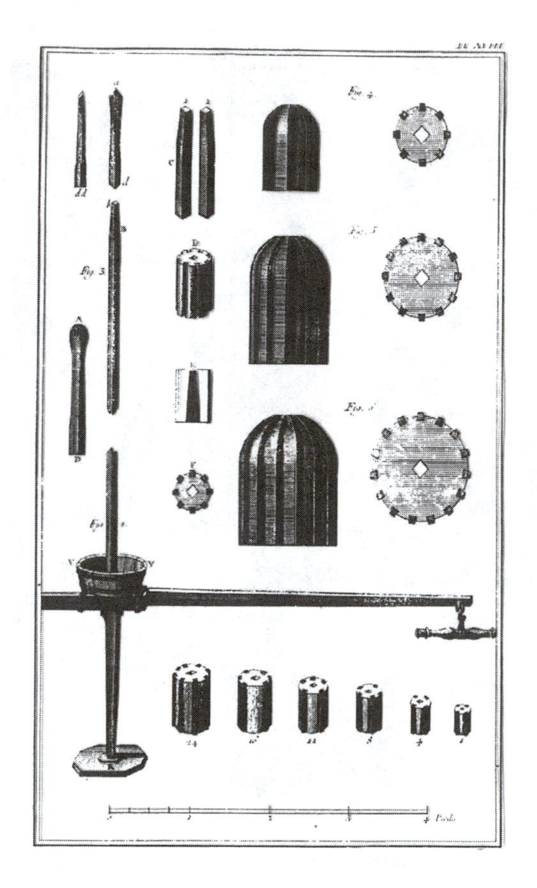

Figure 1.17. Reaming tools used in the apparatus of Figure 1.15, from Diderot[14].

An improvement was seen in Smeaton's boring machine in the Carron Iron Works around 1769 as illustrated in Figure 1.18. As shown in the inset, Figure 1.18b, the reamer head is carried here by a small counterweighted cart that rides on the cylinder walls. So, whereas this design does provide some connection between the tool and the work carrying structure, the accuracy of the tool centerline is still dependent on the unfinished cylinder wall geometry — which in turn is dependent on the quality of the core and casting process/ mold. Note the method for advancing the tool into the work via rope and pulley. A water wheel is, however, introduced as a source of continuous power. Wilkinson finally put all of the elements together in his cylinder boring machine shown in Figure 1.19[15]. The key feature here is that the overshot waterwheel powered tool is mounted on a shaft supported on both ends by the frame to which the cylinder is attached so that the accuracy of the location of the

Figure 1.18a. Smeaton's boring machine, circa 1769.

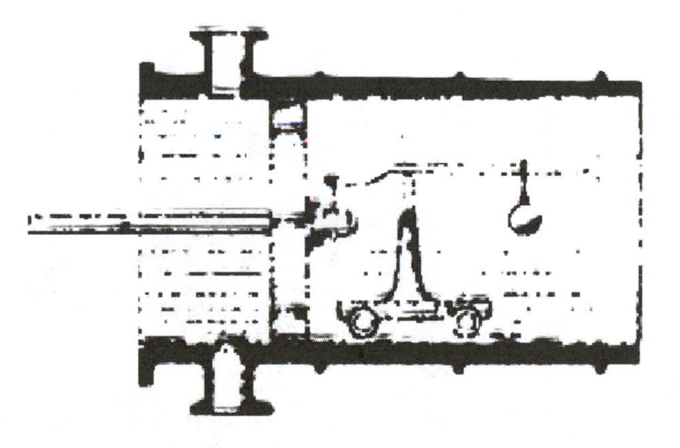

Figure 1.18b. Closeup of tool support mechanism, from Roe[15].

axis of the shaft is not dependent on the unfinished or finished surface of the workpiece. Hence the reamer head rotates with little deviation off axis. As an indication of the importance of these cylindrical elements, Figure 1.20 shows Watt's pumping engine and the use of several large cylinders for the vacuum pump as well as the main cylinder. Scale models of many of these machine tools as well as Watt's engine can be seen at the Henry Ford Museum/Greenfield Village in Dearborn MI.

Figure 1.19. Wilkinson's cylinder boring machine and close up, from Roe[15].

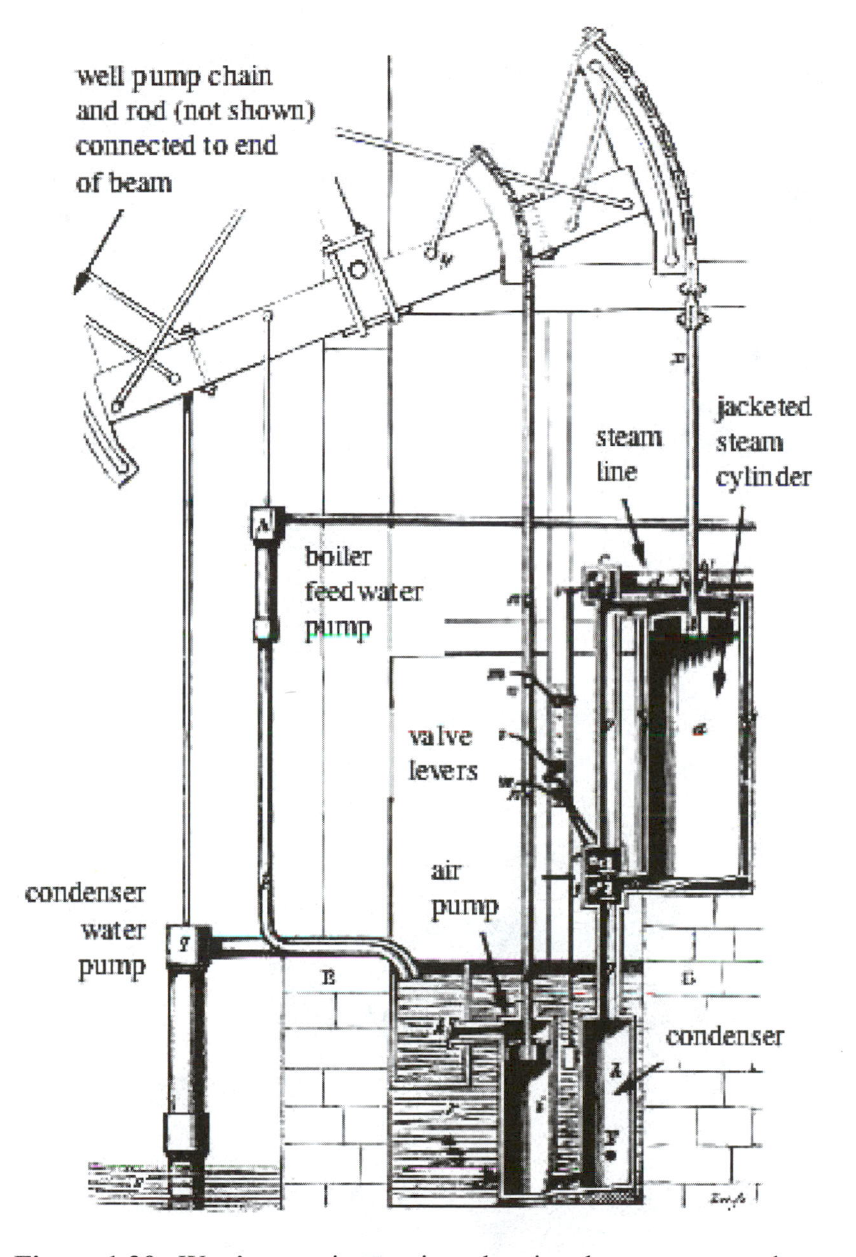

Figure 1.20. Watt's pumping engine, showing the separate condenser to the left below the main cylinder. The vacuum pump is the cylinder to the left of the condenser, which is immersed in water to keep it cool.

Combining machine elements to increase control of the process, automated or mechanized feed of tools and workpieces, and better organization of processes for efficient production, Brunel and Maudslay introduced block making machinery in 1808. The Royal Navy had an almost insatiable appetite for blocks for the many sailing vessels and with this machinery ten unskilled men could make on the order of 130,000 blocks per year. There were five steps in the process to obtain the finished product. Each machine in the process introduced some advances in machine design and construction. Figure 1.21[15] illustrates a mortising machine (akin to a broaching operation) that boasted 150 strokes per minute. Figure 1.22[15] illustrates Brunel's shaping machine for finishing the edges of the block.

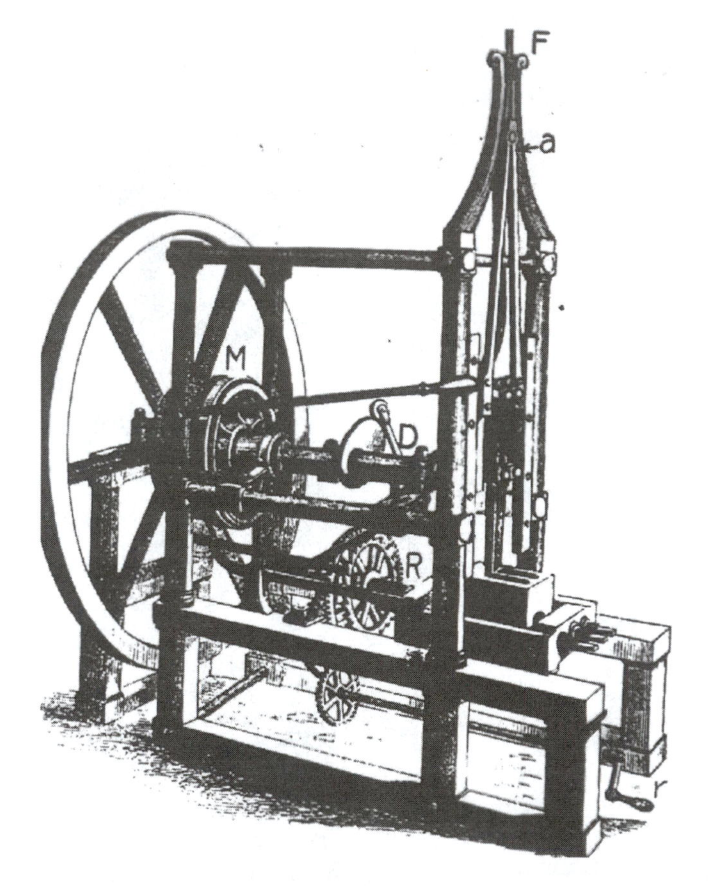

Figure 1.21. Brunel's mortising machine (150 strokes per min.), from Roe[15].

The later machines shown above are able to implement all of the elements previously listed as necessary for controllable manufacture. Further improvements were, of course, necessary but they generally involved refinements and enhancements on these basic elements as illustrated here. The objective of this course is to introduce and explain the science behind these advancements and show how they can be continuously applied to improve the precision of manufacturing machinery and processes.

Figure 1.22. Brunel's shaping machine, from Roe[15].

1.5 Organization of this book

The objective of this introduction was to give the reader an appreciation of where precision manufacturing "fits in" to the field of engineering as well as the critical role precision machine design for manufacturing has had in the technological development over the last several hundred years. We could have followed many other strands of technological development to accomplish the same objective. What follows is an introduction (and review for many) of some of the basic components of precision engineering as they apply to manufacturing. First we will discuss the elements of precision machine design, although this is not a course on precision machine design. We will specifically outline the sources of errors in the performance of machines (remember Taniguchi?). We than step through in more detail some of the most critical of these elements starting with principles of measurement and metrology. As has been oft stated...if you can't measure what you made, then you can't tell if you made it or not! Think about this. Then, in no special order, we will cover thermal errors, techniques for error mapping, compliance errors and form error and finally error compensation and correction. These topics are generic in the sense that the material backing up each one will touch in many different aspects of machine design and machine elements. We then look at the whole machine tool system.

An important next topic, somewhat related to error compensation and control, is on sensors for precision manufacturing. This could be a course on its own but will be covered here from the point of view of the impact of sensor technology on manufacturing. The engineer is charged with the task of reducing the uncertainty in systems introduced by interfaces between machine elements, process elements, materials, etc. Recall the n x m. possibilities for errors of Ayres? These are the "interfaces" we are referring to. Sensors play a major role in this but must be used to reduce uncertainty rather than introduce additional elements for potential failure.

Precision manufacturing ultimately comes down to a process acting on a workpiece. In the last two chapters we cover in some detail machining processes and process planning for precision manufacturing. Although "machining" seems like a rather narrow topic given all of the range of products of interest here, in fact, machining — or rather material removal — is at the basis of either creation of the artifact on the part, or creation of an element or elements of the machinery on which the artifacts are produced. Examples follow which illustrate this — from planarization by loose abrasive lapping to manufacture of steppers for photolithography.

II MACHINE DESIGN FOR PRECISION MANUFACTURING

2.1 Background on machine design for manufacturing

The development of machines over time can be viewed through a number of different lenses. Shirley and Jaikumar[16], for example, refer to a classification of seventeen levels of mechanization of "machines" related to their power and control sources. These developments, or levels, roughly follow progress of man and machine through time. So one sees the development from a person holding a tool at the lowest level, level 2 (level 1 being the person's hand alone) through powered tools to more automated machinery. Finally, at level seventeen, we see a machine which anticipates action required and adjusts itself to provide it in response to some sensor inputs and "intelligence" containing an objective function and means for optimization.

Moriwaki has represented this development in a more engineering-oriented fashion. Figure 2.1, from Moriwaki[17], describes the transition from the machine driven by "predetermined commands" which is much more than open loop — here implying even so-called adaptive control, but control about some pre-determined set of operating conditions based on our best estimate of the required conditions and the existent material, tooling and work geometry circumstances. Crossing the magic dotted line in the figure signifies machines which can make decisions "for themselves." What this rather anthropomorphic term implies is that, based on ambiguous or incomplete information, experience (codified in data bases or process models), as well as an ability to "learn" from conditions experienced while in operation, the machine and controller can process this array of information and determine the best course of action to achieve the objective. The objective is usually the creation of a surface with certain characteristics, artifacts with dimensions within certain tolerances and error of form within other bands. Whether or not one chooses to believe this characterization, the image in the figure does represent the views pertaining to the direction of development of machinery for manufacturing.

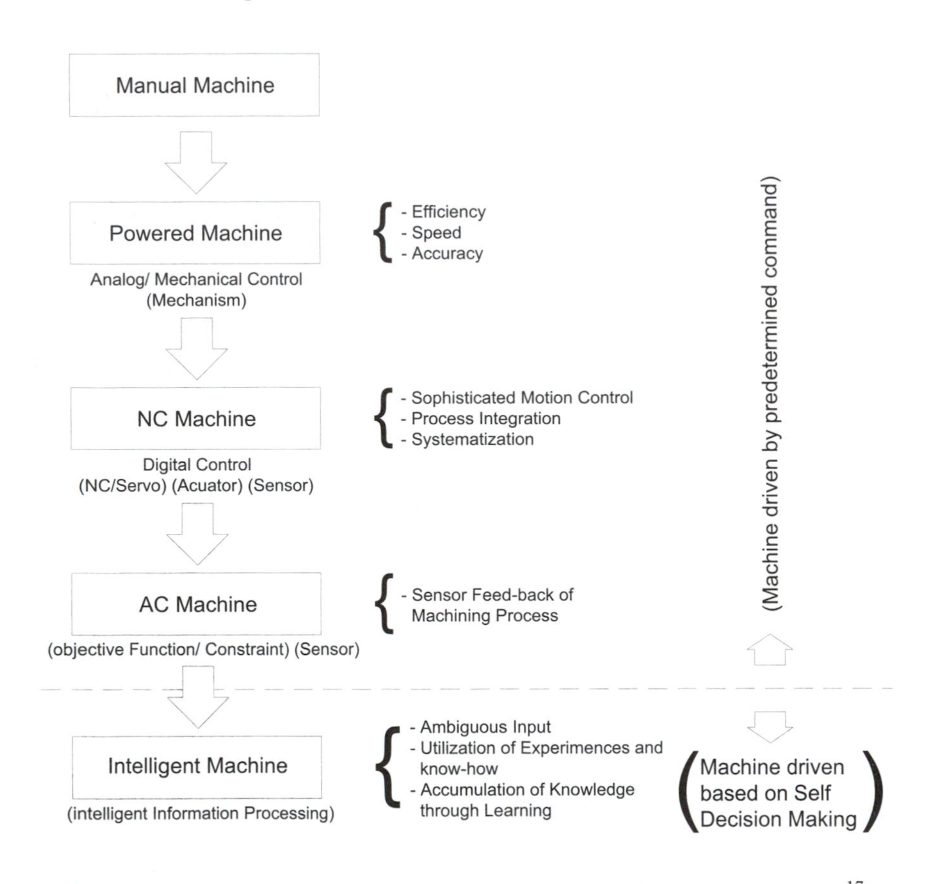

Figure 2.1. History of machine tool development, from Moriwaki[17].

The view outlined in the previous paragraph creates tremendous challenges for the precision manufacturing engineer. It pushes the requirement on "determinism" to the limit as we try to insure the performance of complex mechanisms over ever broadening ranges of performance. Certain design strategies will be employed to insure that determinism is achieved to the extent possible. The "natural enemies" complicating the task are the errors in these mechanisms. These will be introduced after a brief discussion of design philosophy.

2.2 Philosophy of precision machine design

The purpose of this chapter is not to present design philosophies for machines in any detail but, rather, to set the stage for our discussion on precision machinery for manufacturing. This may seen like an arbitrarily fine distinction but there are excellent texts available to which the reader is referred for more on that subject. Specific sources include Slocum[18] and Nakazawa[19] which give very detailed and practical (in the case of Slocum) and more philosophical (in the case of Nakazawa) information on precision machine design. There are many other general texts which cover the principles of design, from identification of functional requirements through project management. The unique features of precision machines or the processes they implement must be considered.

The success or failure of a precision machine can be evaluated with respect to six major items, from Nakazawa[19], all of which will be discussed in more detail. These are:

- dimensional precision
- angular precision
- form precision
- surface roughness
- kinematic precision
- surface layer alterations

These are both elements that must be *designed in* to the machine as well as features or characteristics of machine performance that must be *measured*. Nakazawa describes in some detail methods for insuring that the proposed design solution, building on the functional requirements of the machine or system, can be obtained in the most efficient, mechanically, and cost effective, economically, way. Finally, Nakazawa establishes a set of design principles which are illustrated in the text with specific machine elements or systems. The design principles revolve around the needs for precision machines to meet the four basic functional requirements, Nakazawa[19], of:

- possessing a perfect kinematic reference,
- possessing a perfect kinematic pair which execute perfect movement with respect to the reference,
- being constructed so as to prevent noise (or disturbances) in operation, and
- being able to detect movement accurately.

Some of these principles are derived from basic theories we will be covering later in the text.

Nakazawa's first design principle is the principle of functional independence and states[19]:

"When controllable functional requirements exist, a system in which the functions are independent is preferable to one in which the functions are not independent."

The principle of functional independence was proposed by Suh[20] and applies to a wide range of systems. Nakazawa's second design principle[19] is the principle of total design:

"When constraints exist for certain evaluational items, total design is better than either additive design or combination design."

For example, it may be better to design a wholly new machine tool to meet the six critical characteristics than to modify an existing design or assemble a machine from existing components. Of course, this may cost more initially.

One final reference that is not easily obtainable but offers invaluable insight into the philosophy and practice of design and manufacture of precision machinery is by Moore[21]. Moore's company, Moore Special Tool Company, is arguably one of the finest precision machine makers in the world. Moore Special Tool started in 1924 as a specialty tool making shop in Bridgeport CT making tooling for watch, clock and typewriter plants. Moore approaches the challenge from the point of view of the skilled machine builder and picks up where Nakazawa leaves off. That is, Moore answers the question...okay, so how do we actually build this machine we have so cleverly designed? And, further, how do we prove we built it? Moore emphasizes the need to master what he refers to as "the four mechanical arts:"[21]

- geometry (starting with it's foundation in the flat plane, from which the surface plate evolves and straightedges, and methods of scrapping them)
- standards of length (referring here to the measuring element of a machine tool — the lead screw — from which the machine derives its accuracy)
- dividing the circle (being able to accurately divide the circle is a challenge that has confronted precision machine and instrument makers for centuries, see Evans[1], for excellent background on this.)
- roundness (the performance of these machines is dependent upon the overall accuracy of holes, shafts, balls and other components of the machine.)

We will return to these mechanical arts throughout the course.

2.3 Sources of error - overview

We referred to the sources of errors in precision machine as "natural enemies" of the precision engineer. Recall Donaldson's and Bryan's insistence on determinism in design — that is, the application of sound engineering analysis to overcome the errors in performance of these mechanisms. Taniguchi had referred to these as "systematic errors" with the errors that had no obvious or repeatable clear source as "random errors." It all boils down to how closely (and how much time/money spent) one wants to look for the sources of errors *or* utilize methods of design and manufacture to prevent or minimize the errors. We declare victory over errors when they are either not measurable or measurably small enough to fall below our specifications.

Machine tools, which are a good focus for our discussion as they have all of the critical elements of interest (as well as create all of the critical elements on the workpiece), are basically closed structural frames, Figure 2.2. The spindle, in which the tool is mounted for material removal, is linked to the frame, here comprised of the column, base and table, which supports the workpiece. One can easily imagine the corresponding sketch components for a lathe or other machine. The critical "open" connection in this loop is between the work surface and the tool. Clearly, any error in position between tool and work surface will result in a dimensional error on the part surface (tolerance, form, surface, sub-surface damage, etc.) Thus, anything that contributes to an error in position is of concern us. We will describe the frame of reference for quantifying the errors in position as you can imagine they are both translational as well as well as rotational. And, they will be most troublesome in certain directions, called *sensitive directions* — for example, perpendicular to the surface of a machined part.

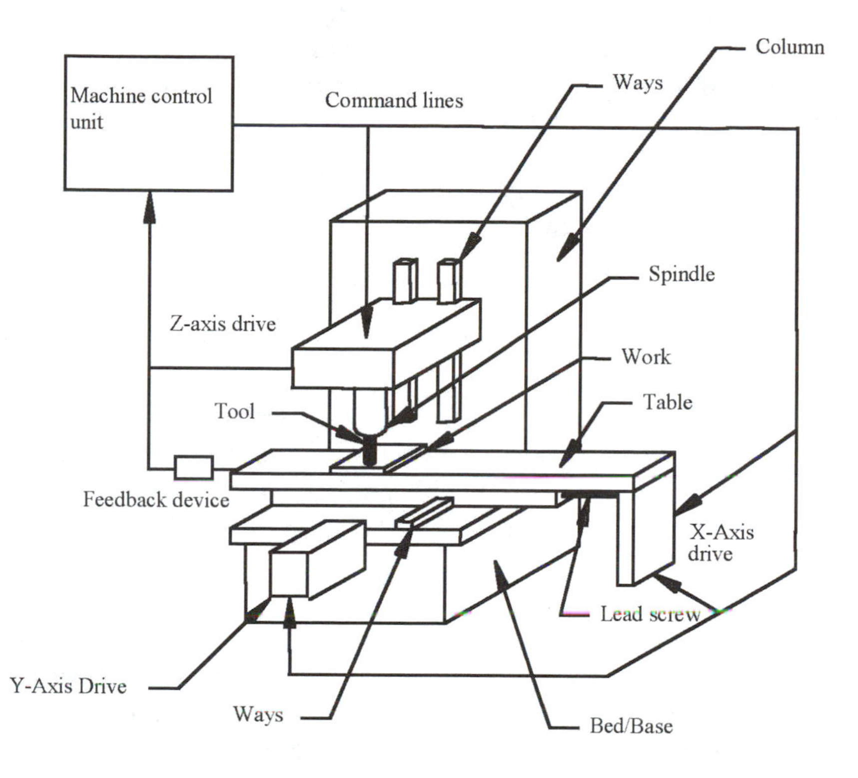

Figure 2.2. Machine tool structural "loop".

In the past, many efforts have been made to characterize the errors in part features, holes and planes, in terms of process parameters. This was done to aid in process planning methodology which generally attempts to map processes on to features for the selection of the minimum set of processes and their sequence of use to create a machined part. Often this is referred to as process capability analysis. Wysk[22] introduced a "process boundary table" which defines for hole and plane producing operations tolerances on dimension and form. These are determined based upon statistical regression fits of data (that is a slope and intercept for linear relationships and exponents and intercepts for nonlinear relationships) based upon intuitive analysis, simulation and/or experimental evidence. Basically, tool position errors for plane generation due to setup or inaccurate measurement of tool length or diameter provide a constant offset or intercept and machining conditions, like metal removal rate, provide a variable input. Tolerances on hole diameters (diameter only, not form such as cylindricity or perpendicularity) are estimated similarly, Scarr[23]. These are of the form

$$\text{Tolerance} = A\,(D)^{n} + B \qquad (2.1)$$

where:
A = coefficient of the process (say drilling)
n = exponent describing the process (sensitivity of process parameters on hole tolerance for a specific diameter)
B = constant describing the best tolerance attainable by the process (and here this could refer to the drill geometry, specifically, and tolerance on the drill diameter as these will have the largest influence on diameter)
D = diameter of the hole

Tool deflection will cause errors in straightness and parallelism so tool length (often normalized by diameter) will appear as a dependent variable. For face milling operations, tool deflection (at a certain tool length but driven by material removal rate — to which tool forces are generally proportional) and the corresponding out of plane deflection of the face of the milling cutter is useful for estimating surface roughness. Think of a rotary lawn more with a bent blade shaft passing over a lawn. The “sawtooth” appearance that results is exactly the same as the surface of a workpiece machined by a deflected face milling cutter. Recently, a number of researchers have developed very sophisticated software programs for predicting these effects in an attempt to better plan the process to meet design specifications; see DeVor[24], for example. Figure 2.3, from the Machine Tool Task Force Study[25], summarizes one prevailing view of the feedback from the process to control machine performance. This, as with most other schemes, still operates at the “pre-determined command” state described by Moriwaki.

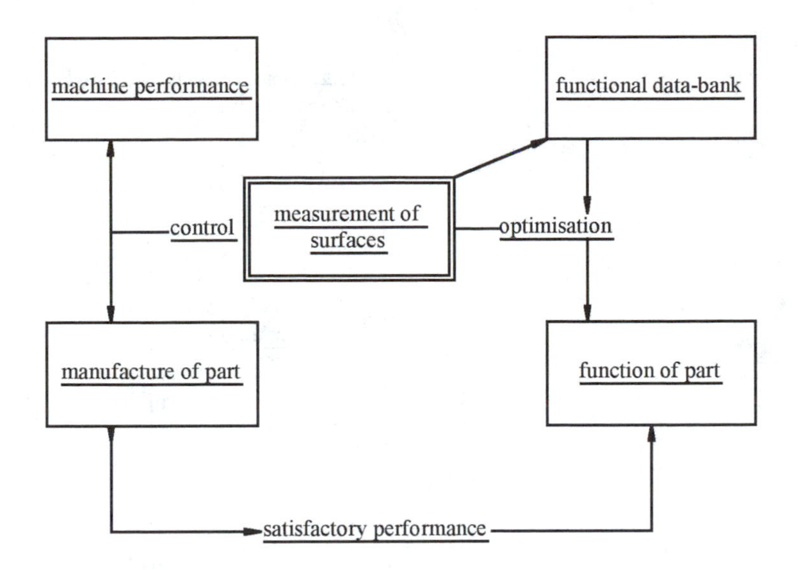

Figure 2.3. Process feedback for manufacturing, from MTTF[25].

As interesting and useful as these estimates of process capability, they are reactive rather than proactive. That is, they try to predict the performance by measuring and modeling the part features. For precision manufacturing, this is seldom effective and can, in fact, confuse the issue by masking interaction effects. Hence, we are interested in the sources of errors and the extent to which we can understand, model and predict the magnitude and direction of their effects. The study by Shirley and Jaikumar[16] also summarized common sources of error in machine tools using the Taniguchi classification of "systematic" and "random." They also included "dynamic" with random but Taniguchi would call this random as well. Classification of errors in machine tools are categorized as mechanical and thermal operational errors with respect to those on the part and those on the machine. They also include operational errors which, basically, cover all other errors from programming the controller to sloppy tool holders to measurement errors as with a coordinate measurement machine. From the point of view of determinism, the systematic errors are most repeatable and predictable while the so-called random/dynamic errors are not. As we will see, most of the errors in their random column are, in fact, predictable (or if not, can at least be bounded).

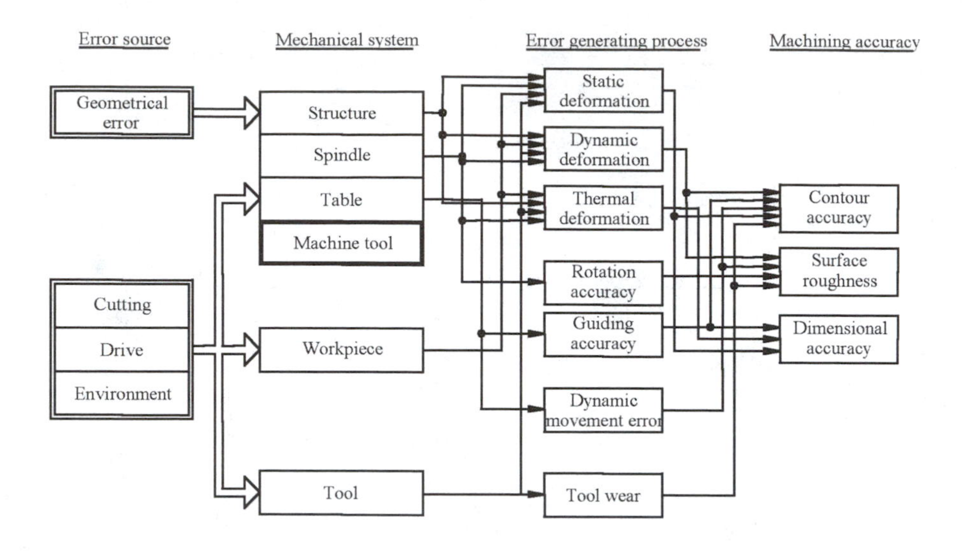

Figure 2.4. Machining error generating process, from Wada[26].

A more instructive view of error sources and their effects is shown in Figures 2.4 and 2.5, both from Wada at a Japanese machine tool engineer's conference[26]. Figure 2.4 constructs a "fault tree analysis" of the source of measurable errors in three of the most critical features on the machined part, contour or form accuracy, surface roughness and dimensional accuracy. It traces the accuracy back to the "process" generating the error, such as static deformation or tool wear, and associates it with the likely mechanical system elements in which the error generation occurs, such as a spindle or table (as part of the machine tool). It includes the other elements of the machine tool loop as well, the workpiece and the tool. We will see that diagrams such as this one, with measures of influence placed on the lines connecting one box to the other, will be the basis of our development of quantitative "error budgets" for machine design later in the notes.

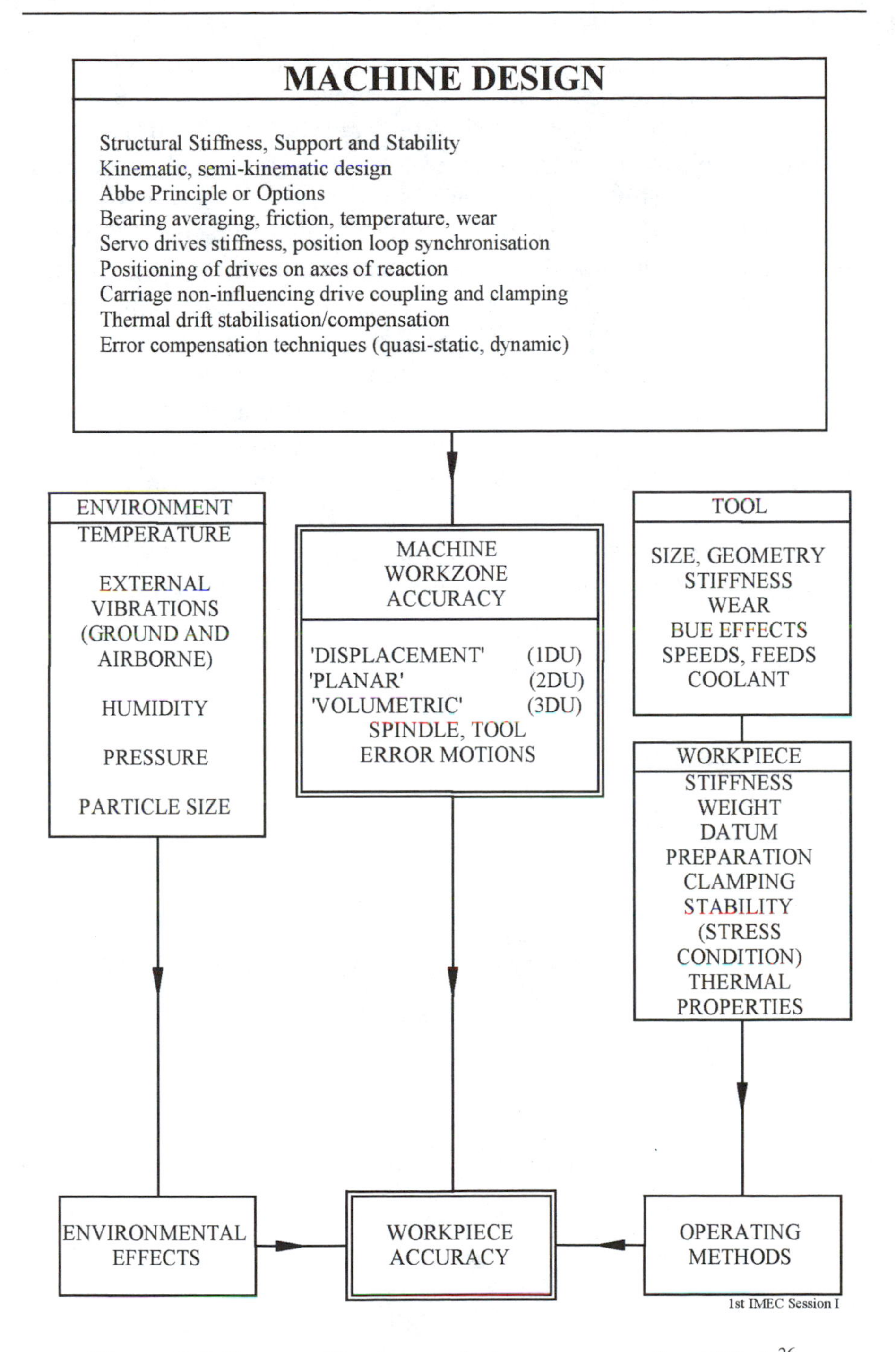

Figure 2.5. Factors affecting workpiece accuracy, from Wada[26].

Figure 2.5 is a detailed view of the specific contributors to workpiece accuracy from the point of view of the machine design (mapped on to the machine workzone accuracy), the environment in which the machine operates, tool characteristics and workpiece characteristics. This figure describes the "to do" list for precision machine tool design for manufacturing. Of interest to us will be the methods of quantifying these contributing sources, estimating their cumulative effect (most can be superposed), and determining how to minimize or eliminate their effects. Process related contributors are listed here under tool and workpiece but this does not give complete treatment to their impact and will be treated in more detail as well. The most significant point to be made from these two figures is that it is possible to trace the dimensional, contour and surface accuracy of a workpiece back to specific machine elements through the error generating mechanisms at work. With that knowledge, we can proactively design machines and processes for precision manufacturing.

III PRINCIPLES OF MEASUREMENT

3.1 Definition of terms – accuracy, repeatability, and resolution

The previous chapter talked extensively about workpiece accuracy with out clearly defining the term *accuracy*. Taniguchi talks of precision, accuracy and smoothness but not repeatability. For our purposes, we are interested in *accuracy, repeatability and resolution*. Different authors on precision engineering have slightly differing views of the definitions of these terms. Let's cover each one in order.

This chapter includes discussions centering on concepts of statistics. It is assumed that the reader is familiar with the basics of statistics, especially Gaussian distributions, *t*-distributions, variance and standard deviation, etc. so those will not be defined in detail here. There are a number of excellent references on this that can be easily found.

3.1.1 Accuracy

When people talk of "machine accuracy" it often implies a broad sense of the performance of a machine. For example, references to machine accuracy may use resolution, repeatability and accuracy in the same definition. We would like to define this term more specifically for the purposes of these notes. We can start by looking at a histogram of the dimension D representing the diameter of a large number of machined parts, Figure 3.1. A sample batch of these parts was measured with a sufficiently accurate instrument. The frequency of occurrence of the measured dimensions, D_i, where i is the specific part measured, are shown as a function of a range of possible dimensions. The distribution shown in Figure 3.1 represents a normal distribution which is not uncommon. To be correct, this distribution is called a *t*-distribution which can be approximated as a normal distribution when the number of samples is on the order of 30 or more. However, there is no reason to be surprised if a different distribution results from this type of test.

On its own, Figure 3.1 represents some basic knowledge about the manufacturing process, often referred to as the "natural tolerance" of the process- that is, the natural range of variation in the dimension from the process. Designers will base the design tolerance on the "handbook" estimate of what that tolerance should be for a particular process. If we estimate the population standard deviation, σ, for the manufacturing process from this data by conventional means, the normally associated design tolerance should be on the order of $\pm 3\sigma$. That is to say, a range of six sigma is what can be expected from the process — minimum to maximum dimension, *D*. Of course, for a normal distribution, this means that some percentage of the machined diameters will fall outside of this six sigma range and, technically, be rejects from the point of view of the tolerance of the design. In fact, exactly 0.27% (or 2700 parts per million) will fail at this tolerance. This last statement is the basis of the process control/quality improvement measured by C_p and C_{pk} since, if the process can be controlled to better than the six sigma range, that is have a smaller natural tolerance range, with respect to the acceptable design tolerance, fewer rejects result. We will discuss C_p and C_{pk} in more detail later.

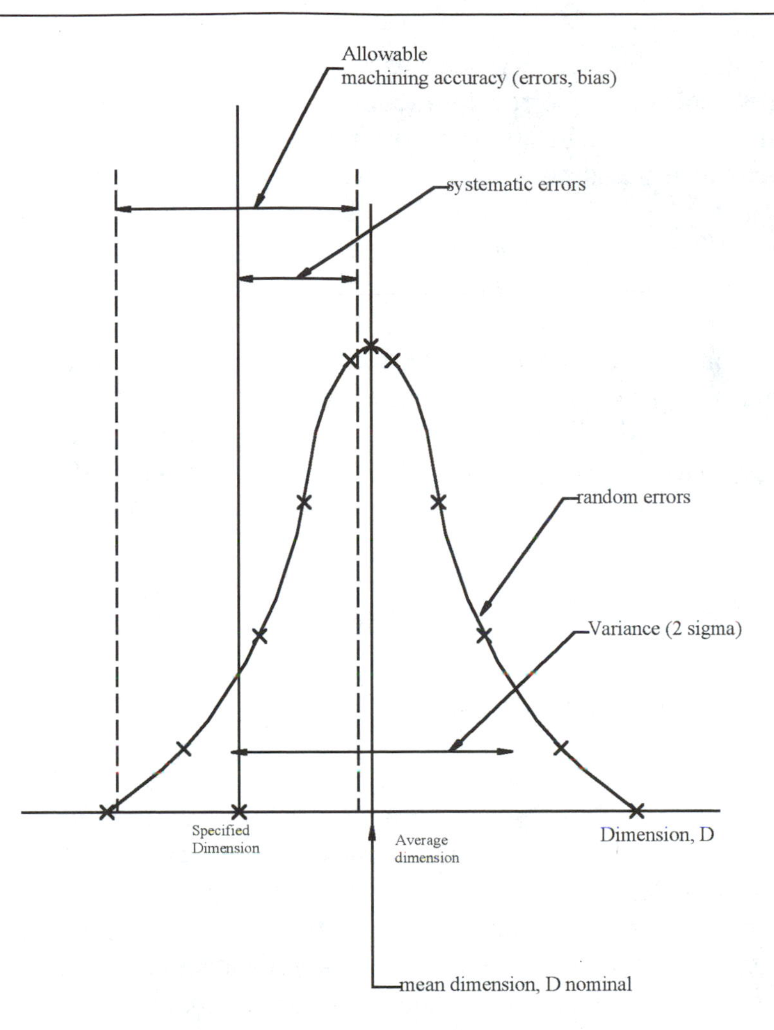

Figure 3.1. Distribution of dimensions of a part machined to diameter D.

Now, the distribution of diameters, D, of the workpiece is caused by something. We could assume that this distribution results from positioning inaccuracy in the tool, which is a machine tool performance issue, all else being under control (all of the other error sources listed in Figure 2.4, for example). Thus, knowing the distribution of D tells us something about the performance of the machine making the part. Another way of getting this information would be to measure directly the motion, or position, of the tool from part to part (operation to operation). This would remove a number of process related sources of error in the distribution and, possibly, more clearly reflect the behavior of the machine tool. For example, any dimensional variation due to tool wear or chip formation would not appear in the tool position data if position is measured on the tool shank. If we measured tool position by looking at the end of the tool, tool wear would certainly be included as a source of variation.

In Figure 3.1 we can also see the mean of the distribution as well as the nominal (desired) dimension. The difference between the nominal or desired dimension and the mean is sometimes called the *bias*, Δ. The smaller the value of the bias, the more accurate the machine or process is. We are calculating this bias as follows:

$$D_{mean} = \frac{1}{N}\sum_{i=1}^{N} D_i \tag{3.1}$$

where N is the number of parts measured.

This can be shown in multiple dimensions as well, Figure 3.2 from Slocum[18]. This figure uses a "bullet-hole" analogy to show the bias in two dimensions as the root mean square distance of all of the holes in one cluster to the "bull's eye" or nominal physical dimension desired. A similar analogy could be shown for volumetric bias, always measured as the root mean square distance of an actual position or cluster of positions, to the origin or nominal position.

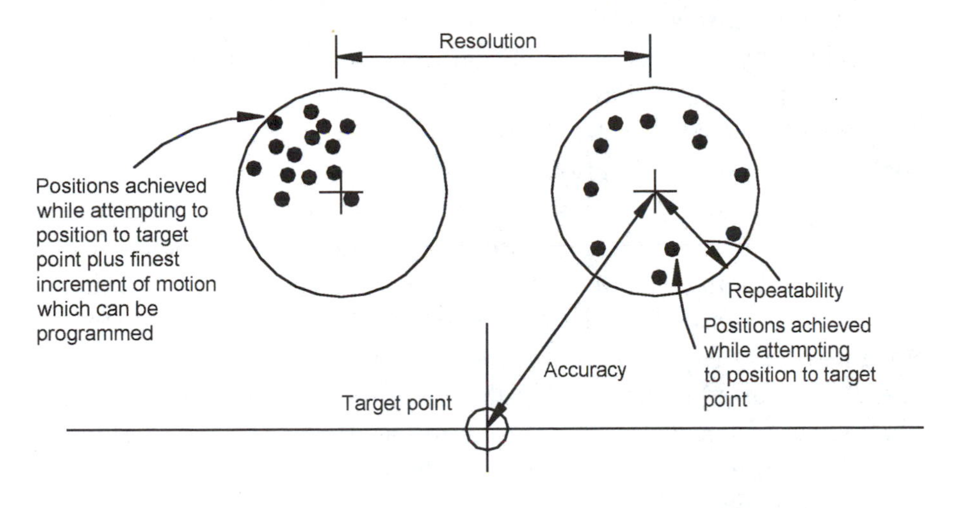

Figure 3.2. Two dimensional representation of accuracy, repeatability, and resolution, from Slocum[18].

3.1.2 Repeatability (or precision)

Repeatability is, as the term suggests, the ability to repeat the same motion or measurement within certain definable bounds. In Figure 3.1, this is indicated by the dispersion of the distribution, or standard deviation, σ. We calculate this in the usual fashion:

$$\sigma = \sqrt{\frac{1}{N-1}\sum_{i=1}^{N}\left(D_i - D_{mean}\right)^2} \tag{3.2}$$

In the figure, the distribution of D_i can represent as well the distribution of positioning accuracy as discussed above. In Slocum's bullet-hole figure this dispersion is represented by the diameter of the circle encompassing all of the holes in a cluster. Obviously, the indication that we are dealing with a marksman rather than a novice would be a tight cluster of holes close to the bull's eye — that is a person that is accurately as well as repeatable or precisely. One can design to a certain precision by insuring that all of the features, here the diameter, fall within a certain distance to the nominal dimension.

As we say earlier, if the process creating the features can be represented by a normal distribution, then the number of parts with dimensions falling within ± 3σ will be 99.7300%; within ± 4σ will be 99.9937%; ± 5σ will be 99.9999%; and ± 6σ will be 100%. We normally assume bi-directional repeatability, as with backlash in a machine tool axis. This is not always the case however. An excellent example can be seen in measuring the positioning accuracy and repeatability of a machine tool slide (axis) with integral covers to keep chips and other swarf off the way elements. In tests in Professor Paul Wright's lab at Berkeley on a Haas vertical milling machine, the repeatability varied depending on the direction of motion of the axis. This occurred because in the direction causing the slides to extend, the extension was smooth with little applied force on the axis but in the opposite direction of motion, the slides bound and caused a significant force resulting in table displacement and position error. This was measured using a laser interferometer.

3.1.3 Resolution

Resolution is, simply, the finest resolvable increment of measurement or motion. This could correspond to the smallest step size on a machine axis controlled open loop by a step motor (for example, in a point to point mode of operation), the smallest increment of motion measurable with an encoder in a closed loop servo motor driven system, or the most significant digit on a metrology system. This could be either a software limit or a mechanical limit. It is often referred to as the *basic length unit* or *control resolution* for numerically controlled machine tools. In Figure 3.2 this corresponds to the measurable spacing between to adjacent clusters of holes. This is a significant point to make here. The occurrence of feature sizes, like dimension here, is such that any value within bounds is reasonable. The resolution of the instrument measuring the dimension will apply some natural "segmentation" with respect to the histogram as in Figure 3.1 — a sort of discretization if you will. This results in the same discretization errors encountered in other systems when going from continuous to the discrete.

3.1.4 Probabilistic measure of accuracy

When bias is found to exist in a machining process or other process by, for example, tracking a dimension (diameter here) as a function of number of parts made on a control chart, the manufacturing conditions are adjusted to compensate for the bias to make the mean on the distribution agree with the nominal, i.e.

$$D_{nominal} - D_{mean} = \Delta \tag{3.3}$$

which should be maintained as small as possible. Of course, $\Delta = 0$ is best. However, errors can occur since we are using the sample mean and not the population mean, μ, as the basis of our correction. Consider the following example from Nakazawa[19].

The probabilistic range over which the difference $\Delta = D_{nominal} - D_{mean}$ will be dispersed can be used as a quantitative measure of accuracy. This is often referred to as the "confidence interval" and can be given by the following:

$$\delta = t(\phi, \frac{\alpha}{2}) \times \sqrt{\frac{v}{N}} \tag{3.4}$$

where $t(\phi, \frac{\alpha}{2})$ is the t-distribution value for $\phi =$ N-1 degrees of freedom and the probability is $1-\alpha$ obtainable from a statistical table. N is the sample size as before, α is the level of significance, and v is the unbiased variance, σ^2. This represents a kind of tolerance band on the mean.

If we consider an example in which 0.95 is chosen as the probability ($1-\alpha$), expressed as having a "95% reliability", so α = 0.05. If we take N = 21 (the dimensions of 21 parts are measured) and v = 4 μm^2 (determined from the sample distribution), we obtain $t(20, 0.05/2) = 2.09$ from the table. Using equation (3.4),

$$\delta = 2.09 \times \sqrt{\frac{4}{21}} = 0.912 \mu m .$$

Thus, the accuracy of the machining process is evaluated as 0.912 μm with a probability of 0.95 (or 95% reliability). That is, μ - *D* will be within ± 0.912 μm with a 95% probability. This is usually expressed as "0.912 μm (95%)".

A similar quantification of precision can be made as follows. Define the dispersion as ε. Precision is proportional to the dispersion. Then using the unbiased variance v from above,

$$\varepsilon = t(\phi, \frac{\alpha}{2})\sqrt{v} \tag{3.5}$$

For this example, with a 95% reliability (α=0.05) and 20 degrees of freedom, the *t*-distribution value is 2.09, so 2.09 x $\sqrt{4}$= 4.18 μm with a probability of 95%. This is expressed as "4.18 μm (95%)". That is, the finished dimension will be disbursed within ± 4.18 μm of the mean value D_{mean} with a .95 probability.

Accuracy and precision can be combined to give a broader definition of precision. If the accuracy as calculated here (δ) and the dispersion (ε) are both known, then τ, equivalent to the error limit, can be defined as

$$\tau = \delta + \varepsilon \tag{3.6}$$

When accuracy and precision can be estimated with sufficient reliability, then precision, broadly defined, can be obtained from the variance of the dispersion of these two quantities, which are treated as independently varying factors, as

$$\tau = \sqrt{\delta^2 + \varepsilon^2} \tag{3.7}$$

3.2 Metrology and measurement

With the above definitions we can now address the issues of determining the accuracy of physical elements of a workpiece or the motion of a mechanism creating those elements. These measurements are, not surprisingly, related to displacement in one direction (1D), surface displacement (2D), and displacement in a volume (3D). McKeown[27] represents these three measurements as shown in Figures 3.3-3.5. In Figure 3.3 displacement accuracy is defined a "Total positioning accuracy" and with one dimensional uncertainty defined as "The maximum uncertainty (systematic plus random) between any two points in a specified line of measurement within a specified displacement." Planar accuracy and volumetric accuracy are defined in their respective figures, Figure 3.4 and 3.5. This is distinctly different from measurement of features such as straightness (1D), flatness (2D), and form (3D) where it is the artifact that is created that is the subject of the measurement and not the motion of the mechanism creating it. Obviously the two are linked. Figure 3.6, from Moore[47], shows examples of one, two and three dimensional errors in machine elements, here the v-ways of a precision machine tool. The classification depends upon whether or not one, two or three dimensions are necessary to represent the error. Figure 3.6c shows the "twist" of a table riding along vee-ways, and demonstrates a rotation of a table about an axis of motion.

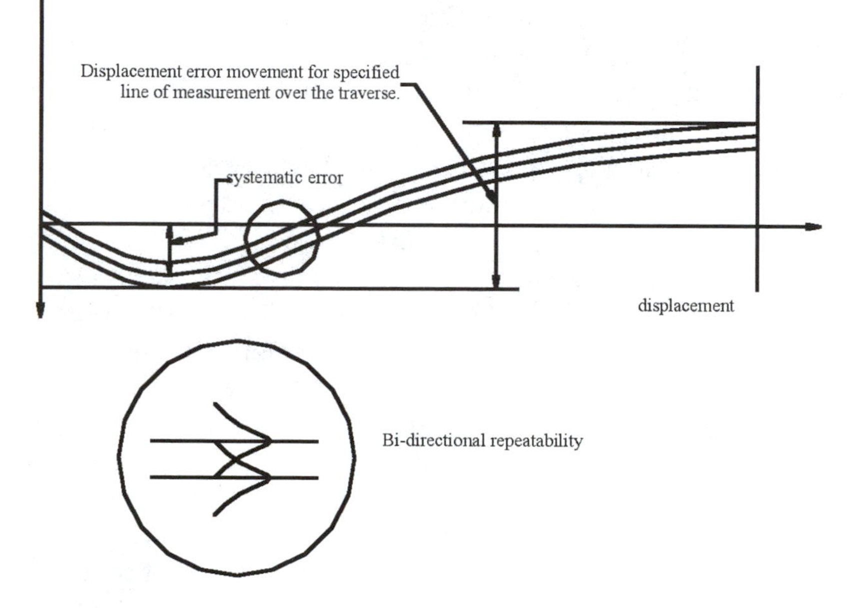

Figure 3.3. Displacement accuracy, from McKeown[27].

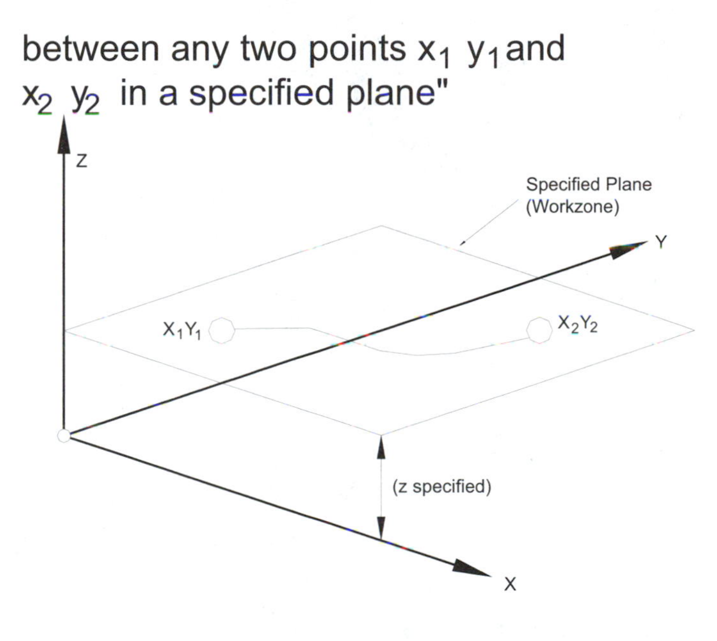

Figure 3.4. Planar accuracy, from McKeown[27].

VOLUMETRIC ACCURACY

[Three dimensional uncertainty (3DU)]
"The maximum errors,

U_{3x} (in x - direction)
U_{3y} (in y - direction)
U_{3z} (in z - direction)

between any two points x_1, y_1, z_1 and x_2, y_2, z_2 in a specified volume"

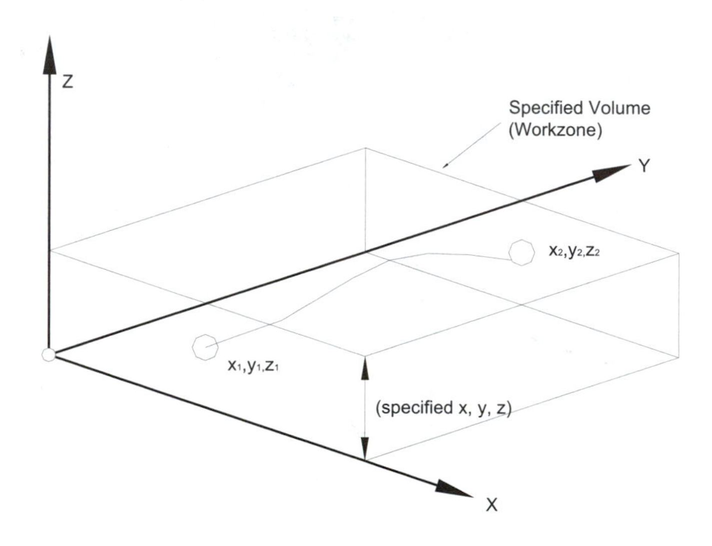

Figure 3.5. Volume accuracy, from McKeown[27].

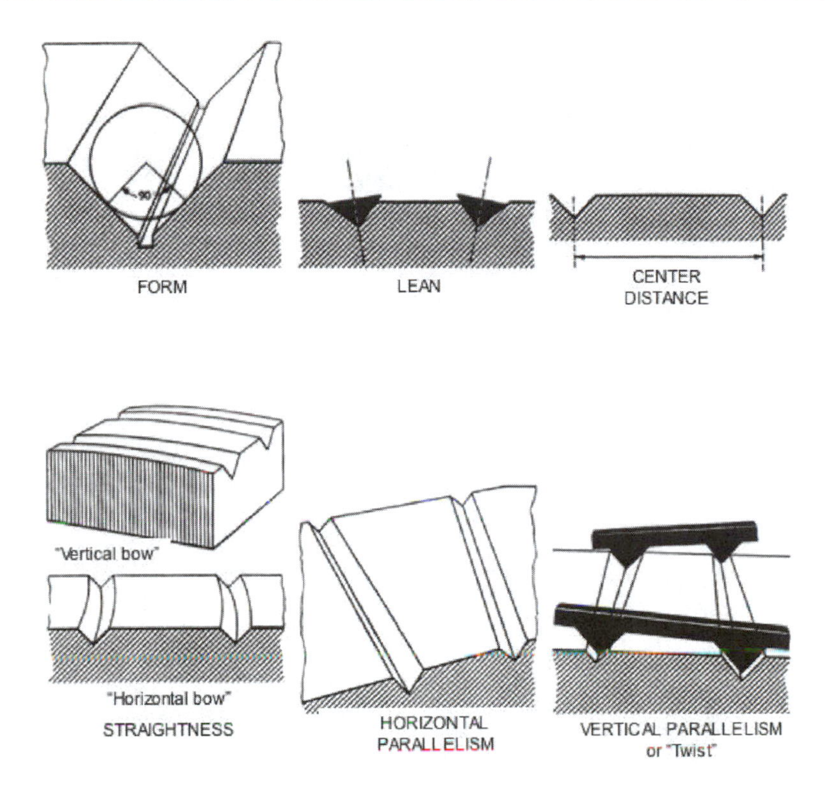

Figure 3.6. Typical machine way errors, from Moore[21].

There is standard nomenclature for defining the possible errors in a mechanical system. A rigid body has six degrees of freedom — 3 in translation: *x, y, and z*; and 3 in angular motion or "tilt" commonly referred to as "roll" , "pitch", and "yaw": θ_{roll}, θ_{pitch}, and θ_{yaw}, Figure 3.7. Measurement techniques have been established to determine the magnitude of errors, with respect to some reference, of each of these degrees of freedom. But how we make the measurement has a big influence on the reliability of the measurement itself.

Nakazawa gives four principles of measurement that point out some of the basis of good measurement or metrology[19] and they are summarized here:

- First Principle of Measurement — When measuring the dimensions of an object, one must know its temperature.

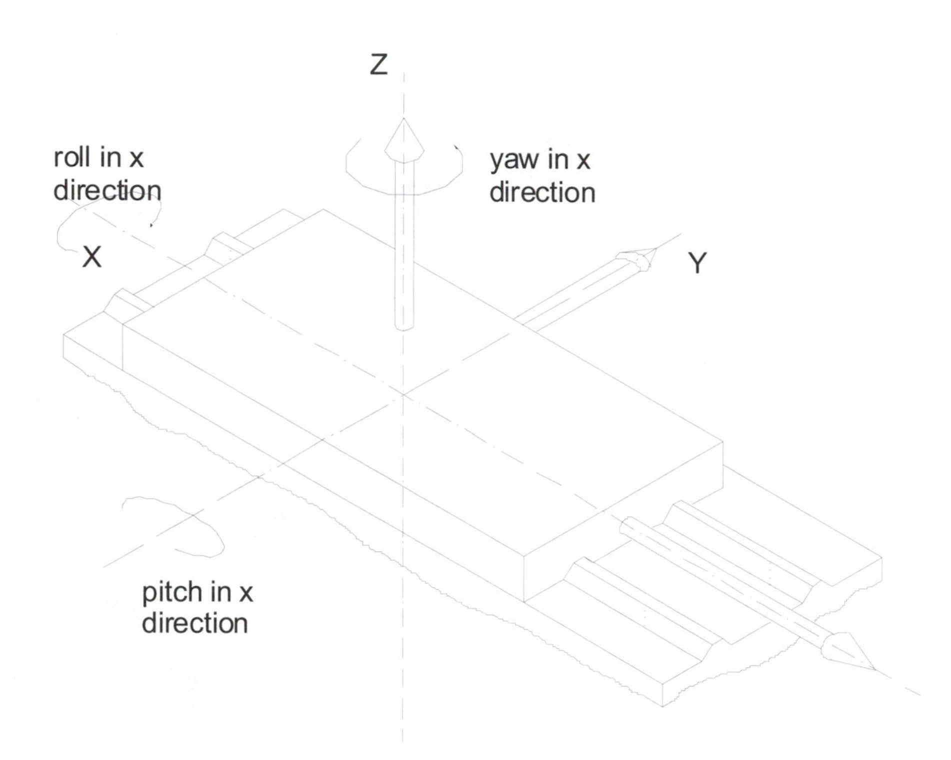

Figure 3.7. Rigid body degrees of freedom.

Coefficients of expansion of materials differ greatly. Referenced to 20°C, steel has a coefficient of approximately 13 (10^{-6}) m/m-°C, while glass is 9 (10^{-6}) m/m-°C, and Invar (a special nickel steel composition) is 0.9 (10^{-6}) m/m-°C. There is a "super Invar" which is specially well suited for precision machine tool applications that boasts a coefficient of linear expansion of -0.01 (10^{-6}) m/m-°C — virtually zero. One can appreciate the errors that are introduced due to linear expansion of machine elements due to temperature differences in the structure. And, if two dissimilar materials are layered (say granite, with a coefficient roughly double that of steel, and steel, any expansion will result in a 2D error). Since most humans can be represented nicely as a 100W light bulb from the point of view of heat sources, just standing next to some instruments or machines is enough to cause a measurable error! We will discuss all of this in more detail later on.

- Second Principle of Measurement — Avoid the application of force. (Non-contact methods of measurement are preferable to contact methods.)

This relates specifically to contact stresses generated by the contact of two bodies under load and, due to the elasticity of the bodies, the resulting deformation. The deflection of the contacting bodies is proportional to the contact force, the equivalent modulus of elasticity of the two bodies and the equivalent radius of contact, recall Hertz. Hence, there will always be some error introduced into the system due to this deflection due to contact. It will vary with surface conditions of the two contacting bodies as well. This too will be discussed in more detail.

- Third Principle of Measurement — The precision of the measuring instrument must be five or ten times higher than the expected precision of the measured object.

This is standard operating procedure for measurements and relates back to our definition of precision and resolution so that, statistically, we can insure a reliable measurement.

- Fourth Principle of Measurement — Be aware of the behavior of the instrument.

All instruments, like machines, exhibit unique behavior which will at some level affect the performance of the instrument. Directional behavior due to screw "windup" or backlash in gears, low speed errors due to stick-slip phenomena, drift due to thermal effects, etc. are the focus here. Skilled machinist or metrologists, having worked with specific instruments for some time, "understand" these peculiarities of performance. And, they use the instruments in such a way as to minimize them by, for example, approaching a measurement point by axis motion in the same direction each time. This is also applicable to machines. Early robots with unsophisticated controllers and encoders for motion would always move to a point for a pick or place operation from the same direction (clockwise, for a polar coordinate machine, for example). This would negate any backlash due to "hunting" around the final position that would introduce inaccuracy in the position if both clockwise and counterclockwise motion was allowed to the final position.

3.3 Abbé's principle

One of the more fundamental issues to be considered in metrology and machine tool control is the amplification of angular errors. Dr. Ernst Abbé, who was one of the co-founders of Carl Zeiss, brought attention to this problem late in the 19th century and his work on this led to what is referred to as the Abbé principle. Abbé observed that

> "If errors in parallax are to be avoided, the measuring system must be placed co-axially with the axis along which displacement is to be measured on the workpiece."[28]

Basically said, the line of measurement should coincide with the measuring line of the instrument. The simplest examples of instrument designs which violate and obey Abbé are the dial caliper and micrometer, respectively. In the case of the caliper, any distortion of the jaws and their tips (the point at which the measurement is made) will result in an error in the reading of jaw displacement on the scale located at the base of the jaws, The micrometer, by contrast, has the scale for measurement in line with the tips of the jaws and no Abbé error results. Nakazawa[19] shows two lathes, one for which the tool post is correctly designed to comply with Abbé and the other violating the principle, Figure 3.8. The effect of Abbé offset is seen on a milling machine in Figure 3.9. Complying with Abbé is consistent with Nakazawa's fourth basic functional requirement of a precision machine — being able to detect movement accurately.

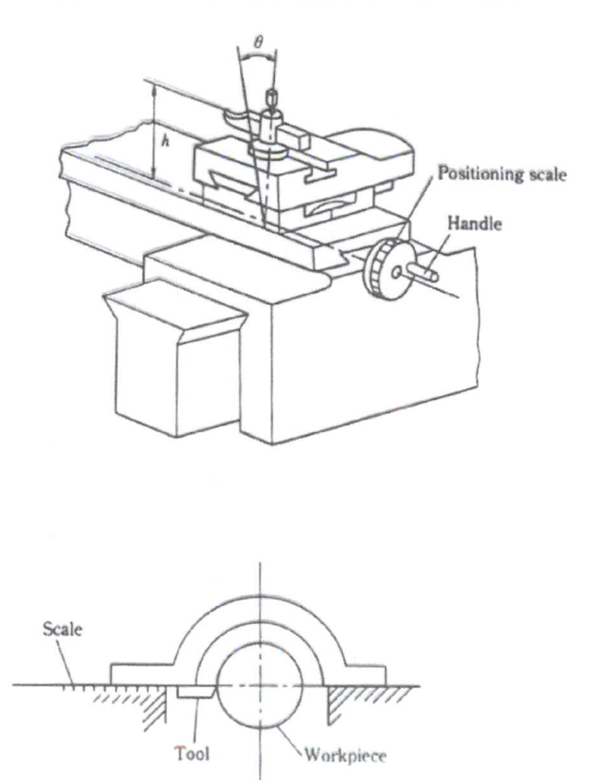

Figure 3.8. Lathe with tool post violating Abbé's principle (a), and a tool post designed in accord with Abbé (b), from Nakazawa[19].

Bryan[28] has amplified and extended the Abbé principle making it more easily understood and, importantly, how to effectively meet the requirements if it is not practical to do so directly. Bryan states that[28]

"The displacement measuring system should be in line with the functional point whose displacement is to be measured. If this is not possible, either the slideways that transfer the displacement must be free of angular motion or angular motion data must be used to calculate the consequences of the offset."

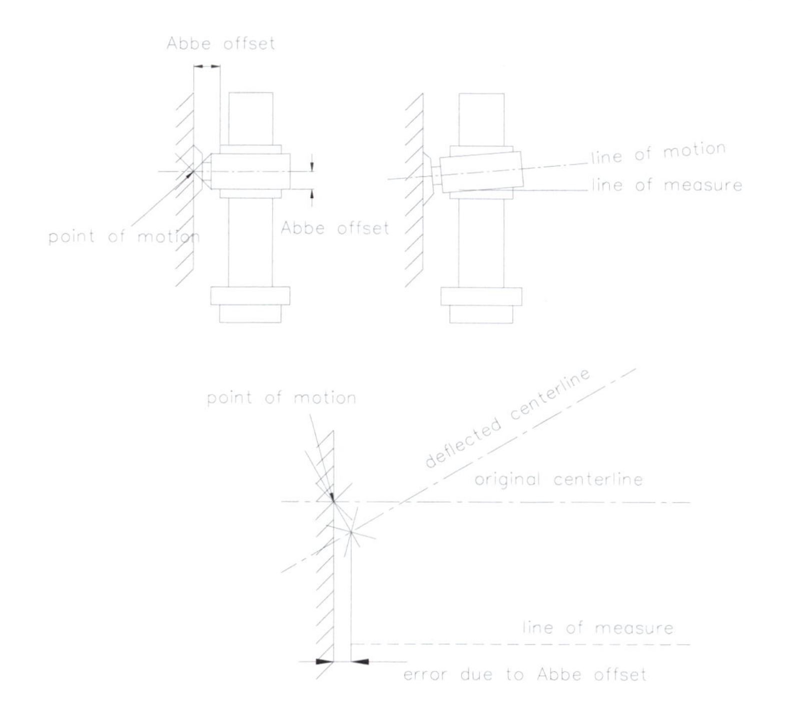

Figure 3.9. Milling machine with Abbé offsets.

In the case of a lathe, the "functional point" is likely to be the end of the tool that contacts the work surface. For a milling machine, the functional point may be the center of the face of the milling cutter creating a surface. Bryan further discusses straightness measurement and re-states the above but replaces "displacement" with "straightness" at each occurrence. So "...the straightness measuring system should be in line with the functional point whose straightness is to be measured..." etc. This is an important addition as, in general, it is generally difficult or impossible to abide by Abbé for all axes of motion on a machine without incurring additional expense. And then, even if Abbé, or Bryan's enhancements, are followed, during the course of operation, other effects may cause non-anticipated distortions or displacements that introduce measurement errors. We will see many opportunities to assess the degree of compliance with Abbé in the material that follows in the rest of these notes.

3.4 Metrology techniques

Not surprisingly, metrology techniques differ according to the particular artifact being measured and the features of that artifact of interest. Recall that we had previously listed six characteristics of a workpiece (or the machine generating it) that are of interest to us and, hence, need to be determined, quantitatively. These were: dimensional precision, angular precision, form precision, surface roughness, kinematic precision, and, surface layer alterations. This section reviews techniques for establishing measures of dimension, angle, form, etc. In many cases, standards have been established to prescribe the method of determining a measurement or how to interpret the results. Surface finish is one such case. In other cases, no standards exist and/or there are no specific methods of measure and the measurement method and analysis depends upon the specific use for the information. Surface layer alterations (or subsurface damage) are such a case for this. Finally, in some cases these measures are more conveniently covered in another section and that will be noted. All of the material below is integral to our study of precision manufacturing and could be discussed from several different points of view.

3.4.1 Measurement of dimension and angle

Measurements are made by comparison of an instrument reading to a standard. This standard may be a length standard, such as the standard meter originally agreed upon in 1875 establishing the length of a meter. In the United States, the inch was defined as 25.4000508 mm. In October 1960, a standard based upon the visible emissions of krypton (Kr) 86 in a certain frequency was adopted. This allowed the replacement of a physically created artifact with a naturally occurring phenomenon — an immutable standard. This standard length was again redefined in 1983 and one meter defined as "...the distance that light travels in a vacuum during a period of 299, 792,458th of a second." The inch is defined by the U.S. National Institute of Standards and Technology (formerly NBS) as "41, 929.399 wavelengths in a vacuum of the reddish-orange radiation corresponding to the transition between levels 2_{p10}-5_{d5} of the unperturbed atom of krypton 86." In practice, we may not have access to a krypton coherent light source but we generally rely on physical length standards such as precision scales, gage blocks (often called end standards) and other measuring instruments. In many cases, the gage blocks are used to calibrate other measuring instruments. Precision scales have lines ruled on them with a spacing of about 20 μm and they are used as references.

Gage blocks, created in the 1890's by Carl Johansson in Sweden (thus sometimes referred to as "Jo" blocks), consist of a set of 102 blocks that in combination allow over 20,000 measurements to be made. The blocks, finely lapped, would adhere when "wrung" together and allow the accumulation of any distance within the 20,000 possibilities. This accumulated length would be used as a reference. Because of the adhesion of the individual blocks, there was no effect of the joint between adjacent blocks. The method of manufacture of these blocks included a sequence of stress-relieving the blocks between processes by alternately exposing the blocks to freezing and heat treatment with periods of "rest" at room temperature in between. The blocks were, thus, exceptionally stable. The blocks were rendered "traceable" by a series of calibrations at the French BIPM (International Measurement Bureau) in Sèvres. The calibration showed that the blocks had the same degree of accuracy as the interferometer used to calibrate them and that adjustment of the blocks only displayed errors less than 0.1 μm. Improvements in the finishing of the blocks enabled them to be wrung together with a force of up to seventeen atmospheres in 1907 and up to thirty-three atmospheres by 1916 (that is, the axial force necessary to pull two blocks apart!)[21].

Interferometry is most widely used for displacement/length measurement today as well as some form measurement. Slocum[18] gives an extremely comprehensive background on interferometry going back to wave theory and the reader is referred to that source for details. Briefly, interference is a result of superimposing two or more waveforms. An interferometer is an instrument or measuring the degree of interference. If we consider two waves, one out of phase (lagging behind the other) by an angle φ, traveling in the same direction and with the same velocity, frequency, ω, and amplitude, A, we can describe their amplitudes, y_1 and y_2, as

$$y_1 = A\sin(kx - \omega t - \phi)$$

$$y_2 = A\sin(kx - \omega t) \tag{3.4.1}$$

Physically, the phase angle means that at any instant of time, t, one of the two waves will be displaced relative to the other along the x axis by a constant amount φ /k. The constant k is the inverse of the wavelength, λ; k = 2π/λ. At any point x one wave will appear to follow the other by a distance θ/ω. If the two waves are superimposed, the resulting wave is of the form

$$y = 2A\cos\left(\frac{\theta}{2}\right)\sin\left(kx - \omega t - \frac{\phi}{2}\right) \tag{3.4.2}$$

This wave has the same frequency as the original waves and, when the phase φ is zero, the amplitude is double that of the original (constructive interference). When the phase φ is π or some integer multiple, the resultant amplitude is zero (destructive interference).

In the application of interferometry to metrology, one monochromatic (ideally) wave or beam of light can be used as a reference and a second reflected from the workpiece or component whose displacement is to be measured. If the two waves travel different paths to get to the same point, they will interfere constructively for certain path length differences to and from the mirror and interfere destructively for others. For path length differences of 0, λ, 2λ, 3λ, ... corresponding phase shifts of 0, 2π, 4π, 6π, ... occur. For path length differences of λ/2, 3λ/2, 5λ/2 ... phase shifts of π, 3π, 5π, ... occur, respectively. Thus, the interferometer is designed to produce a very stable, high frequency observable waveform which, if split, makes up the reference and target beams. Then, observing the passing bands of interference (i.e. passing occurrences of destructive and constructive interference) as the target on the machine component moves, one can determine very accurately the displacement. The use of coherent light sources is employed due to the difficulty of obtaining monochromatic light. A laser is also a good source of monochromatic light. Slocum[18] shows the principal components of an optical heterodyne interferometer (Figure 3.10) and some typical reflectors for mounting on the machine tool (Figure 3.11). Although we are discussing single axis measurements here, it is obvious that the interferometer has applicability to the measurement of form error as well. The implementation of the laser interferometer in machine tool metrology is illustrated in Figure 3.12[25].

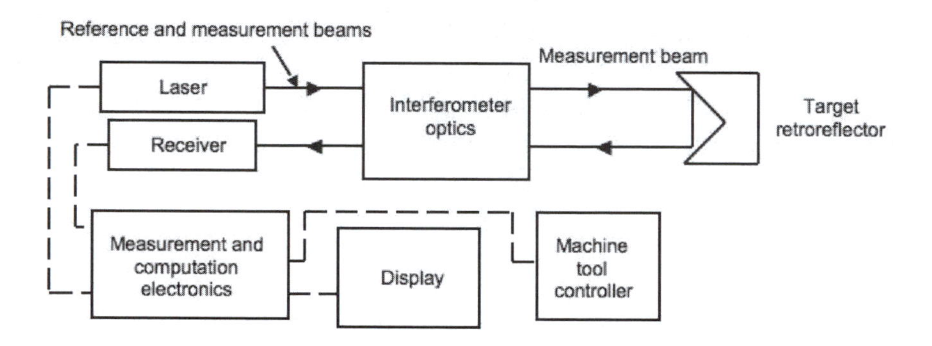

Figure 3.10. Block and flange mount retroflectors (commonly called corner cubes), Slocum[18].

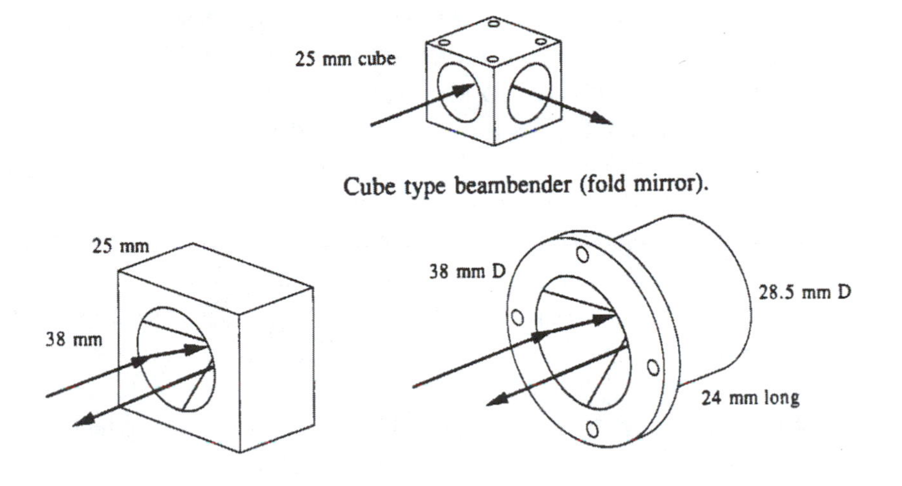

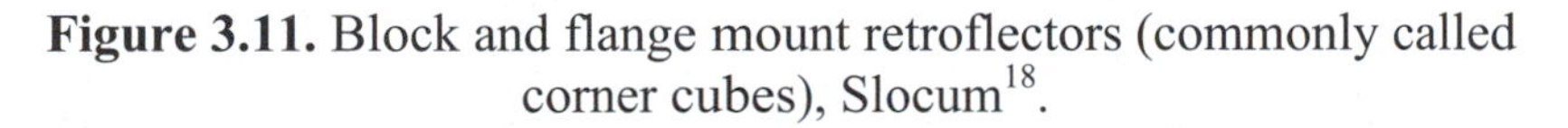

Figure 3.11. Block and flange mount retroflectors (commonly called corner cubes), Slocum[18].

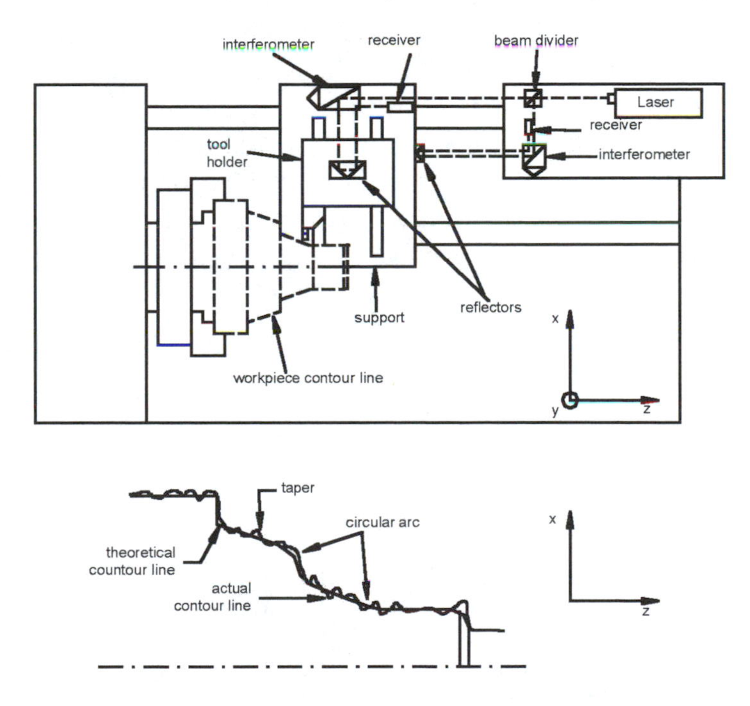

Figure 3.12. 2 axis measurement using a laser interferometer on a lathe, from MTTF[25].

The third "mechanical art" described by Moore[21] is dividing the circle, or more appropriately, measuring angles. Polygonal mirrors, rotary tables, serrated tooth circle dividers and rotary encoders are all instruments for establishing motion through an angle or measuring angles to high precision. We will mention here only the traditional method of using a sine bar to establish an angle. This method is comprised of a sine bar, a block (like a gage block- with a precise length) and an auxiliary flat surface, Figure 3.13, from Nakazawa[19]. The sine bar is designed to act as a gage or angle reference or to make angular measurements of mechanical parts mounted on it. Usually, a stack of gage blocks is set in the correct combination to form length *H* in Fig. 3.13. The corners of the right triangle, *A* and *C*, are hardened gage pins. The inclined surface is hardened, ground and lapped to a high degree of flatness. As it is the ratio of the lengths that is important and not the actual lengths, other length measuring tools may be substituted. Obviously, the system works equally well in SI or English units of measure! The device is fairly reliable at low angles, less than 15°, but becomes increasingly inaccurate as the angle increases. Further, the sine bar only presents the top surface, or that of some mounted workpiece, to the required angle. Another datum, such as a surface plate or machine axis is needed as well as other auxiliary instruments, such as an indicating device, to make the measurements.

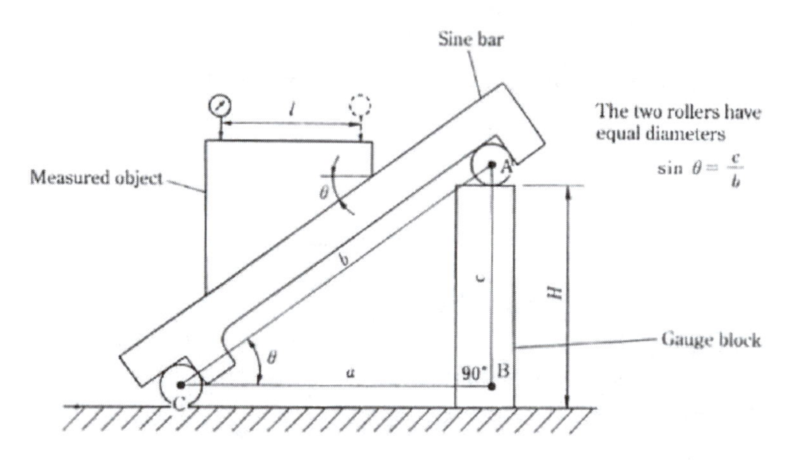

Figure 3.13. Angle measurement using a sine bar, Nakazawa[19].

3.4.2 Measurement of form

Form includes such characteristics as flatness, straightness or curvature, cylindricity, circularity, perpendicularity, parallelism, etc. That is, two or three dimensional aspects of artifacts as opposed to just a dimension, like length or diameter. This is more complicated to measure as we will see as the nature and method of application of the instrument for measurement can substantially alter the accuracy and reliability of the measurement.

Figure 3.14 illustrates some of the basic terminology in form measurement for straightness and roundness. We will look in some detail at elements of the measurement of some of these forms.

3.4.2.1 Straightness

Although interferometry was applied to the establishment of standards of measure and, eventually replaced the autocollimator as the preferred means for measurement of angular errors of form in mechanisms and workpieces), the autocollimator is still in use and serves as a practical instrument or many measurements. The autocollimator is an optical instrument which, if used in conjunction with a reflecting mirror, can accurately measure very small deviations from a datum angle. It is used to measure straightness in a machine's ways in both the horizontal and vertical plane. Moore[21] describes the instrument and its use in some detail. The autocollimator principle of operation is as follows. A target line, placed at the principal focus of an objective lens, is illuminated. The image of the line is reflected back onto a reflector placed on the machine axis by a beam splitter mirror. The image projected on the machine axis is reflected back to the instrument. When the reflector on the machine is tilted through an angle θ, the reflected image is tilted through an angle 2θ and the displacement in the instrument is measured by adjustment with a micrometer. The adjustment "re-zeros" the image as viewed through an eyepiece after each measurement and the angle of tilt can be directly determined. Figure 3.15 shows a schematic of a photo-electric autocollimator, from Moore[21] and Figure 3.16 shows its use in the measurement of errors on a machine way, also from Moore[21].

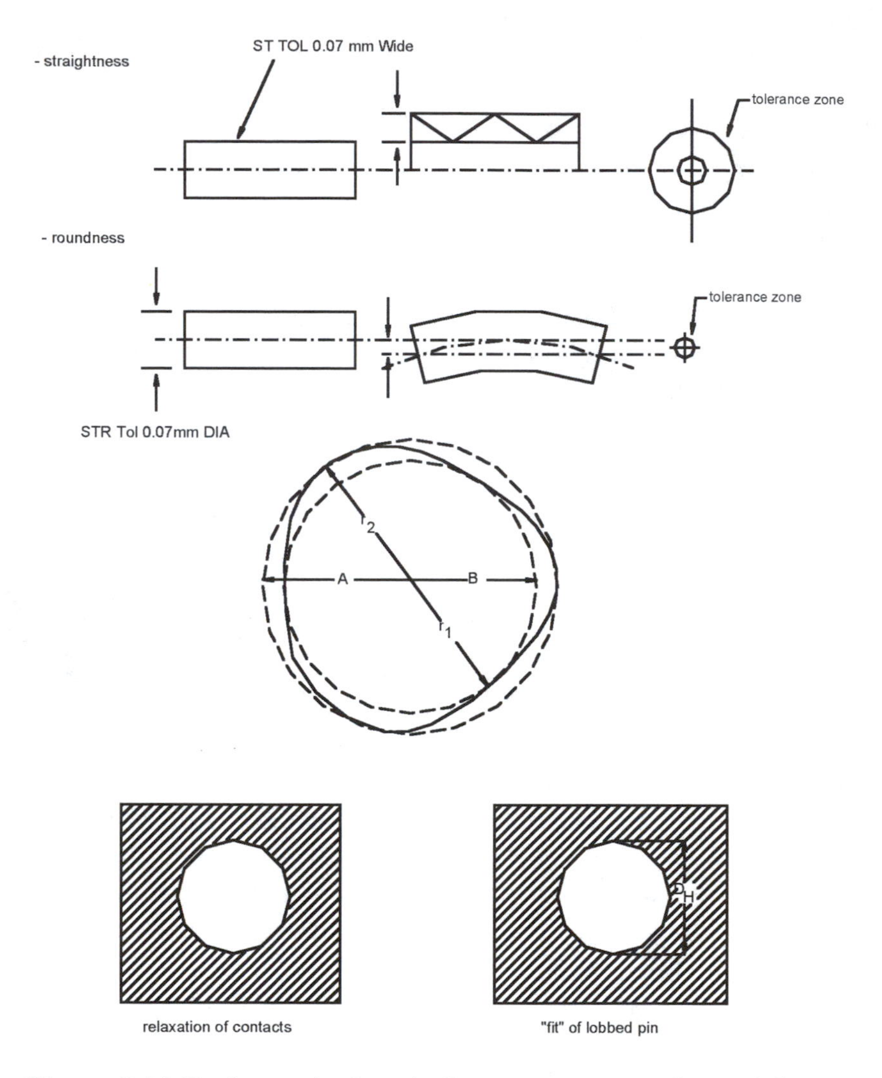

Figure 3.14. Basic terminology in form measurement for straightness and roundness.

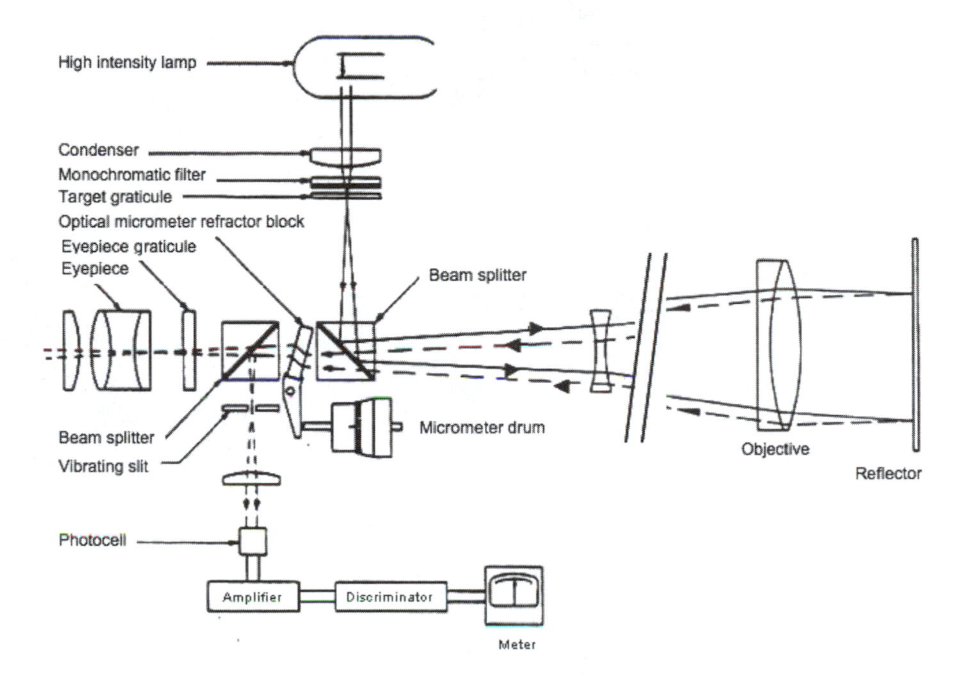

Figure 3.15. Schematic of a photo-electric autocollimator, from Moore[21].

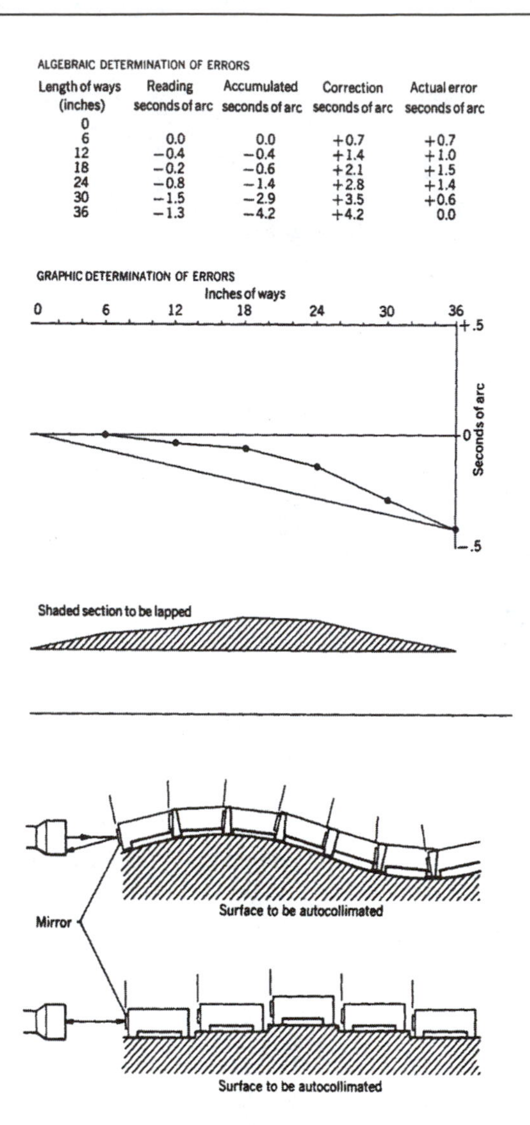

ALGEBRAIC DETERMINATION OF ERRORS

Length of ways (inches)	Reading seconds of arc	Accumulated seconds of arc	Correction seconds of arc	Actual error seconds of arc
0				
6	0.0	0.0	+0.7	+0.7
12	−0.4	−0.4	+1.4	+1.0
18	−0.2	−0.6	+2.1	+1.5
24	−0.8	−1.4	+2.8	+1.4
30	−1.5	−2.9	+3.5	+0.6
36	−1.3	−4.2	+4.2	0.0

Figure 3.16. Implement of an autocollimator in the measurement of surface error on a machine tool way, from Moore[21].

For measurement of surface form errors (as distinct from roughness), we can take advantage of the interferometer described earlier, as well as the autocollimator. As part of a discussion of the Abbé principle, Bryan[28] compares the output of an autocollimator with reflector on the feet of a slide moving over a surface and an electronic displacement indicator indicating between a parallel and a reference surface in various positions with respect to the slide. The purpose of the exercise is to measure the straightness of surfaces (here a slideway) and angular motion and demonstrate the effect of several variables of the measurement system on the results. Bryan points out that people often mistakenly assume that one point on a large mass, like a machine tool, moves in the same manner as other points on the mass. A similar mistake is made for slideway straightness. Bryan proposes a definition for slide straightness as[28]

> "Slide straightness error is the non-linear movement that an indicator sees when it is either stationary and reading against a perfect straightedge supported on a moving slide or moved by the slide along a perfect straightedge which is stationary."[28]

Bryan[28] demonstrates this by an interesting series of experiments. The text of a portion of the reference[49] is repeated here as it best describes these experiments. The figures below illustrate a few of the motions that can occur in slideways and, whereas the motions are exaggerated for clarity, they have all been observed in real machine tools in significant magnitudes.

> "Figure [3.17] illustrates a slide moving along a bed which has a sinusoidal shape. The feet of the slide have a spacing equal to an even number of wavelengths (of the surface undulation). The movement of the slide is free of angular motion, but is not straight. Since there is no angular motion all points move with the same amount of straightness error, and the indicator positions 1, 2 and 3 show the same values. Figure [3.18] is the same as Figure [3.17] except that the straightedge is moving. Here again, the straightness error is unaffected by indicator position."

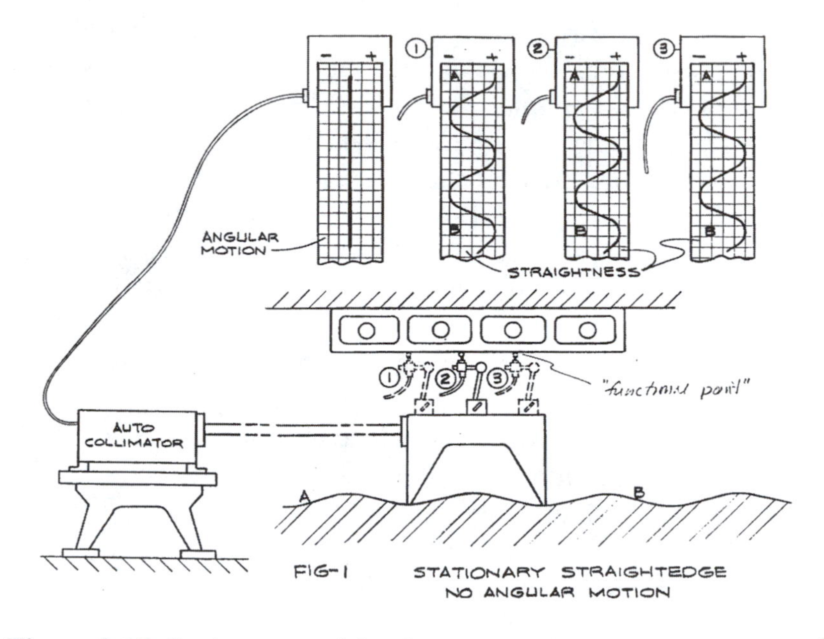

Figure 3.17. Stationary straightedge; no angular motion, from Bryan[28].

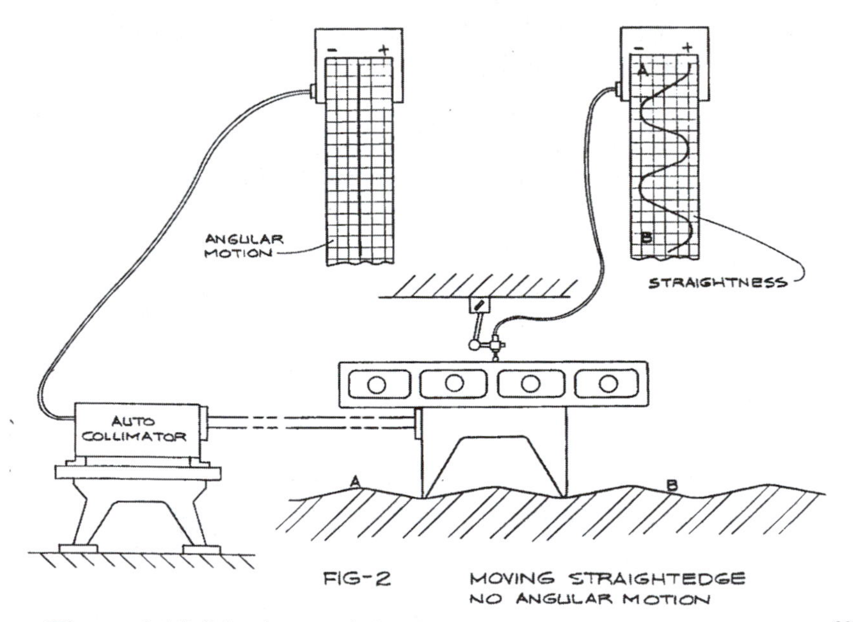

Figure 3.18. Moving straightedge; no angular motion, from Byran[28].

"In Figure [3.19] the feet of the slide have been changed to a spacing equal to a half wavelength longer. The slide now has oscillating angular motion as the autocollimator strip indicates. This condition leads to the rather amazing result that the straightness error depends entirely on the location of the indicator. At position No. 2 there is no error (or second order). Position No. 3 has a different magnitude and phase than position No. 1."

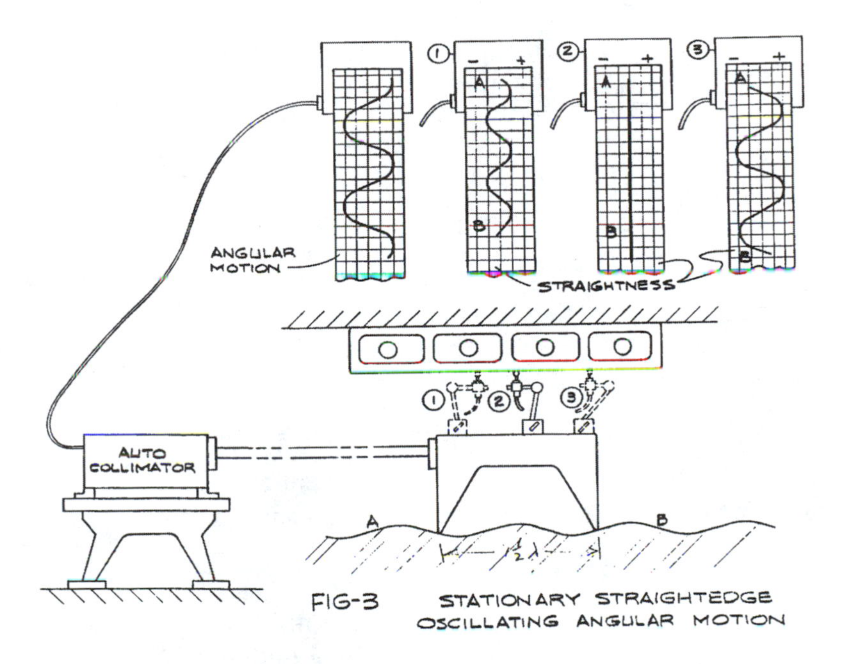

Figure 3.19. Stationary straightedge; oscillating angular motion, from Bryan[28].

"Figure [3.20] is the same as Figure [3.19] except that the straightedge is moving. The straightness error is now a combination of the profiles in Figure [3.19]. When the indicator tip is opposite the centre of the slide there is no error (or second order), when it is at the end of the travel it shows a maximum. The exact shape of the profile is not represented accurately in the illustration. For the purpose of this paper it is sufficient to show that there can be different straightness error profiles depending on the set-up and location of the indicator."

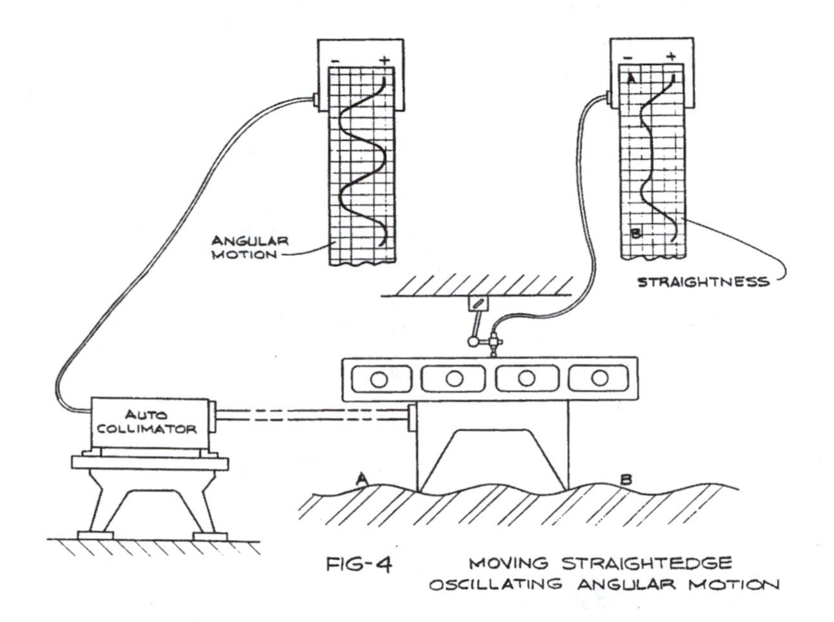

Figure 3.20. Moving straightedge; oscillating angular motion, from Bryan[28].

"Figure [3.21] shows a slide having uniform angular motion. The bedway is circular with a radius of *R*. The autocollimator trace is an inclined straight-line. The difference indicator positions all show the same straight error magnitude *S*, but the inclination of a straight-line passed through the endpoints is different. The general direction of the slide travel changes with the position of the indicator. This can create some terrible confusion when a slide is being adjusted square or parallel with another slide or possibly a spindle."

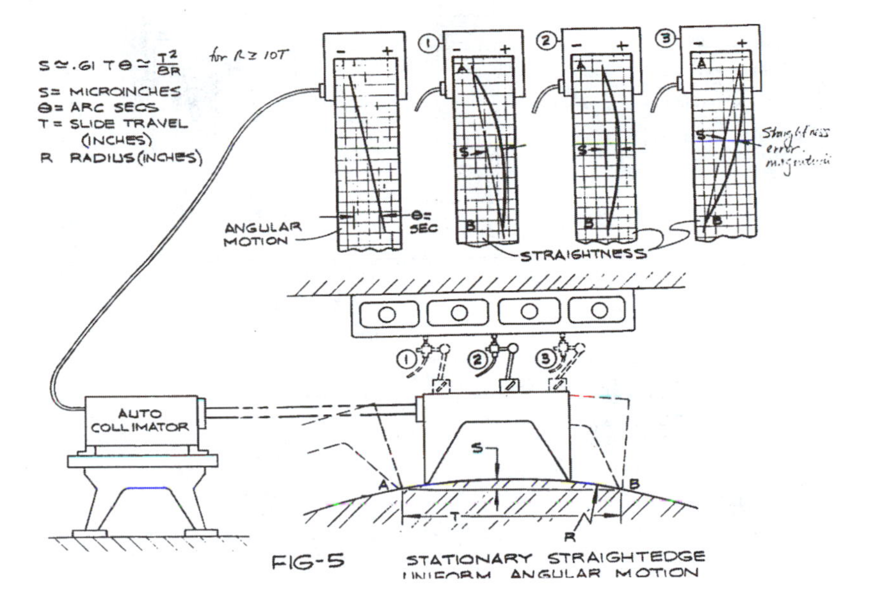

Figure 3.21. Stationary straightedge; uniform angular motion, from Bryan[28].

"Figure [3.21] also shows two approximations for the value of *S* as a function of either the travel of the slide *T* and the angular motion θ, or the travel of the slide *T* and the radius of curvature *R*. These approximations are accurate to a few percent as long as the radius *R* is at least 10 times as large as the travel *T*. These approximations are not relevant to the main subject of the paper but I have found them to be enormously useful in acquiring a feel for the magnitudes involved when dealing with uniform angular motion. Figure [3.22] is the same as Figure [3.21] except that the straightedge is moving. Changing the indicator does not affect the straightness error or the general direction of travel."

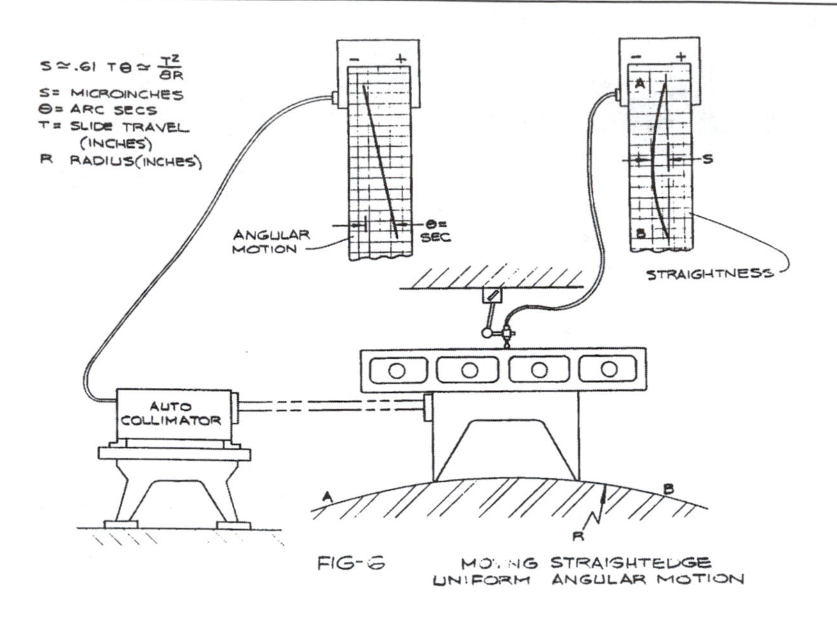

Figure 3.22. Moving straightedge; uniform angular motion, from Bryan[28].

It is obvious from the above that the setup and application of a measuring system greatly affects the results. Also, this discussion reaffirms the importance of adherence to the Abbé principle as the "functional point" in all cases is not in line with the straightness measuring system. In some of the above cases this was not a problem.

Simple straightness measurements can also be accomplished with a taut wire and a microscope fitted with a graticule, Figure 3.23[25]. A detail of the setup is shown in Figure 3.24[25]. With the aid of a four-quadrant photo-diode and a suitably intense and focused light source, one can make a number of straightness measurements, Figure 3.25[25]. The resolution of this system is dependent on the resolution of the photodiode sensor used. In some cases, arrays, with bit-mapped diode outputs, can be used.

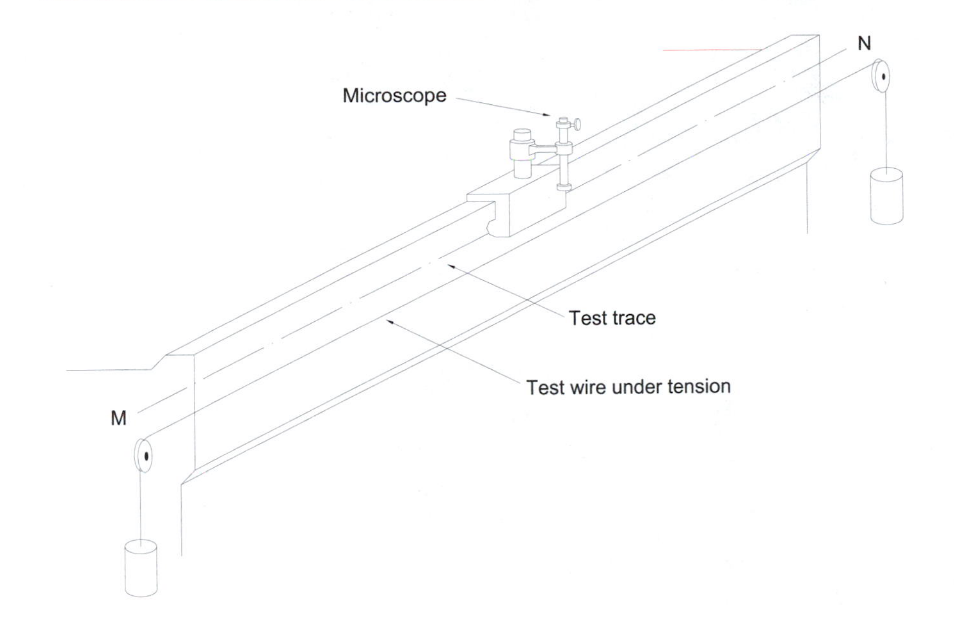

Figure 3.23. Measurement of surface unevenness with test wire and microscope, from MTTF[25].

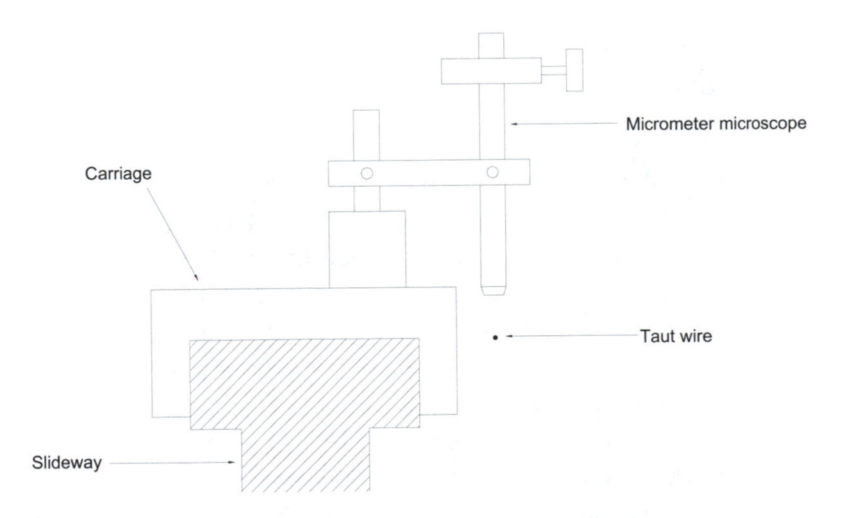

Figure 3.24. Detail of straightness measurement with a taut wire, MTTF[25].

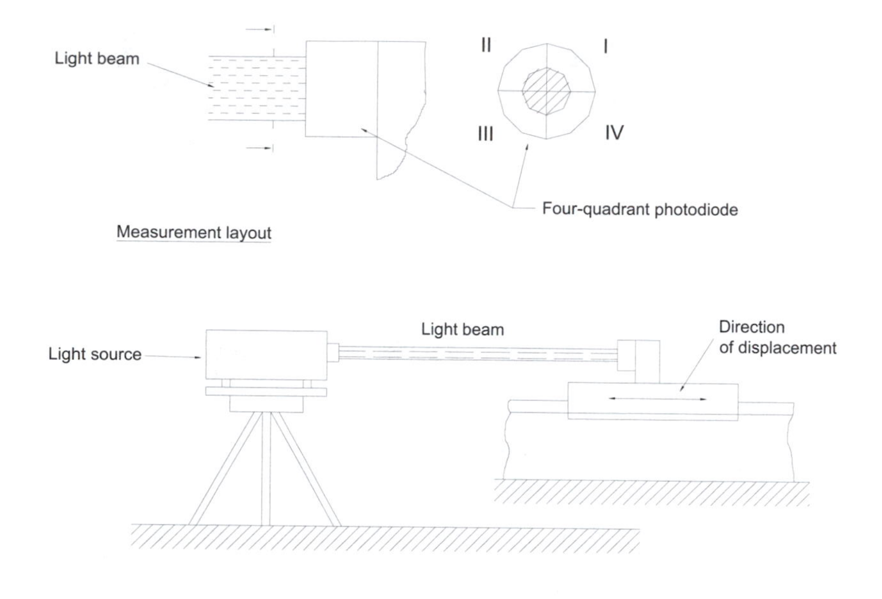

Figure 3.25. Straightness measurement with a 4-quadrant photodiode, MTTF[25].

3.4.2.2 Flatness

For surface error determination, Fizeau interferometry can be applied. In this case, a reference plane, sufficiently flat, is used and the projected light waves from the reference beam interfere with the reflected beams from the work surface. Observing the pattern and fringes of the interference indicates the magnitude and location of surface deviations from flat. Figure 3.26 shows a schematic of the interferometer configuration. Tilting and/or uniform surface irregularity of the "plane" workpiece will result in distortion of the observed interference pattern — called interferograms. Examples of interferograms resulting from various surface contours are given in Figure 3.27.

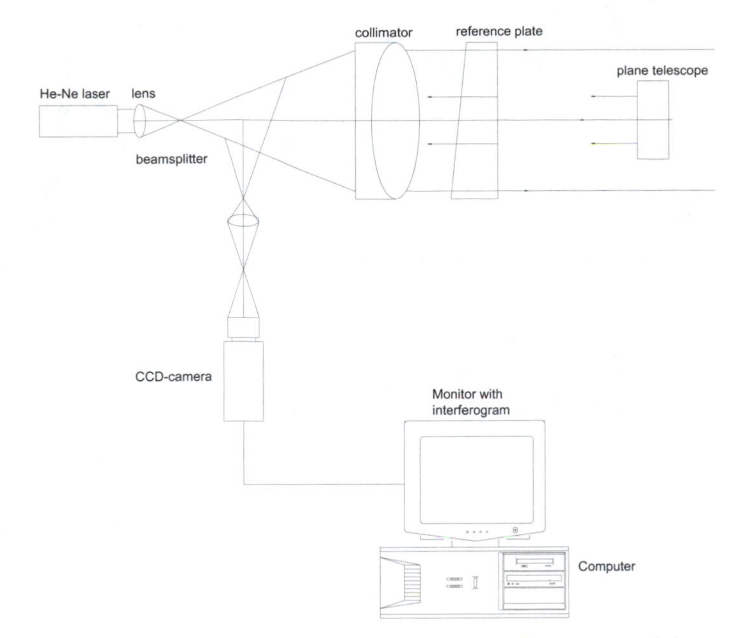

Figure 3.26. Schematic of Fizeau interferometer for flatness Measurement.

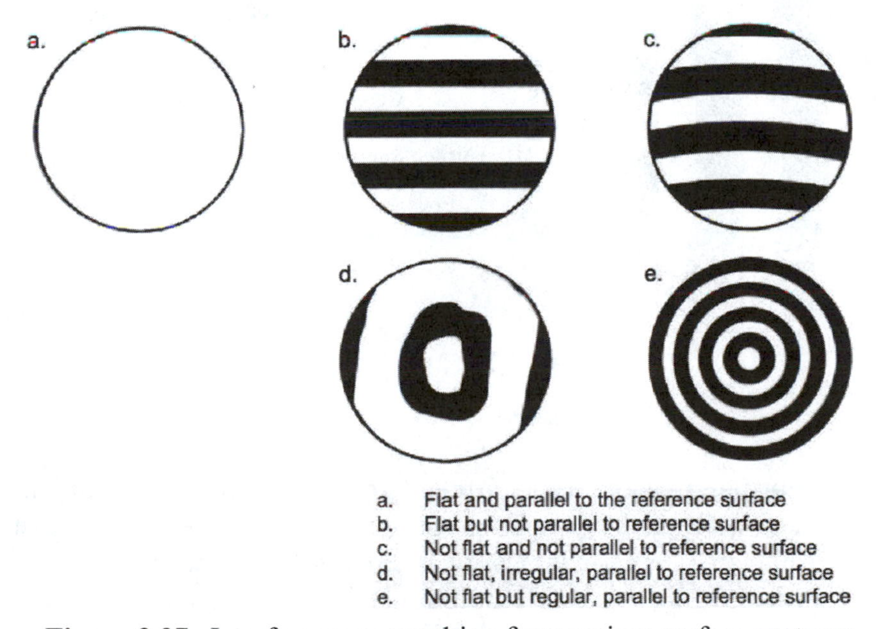

Figure 3.27. Interferograms resulting from various surface contours.

Any surface profile other than "flat and parallel" to the reference surface will result in an interference pattern. The spacing of the fringes corresponds to λ/2 change in the surface profile (where λ is the wavelength of the light source, λ = 633 nm, for a He-NE laser for example). Hence, the interference pattern creates a "topo" map of the surface with the spacing of the contour lines determined by the basic sensitivity of the instrument.

The sensitivity of the interferometer is determined by the wavelength of the light source as well as the angle of incidence of the light onto the worksurface, α, Figure 3.28. Thus, sensitivity, *S*, can be defined as

$$S = \left(\frac{\lambda}{2}\right)\frac{1}{\cos\alpha} \tag{3.4.3}$$

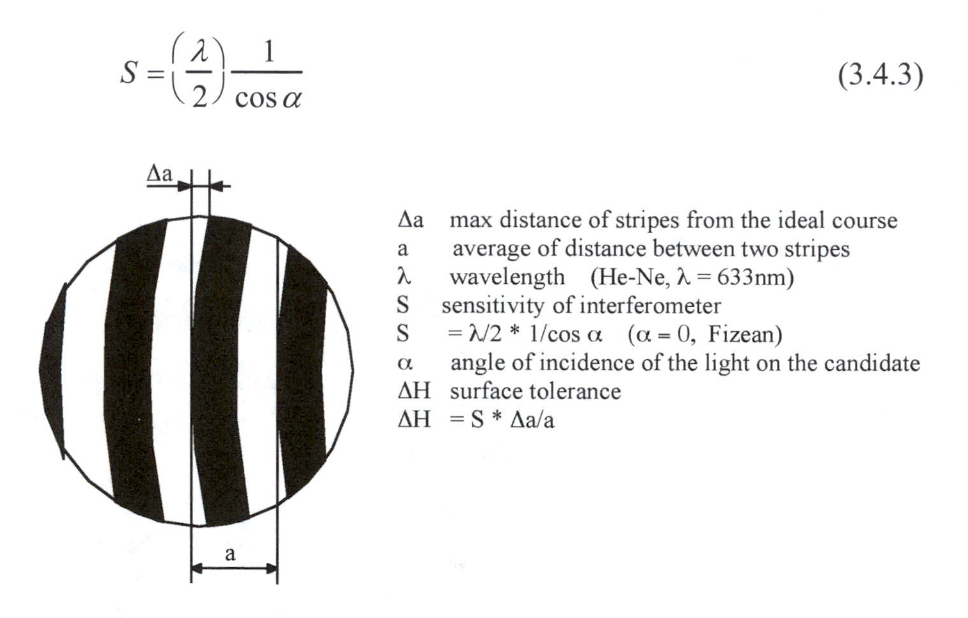

Figure 3.28. Sensitivity of the Fizeau measurement.

For non-planar surface error characterization, the instrumentation used is dictated by the shape of the form to be measured. Most interferometers have provisions for removing tilt of the specimen or worksurface to improve the resolution of the measurement. Figure 3.29 shows an interferogram and 3-D plot of a work surface not parallel to the reference plane. Figure 3.30 shows the same 3-D plot corrected for tilt. The section of the profile in the left-hand side of Figure 3.30 is for the "tilted" worksurface.

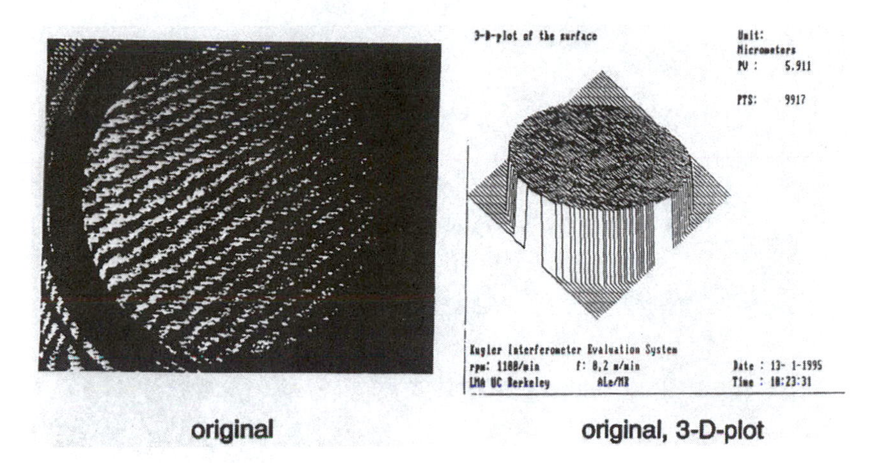

Figure 3.29. Example of interferogram for tilted specimen.

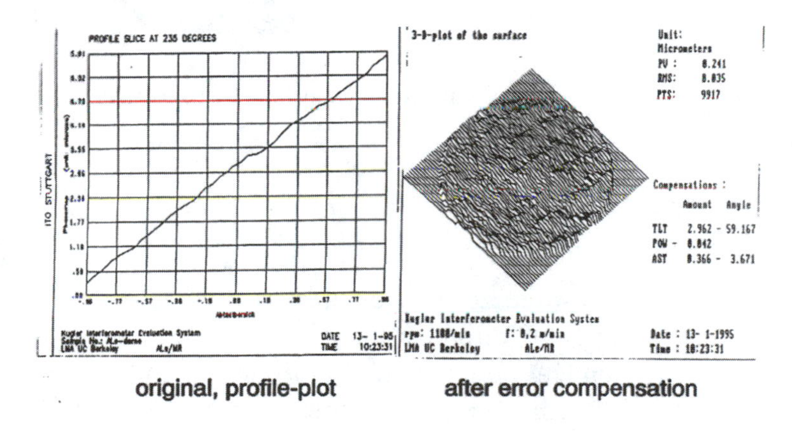

example is flat, but was not parallel to reference plate

Example for error compensation (tilt, power and astigmatism)

Figure 3.30. Correction of the tilt in the specimen.

If one doesn't have a Fizeau interferometer in one's tool box, the use of an optical flat and a monochromatic light source works well for determining surface flatness errors. In fact, this technique has been used for some time and, in the hands of a knowledgeable technician, can be very effective. The application is illustrated in Figure 3.31, from Moore[21].

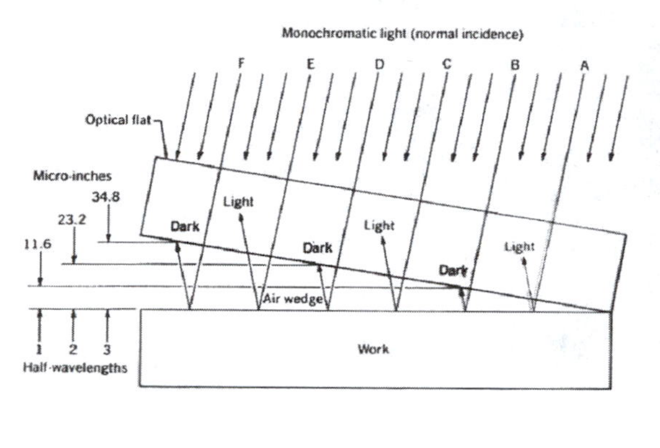

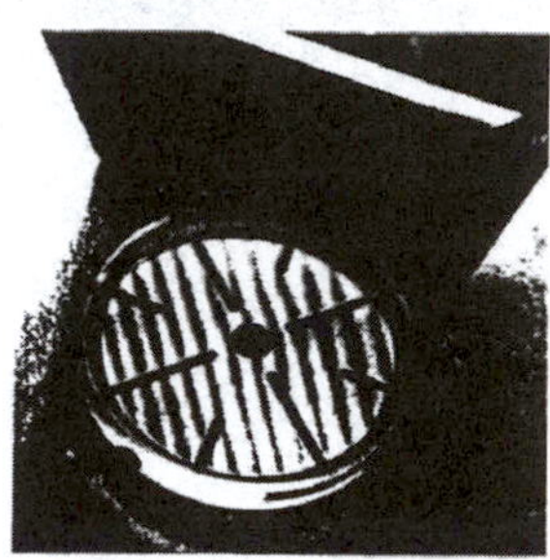

Figure 3.31. Fringe formation with an optical flat, from Moore[21].

3.4.2.3 Roundness

The generation and measurement of roundness of mechanical parts is the fourth "mechanical art" cited by Moore[21]. Location, effective size and fit are all dependent upon our ability to define and measure roundness. Figure 3.32 from Moore[21] illustrates the effect of roundness errors in mating shafts and holes for four situations. Roundness can be measured by a number of methods including:

- diametral method
- limit plug and limit ring gages
- center support method
- V-block
- three point probe
- precision spindle

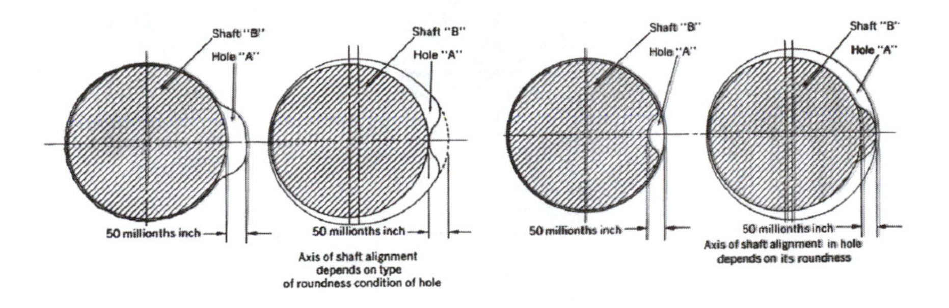

Figure 3.32. Effect of roundness errors in mating shafts and holes for four situations, from Moore[21].

The diametral method employs two parallel surfaces, the platens of a micrometer for example, to measure the diameter. Unfortunately, if the part has an odd number of lobes (due to the manufacturing process, for example, as we will see later) the out-of-roundness of the part cannot be detected. Consider the relaux triangle-shaped part in Figure 3.33. The diameter is the same across any two points of measurement. So the part could meet quite stringent diameter tolerances without being very round. If the part has an even number of lobes, the diametral method will reveal this characteristic of roundness.

The limit plug and limit ring gages require that a separate set of these gages be made for each size of work to be measured. Further, the technique does not allow the measurement of other geometric characteristics such as concentricity. The technique is illustrated in Figure 3.34 for roundness measurement on a shaft. The shaft is rotated in the gage and the inspector looks to excessive swing of the indicator needle.

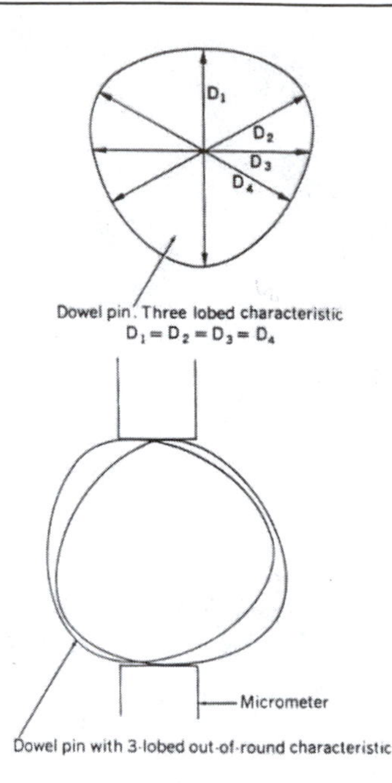

Figure 3.33. Measurement of odd number lobed diameters with the diametral method, Moore[21].

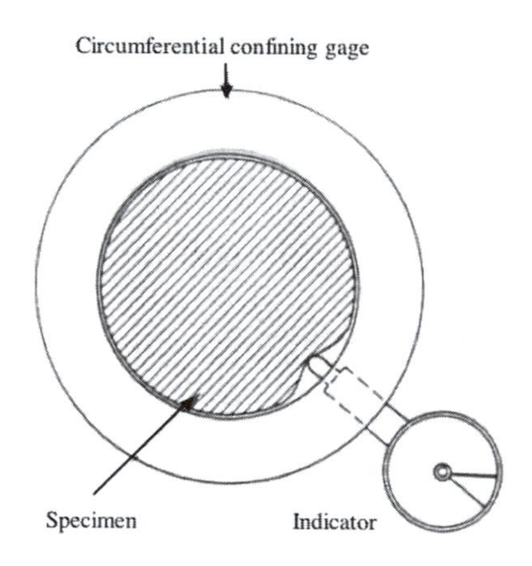

Figure 3.34. Use of limit plug and limit ring gages, Moore[21].

The center support method does exactly that to the workpiece — support it on centers, rotate the part and, using an indicator, note the presence and degree of out of roundness. This works best for shafts and similar workpieces. This can introduce other sources of error into the measurement, for example, because of machine center rotational errors (e.g. surface condition of the centers and center holed). Further, part errors such as lack of straightness, will cause a doubling of the runout effect and appear as a roundness error. We will discuss later some "reversal" techniques that can remove some of these sources of error.

Two V-block methods exist — fixed angle and adjustable angle V-blocks. Depending on the angle of the V-block used and the number of lobes on the workpiece, the method may or may not successfully detect out-of-roundness, see Figure 3.35. For 5 or 7-lobed workpieces no error will be seen for the 60° V-block and micrometer measurement, Figure 3.36. The 60° V-block seldom shows the true error. But this will be true for a V-block of other angles as well as, for example, a 90° V-block will not show the errors in a 7-lobed part. Hence the concept of an adjustable angle V-block which can be set to the appropriate angle once the number of lobes of the part has been determined.

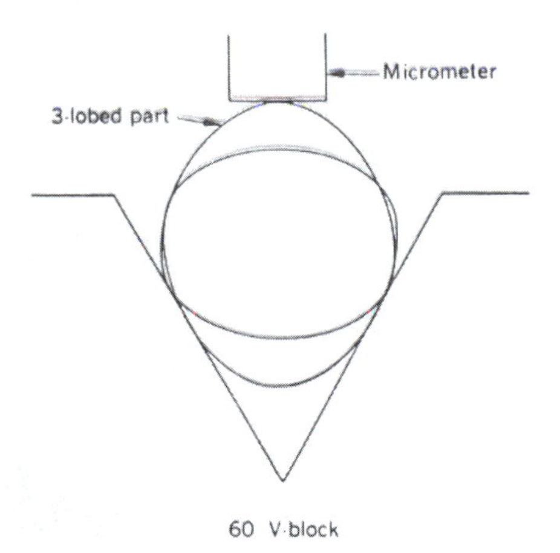

Figure 3.35. Use of a 60°V-block for inspecting a 3-lobedcylindrical part; the out-of-roundness error is exaggerated, Moore[21].

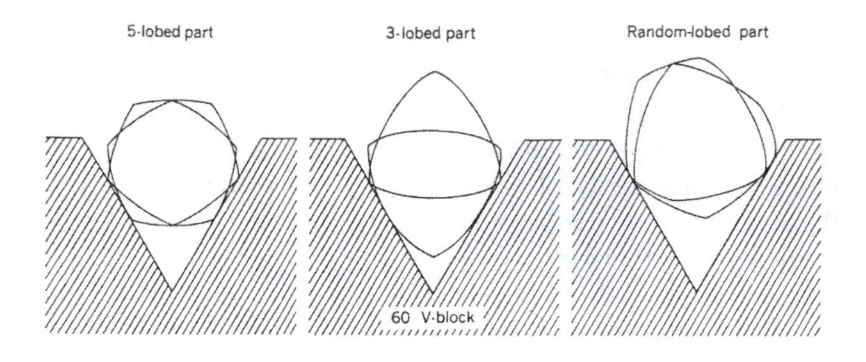

Figure 3.36. Use of a 60°V-block for measuring parts with various numbers of lobes, Moore[21].

The three-point probe works similarly to a 60° V-block method. This is effective for probing effective size when the geometry of the part if is in question. Three probes are arranged 120° apart as with a three jaw inside micrometer, three-jet ring gage for gaging a shaft. However, the same limitations apply to this method as to the V-block method with regard to inaccuracies due to part lobes. Figure 3.37 gives several examples of non-contact probes used in measurement. These probes are often used for a variety of measurements including the applications discussed here.

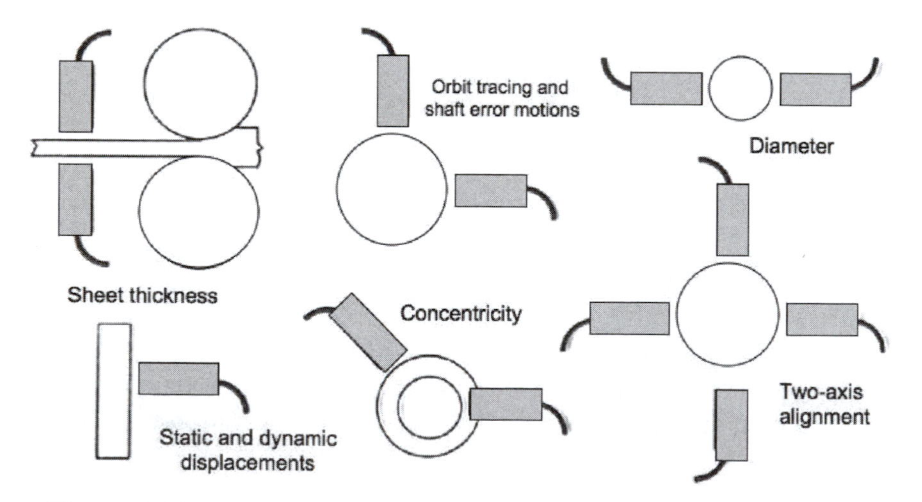

Figure 3.37. Noncontact displacement probes used for measurement, from Slocum[18].

The use of a precision spindle to rotate the part may be the best method for roundness measurement. Either an overhead spindle or a rotating table can be used. Either the probe is rotated around the part (overhead spindle) or the part is rotated on a precision table in the presence of a probe (rotating table). Both methods are illustrated in Figure 3.38, from Moore[21]. The results of these measurements are polar plots of part surface location relative to some reference and by observing these the roundness errors can be assessed, Figure 3.39, from Scarr[23]. If the precision spindle happens to be part of a machine tool, the same methods can be applied to measure the roundness and runout of the spindle, Figure 3.40 from MTTF[25].

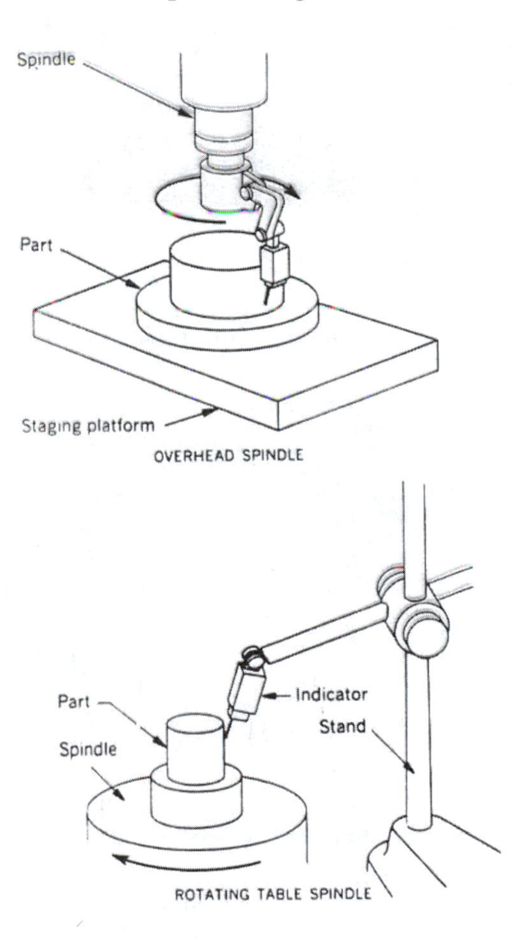

Figure 3.38. Rotating spindle and rotating table methods of roundness measurement, from Moore[21].

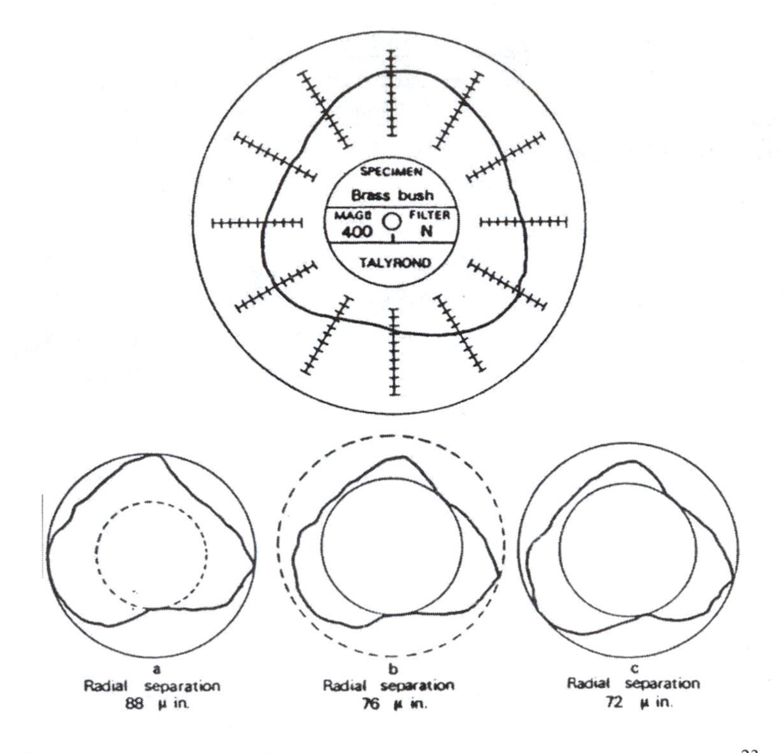

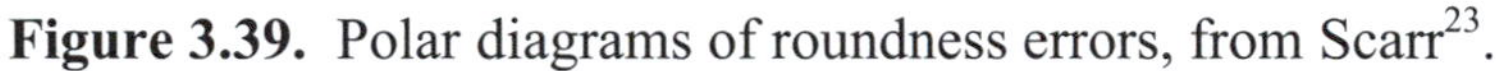

Figure 3.39. Polar diagrams of roundness errors, from Scarr[23].

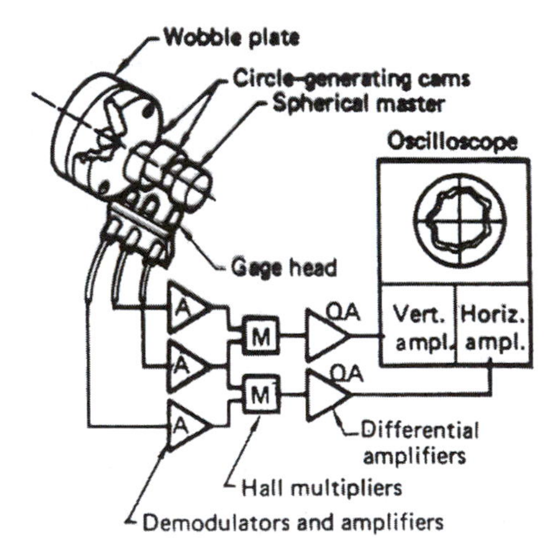

Figure 3.40. Measurement of run-out on a precision spindle, from MTTF[25].

How the error measured by these techniques is quantitatively referenced to is a complex subject. Generally, with respect to the polar plot of roundness error seen in Figure 3.39, some calculation, such as i. *minimum zone circle*, ii. *minimum circumscribed circle and maximum inscribed circle* and iii. *least-squares fit circle*. The three are described as follows. In the minimum zone circle calculation, the out-of roundness is expressed as the smallest difference in radii of concentric circles (inscribed and circumscribed) relative to the shaft or hole center, that is R_{max}-R_{min}, Figure 3.41. The second method is illustrated in Figure 3.42 and consists, for a shaft, of the difference between the radius of the minimum inscribed circle (located at shaft center) and the radius of the minimum circumscribed circle drawn from the same center. For a hole, the measure is the difference in the radii of the minimum circumscribed circle (relative to the hole center) and the radius of the maximum inscribed circle from the same center. The best-fit-circle technique is exactly that. The mean circle is determined either by a true least squares technique or so that the difference in the circle's outer and inner areas bounded by the recorded profile is minimized, Figure 3.43. Scarr[23] gives detail on the techniques of calculating these out-of-roundness measures.

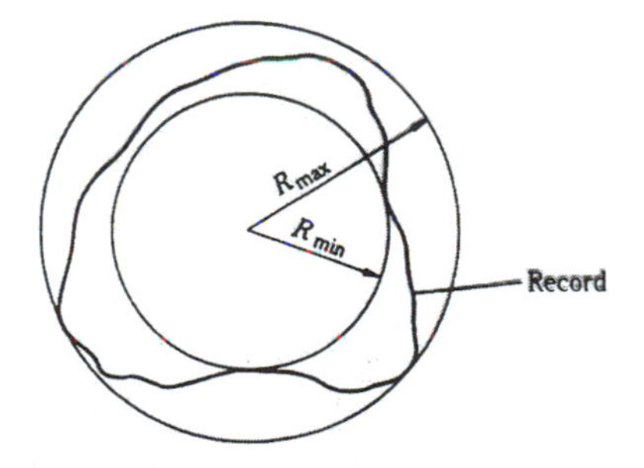

Figure 3.41. Out-of-roundness measured by the minimum zone circle calculation, from Nakazawa[19].

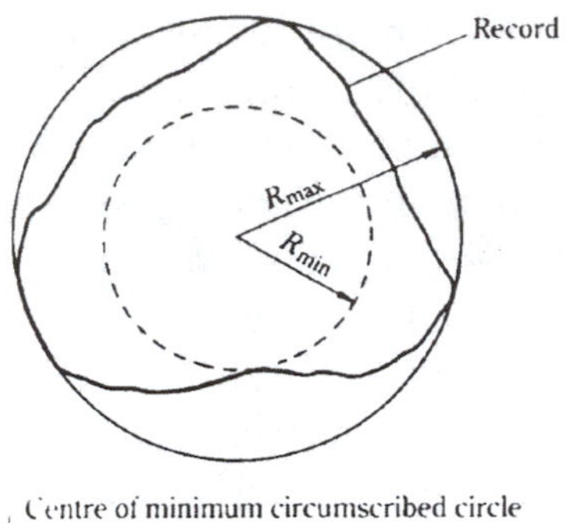

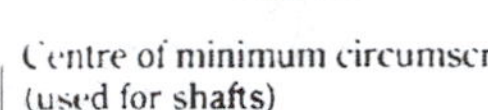

Centre of maximum inscribed circle
(used for holes)

Figure 3.42. Shaft and hole out-of-roundness measured by the minimum inscribed and circumscribed circle method, from Nakazawa[19].

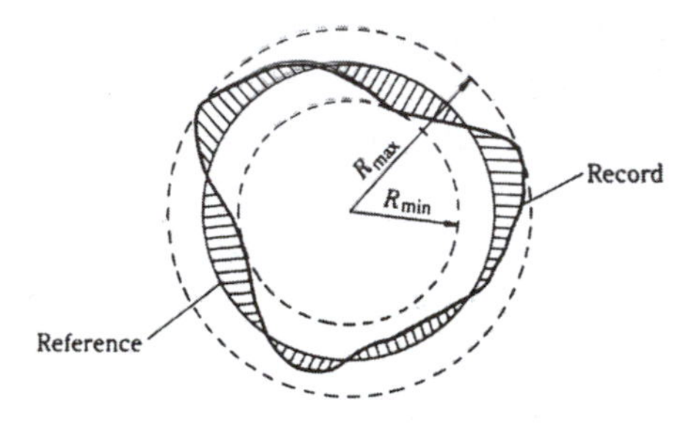

Figure 3.43. Least squares technique for out-of-roundness measurement, from Scarr[23].

For a circle whose roundness is shown as in a polar plot, Figure 3.43, for example, we can calculate a least squares center and circle (i.e. radius). If we have measurements of the radial position of the surface of the artifact (hole or shaft) determined as earlier by, for example, the precision spindle method giving n data points or radii r_i, at equal angular spacings $(r_i \theta_i)$, where

$$i = 1, 2, 3, \dots, n \text{ and } \theta_i = \frac{2\pi i}{n}$$

then we can show that the least squares circle will have a center C, whose rectangular coordinates are (a, b) and whose radius is R, where

$$R = \frac{\sum r_i}{n}$$

and

$$a = \frac{2\sum x_i}{n}$$

and

$$b = \frac{2\sum y_i}{n}$$

Roundness errors can then be established with respect to these least squares measurements as discussed earlier.

Finally, there are machine errors which will cause variations in the profile generated, roundness, for example, from rotation to rotation. Figure 3.44, from MTTF[25], shows the non-repetitive per revolution error motion in a spindle due to a too-low speed in a plain bearing. And this error motion may change with rotation speed of the spindle, Figure 3.45, MTTF[25], measured using the technique of Figure 3.40.

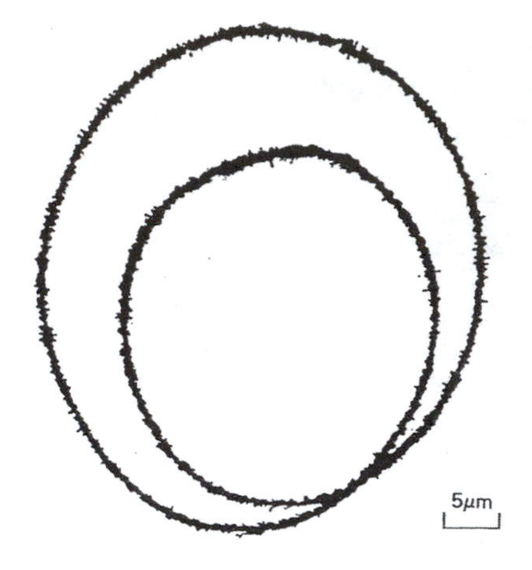

Figure 3.44. Non-repetitive per revolution error motion in a spindle due to a too-low speed in a plain bearing, from MTTF[25].

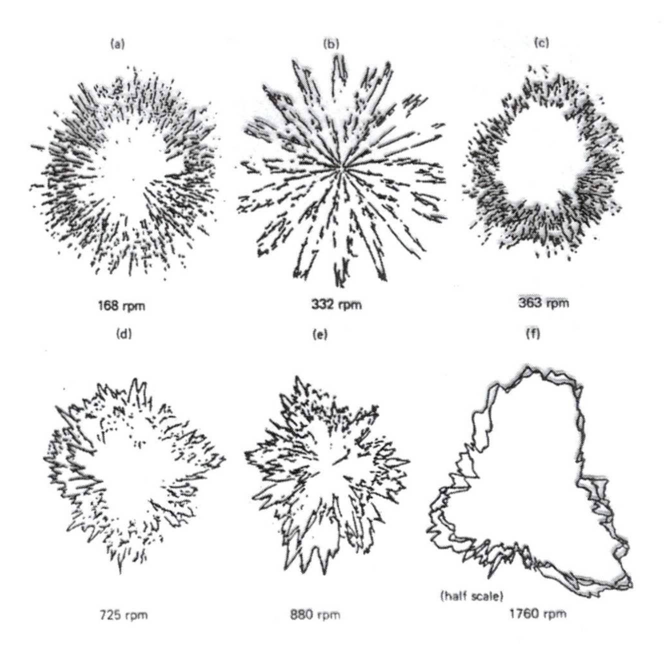

Figure 3.45. Polar error motion plots at different rotational speeds for a precision lathe, from MTTF[25].

3.4.2.4 Other form errors

None of the above measurement techniques will tell anything about the out of plane form errors of the workpiece — cylindricity (e.g. taper), for example. Methods such as the precision spindle/probe techniques can be used for such measurements, usually by allowing multiple measurements at different cross-sections of the workpiece so as to give some "three-dimensional" information. We will see other examples of this in the future but, an excellent illustration is given in Figure 3.46, MTTF[25] showing the three dimensional geometrical deviations of a machined part due to an axis-of-rotation error motion.

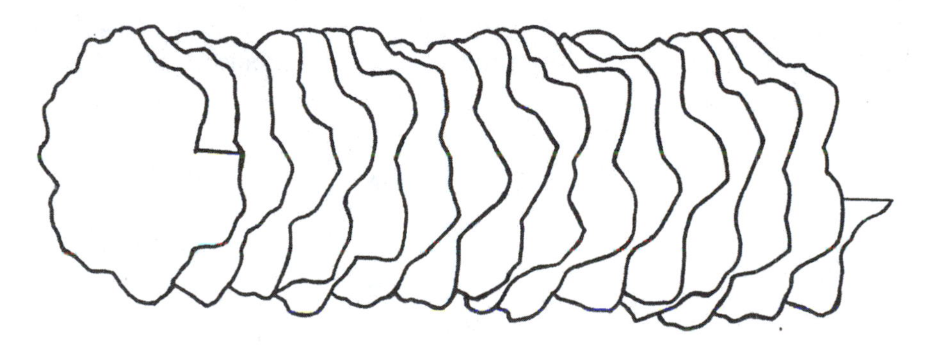

Figure 3.46. 3D geometrical deviations of a machined part due to an axis-of-rotation error motion, from MTTF[25].

3.4.3 Measurement of surface roughness

Although some roughness features are visible in the Fizeau interferogram in Figure 3.30, we usually distinguish between the measurement of roughness and surface profile. In fact, of course, the underlying "waviness" in a roughness profile is the surface profile. And, at the limit, the tolerance on the position of a surface is determined by the roughness. The ANSI Y14.5M standards for dimensioning and tolerancing acknowledge the surface roughness effects in distinguishing between the simulated datum and theoretically exact datum plane, Figure 3.47. In fact, the surfaces of manufactured parts are not perfectly flat for a variety of reasons, some of which we have already spoken. The surface of the part is the “process record” and includes every vibration, shape change, subsurface material defect and effect, etc. Recall the early Edison wax cylinders for recording and playing back music. The wax cylinder was rotated while in contact with a diaphragm-mounted needle fed along its length. The diaphragm was affixed over the end of a horn (remember the master’s voice RCA logo?). When recording, the variation in sound pressure directed against the diaphragm by the horn caused the needle to move and scribe the movements in the soft wax cylinder. On “play back” the needle was perturbed by the hardened wax record, the diaphragm was vibrated generating sound or “music” of a sort. The analogy between this and the creation of a surface of a machined part is exact. Every influence which the machine is exposed to during production of the artifact is recorded, to some degree, based on the “sensitivity” of the configuration to the perturbation, in the surface of the artifact. The measurement and quantification of surface roughness is a subject of study in of and itself and we shall only be able to skim over the surface, so to speak, of this important topic.

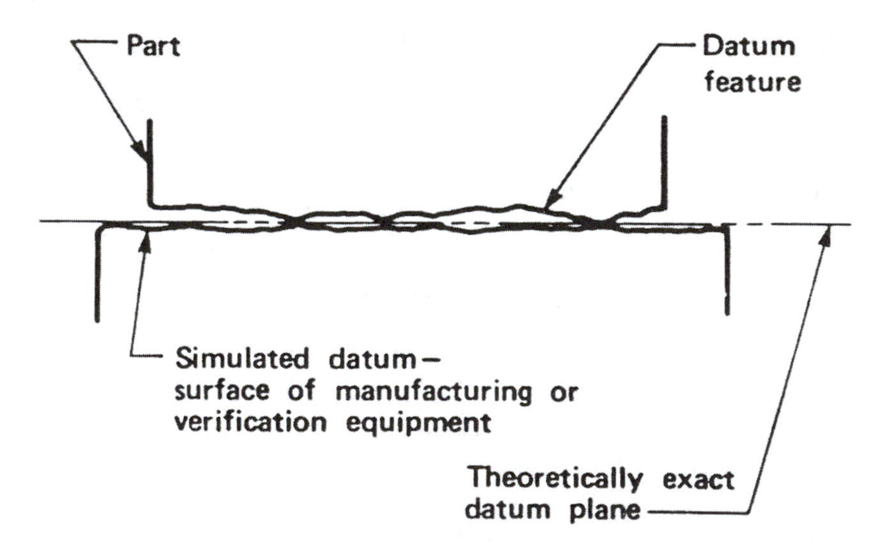

Figure 3.47. Datum feature, simulated datum and theoretical feature, after ANSI Y14.5M-1982 Standard Engineering Drawings and Related Documentation Practices.

The classic terms in surface roughness measurement are illustrated in Figure 3.48. Waviness generally refers to the underlying surface undulation due to, for example, low frequency vibration or the out-of-roundness errors we described in the last section. On top of the waviness is superimposed roughness due to the higher frequency aspects of the tool-work interactions. The roughness is measured over a period of waviness sufficiently long so as the mean value of the roughness can be determined. The exact definitions of parameters for surface characterization is detailed in ANSI B46.1 standards[29]. Importantly, this also includes a cross-reference to the terms used in other countries to define the parameters. Roughness is measured by contact profilometer-type devices that ride on the surface on a skid- which provides the reference for the measurements- or non contact devices such as interferometry-based or, recently, atomic force microscopes. The instrument used is dependent on the resolution needed for the measurement. We will not discuss these instruments here but many excellent references exist.

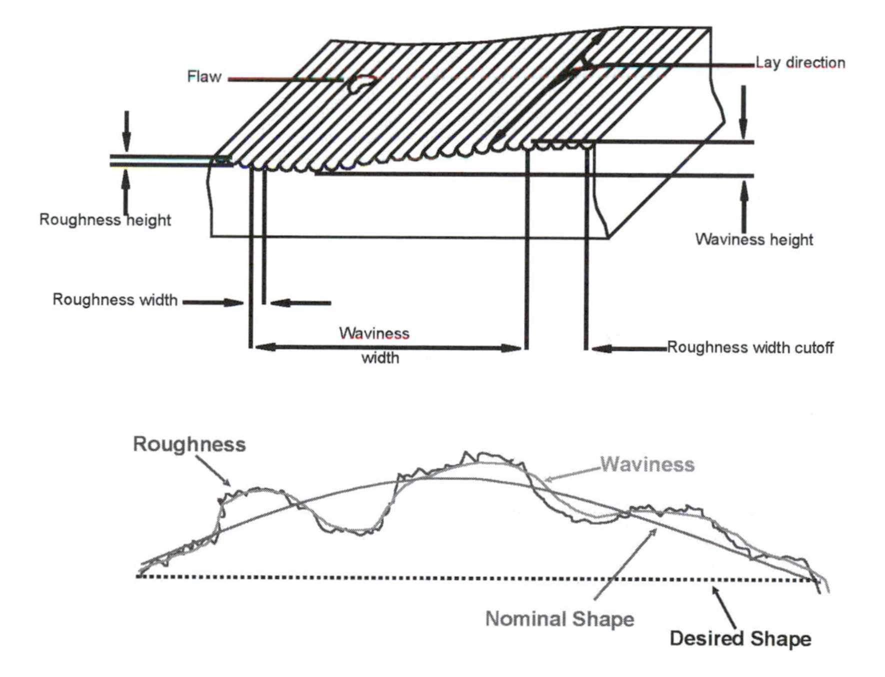

Figure 3.48. Surface roughness terminology.

As a first approximation, roughness is measured as an average, R_a, for arithmetic average or R_q for root mean square (RMS) roughness. These are determined as follows for a profile represented by a series of discrete points *y* representing the profile deviation from the reference:

$$R_a = \frac{1}{L} \int_{x=0}^{x=L} |y| \, dx \tag{3.4.4}$$

where

R_a = arithmetic average deviation from the center line
L = sample length
y = ordinate of curve of the profile

An approximation of the average roughness R_a may be calculated by adding the y increments shown in Figure 3.48 with out regard to sign and dividing the sum by the number of increments, N.

And the RMS roughness is

$$R_q = \left(\frac{1}{L} \int_{x=0}^{x=L} y^2 dx \right)^{1/2} \tag{3.4.5}$$

where

R_q = RMS average roughness

This can be similarly approximated by

$$R_q = \left(\frac{y_1^2 + y_2^2 + y_3^2 + \cdots + y_N^2}{N} \right)^{1/2} \tag{3.4.5}$$

where N = number of samples.

Other measures often referred to include:

— maximum peak-to-valley roughness height, R_y or R_{max}; The distance between two lines parallel to the mean line that contact the extreme upper and lower points on the profile within the roughness sampling length.

— average peak-to-valley roughness, R: The average of the individual peak-to-valley roughness heights, each of which occurs within a defined spacing interval.

— average spacing (wavelength) of roughness peaks, A_r or A_R; This is the average distance between peaks measured in the direction of the mean line and within the sampling length.

— bearing length ratio t_p; A reference line is drawn parallel to the mean line and at a pre-selected distance from it to intersect the profile in one or more subtended. The bearing length ratio is the ratio of the sum of these subtended lengths to the length of the mean line.

— amplitude density function, ADF; This is defined as the probability density of profile heights or amplitudes. Figure 3.49.

— bearing area curve, BAC; This is the ratio of the cut length to the total sample length. The BAC results from increasing the depth of cut from the top to the bottom of the profile. This parameter is comparable to wearing-in or partial finishing of a surface.

— skewness; Skewness is a measure of the symmetry of the profile about a mean line and offers convenient way to characterize load carrying capacity, porosity and characteristics of non-conventional machining processes. A definition of skewness is

$$\text{Skewness} = \frac{1}{\left(R_q\right)^3} \frac{1}{N} \sum_{i=1}^{N} y_i^3 \tag{3.4.6}$$

where y_i is again the profile ordinate at i and N is the number of sample points.

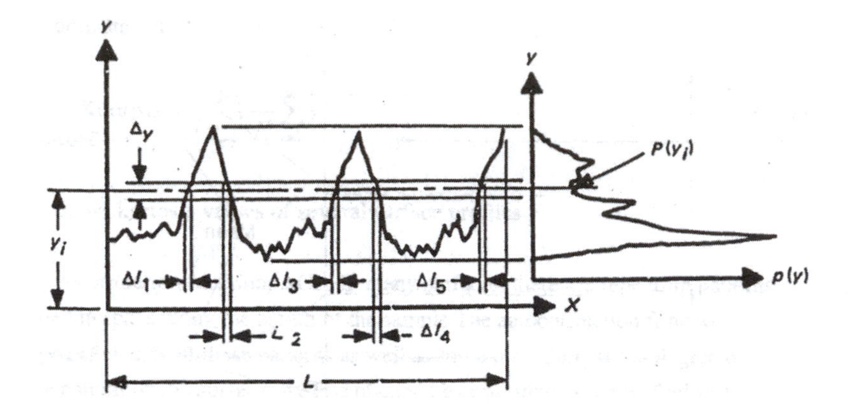

Figure 3.49. Amplitude density function, from ANSI[29].

Figure 3.50 shows skew values for various profiles.

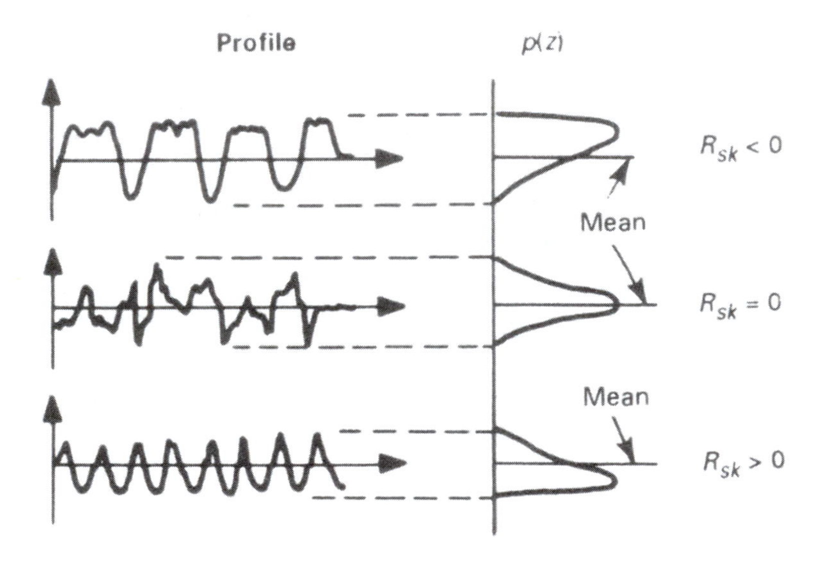

Figure 3.50. Skew values for surface profiles, from ANSI[29].

— kurtosis; Kurtosis is a measure of the amplitude density function sharpness. In addition, it quantitatively describes the randomness of a profile's shape relative to that of a perfectly random surface which has a kurtosis value of 3. A definition for kurtosis in terms of N profile ordinates y_i is

$$\text{Kurtosis} = \frac{1}{(R_q)^4}\frac{1}{N}\sum_{i=1}^{N} y_i^4 \tag{3.4.7}$$

Figure 3.51 shows kurtosis values of several surface profiles.

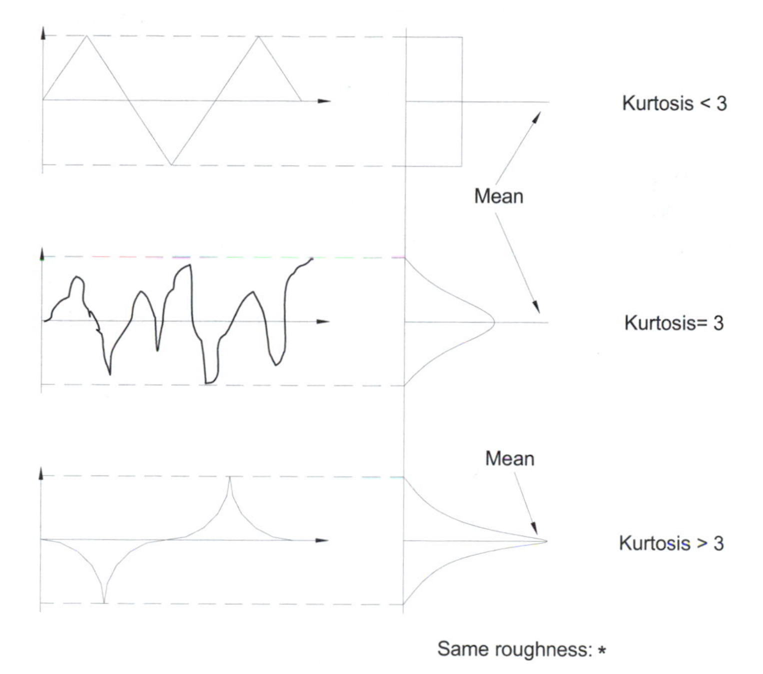

Figure 3.51. Kurtosis values for surface profiles, from ANSI[29].

— autocorrelation function, ACF; In many surfaces, there are repeating patterns that are shifted in space along the length of the sample The autocorrelation function is a measure of the similarity of these patterns as well as some indication of the degree of "shift" in the pattern on the surface. ACF is obtained by multiplying the shifted and unshifted waveforms over the overlapping lengths point by point and calculating the average of these products. At shifted points for which there is a strong correlation the product should be large. Positive or negative correlations can be similarly determined. The ACF can be computed as

$$\mathrm{ACF(m)} = \frac{1}{N-m}\sum_{i=1}^{N-m} y_i y_{i+m} \tag{3.4.8}$$

where N = number of discrete element y_i in the profile and
m = the lag or shift ranging from 0 to N maximum.

With skew and kurtosis we can also calculate parameters of the profile such as mean number of zero crossings, MNZC, as well as the mean number of maxima, MNMA (peaks or valleys). Further, the ratio of the density of extremum (peaks or valleys) to density of zero crossings can give us a measure of the degree to which the surface is isotropic. These terms are all important depending upon the eventual use of the surface and the precision of the application.

The basic model for the generation of an ideal surface with a tool is shown in Figure 3.52. For a tool with a corner radius r_e used in a turning operation with feed, f, we can estimate the surface roughness, R_a as

$$R_a = \frac{0.0321 f^2}{r_e} \tag{3.4.9}$$

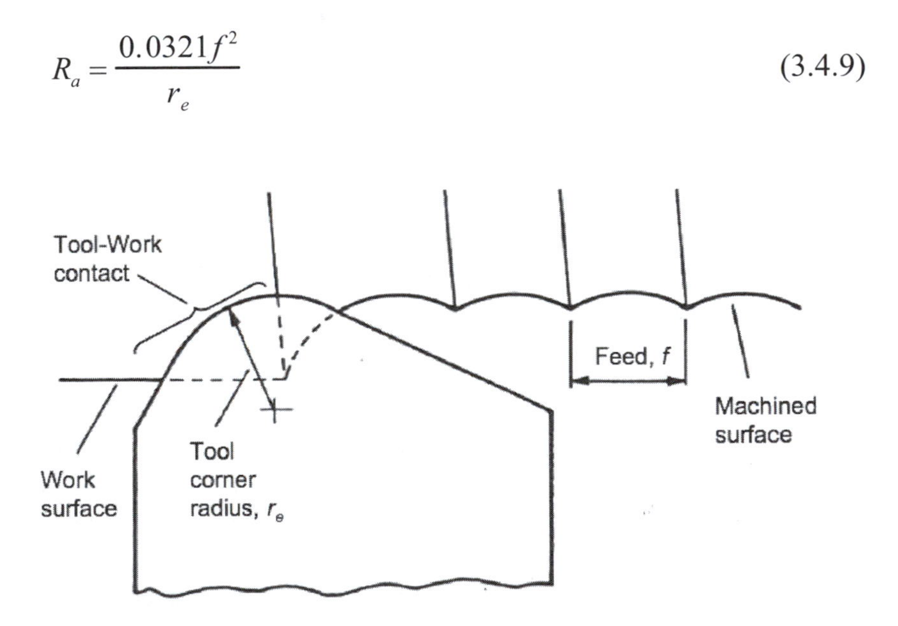

Figure 3.52. Idealized model of surface roughness for a tool with a rounded corner.

The resulting profile, depending upon the machining conditions, depth of cut, etc., can look quite similar to the profile in Figure 3.48. We will talk more about the influence of machining parameters on the surface generated in Chapter 10. The condition of the tool, commonly referred to as "tool wear" has a significant influence on the surface generated. The ideal surface in Figure 3.52 is due to an ideal tool profile. When the tool contacts the workpiece and as the tool condition deteriorates (i.e. the tool wears) the surface degrades from ideal. Other effects such as built up edge on the tool will affect the surface. A series of illustrations is shown in Figure 3.53. Note the change in shape of the profile and the amplitude distribution. Figure 3.54, from MTTF[25] shows a similar sequence of profiles, this time indicating as well the power spectra of the data with a strong periodicity due to the feed marks.

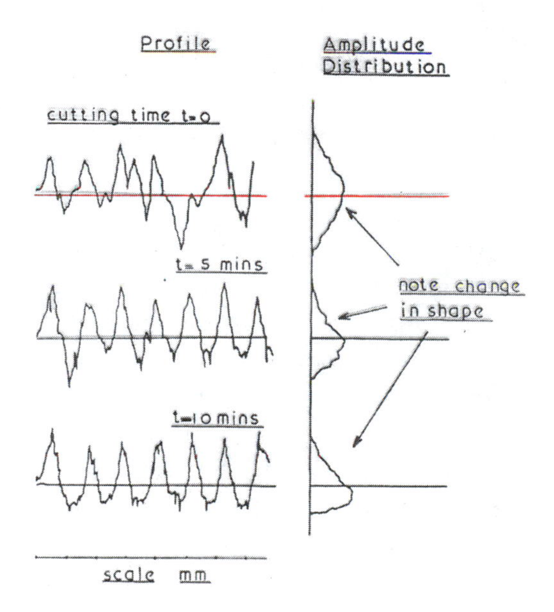

Figure 3.53. Effect of tool wear on surface profile, from MTTF[25].

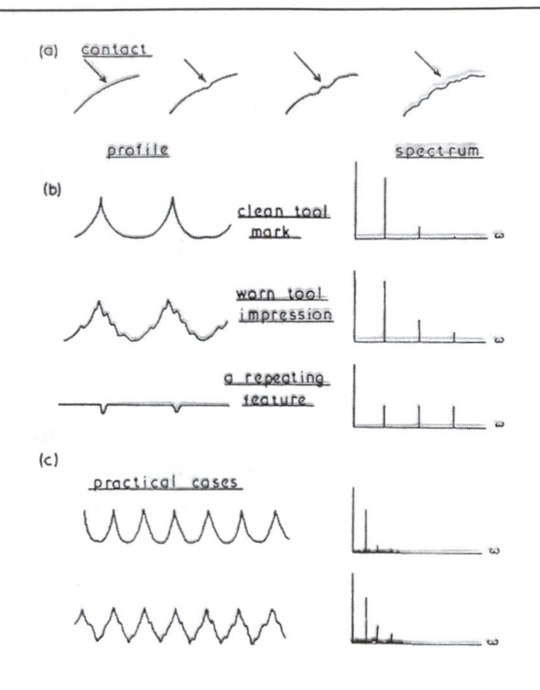

Figure 3.54. Effect of tool wear on surface profile and spectra, from MTTF[25].

The deterioration of the tool profile, meaning the profile goes from a carefully defined profile to one with superimposed nicks, dings, fractures, build-ups, etc. causes the surface profile to reflect the new shape of the tool. That is, each tool profile "error" shows up in the surface profile or, stating it another way, each (or many or the) feature in the surface profile is traceable back to the tool profile. The two surface profiles seen in Figure 3.55, from MTTF[25], are due to such a tool profile deterioration, here due to a built up edge. Note the effect on the autocorrelation and power spectra of the profile data. In Figure 3.56, from MTTF[25], another effect is illustrated, that of machine vibration, on the surface. The careful reader will begin to notice that a lot of these effects look the same in the frequency domain or expressed as other parameters. The "dissection" of a surface and the identification of the causes of its many characteristics are not straightforward and there are no magic analyses that can accomplish this. That is why an understanding of the genesis of the surface in terms of machine, tool work material and process conditions is so important.

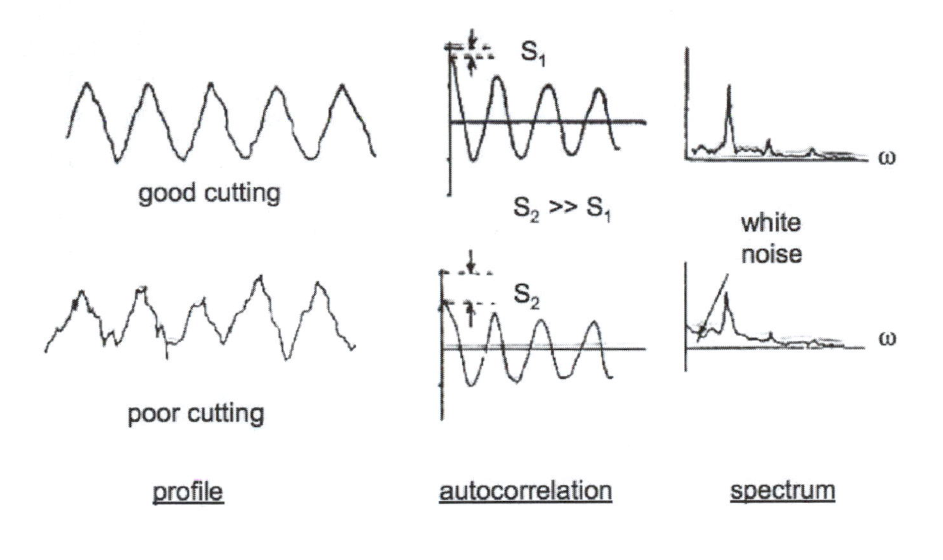

Figure 3.55. Use of random process analysis to reveal built up edge on the tool, from MTTF[25].

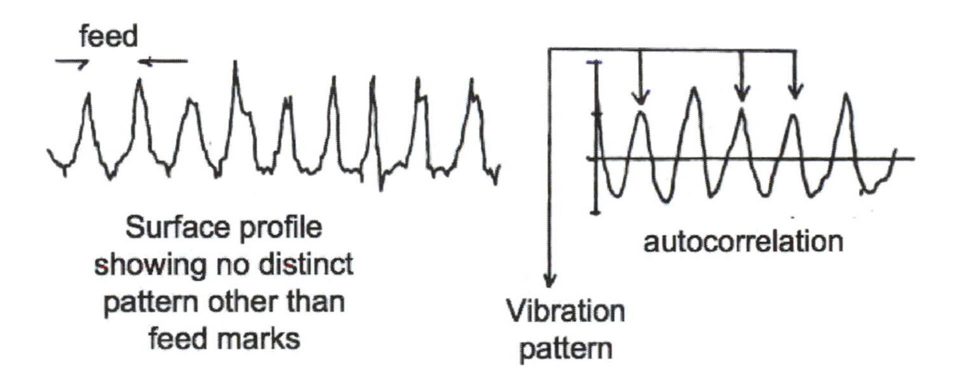

Figure 3.56. Effect of machine vibration on surface profile, from MTTF[25].

Finally, this phenomenon of surface generation dependent on tool profile is not restricted to single point turning. Obviously for multiple point tools it is more complicated. Often, the researcher attempts to isolate on of the tool points and then, knowing its path and impact on the surface, use superposition to simulate the surface generated by a multi-point tool. We will see examples of this later in milling surface generation. For grinding, where the "tool" is an abrasive grain of poorly determined shape and orientation the deterioration of the grain shape will influence the surface, Figure 3.57, from MTTF[25].

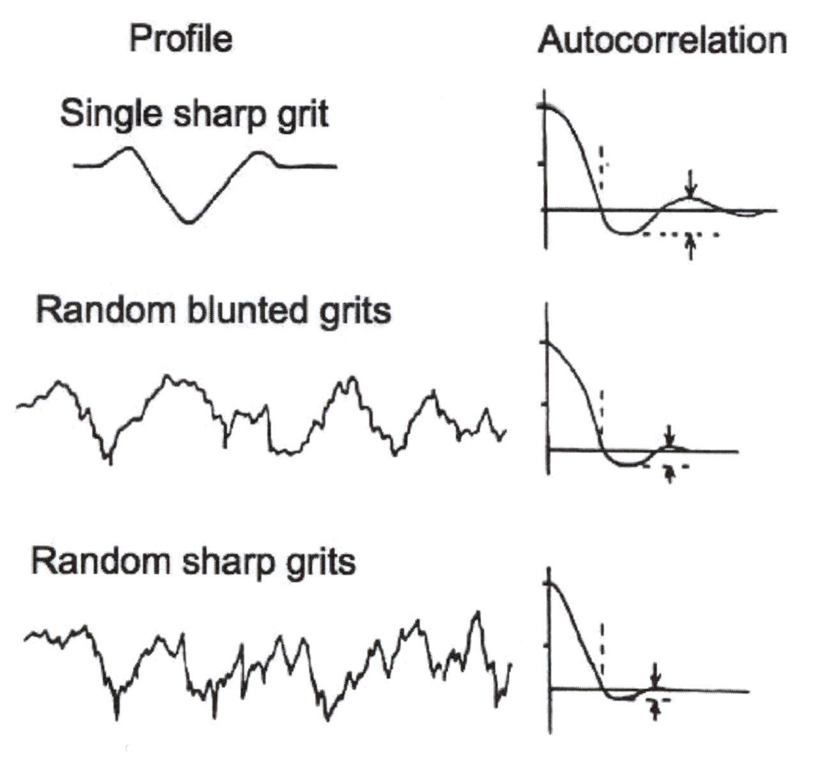

Figure 3.57. Use of autocorrelation function to characterize the effect of various grip shapes on the surface profile generated, from MTTF[25].

3.4.4 Kinematic precision

Kinematic precision refers to the precision in position, velocity and acceleration of an element of the machine during operation. They are in a sense the dynamic measures of the machine's performance whereas the straightness, etc. are considered as steady state measures. Of course, when we start talking about thermal errors, there will be a dynamic effect there but at considerably lower frequencies (several orders of magnitude lower, in fact.) Our discussion and illustration of rotational precision earlier could be covered as well in this section under kinematic precision. Accelerometers and vibrometers are other tools applied here.

This is most difficult to determine for volumes (i.e. three dimensional measurements). The "instrument of choice" for this type of measure is the double ball bar (DBB) or laser ball bar (LBB), Figure 3.58, from Nakazawa[19]. With this, and a prescribed test pattern in the machine tool work volume, the comparison between the desired position and the actual as measured by the instrument (either Moiré scale or laser interferometer).

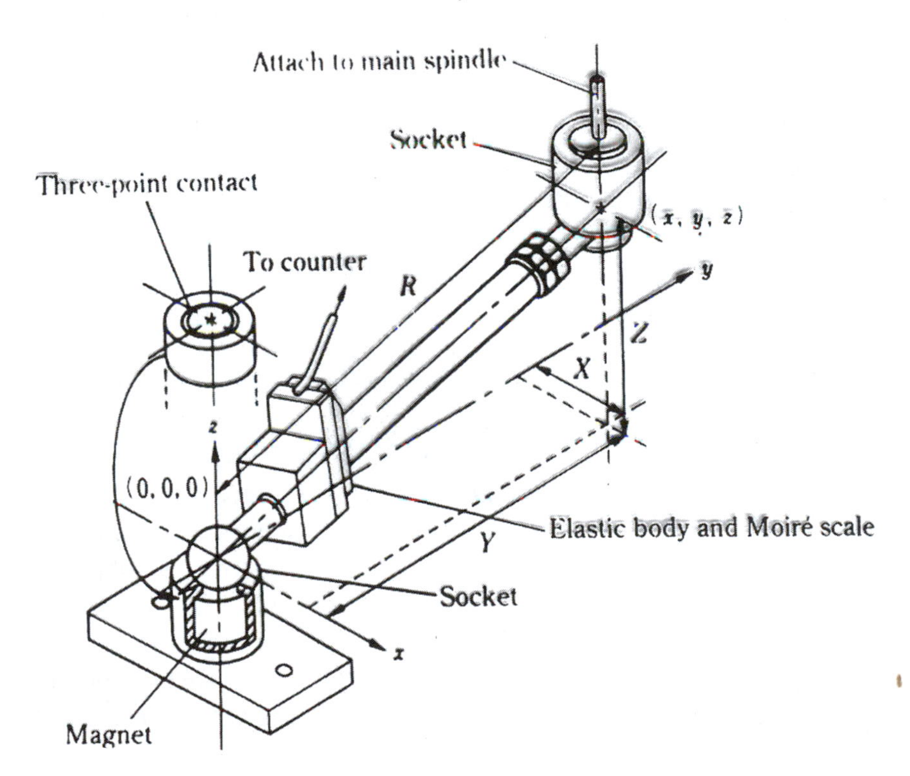

Figure 3.58. Double ball bar, from Nakazawa[19].

3.5 Subsurface damage

This topic deserves some special attention as it is commonly ignored when measurement issues are discussed. However, it can be ultimately the most important when the final performance of the workpiece or machine is evaluated. We will be going into aspects of machining processes in much more detail in Chapter 10 but it can be said that the passing of a tool over a surface as in a machining operation will both remove material (generally) and leave a residual effect on the surface. Just as the surface finish after machining is influenced by the magnitude of the chip removed (recall equation 3.4.9 above in which the feed, f, is approximately the uncut chip thickness — thus smaller feed/chip yields a smoother surface) the magnitude of the surface effect is proportional to chip removed. The contact stress of the tool against the work surface on the clearance face of the tool due to elasticity will result in some residual stress in the workpiece. Similarly, the degree of finish of the surface against which the tool contacts will affect the final surface.

These effects are summarized in Figure 3.59, from Ohuchi[30]. Although we could construct other sets of relationships, this nicely summarizes a key set of interactions. Consider first the surface roughness due to a previous machining operation. Figure 3.60 shows the cross sections of four surfaces machined by four quite different processes, from a rough shaping operation to a fine lapped surface. It is clear from this that the level and degree of surface damage differs in each case. And, any subsequent process, such as a finishing operation on the ground surface, will have to remove a substantial amount of material in order to remove the effects of the last process if that is important for the final use of the surface. Figure 3.59 forces us to look both on the surface, for oxide layers and surface films as well as deeper into the surface for other effects of the surface creation process, like crystallographic orientation and grain size and degree of plastic deformation. On of the major concerns is the release of residual stresses by a machining operation and the resulting distortion (recall Johansson and his gage blocks.)

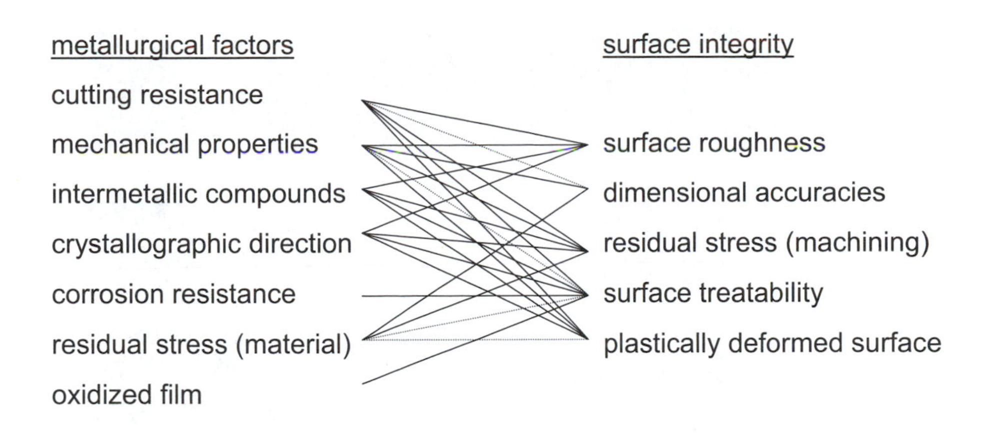

Figure 3.59. Relationship between metallurgical factors and surface integrity, from Ohuchi[30].

A cross section of a typical surface with variations of material from the lowest bulk or undisturbed material to the surface is shown in Figure 3.61, from Nakazawa[19]. Nakazawa identifies the following layers, from the exposed surface with, likely, a lubrication or oxide film on top

- amorphous (or Beilby) layer approximately 1 μm thick
- fibrous layer approximately 10 μm thick
- plastically deformed layer usually 1-2 grain diameters in thickness
- transgranular slippage layer (thickness depends upon on the manufacturing process)
- base material unaffected by the process

The first three (amorphous, fibrous and plastically deformed) are usually referred to as the “affected layer”.

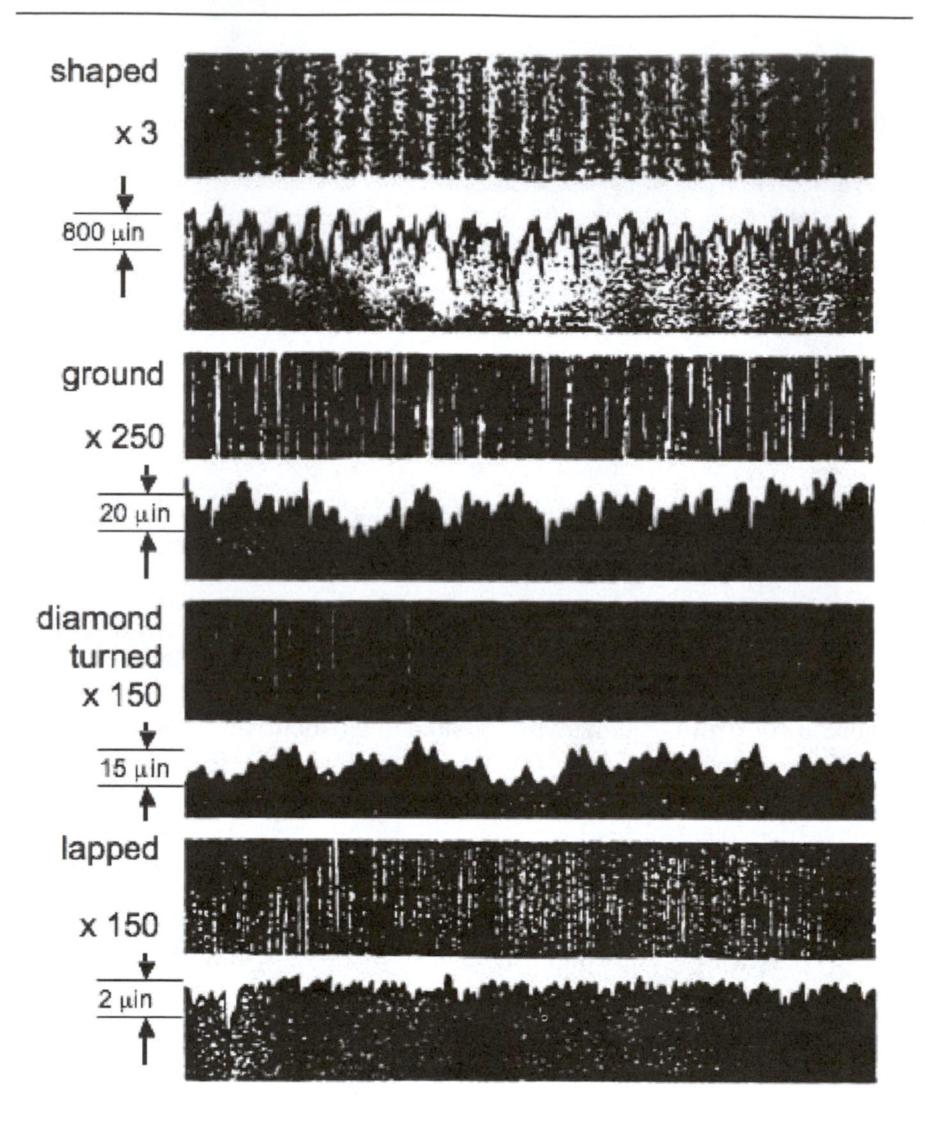

Figure 3.60. Photomicrographs showing plan view, and graphs showing cross-section (with exaggerated scale of height) of typical machined surfaces.

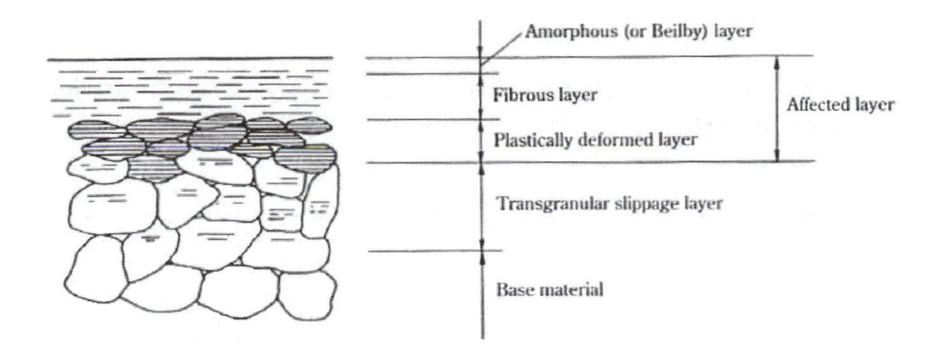

Figure 3.61. Cross-section of work affected layers in a typical machined workpiece, from Nakazawa[19].

The determination of the size and extent of the affected layer is extremely complex for exactly the reasons shown in Figure 3.59. For a specific process operating at a specific energy of the machining process, depending on the material and its composition and previous processing history, the thickness of this layer can vary a lot. We will not try here to lay out a definitive set of rules (if they exist) for this analysis but hope only to review some of the significant elements to be considered. We can address only the FeC system for starters, cast iron, iron and steel, for example, Figure 3.62, from Gronsky[31]. We know that depending on the %C and thermal history of an alloy the grain structure can differ substantially. Figure 3.63 shows, for a low carbon steel, the range of grain distortion due to different percentages of reduction by a rolling operation for a hot rolled material. Although we cannot judge the depth of the layer (although it can be calculated if process conditions were known!) we can see the tremendous variation in grain geometry. This would correspond to the plastically deformed layer in Figure 3.61. Higher alloy content, more C for example, will under proper heat treatment create a finer grain size or a more homogeneous layer. Heat treatment (the temperature and time at temperature as well as quench rate) also significantly affect the subsurface.

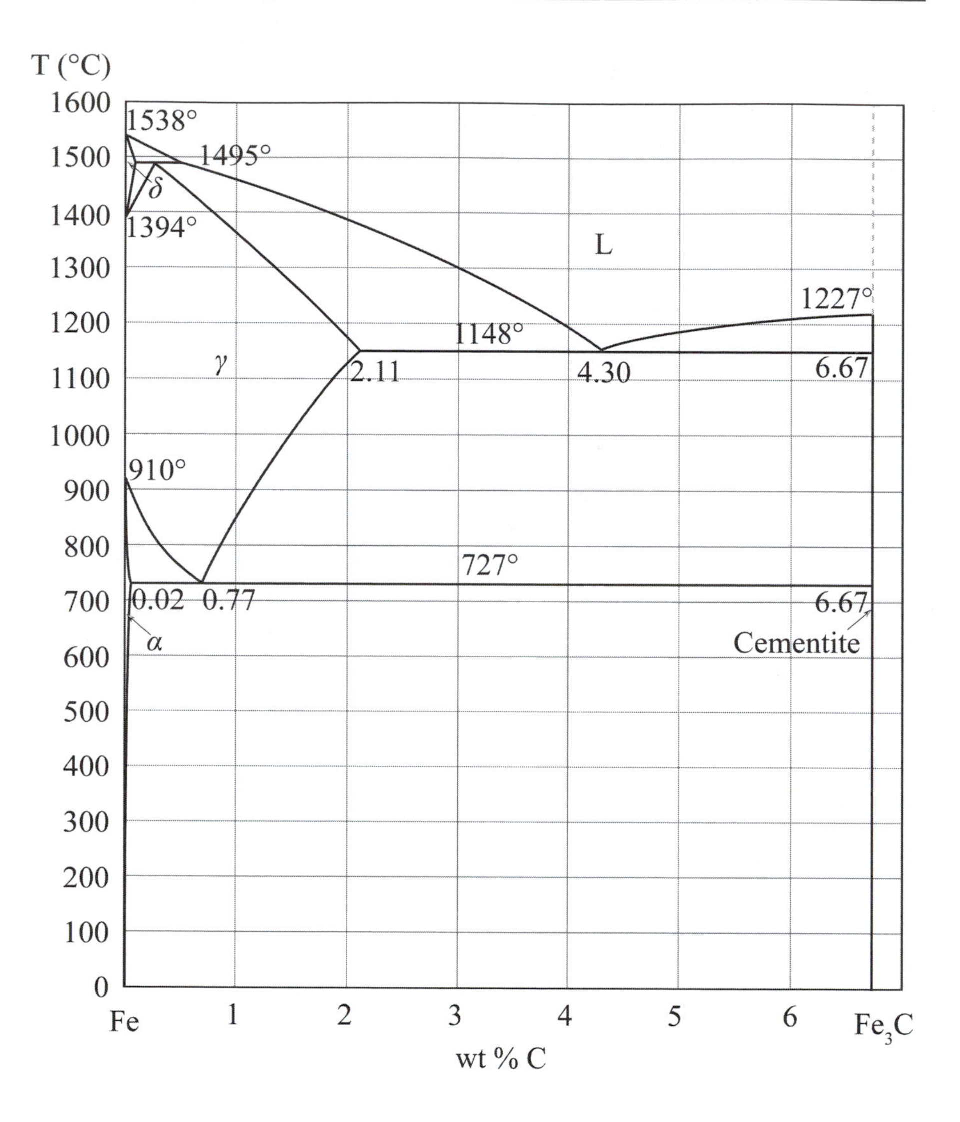

Figure 3.62. Fe-C phase diagram, from Gronsky[31].

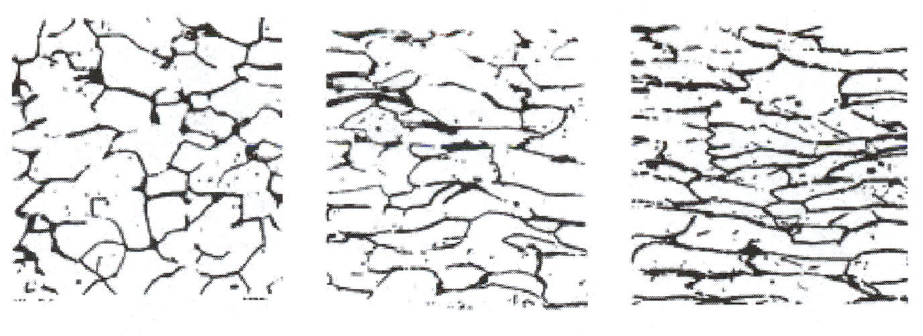

Figure 3.63. Illustration of Effect of percentages of reduction by cold rolling on 1008 steel (Magnification 1000X).

For higher percentages of C (2% or so and above) and Si added we get a variety of cast irons depending on the exact composition and heat treatment/processing during fabrication. The microstructures of these are impressively different from carbon steel, Figures 3.64 and 3.65. The figures show the heterogeneous nature of cast iron due to flakes of graphite and a distinct dendritic solidification pattern in gray iron and the graphite nodules in a steel band all in pearlite matrix in nodular or ductile cast iron. Although these materials are not often plastically deformed as steel alloys are, the resulting structure after various manufacturing processes can be quite interesting and complex.

a. Distribution of graphite flakes

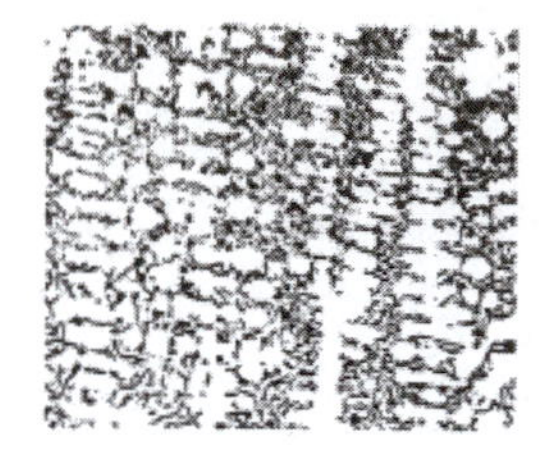

b. Distribution showing dendritic segregation and orientation

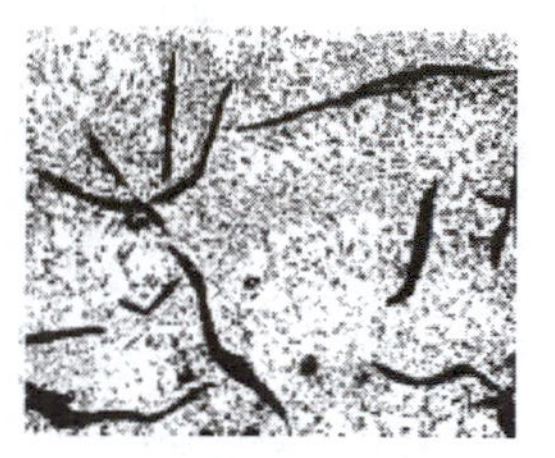

c. Annealed gray iron showing graphite flakes and dispersion of fine spheroidized carbide particles

Figure 3.64. Schematic showing types of graphite flakes in gray iron, structures in class 20 gray iron.

a. Ductile iron (nodular) as cast showing graphite nodules; "Grade 6" 105 nodules per square millimeter

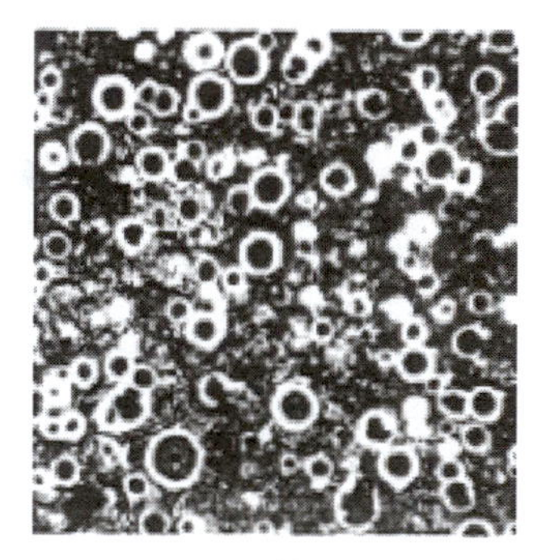

b. Perlitic ductile iron as cast showing graphite nodules in surrounded by free ferrite ring in a matrix of pearlite

Figure 3.65. Schematic of graphite nodules and structures in nodular cast iron.

Let's switch to the non-ferrous system of Al-Si for a final example, Figure 3.66, from Gronsky[31]. This is a very popular metal series and finds a lot of use in manufacturing from die casting to engine blocks. This is an alloy that is precipitation hardened and, like in cast iron where the graphite comes out of solution on cooling, shows a wide variety of structures, Figure 3.67, specially when hot worked. The figure gives some sense of the influence of the as cast material as well as metal working on the structure.

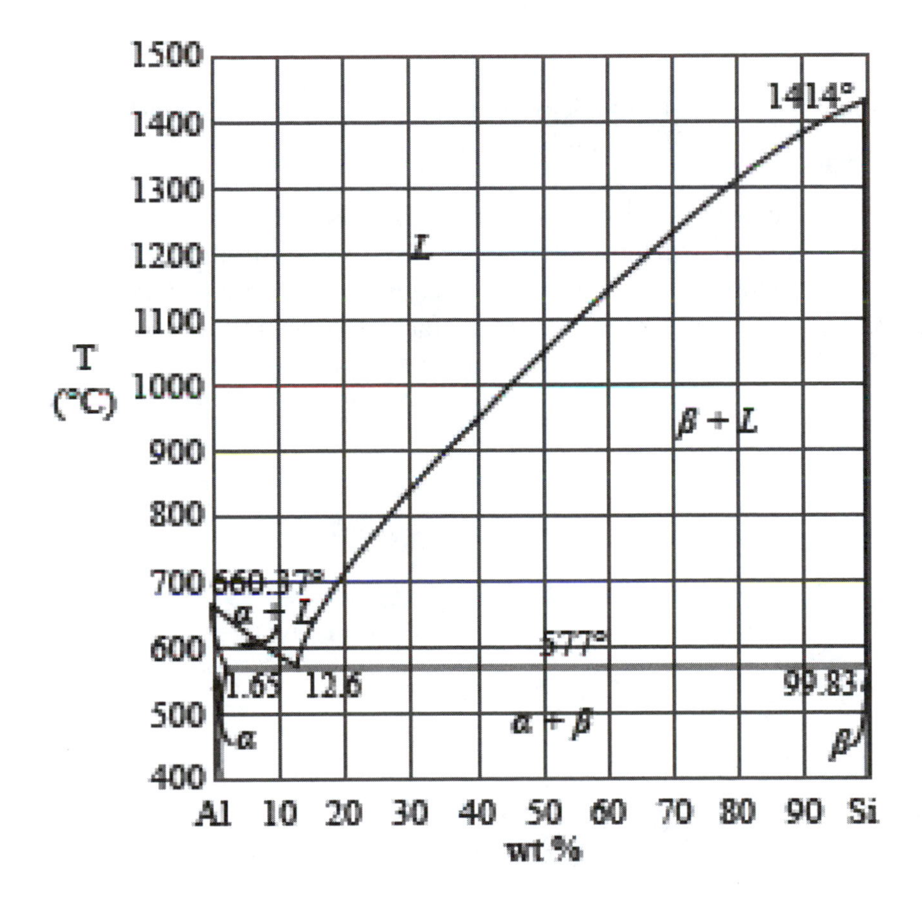

Figure 3.66. Al-Si phase diagram, from Gronsky[31].

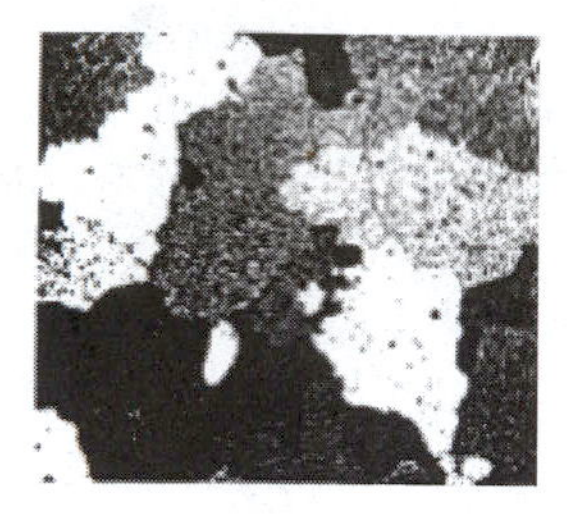

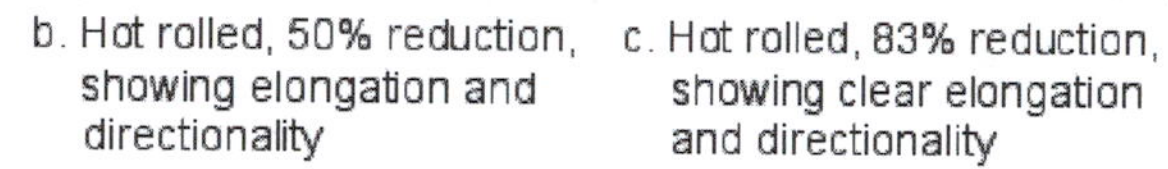

a. Equiaxed grains in ingot of material

b. Hot rolled, 50% reduction, showing elongation and directionality

c. Hot rolled, 83% reduction, showing clear elongation and directionality

Figure 3.67. Schematic of microstructure for Aluminum Alloy 7039 with various processing.

What we draw from these examples is the need to be aware of the "pedigree" of the material we are using. Variations in alloy content and processing history can cause a material subjected to seemingly the same precision manufacturing operation to behave quite differently — often yielding unsatisfactory results.

IV MECHANICAL ERRORS

4.1 Introduction

Precision manufacturing processes rely, primarily, on mechanical devices and structures as the basis of their operation. We saw the machine tool structural loop connecting the cutting tool to the work-piece through the spindle, machine overarm, column, base and table and through the work fixture. Shirley and Jaikumar[16] identified error sources with respect to part and machine contributions to *systematic* and *random/dynamic* errors. These systematic errors were detailed in Figure 2.3 and the environmental and process influenced (often dynamic or random) were shown in Figure 2.4. Of course, the way a machine behaves in the production of a surface or a form on a work-piece reflects the combination of a number of these sources of error. This chapter will review the mechanical error sources in machines. Chapter V will do a similar review but from the point of view of thermal error sources in and out of machines. Then, Chapter VI will describe a methodology for combining the particular error sources with the particular machine structure or configuration to estimate the limits on the machine accuracy.

Mechanical errors can be associated with the part, the machine or the operation (process). Similarly, they can be described as systematic/static in nature (i.e. contributing to machine bias) or random/dynamic (i.e. "non stationary" and more difficult to quantitatively describe). The errors are introduced by a variety of causes generally summarized as:

- compliance
- property variation
- motion error
- setup errors
- programming errors
- measurement errors
- process characteristics

There is one chapter devoted to compliance as it is the most pervasive source of error incorporating machining, processing, and fixturing forces, part/machine section changes, and part/machine material property variation. We are dealing with components that are elastic. Hence, the application of force, as by a cutting tool, or load, as by the mass of a moving element will cause elastic distortion. And distortion of any element of the structural loop (as defined above) is a source of error as it will alter the relative position of the tool to the work surface.

Motion errors are the roll, pitch and yaw of axes along with the positional errors and squareness, straightness and parallelism of axis motion due to machine element behavior. Positioning errors due to incorrect motion commands (due to incorrect programming) are not the same but are still errors. Similarly, setup errors, that incorrectly position the tool tip in the tool holder, or fail to square the fixture with the machine axes are mechanical errors. Generally all of these errors listed above are relatively "static" and measurable.

Dynamic errors are due primarily to motion and the variations of forces displacements and acceleration due to that motion. Thus, the vibration induced by a rotating spindle and its excitation of a machine structural element may vary with speed and be hard to quantify. Similarly, distortion of the cutting tool or fixtures fall in the category if not clearly repeatable. The process changes due to work material variation (chip formation and the changes in machining forces that result, or burr formation and the resulting dimensional error, are two examples) are very difficult to quantify. Phenomenon such as built up edge (BUE) on the tool when machining certain materials under certain conditions fall into this category as well. BUE has an undesirable affect on the part dimension and surface finish. Whereas the conditions under which it occurs can be approximated, during the manufacture of typical workpieces it cannot be fully predicted. Similarly, the variation in insert geometry for an indexable cutting tool due to progressive tool wear or manufacturing variations, or tool axis location for a collet-type tool holder, or run out in a milling cutter, are all likely elements contributing errors but not quantifiable *a priori* in a manufacturing system.

4.2 Errors due to machine elements (excluding bearings)

Machine structural components and their orientation and relative motion are major contributors to mechanical errors. We need to switch our perspective from the measurement of the form error of a workpiece, as seen in our discussion of roundness errors, for example, to the action of the machine elements that create the error of form in the first place. And, we need to keep in mind the metrology techniques appropriate for measuring the element behavior. Figures 3.3-3.5 illustrated one, two and three dimensional errors in machine elements. With a tool in a tool holder fixed onto the table riding on these axes, it is easy to visualize the resulting surface and the relative error between its ideal and actual location. We also discussed several sources of machine errors with respect to Abbé errors in the last chapter. We will see other examples of machine element errors throughout the notes as they are impossible to separate from the general discussion and we will learn how to quantitatively estimate the magnitude of these kinds of errors in Chapter VI. The directions in which errors occur for a conventional machine tool are shown in Figure 4.1, from McKeown[27]. Means of testing the element errors are reviewed in Figure 4.2.

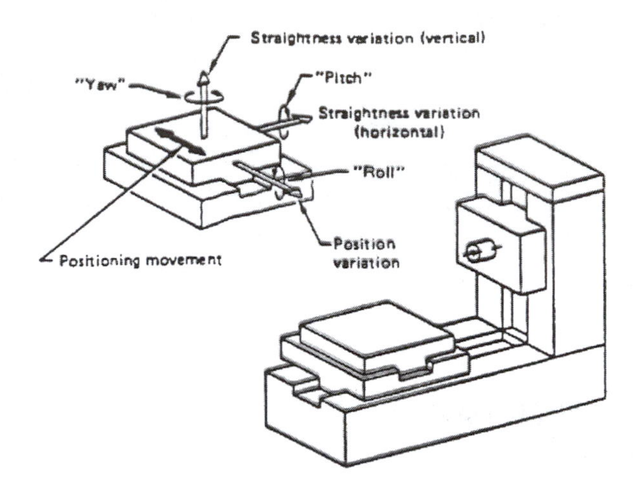

Figure 4.1. Directions in which errors occur in machine tools, from McKeown[27].

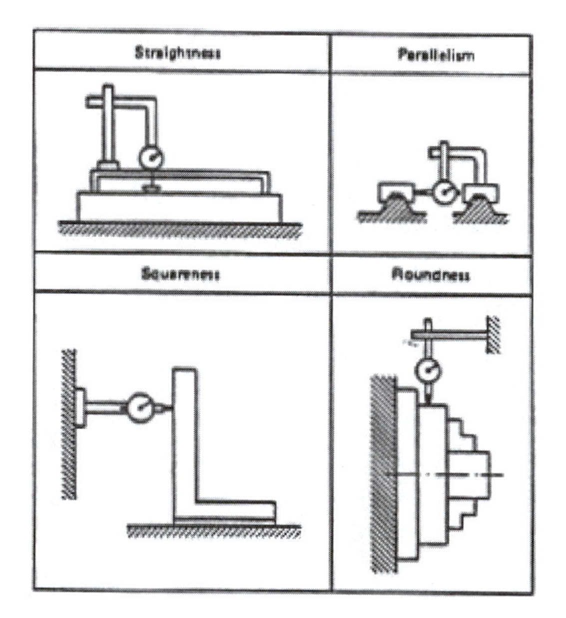

Figure 4.2. Geometric testing criteria, from MTTF[25].

Non-parallel ways on which a machine tool component rides, a spindle for example, will introduce a positional error in the endpoint of a tool driven by the motion. Figure 4.3 illustrates the effect of non-parallel motion caused by straight but non-parallel guideways positioning a spindle above a worktable. A positional error of $\overline{O_1 O'}$ occurs for a vertical column motion of *h* at a radius *r* from the column. Clearly, increasing *h* will increase the error. And, the displacement measuring hardware is invariably mounted on the column of the machine introducing a substantial Abbé error.

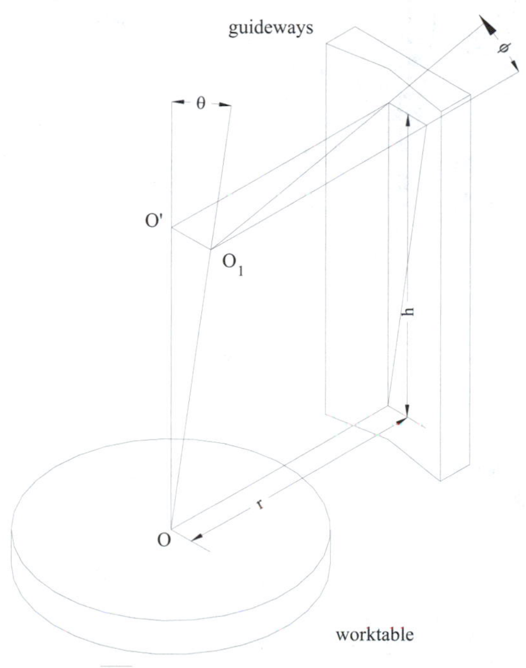

Figure 4.3. Effect of non-parallel ways on machine positional error in a vertical milling machine.

The example in Figure 4.3 was for a vertical spindle milling machine. Similar effects are seen due to "crosswind" in a lathe due to lack of flatness between front and rear slideways, Figure 4.4, from Scarr[23]. Here the crosswind present is defined as being the vertical displacement of one end of the front slideway with respect to the rear slideway (here a "vee" and a "flat", respectively). The effect of this, as the saddle is traversed along the slideway, is to move the cutting tool radially with respect to the workpiece.

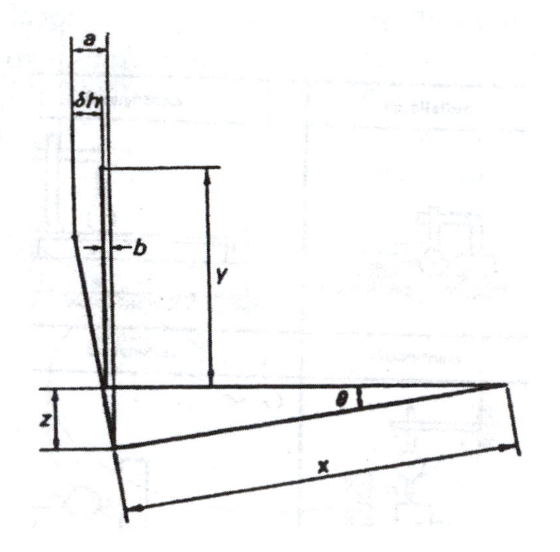

Figure 4.4. Effect of non-parallel ways on machine positional error in a lathe, from Scarr[23].

With respect to Figure 4.4, we can analyze the effect of the cross-wind as follows:

Let δh = radial movement of the cutting tool
x = distance between the vee and flat sideways
y = the height of the tool above the slideway
z = crosswind present between the vee and flat slideways.

Then

$$\delta h = a - b$$

$$= y \sin\theta - \frac{z^2}{2x}$$

$$= \frac{yz}{x} - \frac{z^2}{2x} \qquad (4.1)$$

It can be seen that the relationship between δh and z is nonlinear and the effect of crosswind on the geometric shape of the workpiece is to produce a non-linear taper as shown in Figure 4.5.

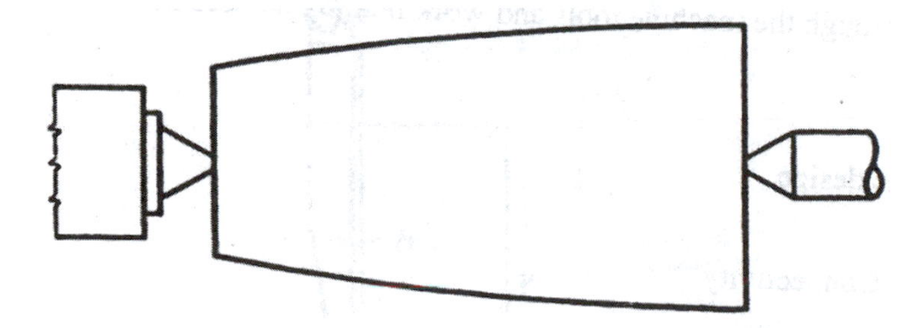

Figure 4.5. Non-linear taper produced by crosswind, from Scarr[23].

Finally, as an example we consider the basic distortion, due to load or uneven foundation, of the machine tool, Figure 4.6. Here the machine tool is supported on conventional distributed foot pads adhesively, or mechanically, fastened to the factory floor. While this keeps the machine from moving during operation, it is a non-kinematic mounting (meaning the machine base is over constrained) so any distortion of the floor (due to swelling of concrete during a "wet" season, for example) will cause the base to distort. This distortion will be seen as part form errors due to the Abbé offset error illustrated. In fact, kinematic mounting and coupling techniques can prevent the introduction of many errors in workpieces through the machine tools and work holding devices as we shall see in the next section.

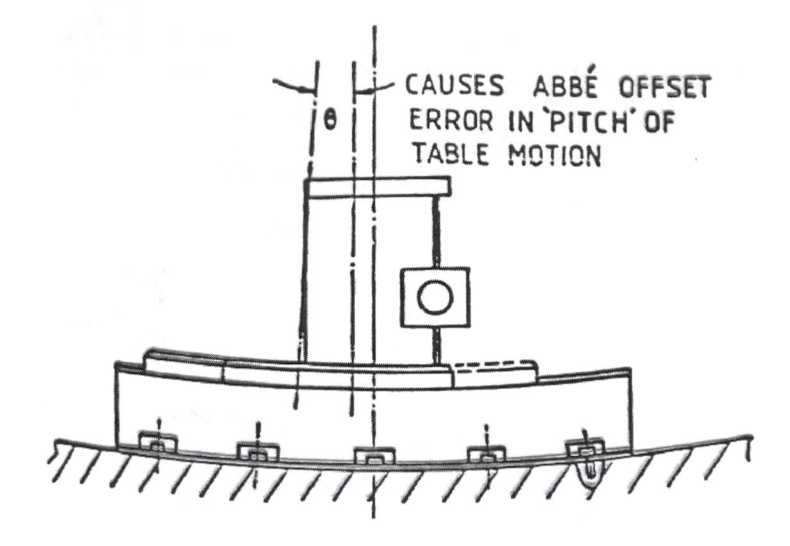

Figure 4.6. Foundation movement causes distortion of the machine structure due to uneven support, from MTTF[25].

4.3 Kinematic design

4.3.1 Connectivity

The real subject of this section could be titled "connectivity between structural elements" — for example between a spindle and the frame or bed of a lathe. There are two basic methods of connection of structural elements-kinematic design and elastically averaged design. The features of each are as follows:

- kinematic design — deterministic, less reliance on manufacturing of the components, limit to the load capacity and stiffness.
- elastically averaged design — nondeterministic, heavier reliance on the particulars of the manufacturing process, no limit on the load capacity and stiffness.

Elastically averaged designs are referred to as over-constrained. As such they can suffer from the problem illustrated in Figure 4.6. They are also susceptible to warping due to uneven thermal expansion. On the other hand, kinematic designs have concentrated loads at points and can experience local deformation. As a general rule, it is preferable to replace excessive surface and line contact with point constraints. Figure 4.7 shows a kinematic three point support of the machine tool in Figure 4.6 eliminating the problem of machine base distortion due to uneven foundation support. While the machine may not stay level, it will not see any distortion of its frame due to the foundation support. We now look at kinematic design in more detail.

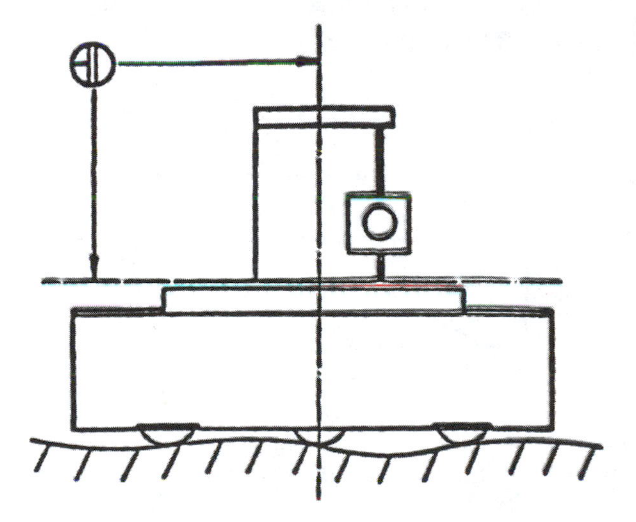

Figure 4.7. Kinematic support of machine tool does not distort the machine structure, from MTTF [25].

4.3.2 Kinematic elements

The six degrees of freedom of a body in space are illustrated in Figure 4.8a. The constraints necessary to kinematically locate the body with respect to the six degrees of freedom are shown in Figure 4.8b. Each contact is in theory a point contact. And, in general, each successive point removes one degree of freedom. One point holds the body with respect to translation in x; two points hold the body with respect to translation in y and rotation about y; three points hold the body with respect to translation in z and rotation about x and z. It is assumed here that the mass of the body is sufficient to hold it against the location points. Lord Kelvin in the mid-19th century and, later, Clark Maxwell, developed the theory and principles of kinematic design. Briefly, these are

> "when a rigid body is constrained by more than the necessary minimum number of constraints, the *redundant* constraints will cause strain which results in distortion, wear, and higher than necessary cost to achieve a specified precision."

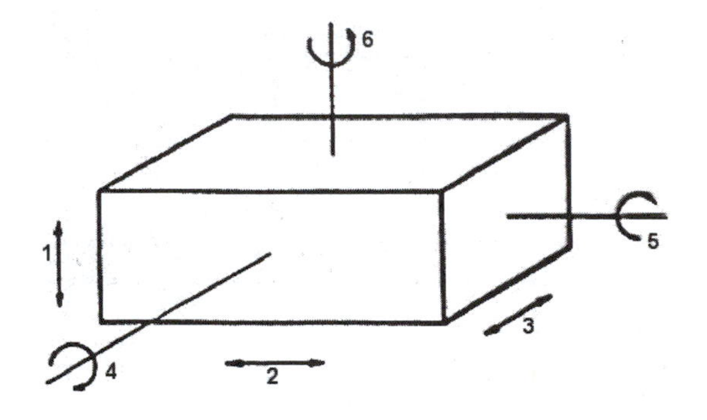

Figure 4.8a. Six degrees of freedom of a body in space.

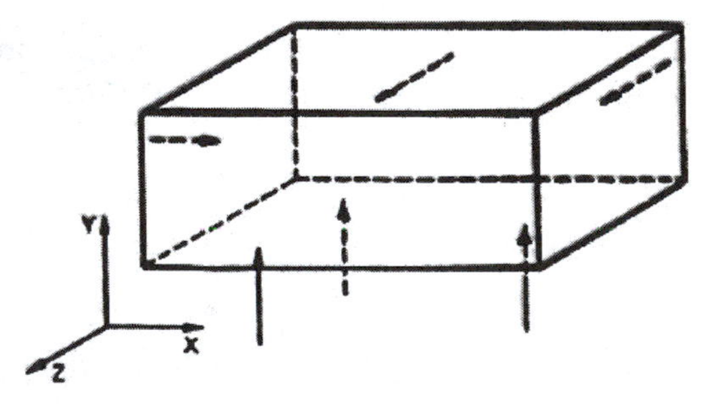

Figure 4.8b. Kinematic location of a rectangular body in space.

Good kinematic design seeks to utilize only the minimum number of degrees of freedom to constrain the motion of a body.

Thus six and only six constraints in the correct positions are necessary to define and fully locate a rigid body positionally with respect to a fixed frame of reference. The fundamentals of fixture design for manufacturing and assembly are based on the same elements of kinematic location. The closure force for the application of kinematic design can be gravity, as noted above, or spring loading or clamping, but the latter must be applied through the kinematic contacts. We will see an example of this in a bit.

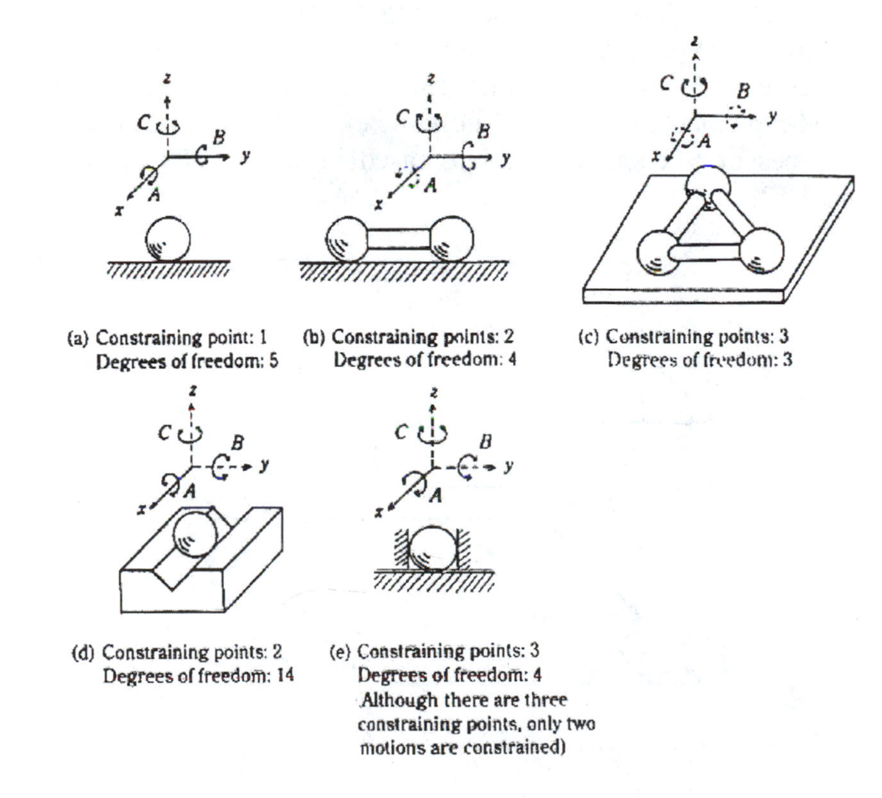

Figure 4.9. Illustrations of a number of constraining points and degrees of freedom.

The elements of kinematic constraint can be broken down to ideal constraining points and degrees of freedom as shown in Figure 4.9. These ideal constraints, like the ball on a surface for one degree of freedom constraint, have practical counterparts in machine elements. The ball and groove in Figure 4.9d is an example of a common machine element. We can look first at static kinematic location as in a fixture or measurement jig. Figure 4.10 illustrates what is commonly referred to as a Kelvin coupling or clamp — a device for locating one machine element with respect to another in all degrees of freedom. It is comprised of the ball on plate (one degree of freedom), ball in vee groove (two degrees of freedom), and ball in trihedral nest (three degrees of freedom) elements. The trihedral nest can be constructed in various ways — the three nested balls as in Figure 4.10, three machined intersecting radial grooves, or a trihedral hollow (as by electrical discharge machining or a coining operation, Figure 4.11. The "Stewart platform" recently invigorated in the hexapod-type machine design is a classic example of a machine with six degrees of freedom movement based in six parallel actuators, Figure 4.12.

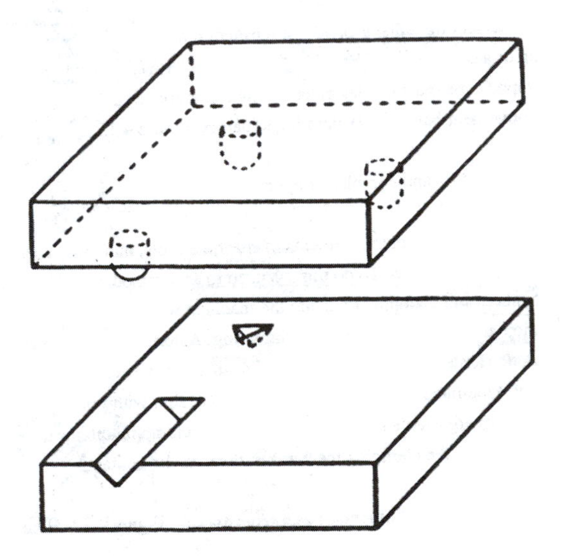

Figure 4.10. Illustration of a Kelvin coupling or clamp.

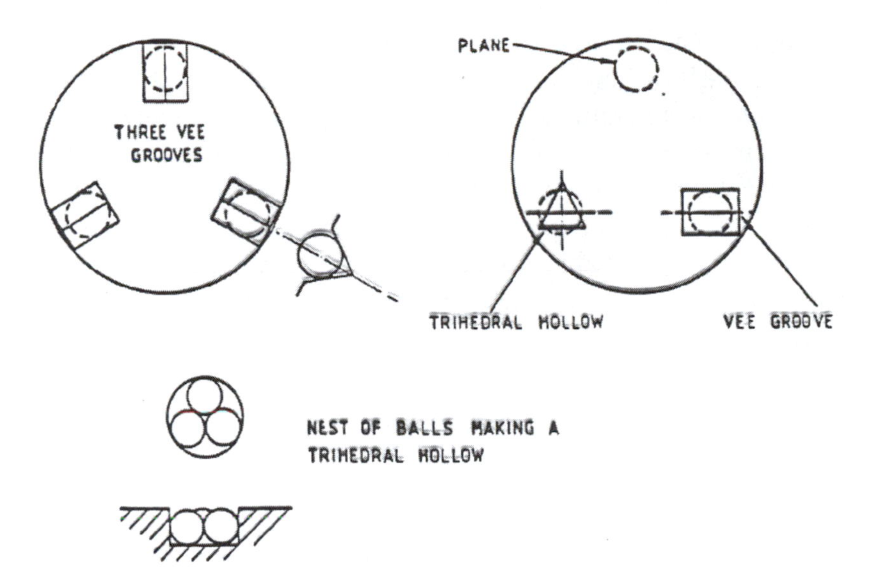

Figure 4.11. Two forms of a kinematic clamp (no degrees of freedom, six point contact), from McKeown in MTTF[25].

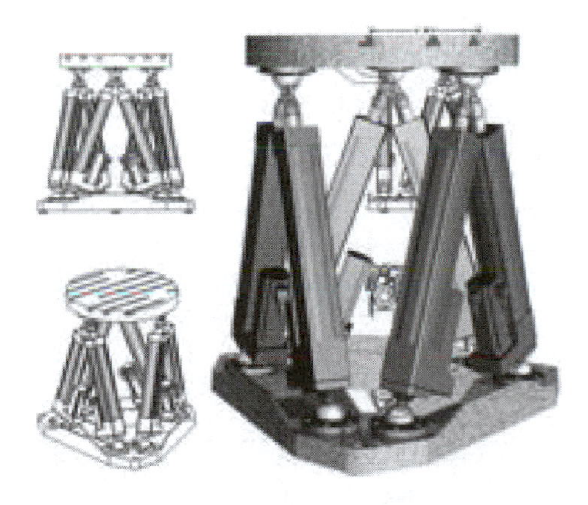

Figure 4.12. Hexapod-type machine tool based on the Stewart platform concept, from Fortier[32].

4.3.3 Contact and complex support

We would like to construct the elements of machine tools, including ways and axes from kinematic design principles. To do this we need to accommodate motion of elements relative to others and clamping to insure the machine will not disassemble itself under the loads, motion and accelerations of manufacturing. As previewed above, any clamping or loading of one element to another to provide positive contact and insure stability should be done through kinematic contacts. One such clamp design with point contact is shown in Figure 4.13. This adjustable kinematic ball seat with clamp maintains the kinematic ball in cone design and has the clamping mechanism through the contact.

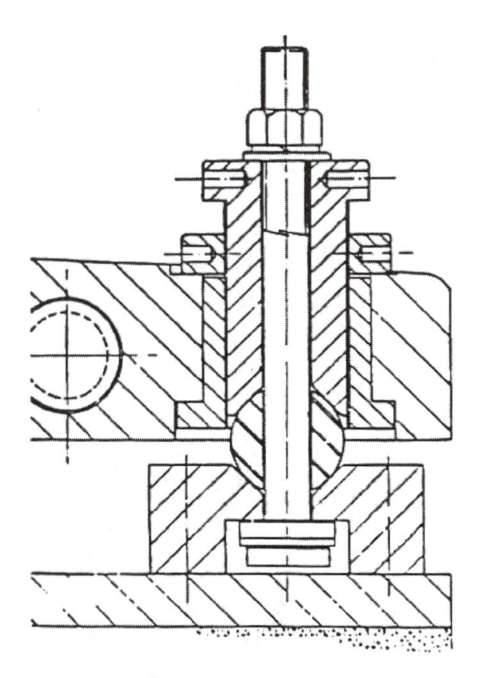

Figure 4.13. Adjustable kinematic ball setting with clamp, from McKeown in MTTF[25].

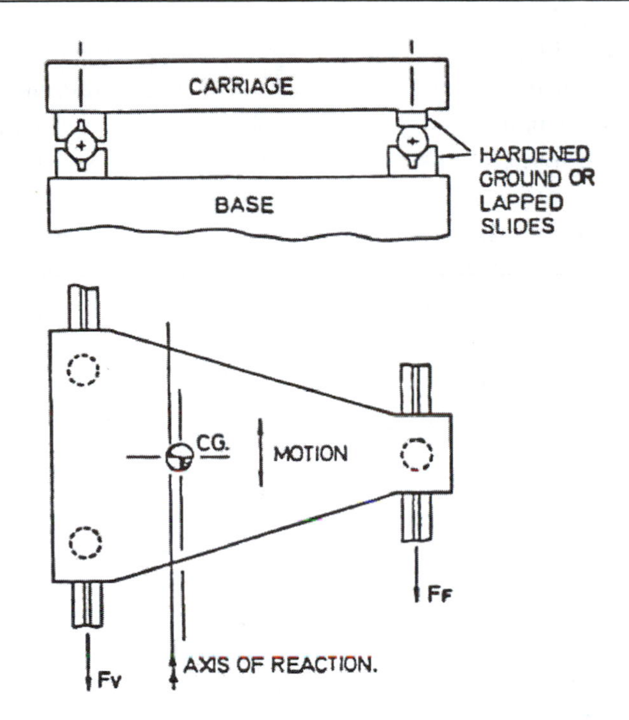

Figure 4.14. Linear kinematic ball slide (one degree of freedom), from McKeown in MTTF[25].

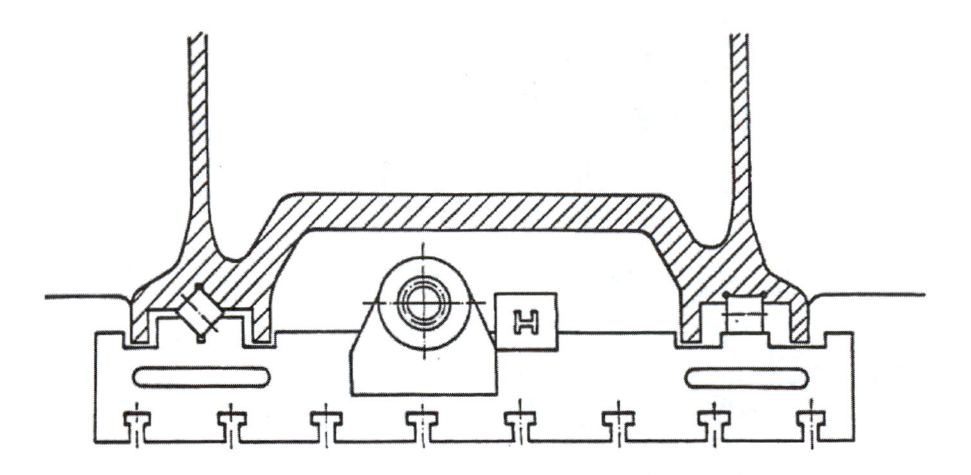

Figure 4.15. Precision roller guideway in a semi-kinematic configuration, from McKeown in MTTF[25].

Axes can be constructed of kinematic elements as well, Figure 4.14. Here we see the vee slot and balls used to kinematically support the table. It is necessary to carefully fix the center of gravity of the load on the table to insure that the assembly is stable. Also note the axis of reaction necessary to insure smooth linear translation. Although Figure 4.14 illustrates an ideal but usable mechanism, traditional machine tool design relies on more robust designs that tend to relax the kinematic configuration, Figure 4.15. This shows the use of conventional bearing elements in place of the balls. These elements are shown in more detail in Figure 4.16. The error due to ball inaccuracy is illustrated along with relative velocities of table and ball carriage.

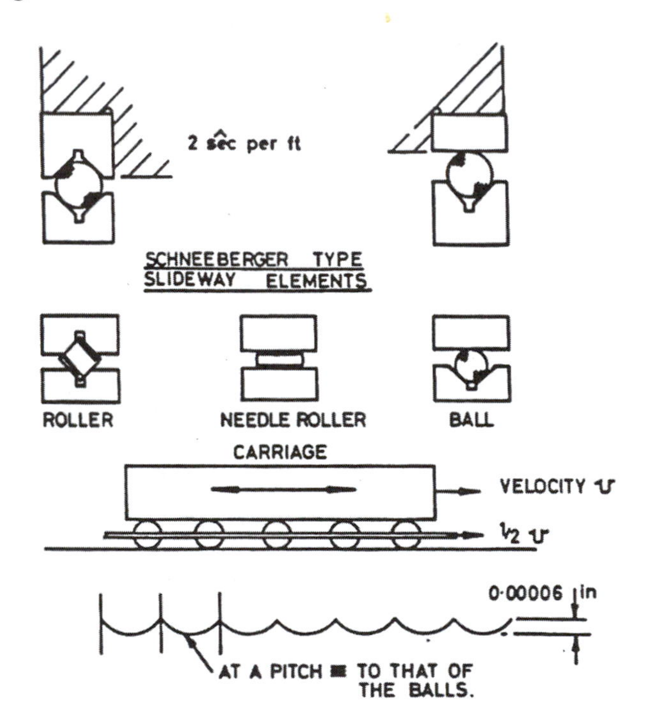

Figure 4.16. Bearing elements, from McKeown[27].

Figures 4.17-4.19 show other configurations of semi-kinematic slides with bearings including one, Figure 4.19, with a table "hung" for mounting a machine element with a vertical spindle or axis of rotation. Exact kinematic vertical supports are possible although complicated. Generally, the most stable is as described in Figure 4.20, from Slocum[18]. Other less stable configurations are shown in Figure 4.21.

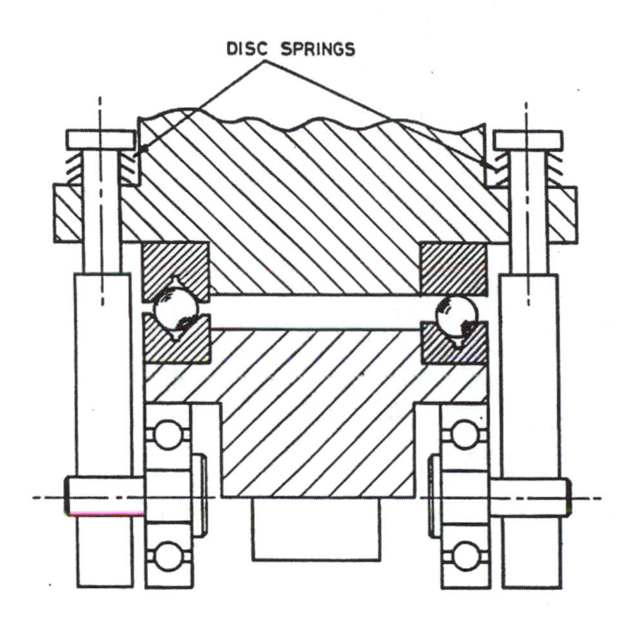

Figure 4.17. Typical mounting configuration for Schneeberger bearing, from McKeown[27].

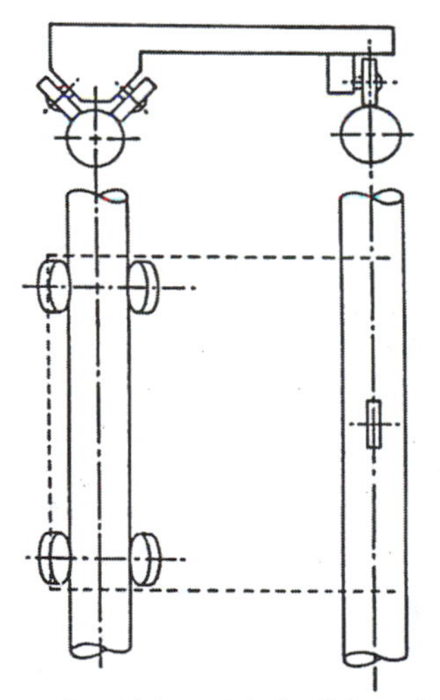

Figure 4.18. Kinematic slide with ball bearings as wheels, from McKeown[27].

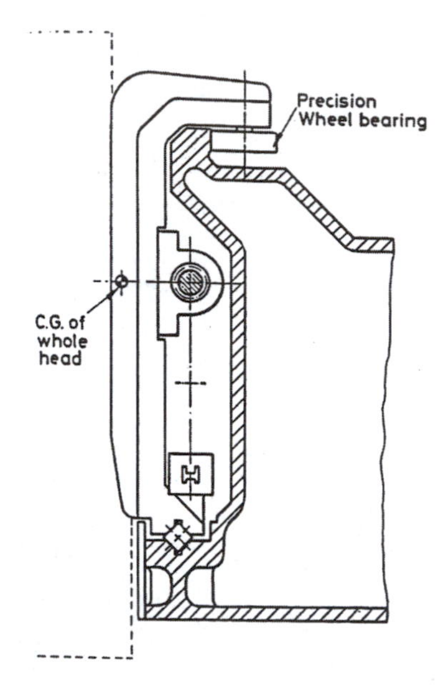

Figure 4.19. Carriage guiding system with "hooked on" configuration, from McKeown[27].

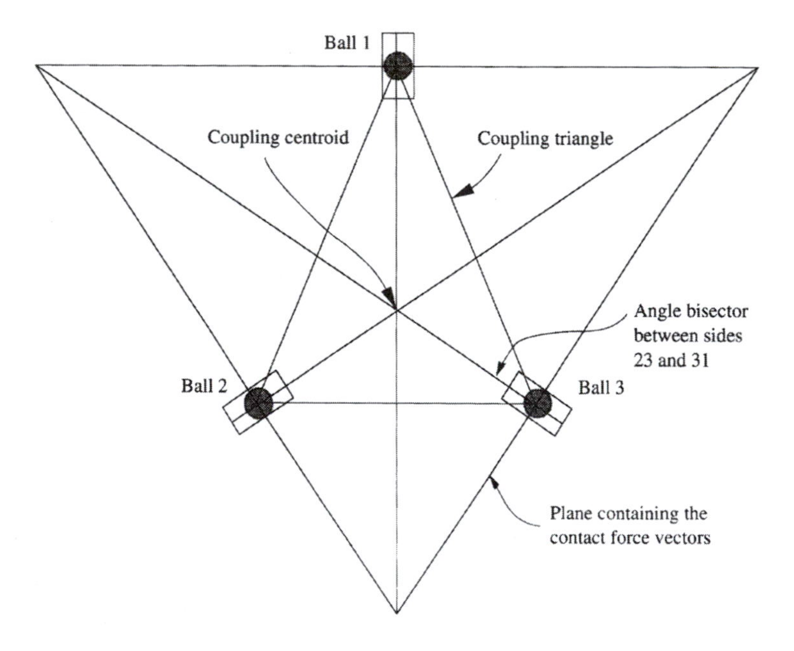

Figure 4.20. Stability in three-groove kinematic couplings, from Slocum[18].

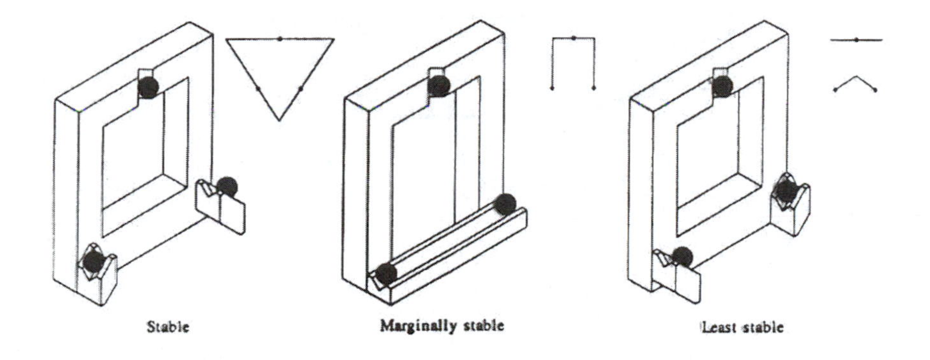

Figure 4.21. Use of analysis of planes of contact force vectors for assessing the stability of three groove kinematic couplings, from Slocum[18].

Although not discussed here, an important consideration is driving the moving axis. This must be done in such a way as not to distort the motion or induce additional forces on the mechanism. This can be accomplished most readily with flexible connections such as flexible nuts or flat springs or hydrostatic nuts.

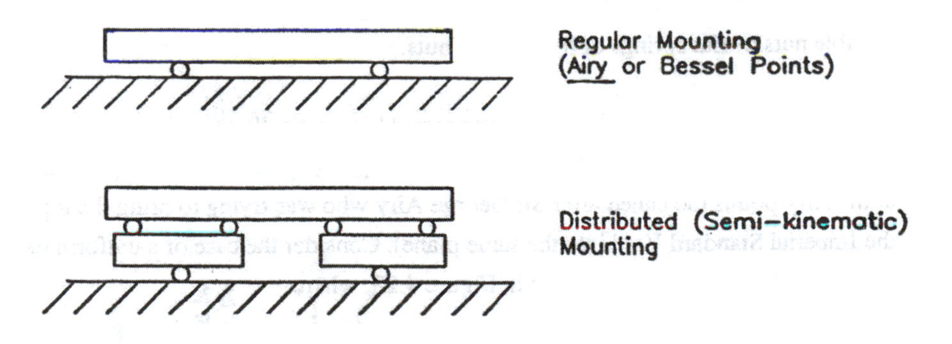

Figure 4.22. Illustration of a 2D "wiffle tree" type mounting used on length standards.

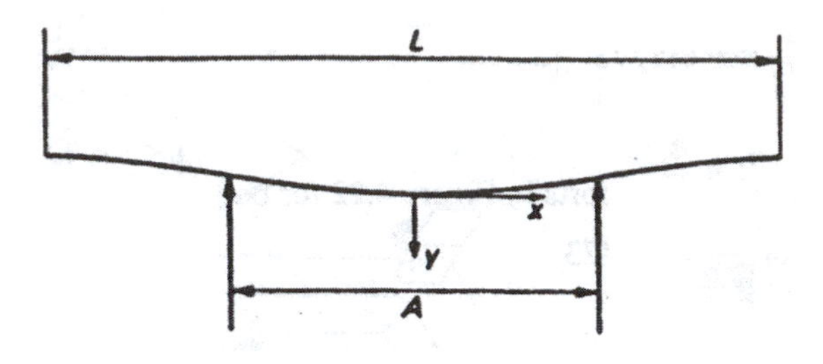

Figure 4.23. Deflection of a uniform bar symmetrically supported.

Finally, large surfaces (planes) can be supported kinematically to maintain their insensitivity to disturbances, as seen in Figure 4.22. In the figure, the kinematic supports are located at the Airy points (so called after Sir George Airy who was trying to bring the top faces of the Imperial Standard Yard into the same plane). Consider the case of a uniform bar symmetrically supported as shown in Figure 4.23, where:

E = Young's modulus
I = second moment of area of bar ($\pi d^4/64$ for a circular section of diameter, d)
w = weight of the bar per unit length

Then between the point of support and the end of the bar the slope of the bar dy/dx is given by:

$$\frac{dy}{dx} = \frac{w}{6EI}\left[\frac{L}{8}(L^2 - 3A^2) - \left(\frac{L}{2} - x\right)^3\right]. \tag{4.3.1}$$

In order to configure the bar so the slope at the ends (i.e. when x=$L/2$) is zero, a value of A/L has to be chosen to make $dy/dx = 0$. That is, A/L is chosen so that $l^2 - 3A^2 = 0$. Thus,

$$\frac{A}{L} = \frac{1}{\sqrt{3}}, \text{ or } 0.5773 \tag{4.3.2}$$

So, the position of the supports in Figure 4.22 for both configurations meets the requirements for $\frac{A}{L} = 0.5773$.

This concept can be applied to more complex assemblies of planar surfaces as for mirror segments for the Keck telescope, Figure 4.24. Figure 4.25 shows this concept applied to the support of a large flat surface. Check your windshield wiper mechanism next time you are driving and notice how this approach insures that the blade maintains contact even with a changing windshield curvature.

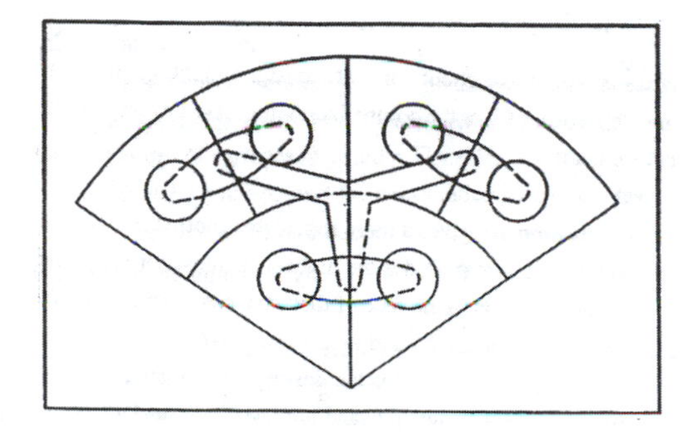

Figure 4.24. "Wiffle tree" mounting for mirror segments on the Keck telescope.

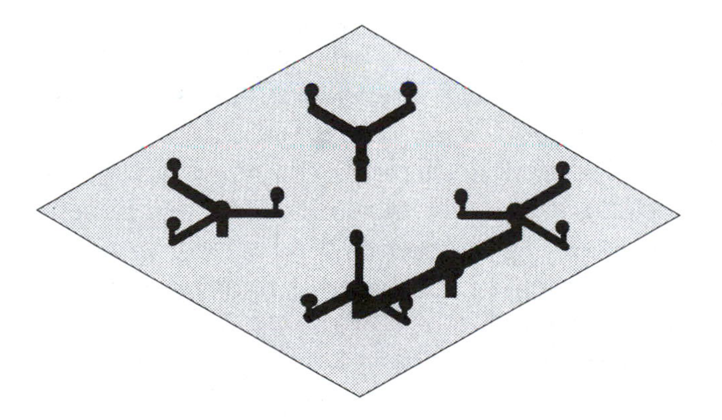

Figure 4.25. Three point mounts with ball joints in a "wiffle tree" arrangement for supporting a large flat surface, from Slocum[18].

4.3.4 Summary of kinematic design

To summarize the major points about kinematic design/mounting and its application to machine elements we can look at the comparison with elastic averaged mounting. The key concerns are:

- which approach gives a more repeatable motion (or assembly)?
- which approach provides the highest stiffness for ideal one degree of freedom motion in the face of disturbances from the environment, machine action (rotation, etc.) and load variations?
- which approach maintains the desired performance and characteristics best over the typical range of usage and wear conditions?
- what is the relative cost of manufacture for the same level of performance?

In general, when applied to pairs, kinematic design will improve kinematic precision. If the center of gravity and axis of reaction concerns are met, kinematic design will reduce the possibility of undesirable motion, such as rotation in plane. Long ways will also help minimize this. Kinematic designs in connections and couplings minimize errors in mating surfaces due to manufacture. Sources of play (as in transmission of drive screw non-linearity to an axis of motion) must be minimized realize the highest precision in the kinematically designed mechanism. The point contacts are ideal in a kinematic design, large loads transmitted through these contacts can result in substantial compliance and wear and, in this case, elastic averaging by the use of larger contacts is advantageous. And, when ever possible, the number of constraints should be reduced to the minimum achievable.

The challenges of structural distortion due to mounting and coupling in kinematic mechanisms are discussed in the next section.

4.4 Structural compliance

We concern ourselves with compliance on two very different scales — micro and macro. That is to say, on a level of point interaction between elements of a mechanism as described in the last section due to, for example, Hertzian contact, which we call micro. And, on a level of structural interaction between, say a machine element and the foundation, which we call macro. Although the micro elements are certainly present in the macro problem, they are small by comparison.

4.4.1 Microscale compliance

For micro scale compliance we concern ourselves with the point contact between two objects which, ideally, depends on the geometry in the contacted region...and this depends on the algebraic sum of curvatures of surfaces in contact. Consider two spheres of radii R_1 and R_2 with modulii of elasticities E_1 and E_2, and Poisson's ratios μ_1 and μ_2, respectively. Then the equivalent radius of contact of the interaction, R_e, can be defined as (for general curved body contact with major and minor radii):

$$R_e = \frac{1}{\dfrac{1}{R_{1\,major}} + \dfrac{1}{R_{1\,\min or}} + \dfrac{1}{R_{2\,major}} + \dfrac{1}{R_{2\,\min or}}} \qquad (4.4.1)$$

Here a concave surface is considered positive and a convex surface considered negative; a plane surface is considered to have an infinite radius. For spheres, the two radii, major and minor, are equal so only one need be considered for each sphere.

The equivalent modulus of elasticity, E_e, can be represented as follows:

$$E_e = \frac{1}{\dfrac{1-\mu_1^2}{E_1} + \dfrac{1-\mu_2^2}{E_2}} \tag{4.4.2}$$

The contact area of surfaces with an equivalent radius of contact R_e is defined in terms of the equivalent circular contact area between two bodies, a, as:

$$a = \left(\frac{3FR_e}{2E_e} \right)^{1/2} \tag{4.4.3}$$

where F = contact force between the two spherical bodies, and the deflection, δ, due to the load and area of contact is defined as:

$$\delta = \frac{1}{2} \left(\frac{1}{R_e} \right)^{1/3} \left(\frac{3F}{2E_e} \right)^{2/3} \tag{4.4.4}$$

Further, we can calculate the contact pressure, q, under these conditions as:

$$q = \frac{1}{\pi} \left(\frac{1}{R_e} \right)^{1/3} \left(\frac{3E_e^2 F}{2} \right)^{1/3} \tag{4.4.5}$$

or

$$q = \frac{aE_e}{\pi R_e} \tag{4.4.6}$$

Under these circumstances, a force of 1 N on a 3mm diameter steel ball resting on a flat steel surface (giving an equivalent elasticity of 110 GPa and equivalent radius of .75 mm for an equivalent diameter of contact of 21.7 μm) results in a deflection (deformation), δ = 31 μm. Interestingly, if the load is increased by 10% this results in a 6% increase in deflection, or approximately 0.021 μm. This is called the stiffness of the system and, here, for a 1 N load and 31 μm deformation the stiffness is .032 (10^{-6}) N/m. We will see this measure for other machine systems later in the notes.

To avoid surface damage (recall our discussion about the interaction of two surfaces and the contact area at asperities?) the contact pressure, q, should be less than $\tau_{\max}$ for the surface. Brinnelling will occur for contact pressures greater than this. If $\tau_{\max}$ is one half the maximum tensile stress ($\sigma_{\max} / 2$) and $\tau_{\max} = q/3$ then we should insure that the maximum Hertzian contact stress, $q_{max} = \frac{3}{2}\sigma_{allowable-tensile}$. This is most serious for contact probes for measurement instruments but can also affect the precision of fixtures and other machine elements. This was the basis of the earlier comment under the discussion of metrology that non-contact techniques are preferred.

4.4.2 Macroscale compliance

Bodies supported in either symmetric or non-symmetric, kinematic or non-kinematic fashion will all exhibit distortion due to their compliance (elastic) behavior. Materials and/or structures designed for enhanced stiffness (as with an H cross section) or of concrete resin composition may have lower compliance but all have a measurable compliance due to structural deflection. Slocum notes[18] that a symmetrical body in which the mounts have the same axes of symmetry will, ideally, have the same symmetric distortions. Overconstraint of mounts (i.e. non-kinematic) can result in significant nonsymmetrical distortion due to "forced geometrical congruence." And, if a kinematic mounting is used, its deterministic nature enables the computation of the amount of distortion, which may not have the same axes of symmetry as the body, and then to design the system to maintain the distortion below a threshold.

The problem of machine distortion is enhanced by the physical size of most machines, the mass of their moving elements and the unfortunate fact that they are rarely supported kinematically. More typically, large machines are resting on pads spaced around the perimeter of the machine on a thick concrete slab. In such a way, any movement in the concrete due, for example, to thermal expansion or absorption of water and subsequent expansion, which usually occurs non-uniformly, will cause differential movement of the machine tool and distortion of the ways, etc.

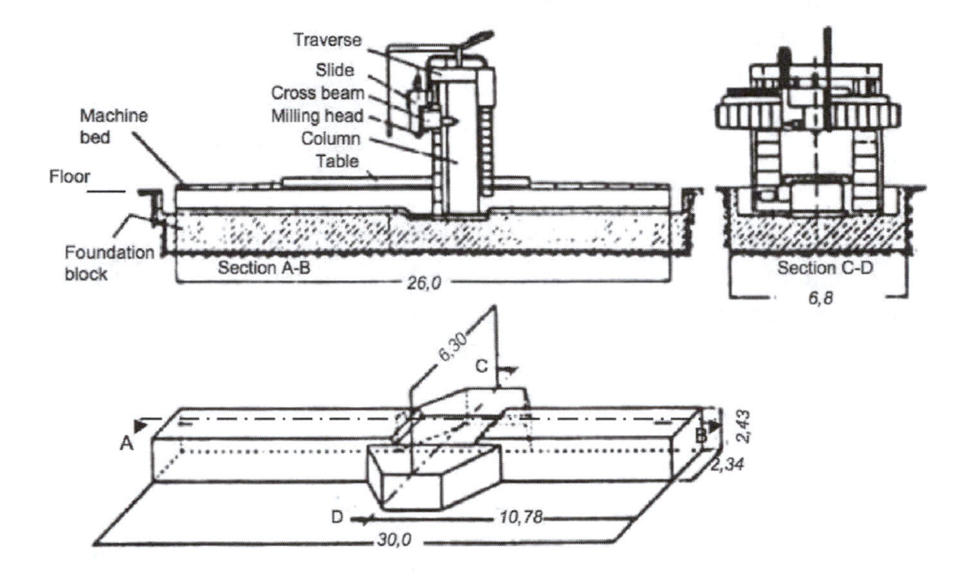

Figure 4.26. Sketch of a planer mill, from MTTF[25].

An extreme example of this is now described with respect to a planer mill shown in Figure 4.26; the sketch shows the machine and its foundation. In Figure 4.27 is shown a model of the machine and it's foundation from static analysis. The figure shows the machine-foundation elements supporting the machine. The major source of static deflection of the machine bed is due to movement of the mass of the workpiece and table over the bed of the machine tool, Figure 4.28. The two diagrams show the machine deformation under the static load of the table at two locations. The graph shows compliance (inverse stiffness) as a function of position in μm/kN. Kinematic support of the machine tool, as seen in Figure 4.7, can substantially prevent this.

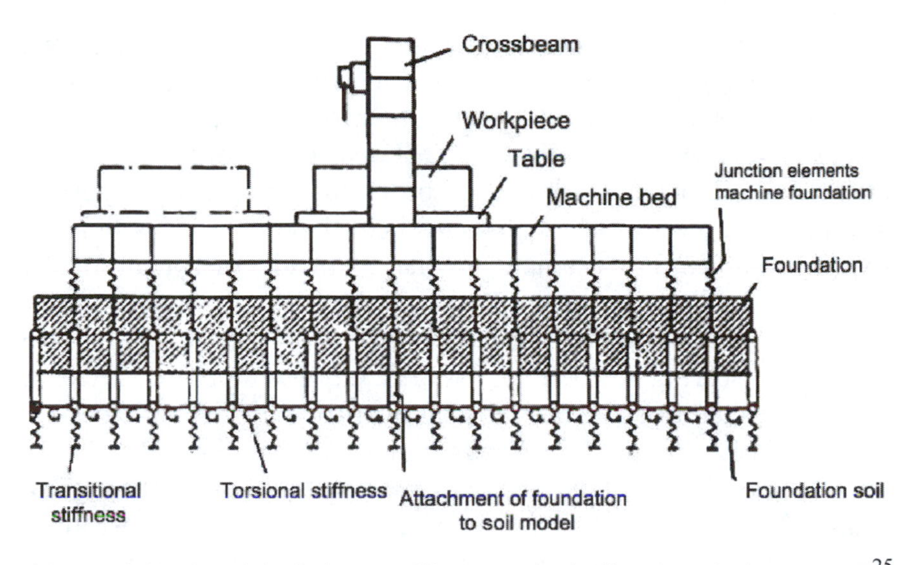

Figure 4.27. Model of planer mill system including foundation, MTTF[25].

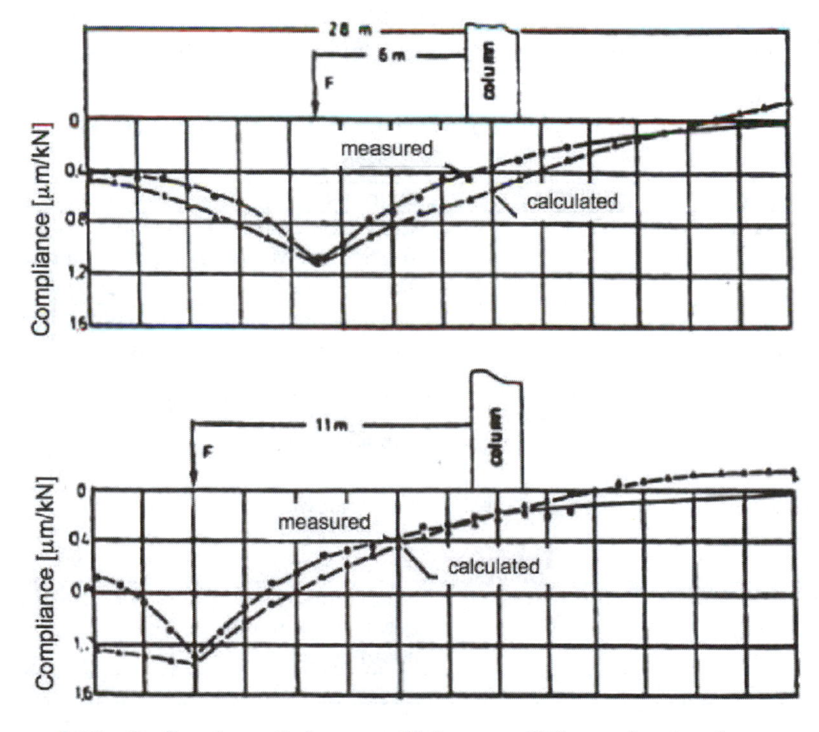

Figure 4.28. Deflection of planer mill for two different load points; measurement (upper trace) and calculation (lower trace), MTTF[25].

To effectively design for minimization of compliance, one must consider the stiffness of various machine elements. Some of these are more quantitatively known than others, Table 4.1. If the stiffness characteristics are well understood, the information can be applied in computer-aided design programs for machine design. Unfortunately, most machine elements behave nonlinearly with some process variables. For example, we have already seen that spindle runout varies with rotational speed, Figure 3.39, although this may not be due solely to change in compliance. In Figure 4.29 is shown variation in bearing stiffness with speed and assembly clearance. A change in a factor of two can be seen over the range of 400 to 1400 rpm. Similarly, Figure 4.30 shows contact deflection under conditions of varying contact pairs. It is interesting to note that the highest stiffness is for combination 1 (lapped cast iron on lapped cast iron). Simply changing one of the contact pairs, for example, curve 2, to ground cast iron, reduces the stiffness significantly. Finally, stiffness of the individual machine structural elements can be substantially affected by reinforcement and other design elements, Figure 4.31.

Table 4.1. Survey of characteristic stiffness values, MTTF[25].

Category	Stiffness	Usefulness in CAD
Machine tool joints	known, but only specific knowledge of static stiffness	by means of special finite element programs
Plain slide ways	some specific knowledge	by means of special finite element programs
Roller bearings and guideways	much knowledge about static, some about dynamic	useful
Hydrostatic bearings and guideways	much knowledge about static, some about dynamic	useful
Aerostatic bearings and guideways	much knowledge about static, some about dynamic	useful
Dampers	-	-

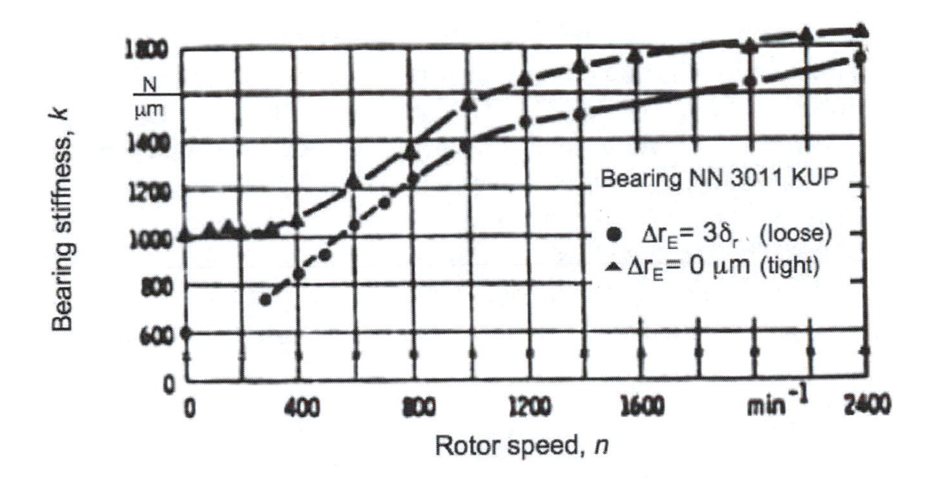

Figure 4.29. Bearing stiffness as a function of speed and assembly clearance, from MTTF[25].

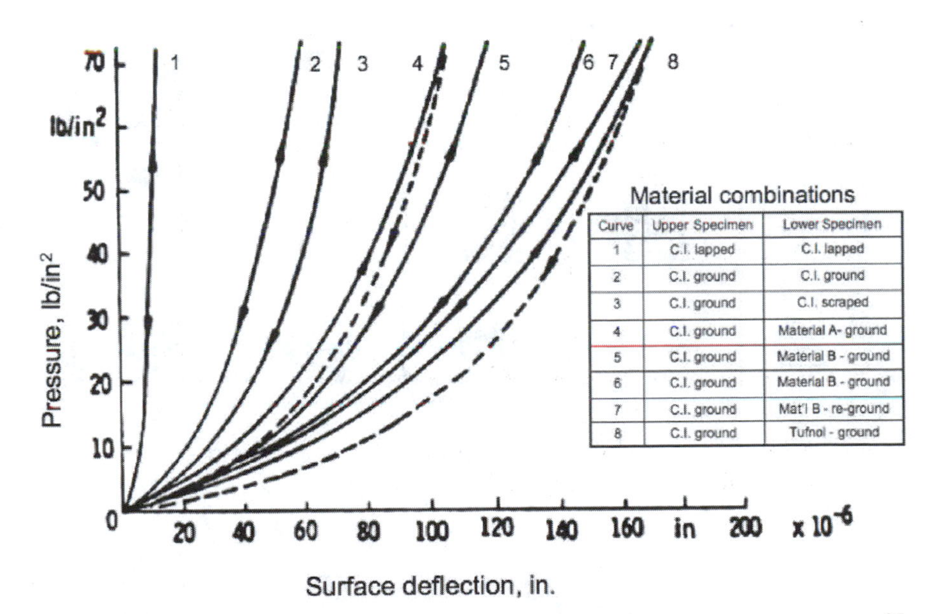

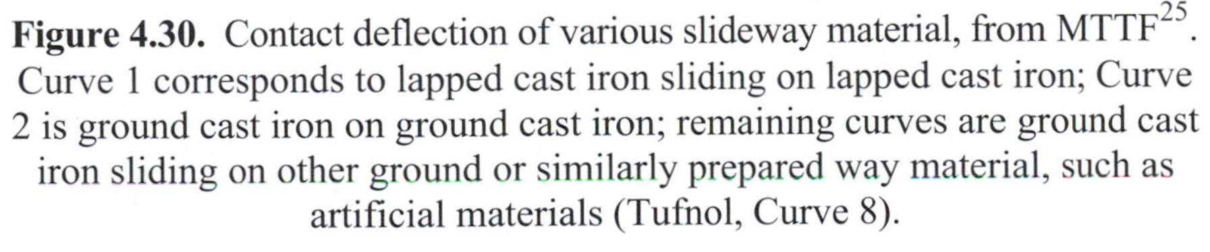

Figure 4.30. Contact deflection of various slideway material, from MTTF[25]. Curve 1 corresponds to lapped cast iron sliding on lapped cast iron; Curve 2 is ground cast iron on ground cast iron; remaining curves are ground cast iron sliding on other ground or similarly prepared way material, such as artificial materials (Tufnol, Curve 8).

When the machine elements are assembled into a machine tool their individual behavior contributes to the overall behavior of the machine tool, Figure 4.32. The curve shows the variation in x-displacement with static load in a direction similar to that observed in machining, orthogonal to the spindle. This is not the primary sensitive direction with respect to surface finish but does affect part accuracy. The slope of this curve is compliance, μm/kN. Notice the significant change in compliance with load. In some cases, counterweights can be used to offset this change in compliance with position or load, Figure 4.33.

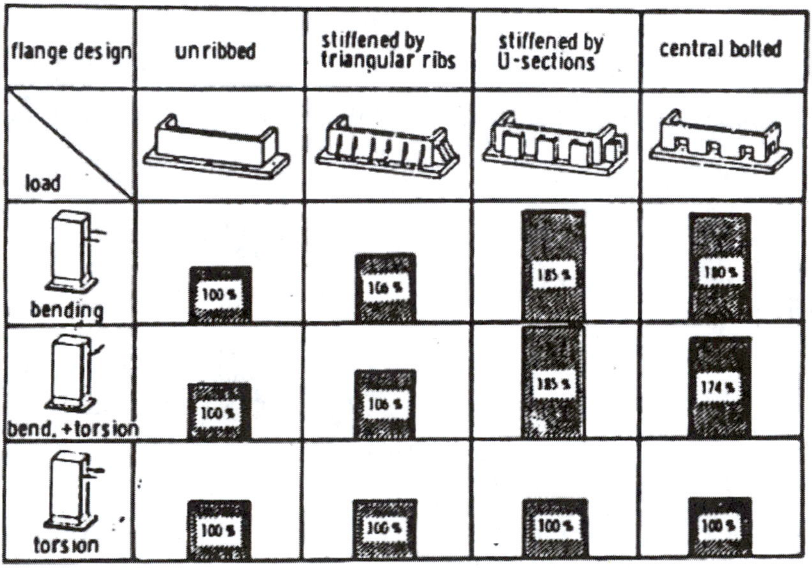

flange design / load	unribbed	stiffened by triangular ribs	stiffened by U-sections	central bolted
bending	100 %	106 %	185 %	180 %
bend. + torsion	100 %	106 %	185 %	174 %
torsion	100 %	100 %	100 %	100 %

Figure 4.31. Effect of different flange designs on the stiffness of machine tool column under various load conditions, from MTTF[25].

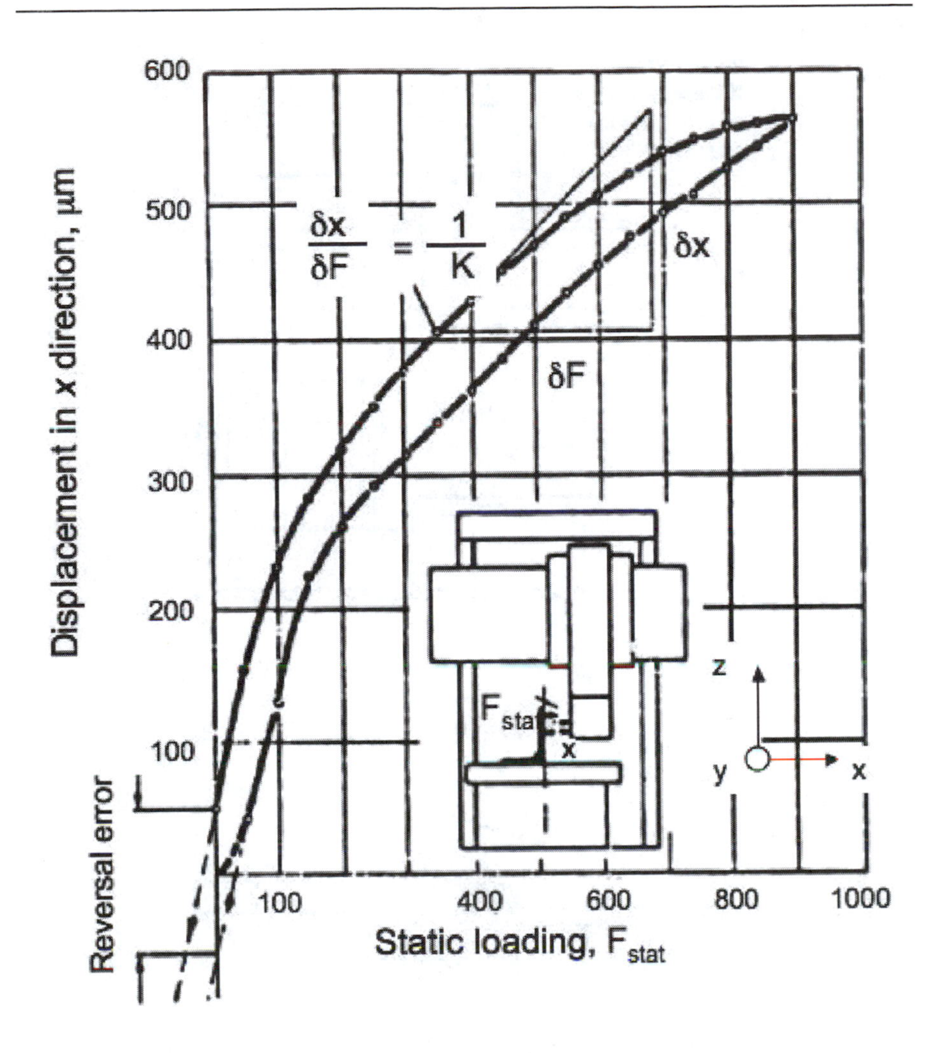

Figure 4.32. Deformation characteristics of a vertical boring mill, from MTTF[25].

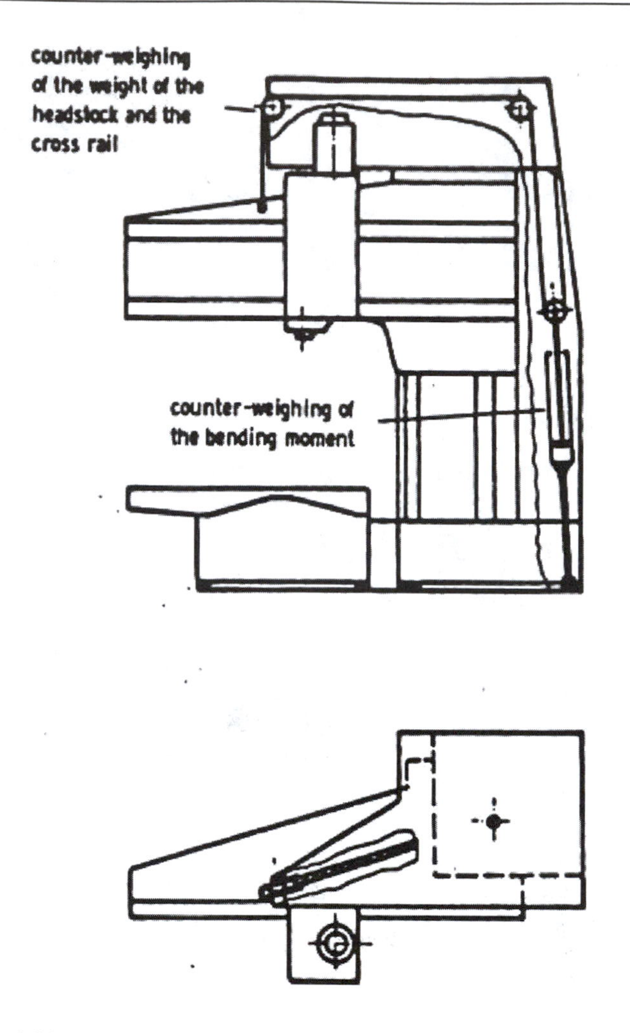

Figure 4.33. Vertical milling machine with counter-weight, from MTTF[25].

We will talk in much more detail about the effect of compliance on the part precision in a later chapter. Similarly, we will discuss the source and impact of dynamic process effects on machine distortion.

4.5 Bearings and spindles

4.5.1 Bearings

We will review here some of the basics of bearings so that we understand the principles of design and utilization, especially from the point of view of mechanical errors. This will be only a cursory review. For more details the reader is referred to standard texts in the field and, for the best information relative to precision machine design, to Slocum[18]. Chapters 8 and 9 of Slocum cover contact and noncontacting bearings, respectively in tremendous detail.

The purpose of bearings is to insure smooth motion of one element relative to another. We would like to insure both minimum friction as well as constant friction under differing conditions. Table 4.2 compares the characteristics of several bearing types:

- hydrostatic
- rolling
- air bearing, and
- magnetic

Table 4.2. Comparison of bearing characteristics, from Furukawa[33].

	Rolling bearing	Hydrostatic bearing	Air bearing	Magnetic bearing
Speed characteristics				
Maximum speed				
Range	35×10^4	30×10^4	200×10^4	$200\text{–}300 \times 10^4$
DN-value*				
Load characteristics				
Maximum load capacity	Large	Large/intermediate	Small	Intermediate/small
Static/dynamic rigidity	High	High/intermediate	Low/intermediate	Intermediate/low
Kinematic characteristics				
Rotational precision	Intermediate/low	High/intermediate	High	Intermediate
Axial direction precision	0.2–1.0 μm	~0.05 μm	0.05 μm	~0.5 μm
Thermal characteristics				
Heat generation	High	High	Low	Intermediate/low
Heat dispersal				
Average temperature	+10–20 °C	+10–20 °C	+2–5 °C	+5–20 °C
	Low-speed, high-output spindle (Rolling and Hydrostatic)		High-speed, low-load, high-precision spindle	High-speed spindle

* DN value = shaft diameter (mm) × speed (r.p.m.)

Not included in the table but an important type of bearing is

- sliding

which could be considered as a class of hydrostatic bearing. Sliding contact, rolling element (and flexural type) have contact between the elements. Hydrostatic, aerostatic (or airbearing) and magnetic have no contact between the elements. Hydrostatic bearings are so called as the fluid film is created by fluid pressure in the area between the bearing elements. Hydrodynamic bearings have a fluid film due to relative motion of the elements. The resulting coefficients of friction, μ, for these bearings differ tremendously. For sliding bearings with little of no fluid film (i.e. contact) μ can vary from 0.01 - 0.10 plus the potential for wear exists. Hydrostatic bearings (as well as hydrodynamic) have μ values in the range from 0.001 - 0.006 for oil and 10^{-7} to 10^{-8} for air as the fluid. Rolling contact bearings have μ in the range of 0.02 - 0.04. Figure 4.34 shows graphically the variation of coefficient of friction with the dimensionless quantity $\eta N / p$, sometimes referred to as the characteristic number. Here η is the fluid viscosity (cP), N is the rotational speed (rps), and p is the average pressure on the plane of projection (N/mm^2).

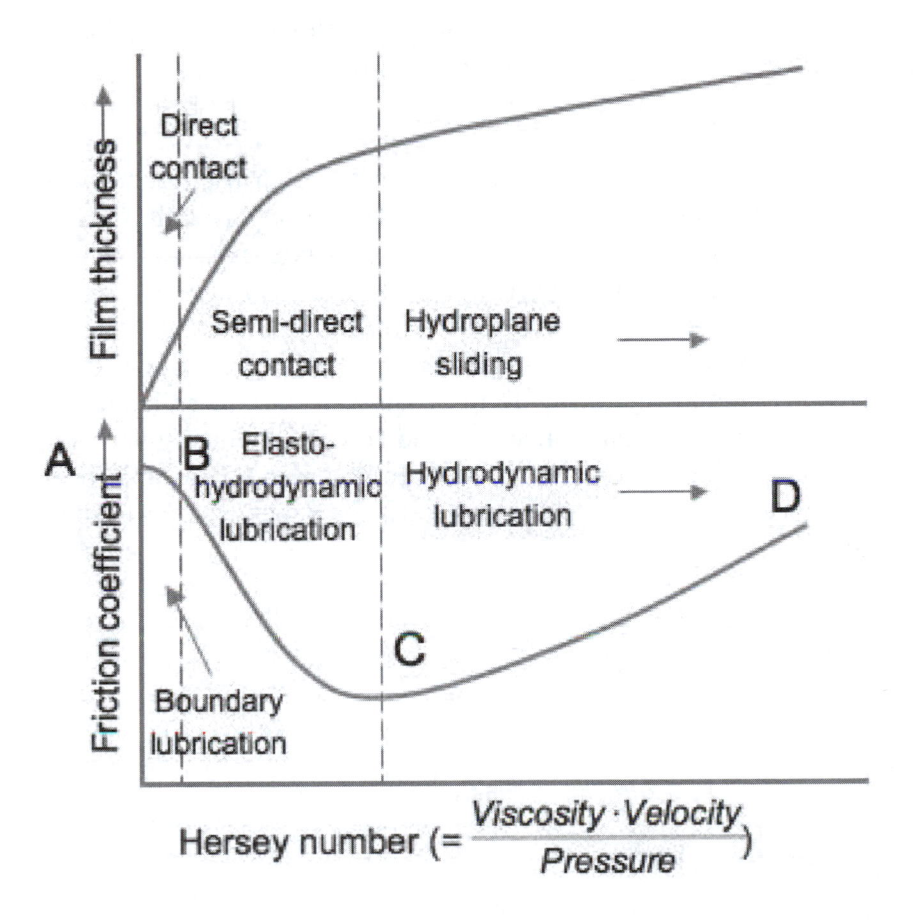

Figure 4.34. Stribeck Curve: Film thickness and coefficient of friction μ vs. Hersey number.

Three distinct regions of friction behavior are apparent depending, primarily, on the thickness of the fluid film between elements. Between A and B on the figure the thickness of the fluid film is sufficiently thin that there is solid to solid contact — i.e. the film thickness is less than the surface roughness. Partial asperity welding can occur here and we refer to this as the boundary lubrication state. Between B and C represents an intermediate stage, referred to as mixed lubrication. Between C and D is hydrodynamic lubrication — the fluid film thickness is greater than the roughness and no surface to surface contact occurs. Lubrication in sliding guides corresponds to the region between A and C whereas hydrostatic and aerostatic bearings operate in the hydrodynamic lubrication state between C and D.

At very slow speed in the presence of high friction a "stick-slip" phenomenon can occur. This is often a problem with machine tool axes during complex motion where one axis may slow to zero velocity, change direction and then accelerates (as in a two axis circular motion). This can be avoided, or at least minimized, by using material pairs in bearings that have an increasing frictional resistance (force due to friction) with increasing speed rather than the inverse. Then, speed reduction will yield lower friction. Figure 4.35 shows the frictional properties of a three commercial castable bearing materials that exhibit both characteristics depending upon speed range.

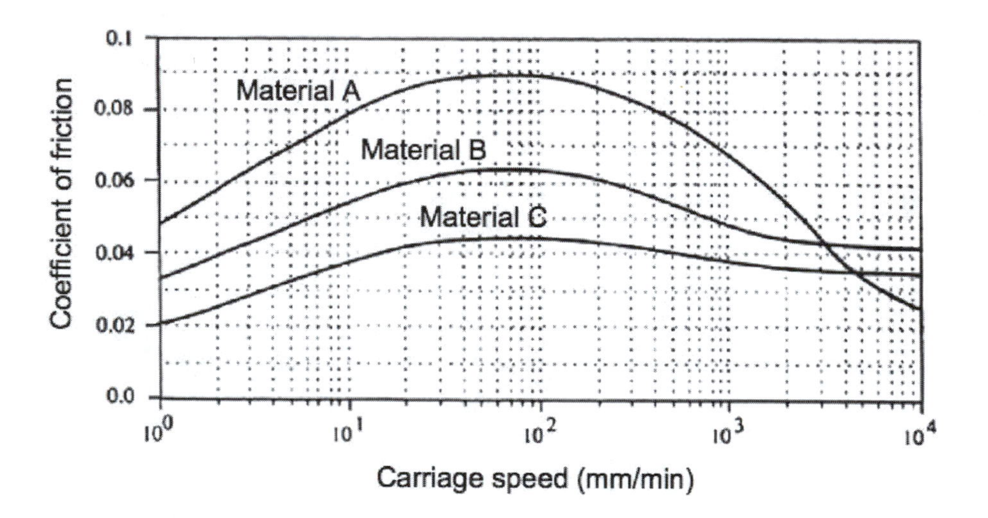

Figure 4.35. Frictional properties of a castable bearing material.

We include here a list of considerations in bearing design that have an influence on performance in a precision engineering sense. Recall that bearings (in ways and in spindles) are significant sources of error due to compliance etc. Some of these concerns have already been discussed and some will be discussed in more detail. Considerations include:

- speed and acceleration limits
- instability at high speed
- coefficient of friction variation
- thermal effects
- range of motion
- applied loads
- contact areas/lines
- constancy of motion (static or dynamic)

As we saw with spindle speed variation, the performance of most bearings is very dependent on operating conditions. With respect to the precision of the mechanism in which the bearing is used, we should consider the following:

- accuracy
- accuracy of the motion of the components supported
- lateral deviation/repeatability as part of a system under servo-control (backlash, hysteresis, etc.)
- repeatability
- similar to accuracy (component and system)
- resolution
- friction level/smoothness of motion
- surface finish, shape accuracy
- short travel ranges (where flexural bearings are generally preferred)

Often, bearing are pre-loaded to eliminate problems with backlash of excessive compliance with the following effects:

- preload
- nonlinear deformation
- higher stiffness
- reduced backlash, increased repeatability to a point
- higher the preload, greater component deformation (Hertzian stresses), higher friction, lower resolution and repeatability

Figure 4.36 from Slocum[18] illustrates the variation in compliance in rolling element bearings for different designs. An additional row of balls in the bearing reduces the amplitude of compliance but increases the frequency of variation.

- stiffness
- pre-load/stiffness tradeoff
- vibration and shock resistance
- no mechanical contact between bearing components (or)
- pre-load high enough to guarantee resultant varying stress load is small (<10%) compared to that of pre-load
- sliding contact and fluid bearings best

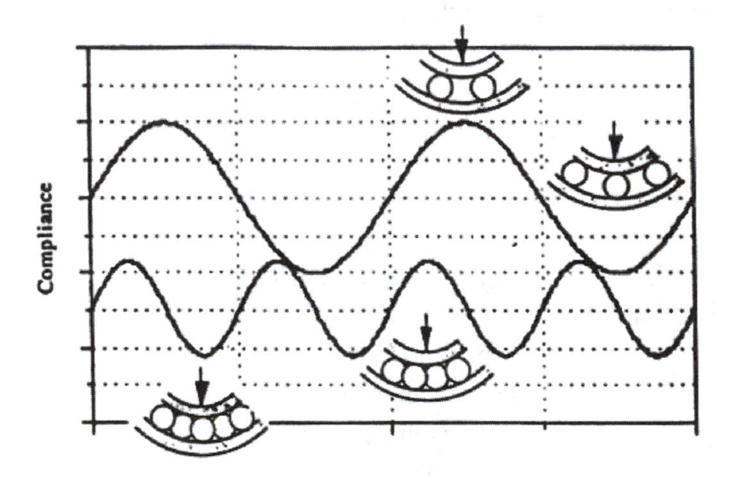

Figure 4.36. Comparison of compliance differences for single and double-row bearings, from Slocum[18].

Figure 4.37 shows the relative effects of bearing preload on load distribution for a range of applications for which the preload varies. Usual classes of preload include: i. heavy/medium — which is good for shock and vibration loading conditions, overhanging offset loads such as may be found in heavy machining conditions generating large cutting forces; ii. medium/light — for lower shock and vibration conditions and machining applications; iii. light — for small vibration environment; iv. very light — precision machines and no overhanging loading environments; and v. very light but with clearance — for no load conditions where thermal effects are not important.

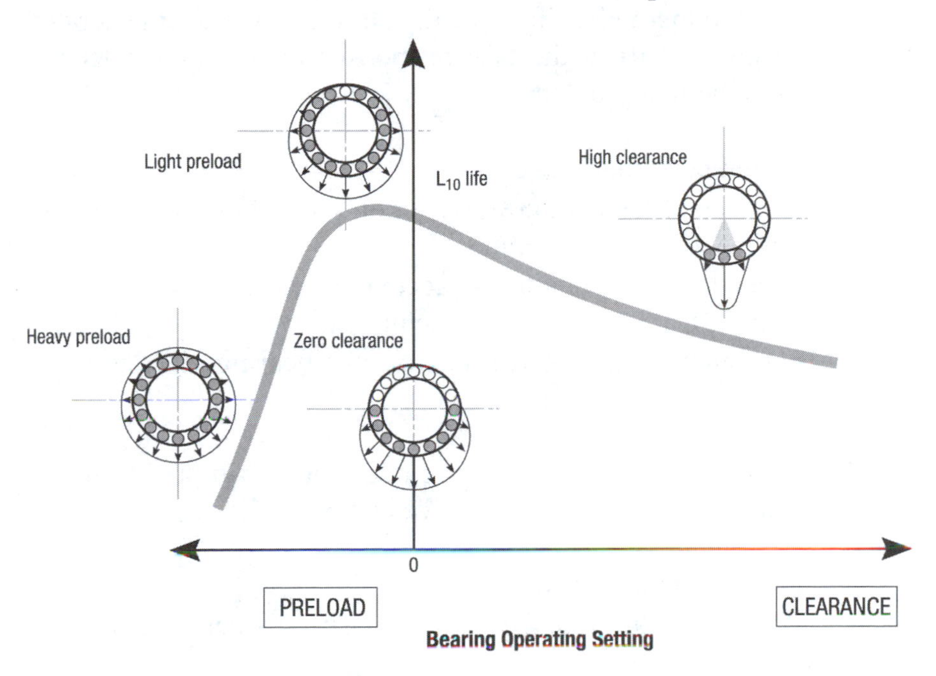

Figure 4.37. Effect of bearing preload on the region of the bearing that effectively acts to support the load, courtesy of The Timken Company.

Bearings can introduce damping in a structure depending on the type of bearing used:

- damping capability
- sliding contact and fluid bearings best because of viscous nature of film
- bearing maybe only damping in the system
- friction
- stick-slip friction (high static with low dynamic friction) causes tracking errors about zero velocity and limit cycling in servos
- sliding or dynamic friction helps damp out motion
- sliding friction causes heat

Bearings by their design and utilization consume energy which appears primarily as heat in the mechanism. In addition to the influence on the thermal stability of the structure, the heat generated by the bearing can affect the bearing performance thus further aggravating the thermal problems:

- thermal performance
- influence of temperature on friction properties and, thus, dynamic behavior of machine
- influence of temperature on accuracy, repeatability and resolution; effect of differential expansion
- heat transfer characteristics across the bearing; does it act as an insulator r conductor?

In addition, general concerns span the range of sensitivity to operating environment to ease of maintenance and reliability:

- environmental sensitivity
- influence of dirt, moisture, etc. on bearing's performance
- sealability
- can environment be effectively sealed out
- size and configuration
- size/strength ratios or size/performance ratios
- weight
- weight/strength ratios or weight/performance ratios
- support equipment (to keep the bearing operating)
- upkeep, lubrication, dry air, pressure regulation
- maintenance
- ease of service
- frequency of service
- what are MTBF and MTTR values; availability
- material compatibility
- interaction/reaction of bearing materials to other materials (differences in thermal expansion, fretting, corrosion, etc.)
- mounting requirements
- bearing mounts/supports
- required life

- cyclic duty/failure probability/service

Finally, design and manufacturing considerations are important:

- availability
- special purpose of "off-the-shelf"
- "designablity"
- ease of employ whether standard or special design
- manufacturability
- cost to manufacture components
- cost to assemble, align, adjust preload, etc.
- cost
- cost/performance or cost/quality ratio

Many of these considerations relate as well to the whole machine system and are not unique to bearings.

Nontraditional bearings include magnetic bearings as a low friction (almost frictionless) bearing for spindles and shafts in precision mechanisms, Figure 4.38, from Slocum[18]. These are especially useful at high rotational speeds as utilized increasingly in high speed machining applications.

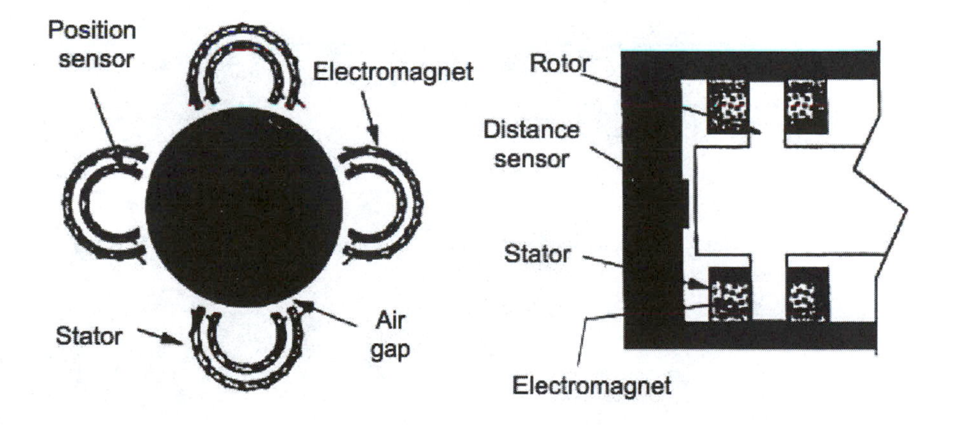

Figure 4.38. Magnetic bearing configurations for radial and thrust loads, from Slocum[18].

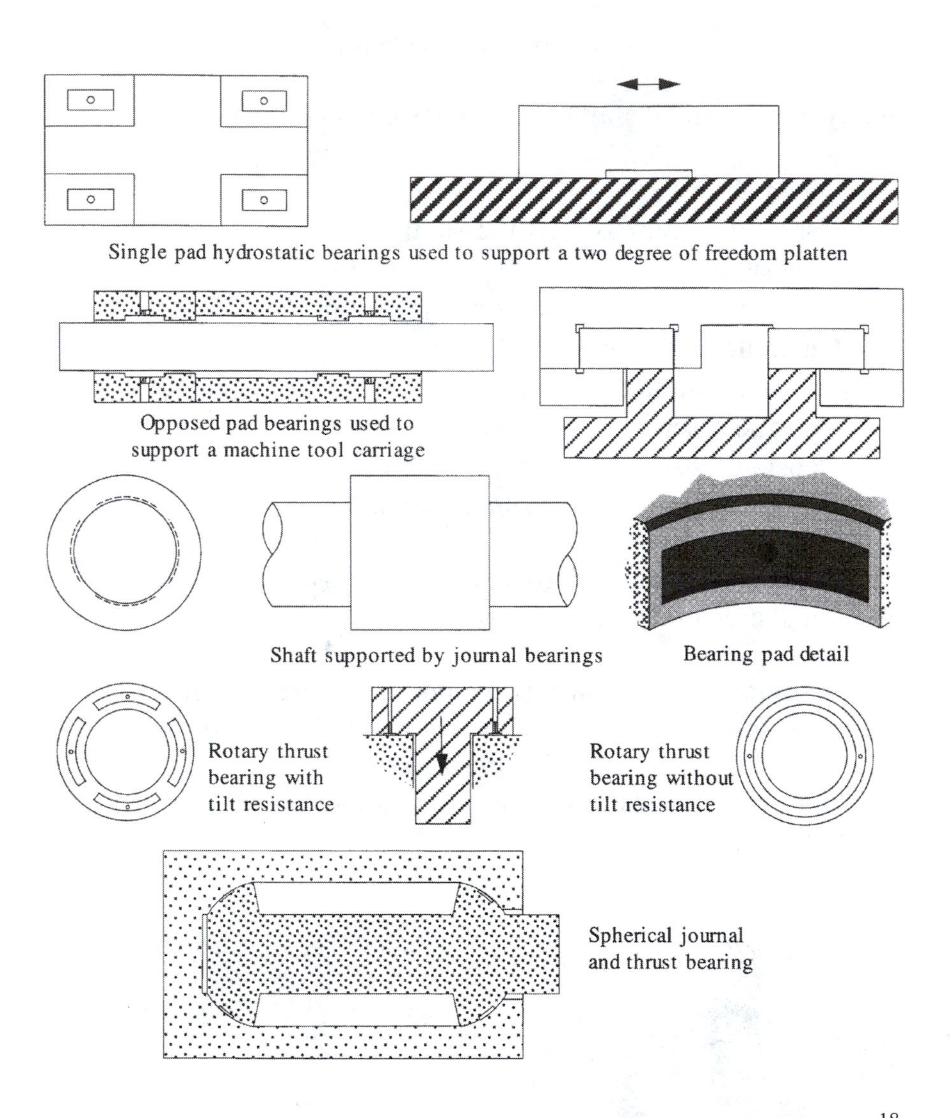

Figure 4.39. Various hydrostatic bearing configurations, from Slocum[18].

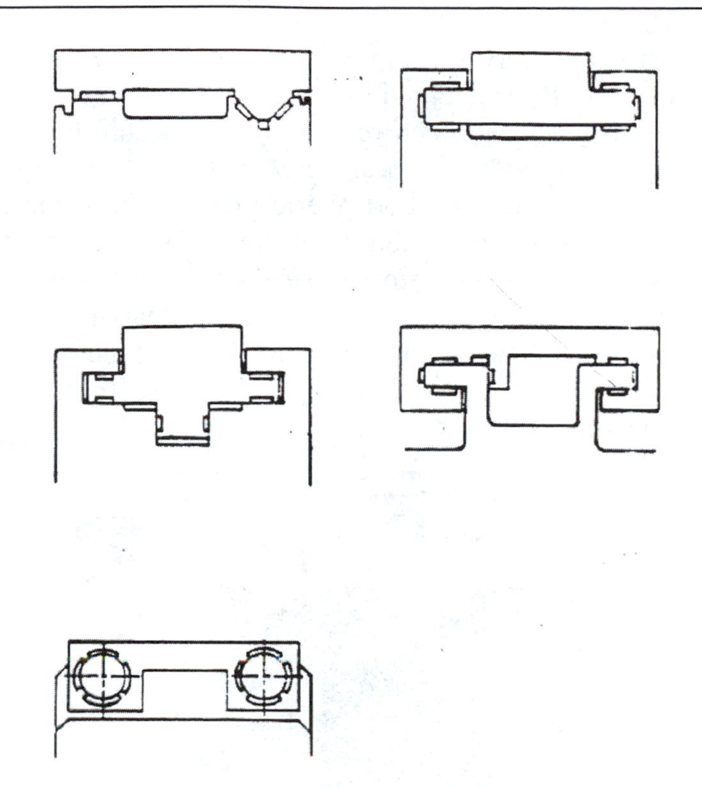

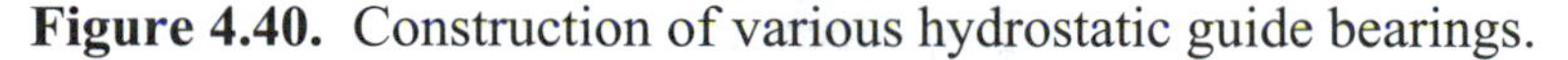

Figure 4.40. Construction of various hydrostatic guide bearings.

Hydrostatic bearings have a traditional wide application in both linear and rotational motions. Figures 4.39 and 4.40 review a variety of hydrostatic bearing configurations. Figure 4.40 shows several practical configurations of hydrostatic guide bearings for machine axes.

4.5.2 Aerostatic bearings and spindles

Figure 4.41 from Slocum[18] illustrates the construction of a hydrostatic annular thrust bearing. The fluid restrictors insure a balanced and uniform flow of fluid into the pads in the bearings. Balance is needed to insure equal and opposite pad pressures for true running and stiffness (i.e. maintaining a fixed nominal gap between bearing elements.) The basic design and operation of air bearing spindles and slides is straightforward, Figure 4.42, also from Slocum[18]. Gas at supply pressure P_0 is admitted to the bearing clearance through an orifice/restricting device which regulates the pressure to P_d from a supply. As the gas enters the bearing clearance and eventually exhausts it reduces further to atmospheric pressure P_a at the bearing outlet. Changes in the restriction or clearance between the bearing elements affect the downstream pressure P_d for a given restriction and thus allowing a higher load (for a smaller clearance) to be carried by the bearing. A higher clearance has the opposite effect. There is always a trade off between load capacity and stiffness in the design of the bearing.

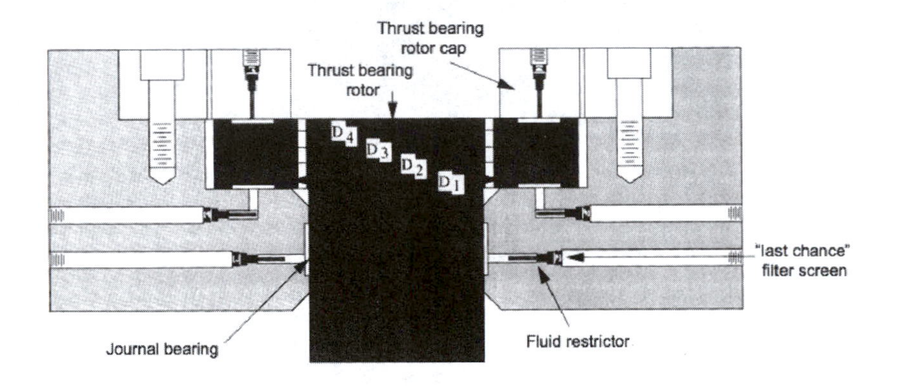

Figure 4.41. Annular thrust bearing used to support bidirectional thrust loads on a shaft, from Slocum[18].

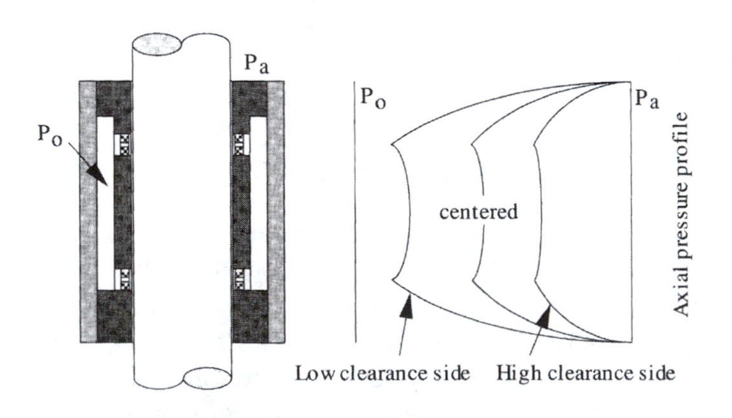

Figure 4.42. Principle of air bearing operation, from Slocum[18].

Air bearings are used in high precision machine tools where high speed and low friction are needed. Figure 4.43 shows an application with an integral drive motor of a spindle for a grinding machine based on aerostatic journal bearings. And, Figure 4.44 shows an application of an air bearing machine tool axis. Consideration must be given to driving the table so as not to induce unnecessary load on the bearings and, importantly, for both axes and spindles consideration must be given to the rest positions of the elements when the air pressure is reduced or shut off.

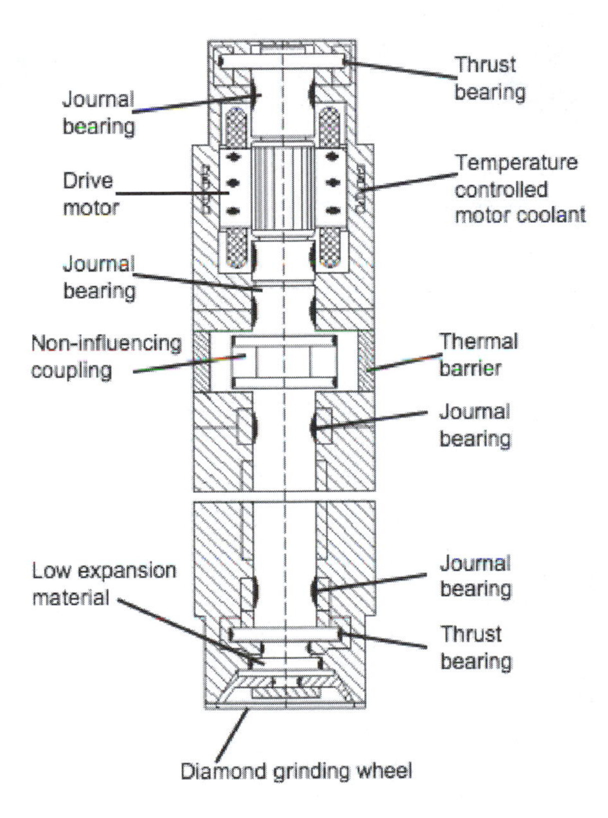

Figure 4.43. Air bearing grinding spindle.

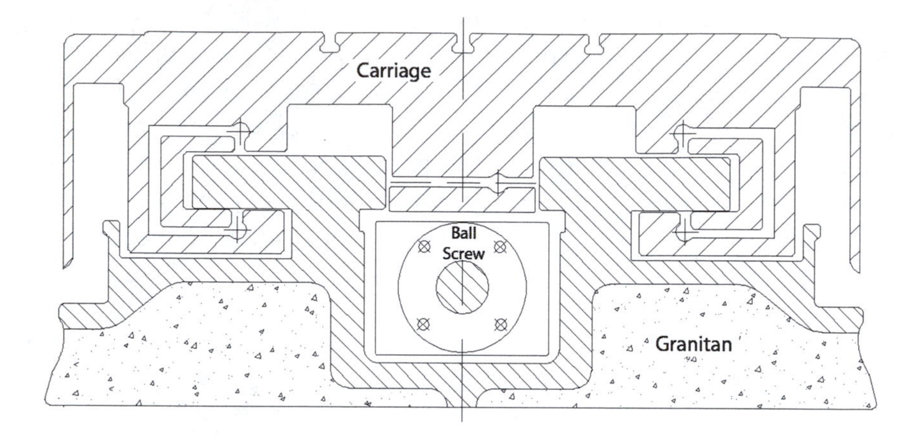

Figure 4.44. Schematic of air bearing X axis worktable.

V THERMAL ERRORS

5.1 Background on the thermal error problem

In our earlier general discussion on sources of error in precision machines we referred to Figure 2.4, which described the various sources and their inter-relationship with the mechanical system components and the resulting impact on part accuracy. In this chapter we address specifically thermal errors which are "the largest single source of errors in close tolerance work."[34] In fact, some place thermal errors at some 40% of the machine tool errors. These errors are, for machine tools, due to a combination of effects including:

- room environment
- coolants
- people
- machine itself
- machining process

and they transfer energy to the structure, tooling, workpiece and metrology or scales via conduction, convection and radiation. Figure 5.1, from Bryan[34] illustrates the sources and transmission of thermal energy in a precision machine tool. It is sometimes straightforward to determine the contribution (thermal) to the environment of motors and other energy consuming (and hence thermal sources due to losses) devices. Other sources are more difficult to determine..... friction effects in slide ways or chip temperature effects in conventional metal cutting are both common sources more difficult to characterize, The average human being working next to a machine tool appears to be the thermal equivalent of a 150 watt light bulb (some of us are a few more watts than others!). All these effects need to be accounted for at some level as they will affect the process in a measurable way. The importance of adjusting the measurement of a workpiece to the reference standard temperature of 20°C (68°F) is pointed out as is the need to allow machines, tooling and work pieces to adjust to the present environment before proceeding with the operation.

Isolation, compensation and minimization or removal of the thermal energy sources are all practical means to reduce the impact of thermal effects. These are summarized as "avoidance" and "compensation" in Figure 5.2 from MTTF[25]. Avoidance is proactive in the sense that a model or understanding of the error source is used to affect the process to minimize (or avoid) the error by eliminating the source of the error or alter the process whereby the source is an error. Compensation, which can be used with avoidance, is to feedback a measurement of the error and offset it in process through some suitable control loop or additional input to, in effect, cancel the error. In either case, some understanding of the sources of thermal effects and their sensitivity to process parameters is necessary. And, the dynamics of the process play an important role as we will see. Some thermal effects occur over a significantly long period of time such that, if the process is well setup, they may not interfere with the process at all.

Ultimately, thermal effects are seen in distortion of the machine tool, Figure 5.3. Sources of heat inducing thermal errors in this figure include bearings, transmission and hydraulic oil, gears and clutches, pumps and motors, guideways, cutting tool and chips as well as external sources. Approaches to the minimization, or elimination, of these thermal error sources are in some cases rather obvious. In other cases it is more complicated. For example, the recent move towards reduction or elimination of coolant in machining has created a flow of hot chips landing on machine components. In the past these were cooled by the fluid as part of the process. Now, machine re-design to allow chip flow off the machine components is needed.

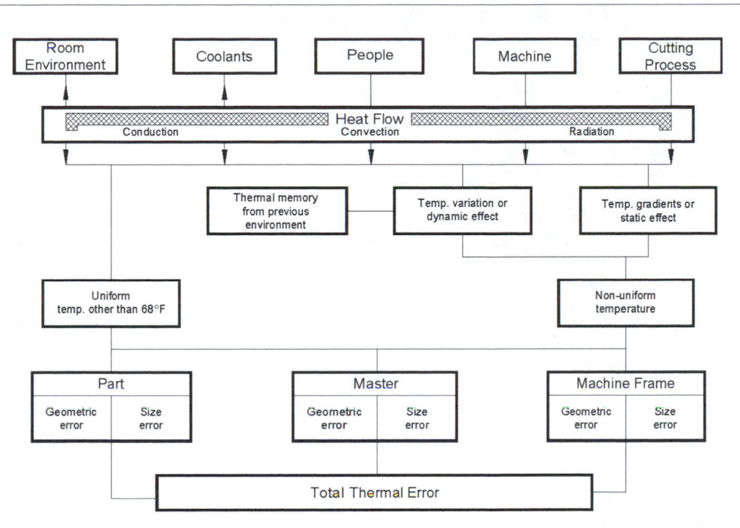

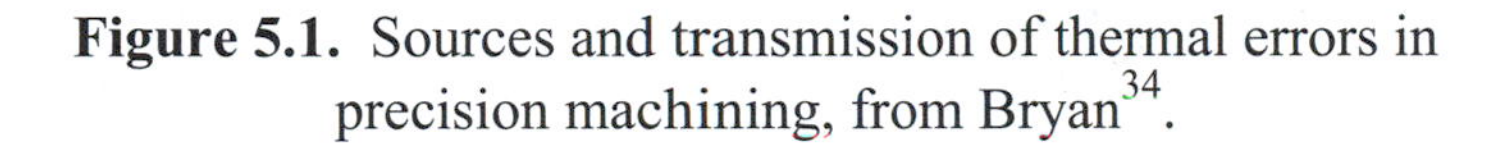

Figure 5.1. Sources and transmission of thermal errors in precision machining, from Bryan[34].

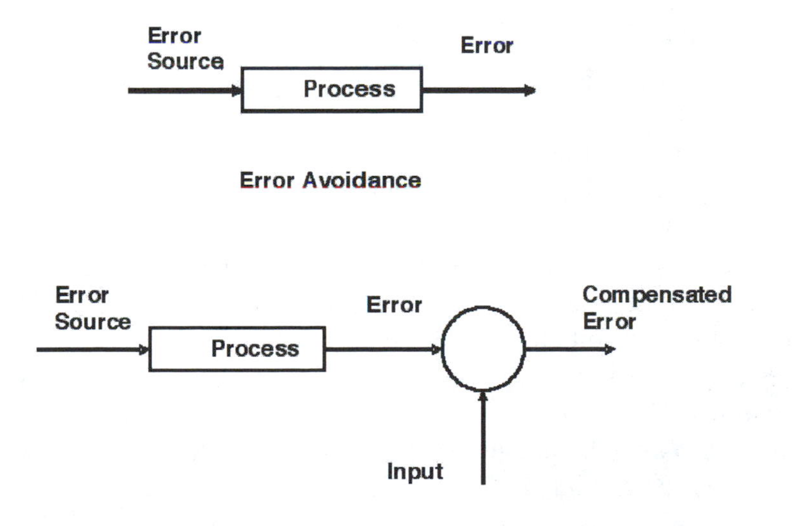

Figure 5.2. Error avoidance and error compensation for thermal error correction, from MTTF[25].

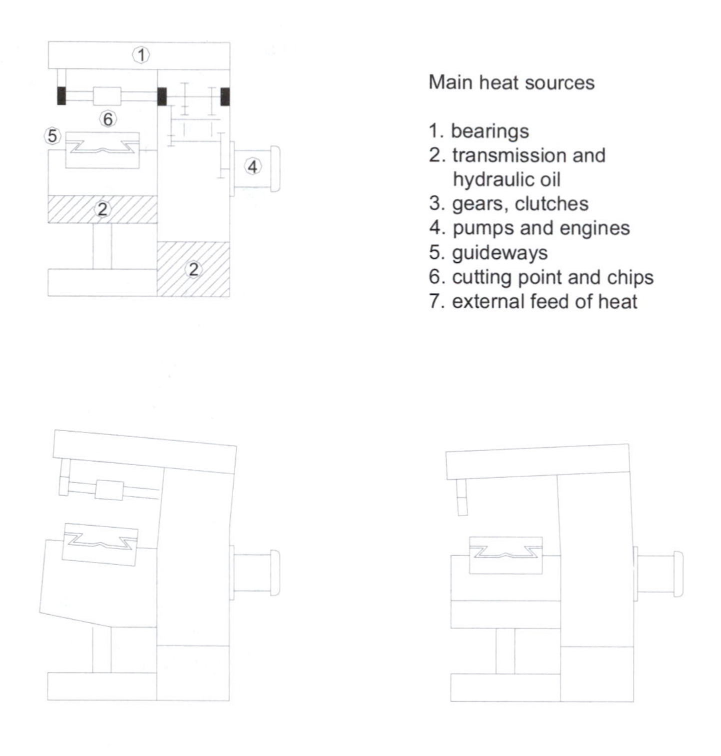

Figure 5.3. Example of deformation arising from thermal conditions in a milling machine, from MTTF[25].

A large portion of the material in this chapter is taken from a tutorial on thermal effects in precision engineering developed and taught by Dr. Kenneth Blaedel of Lawrence Livermore National Laboratory (LLNL) in Livermore (Dr. Blaedel, now retired from LLNL, was part of the Precision Engineering Division at LLNL and, along with many others whose work is cited in this book (J. Bryan, for example), was a major contributor to the precision engineering community. He has allowed the use of this material in this book.) On several occasions he presented this material as part of short courses and lectures at Berkeley. The figures Dr. Blaedel used are presented along with text and additional comments and figures added by the authors.

5.2 Thermal effects in precision engineering

The simple fact is that almost all the energy (or power) supplied to a machine tool is converted to heat. It could be seen in Figure 5.3 that typical sources such as motors driving axes, the process itself (chip formation, for example) as well as external heat sources from workers nearby, sun shining into a work area and onto a machine component or the cyclical heating and cooling of a building due to the HVAC systems all play a role in the thermal "reaction" of the machine. The resulting temperature field in the machine, and the consequent thermal distortion, both result in errors in the workpiece being machined or the part being measured. This is primarily true when the thermally induced distortions act along sensitive directions of the machine relative to the workpiece. Although for some levels of precision the thermal behavior may be ignored (due either to lack of requirements for precision or time constants of the process relative to thermal time constants), in general, when the accuracy of the workpiece becomes demanding the consequence of thermal effects dominates. For example, in Figure 5.3 the predominant distortion is due to the heating of the column of this horizontal milling machine causing tool-work relative errors. This will occur after some time due to the massive casting comprising the machine tool structure.

A change in temperature results in a change in dimension for most engineering materials. There are different ways to estimate the impact of temperature changes on a material. Figure 5.4 gives the usual thermal coefficients for a range of materials commonly used in precision machinery. On a basic level, temperature changes induce thermal elastic strains, ε_T that are proportional to the product of the coefficient of expansion, α, of the material and the temperature change, ΔT, the material is exposed to

$$\varepsilon_T = \alpha \, \Delta T \tag{5.1}$$

Multiplying the resulting strain by the length of the element experiencing the temperature change gives the actual change in length (in this case) of the machine element. Common solutions to minimize thermal expansion include choosing materials that have low coefficients of expansion (like Zerodur with a coefficient of 0.05 μm/m/°C, that is, very small, as opposed to more common materials like cast iron with a coefficient of 11 μm/m/°C – over 200 times as high). This is a reasonable approach for relatively static thermal situations. If thermal transients are of concern (that is establishing the effects of thermal transients in a structure), then investigating the thermal diffusivity may be of use.

Material	ν	E (GPa)	ρ (Mg/m3)	K (W/m/Co)	Cp (J/kg/Co)	K/ρCp (10-6m2/s)	α (μm/m/Co)
Aluminum (6061-T651)	0.33	68	2.70	167	896	69	23.6
Aluminum (cast 201)	0.33	71	2.77	121	921	47	19.3
Aluminum oxide (99.9%)	0.22	386	3.96	38.9	880	11.2	8.0
Aluminum oxide (99.5%)	0.22	372	3.89	35.6	880	10.4	8.0
Aluminum oxide (96%)	0.21	303	3.72	27.4	880	8.4	8.2
Beryillium (pure)	0.05	290	1.85	140	190	398	11.6
Copper (OFC)	0.34	117	8.94	391	385	114	17.0
Copper (free machining)	0.34	115	8.94	355	415	96	17.1
Copper (bery.-copper)	0.29	125	8.25	118	420	34	16.7
Copper (brass)	0.34	110	8.53	120	375	38	19.9
Granite	0.1	19	2.6	1.6	820	0.8	6
Iron (Class 40 cast)	0.25	120	7.3	52	420	17	11
Iron (Invar)	0.3	150	8.0	11	515	2.7	0.8
Iron (Super Nilvar)	0.3	150	8.0	11	515	2.7	0
Iron (Nitralloy 135M)	0.29	200	8.0	4.2	481	1.1	11.7
Iron (1018 steel)	0.29	200	7.9	60	465	16	11.7
Iron (303 stainless)	0.3	193	8.0	16.2	500	4.1	17.2
Iron (440C stainless)	0.3	200	7.8	24.2	460	6.7	10.2
Polymer concrete	0.23-0.3	45	2.45	0.83-1.94	1250	0.27-0.63	14
Zerodur	0.24	91	2.53	1.64	821	0.8	0.05
Silicon carbide	0.19	393	3.10	125	-	-	4.3
Silicon nitride (hip)	-	350	3.31	15	700	13	3.1
Tungsten carbide	-	550	14.5	108	-	-	5.1
Zirconia	0.28	173	5.60	2.2	-	-	10.5

Figure 5.4. Basic material properties for a range of materials used in precision machinery, after Slocum[18].

A measure of thermal diffusivity, D = k/cρ (with k = thermal conductivity, c = specific heat and ρ = density) indicates the relative rate of settling of transients – faster is better to insure stability. A further indication of a materials resistance to thermal effects is to divide the thermal diffusivity by the coefficient of thermal expansion, α. That is, D/α. And, a larger value is better (faster settling and lower expansion).

An effective means of assessing the relative position of materials in terms of their thermal characteristics is to plot the linear expansion coefficient as a function of thermal conductivity, Figure 5.5, below. The graph includes a wide variety of materials including those that are not commonly used in precision machines, from Ashby[35].

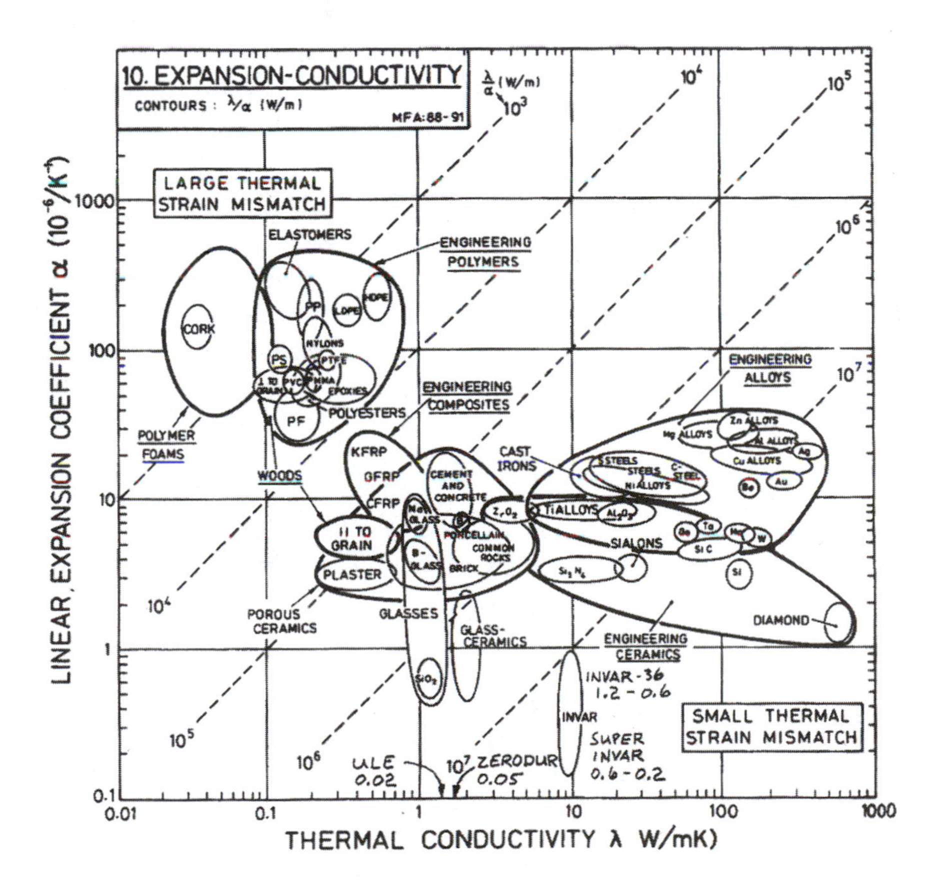

Figure 5.5. The thermal expansion coefficient plotted vs. thermal conductivity, from Ashby[35].

Interestingly, different phases of materials can have different thermal responses. For example, among common cast irons (shown only as "class 40 cast" in Fig. 5.4) have different coefficients of expansion depending on the exact structure of the cast materials. Recall the discussion of material property variation at the end of Chapter 3. Cast iron is (or was) a popular material as it is easy to form into intricate shapes as well as it has excellent damping characteristics (to internal vibrations). Two common cast irons are white cast iron and gray cast iron (and both can occur in the same piece depending on the cooling rates of different parts of the piece – surface cools faster (hence white cast iron) and the inner part cools slower (hence gray cast iron)) have the following corresponding coefficients of expansion, $\alpha_{\text{white iron}}$ = 9.45 μm/m/°C and $\alpha_{\text{gray iron}}$ = 11.0 μm/m/°C are typical. Hence, a machine element comprised of both types of cast irons will have differing response at differing temperatures due to the differing coefficients.

By international agreement, an object's dimensions are given only at 68°F or 20°C, Figure 5.6. Metrology systems must be calibrated at this temperature and applied in a controlled environment or errors will occur. Machinists have always made differential expansion corrections based on the actual temperature, but they often overlooked the fact that such corrections are not always exact.

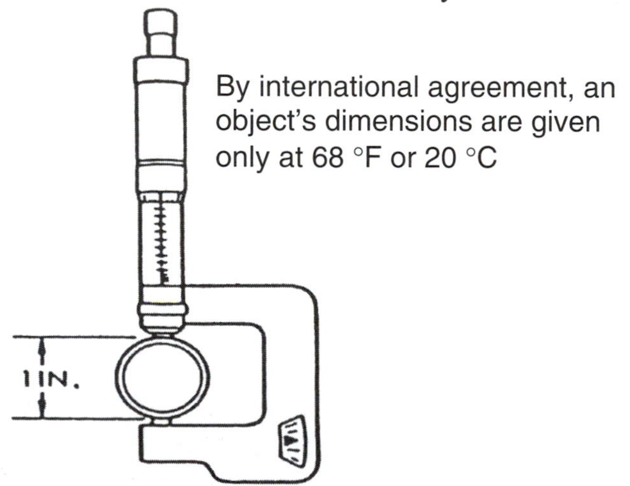

Figure 5.6. American national standard temperature and humidity environment for dimensional measurement.

So, how do we determine if a thermally induced error is large enough to be of concern. We can calculate a "thermal error index' (TEI), given by standard ANSI B89.6.2[36], as:

$$TEI = ((NDE + UNDE + TVE)/\text{Total Permissible Error}) \qquad (5.2)$$

where: NDE = nominal differential expansion
UNDE = uncertainty of NDE
TVE = temperature variation error

It is often very difficult to determine the precise value of a coefficient of thermal expansion for a material under the conditions of analysis for each part. The effect of this lack of precision is called Uncertainty of Nominal Expansion, UNE, according to ANSI B89.6.2[36]. The uncertainty of length measurement due to uncertainty of UNE is zero at 20°C. It should also be pointed out that errors associated with the measurement if the 20°C temperature will also lead to length measurement uncertainty! The magnitude of uncertainty in these coefficients varies according to the material. It is about 10% for gauge block steel and up to 25% for other materials[37]. Differences between the actual coefficients of thermal expansion and the "nominal" coefficients occur because of experimental errors, anisotropy, variations in chemistry, cold work and other mechanical effects and the state of heat treatment of the material. Figure 5.7 clearly illustrates the impact of temperature variation in the drift in dimension of a gage[37]. The correlation between temperature swings and gage "growth" are clearly seen.

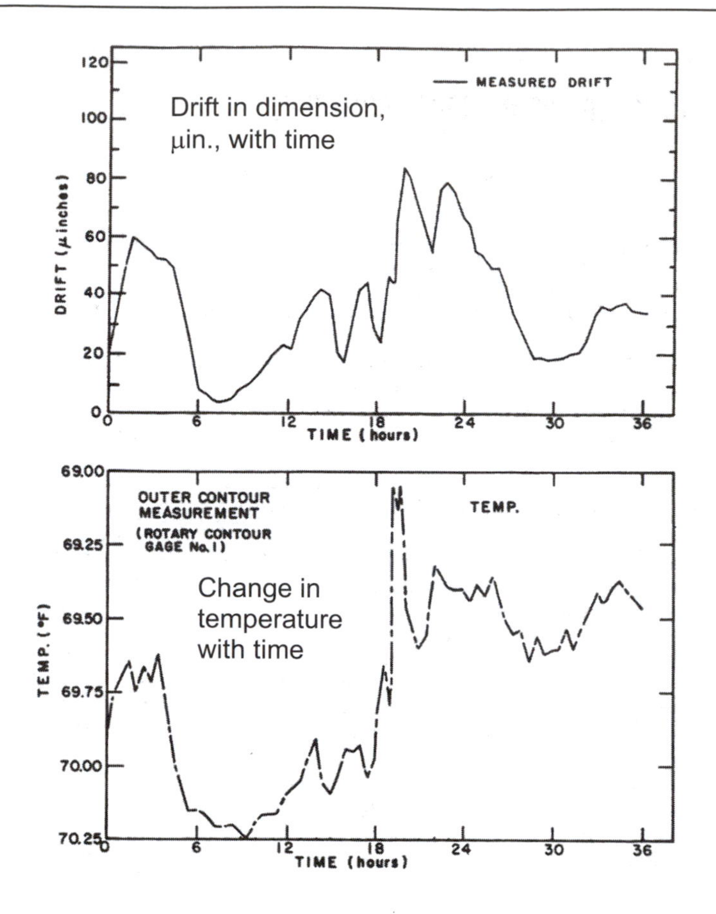

Figure 5.7. Thermally induced drift in a 15 inch Sheffield rotary contour gage with steel part, from Bryan[37].

Several examples of applications of the TEI discussed above are shown throughout this chapter.

EXAMPLE 1: Consider a case of mastering a steel comparator with a 100 mm steel gage block in a 21°C average environment (that is, the comparator is "calibrated" if you will against the steel block but at a temperature that deviates from the standard temperature by + 1°C). An aluminum part of 100 mm nominal length is then measured using the comparator at about 21°C. If the measurement is to be accurate to 3 µm, what is the Thermal Error Index (TEI)? That is, what fraction of the total permissible error has been introduced by mastering and measuring at a temperature other than 20°C?

We calculate the TEI as before:

TEI = ((NDE + UNDE + TVE)/ Total Permissible Error) x 100%

The nominal differential expansion (NDE) = $[NE]_{part}$ – $[NE]_{master}$, where NE (nominal expansion) is calculated from k_e (estimate of the average coefficient of thermal expansion), length of the object, L, and the temperature at which the measurement is made, T, as

$[NE]_{part} = k\,L\,(T\text{-}20°C) = 23.7(10^{-6})\ K^{-1}\ x\ 0.1\ m\ x\ 1°K = 2.37\ \mu m$
$[NE]_{master} = 11.2(10^{-6})\ K^{-1}\ x\ 0.1\ m\ x\ 1°K = 1.12\ \mu m$

thus, NDE = 2.37 μm – 1.12 μm = 1.25 μm (due entirely to the difference in k_{Al} and k_{steel})

UNDE = $[UNE]_{part}$ + $[UNE]_{master}$

where UNE $= (\alpha - \kappa)/\ \alpha\ x\ NE$
$= (\alpha - \kappa)\ x\ L\ x\ (T - 20°C)$

with UNE = Uncertainty of Nominal Expansion
α = actual average coefficient of thermal expansion (CTE)
κ = estimate of the average CTE

so $[UNE]_{part} = 0.55(10^{-6})K^{-1}\ x\ 0.1\ m\ x\ 1°K = 0.05\ \mu m$
$[UNE]_{master} = 0.50(10^{-6})K^{-1}\ x\ 0.1\ m\ x\ 1°K = 0.05\ \mu m$

Then, UNDE = 0.05 μm + 0.05 μm = 0.10 μm

We also need the temperature variation error (TVE) which can be obtained from a drift check (see Figure 5.9). We assume a TVE = 0.12 μm for this example.

Then,

TEI = (1.25 μm + 0.10 μm + 0.12 μm) / 3.0 μm *x* 100% = 49% (!)

This means that we have given up almost half of the allowable tolerance of 3 μm just by performing the mastering and measuring at 1°K higher than the standard.

A system view of accuracy (for example, an error budget) requires an accurate assessment of the thermal error index. Figure 5.8 below is an example of error budget data showing the contributions of various elements to the "inner surface measurement" of a workpiece. The values for many of the elements listed are based on an "average" of the measured variation over some time. A typical example of the temperature data variation over time is shown in Figure 5.9. The left column shows specific samples of temperature at a location on the machine tool and the calculated mean and standard deviation. However, these two values do not show the reality of the situation. For example, the right figure in Figure 5.9 is a real time trace of the length variation (right side trace of the chart) and the room temperature variation (between 67 and 71°F) in the left side trace of the chart. As easily seen, the variations in length are in sync with the variations with the room temperature. Hence, the "randomness" implied by the standard deviation shown in Figure 5.9 is really due to systematic errors, and easily correctable.

	P-V Magnitude (nm)			
	Without Geometry Compensation		With Geometry Compensation	
	Y	Z	Y	Z
LASER INTERFEROMETERS				
Frequency stability	15.0	30.0	15.0	30.0
Resolution	2.5	2.5	2.5	2.5
Index of refraction	12.5	25.0	12.5	25.0
Optical, electronic factors	5.0	5.0	5.0	5.0
ROTARY TABLE				
Total radial motion (including tilt)	130.0		130.0	
Total axial motion (including tilt)		130.0		130.0
MACHINE GEOMETRY				
Y-slide straighness		500.0		100.0
Lower z-slide straightness	500.0		100.0	
Lower z-slide pitch		250.0		25.0
Squareness				
Lower z-slide to y-slide	250.0	250.0	75.0	75.0
Rotary table to y-slide	250.0	250.0	50.0	50.0
THERMAL EFFECTS				
0.05 degree C gradient--x-direction				
0.05 degree C gradient--y-direction				
0.05 degree C gradient--z-direction		187.5		187.5
0.05 degree C overall change	420.0		420.0	
Rotary table drive motor		75.0		75.0
Part	120.0	120.0	120.0	120.0
LVDT PROBE				
Electronic noise	25.0	25.0	25.0	25.0
Linearity	25.0	25.0	25.0	25.0
PROBE TIP				
Size	18.0	18.0	18.0	18.0
Contour	50.0	50.0	50.0	50.0
GAUGING FORCE	25.0	25.0	25.0	25.0
MASTERING	50.0	50.0	50.0	50.0
DATA ACQUISITION	25.0	25.0	25.0	25.0
Arithmetic Sum: $E_A =$	1923.0	2043.0	1148.0	1043.0
RMS Sum: $*E_R =$	221.9	207.9	139.7	91.4
Mean $((E_A + E_R)/2)$: $E_M =$	1072.5	1125.4	643.8	567.2
Predicted Maximum Error: $((E_{MY}{}^2 + E_{MZ}{}^2)^{1/2})$	$E_T = 1.55\ \mu m$		$E_T = .86\ \mu m$	

*Calculations for RMS values assume a uniform probability density for each error about its mean value.

Figure 5.8. Table of final error budget for inner surface measurements, from MTTF[25].

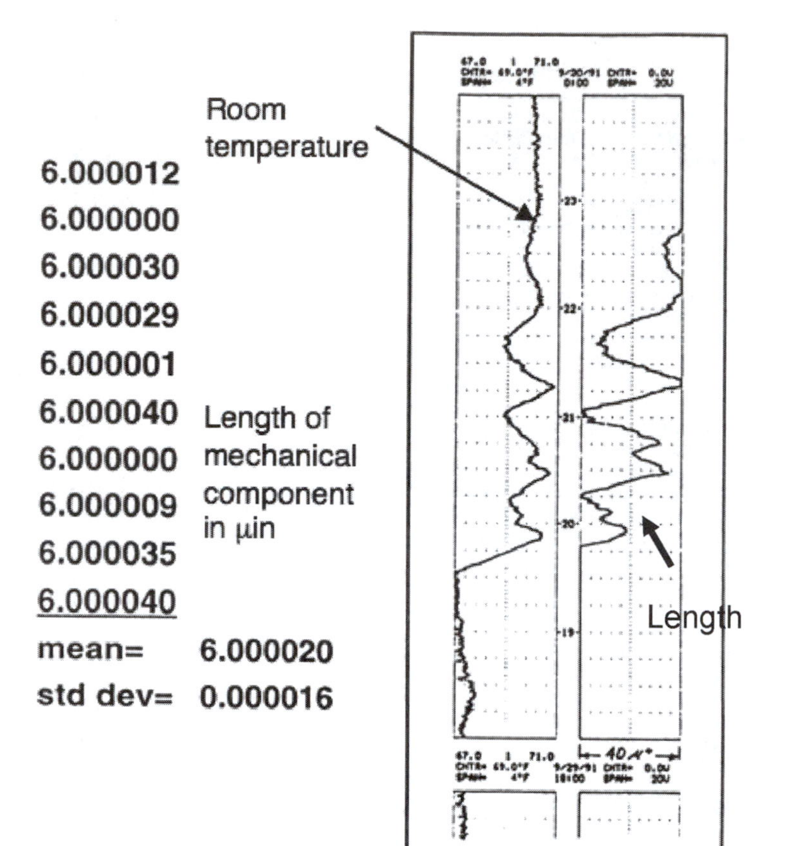

Figure 5.9. Data trace of length variation (in μin) over time and calculated "random" effects.

5.3 Determining the effect of temperature other than 20ºC

Materials will distort when they are subjected to energy sources that result in differential thermal environments. This is especially important to understand for components of precision machines and gages. The resulting deformation can be analyzed considering a number of standard "thermal property groups." These are summarized in Table 5.1 for different thermal source characteristics and design objectives.

5.3.1 *Free and constrained bodies*

Here the distortion of the component (as the "gauge block", Figure 5.10, below) will be proportional to the thermal expansion coefficient, α, and the change in temperature.

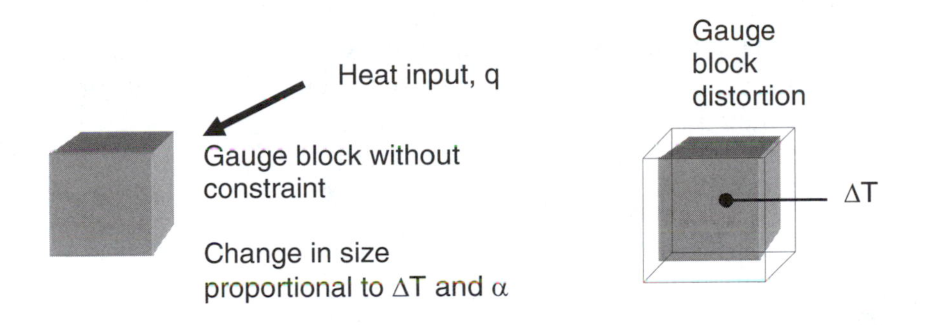

Figure 5.10. "Unconstrained" gauge block distortion when heated.

The deformation does not usually result in any residual stress as long as the body is uniformly heated throughout with out large temperature variations within. Thus, the shape is uniformly increased (or decreased if cooled) without distortion of the form (that is, a cube remains a cube).

If the component is constrained in any direction, for example by being connected to other components that are less flexible or compliant, substantial distortion in the component can result, Figure 5.11. Again, assuming that the component is uniformly heated throughout, it will deform along free directions proportional to the materials change in temperature, modulus of elasticity and thermal coefficient of expansion. In addition, if the yield stress of the material is exceeded (in tension or compression) permanent deformation can occur resulting in a residual stress upon unloading (or return to ambient), Figure 5.12. In the figure the component is "bowed" proportional to the induced stress due to expansion due to the rise in temperature. The upper surface length, $l_{top} > l_c$ of the neutral axis. The inverse is true for the lower surface due to compression. The stress is due to strain induced by expansion coupled to the coefficient of thermal expansion by the modulus of elasticity, E.

$$\sigma = E \, \alpha \, \Delta T \tag{5.3}$$

for $\sigma > \sigma_y$ plastic deformation usually occurs resulting in permanent residual stress. Hence, to minimize the effects of thermal distortion in components, it is necessary to keep E and α low. Table 5.1 serves as a designer's guide on which material parameters to account for in the design when thermal sensitivity is an issue.

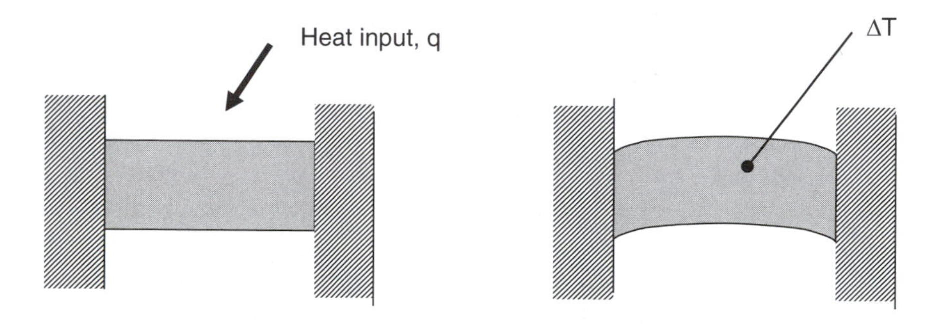

Figure 5.11. Constrained component under thermal distortion when heated.

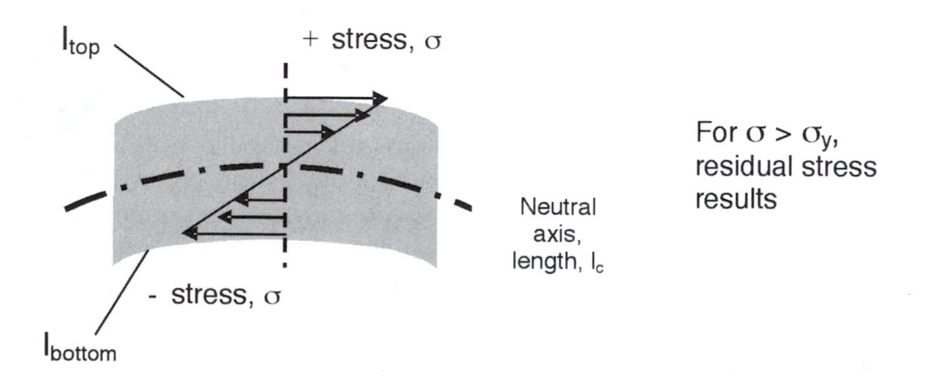

Figure 5.12. Distortion of a constrained component.

Table 5.1. Summary of thermal property groups and design objectives.

Thermal source and characteristics	Design objective		
	Freely supported	Minimally constrained	Heavily constrained
Environmental (uniform heat source)	low α	low $E\,\alpha$	low $E\,\alpha$
Conduction $q = k\,A\,\Delta T/L$		low α/k	low $E\,\alpha/k$
Non-steady $\frac{\partial\theta}{\partial t} = \frac{k}{c\rho}\frac{\partial^2\theta}{\partial x^2}$			
continuous	higher $k/c\,\rho$	low $E\,\alpha$	low $E\,\alpha$
burst		low $\alpha/c\,\rho$	low $E\,\alpha/c\,\rho$

EXAMPLE 2: In this second example we consider a slideway composed of materials with different thermal expansion, Figure 5.13.

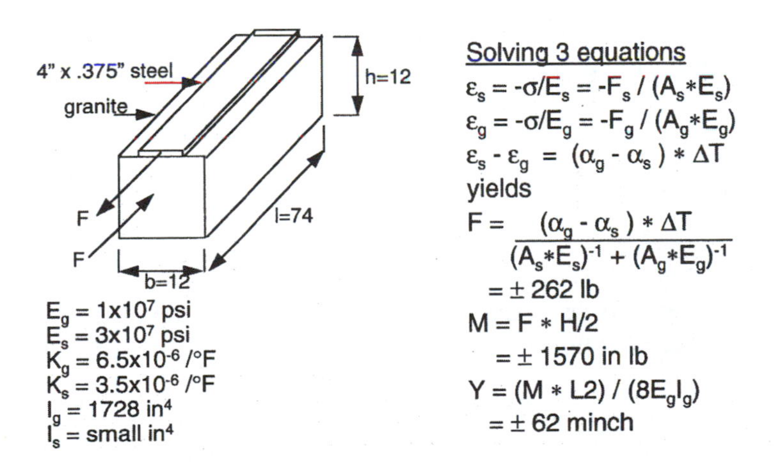

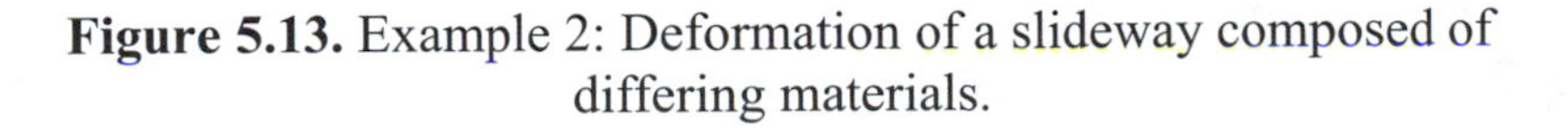

Figure 5.13. Example 2: Deformation of a slideway composed of differing materials.

The slideway has a manufacturer's specification (advertised) of having a vertical non-straightness of about ± 163 μ inch in 74 inches of travel, Figure 5.14. For a temperature variation of ± 2 °F, approximately how much does the simply supported beam sag at mid span? The details of this problem are given in Figure 5.13.

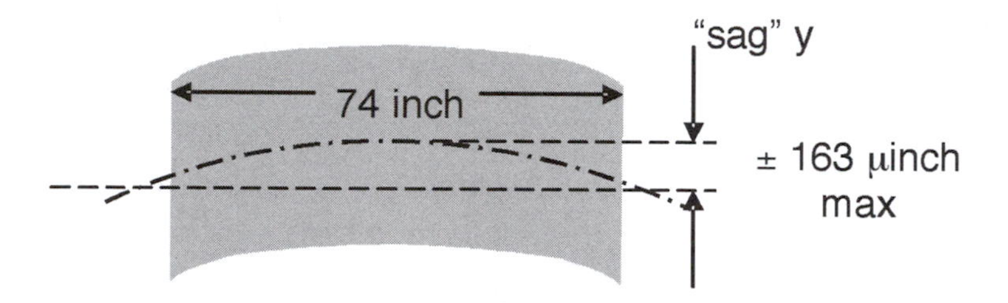

Figure 5.14. Definition of "vertical non-straightness" in Example 2.

The material properties of the two slideway materials, steel and granite are shown (modulus of elasticity, E, coefficient of thermal expansion, α, and moment of inertia). The strain induced in each different material as a result of the two degree F temperature variation is first calculated, ε_{steel} and $\varepsilon_{granite}$, based on the force induced due to the differential expansion. The expression for force as a function of CTE's, ΔT, and the other mechanical properties is shown in Figure 5.13. The resulting deflection (or sag), *y*, is calculated to be ± 62 μ inch. With this the TEI is already 38% (calculated as 62/163 *x* 100%) of the allowable non-straightness specification and this does not yet include any UNDE.

5.3.2 Effect of spatial temperature gradients

A spatial temperature gradient in a component, due to conduction from a heat source for example, creates a special condition. Conduction is governed by the following relationship for heat flow, q:

$$q = k\,A\,\Delta T/\,L$$

where k is the conduction coefficient (in W/m/°C), A is the cross section area of the component, L is the length of the component and ΔT is the temperature gradient along the element, Figure 5.15. The average temperature rise in the beam is $\Delta T/2$. Accordingly, the strain (change in length per unit length) due to the thermal gradient is:

$$\varepsilon = (\Delta T/2)\,\alpha \tag{5.4}$$

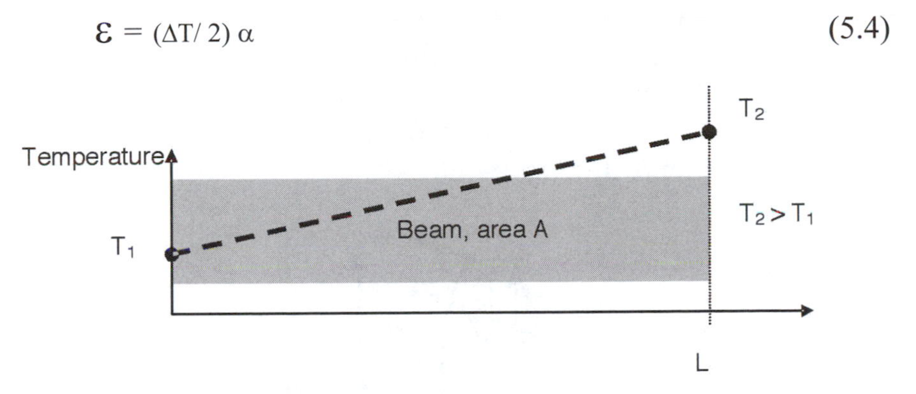

Figure 5.15. Thermal gradient in a beam element.

In this case, one prefers a material with a low α/k factor if the element is unconstrained or minimally constrained, or a low $\alpha E/k$ factor if the element is heavily constrained. Let's look at an example of a frame exposed to a temperature gradient, Figure 5.16.

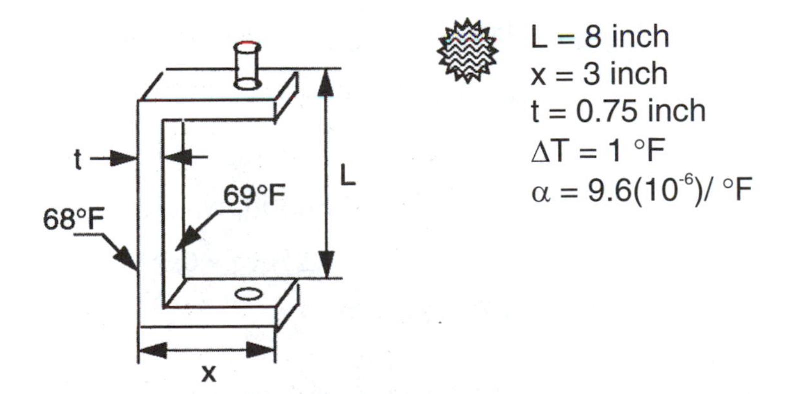

Figure 5.16. "C-frame" comparator details (with sun heat source shown).

The specifics of the comparator are shown in the figure. One side of the comparator "spine" is at 69°F because it is facing the sun and the other side is at 68°F. We are trying to determine what error is introduced in the measurement made by the comparator if it were mastered in a uniform 68°F environment. The sensitive direction with respect to the measurement is along the "L" dimension. The error is determined as shown in Figure 5.17. The radius of curvature of the spine of the comparator is determined by considering the expansion of the front and back surface of the spine due to the temperature difference. Finally, the total error, δ, is 346 μ inch considering an average temperature difference of 0.5°F.

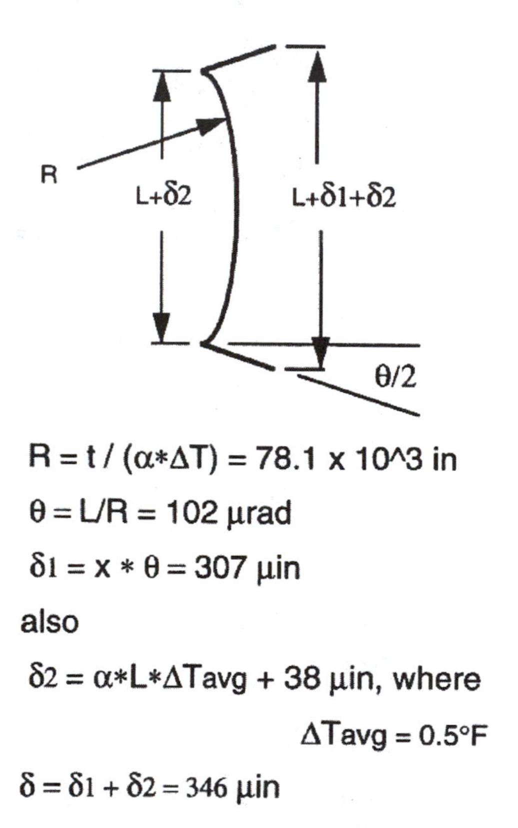

Figure 5.17. Calculation of error due to temperature gradient in a comparator.

5.3.3 *Effect of temperature transients: soak-out time and sinusoidal response*

Machine components are often subjected to some kind of transient thermal input due to varying production schedules, varying tool activity during the production of a workpiece, or varying environmental conditions (cyclic variation of room temperature, for example.) These are called non-steady disturbances due to fluctuating heat sources and the change in temperature of a point over time as a result of spatial variation in temperature is governed by the following relationship:

$$\frac{\partial\theta}{\partial t} = \frac{k}{c\rho}\frac{\partial^2\theta}{\partial x^2} \tag{5.5}$$

where c is the specific heat, ρ is density, θ is the temperature at a distance x along the length of the element, Figure 5.18. The temperature change in the component is governed by the thermal diffusivity k/cρ — higher is better for fast response (meaning a tendency to a more uniform temperature distribution throughout the element). This also depends on the power input of time factor. For a long (or continuous) heat input it is preferable to have low αE factor in the element (thus low distortion for uniform temperature difference throughout). For a short (or burst or instantaneous) heat input, it is preferable to have a low α/cρ for unconstrained elements and a low αE/cρ for constrained or clamped components.

The "thermal memory" of a part requires that it be allowed a certain amount of time to reach thermal equilibrium with its environment when it is moved to a different environment. During this soak-out period the part will be changing size and possibly form and any attempts to measure it before reaching equilibrium can result in serious errors. It can often be a bottleneck in a manufacturing system due to the time required for parts to reach equilibrium. The time involved for soak-out will depend on the "fluid" in which the part is immersed. If still air, the soak-out time for large machined components can approach several days. If moving air, air of different humidity or a liquid coolant is used (water, for example), the time can be substantially shortened. Cooling rates are a function of the thermal properties of the fluid.

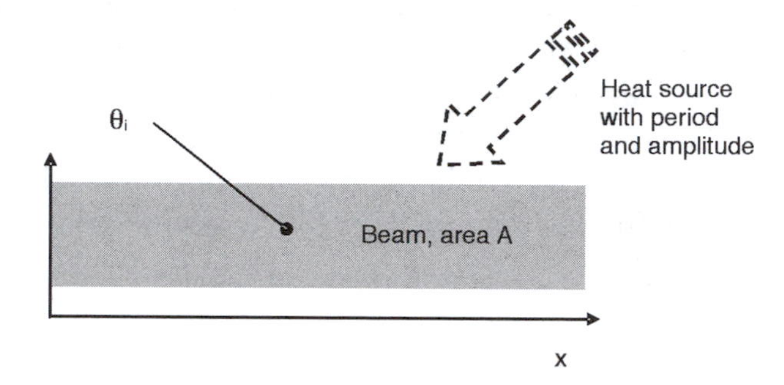

Figure 5.18. Geometry of element subjected to fluctuating heat source.

Bryan[6] relates some experiments done on soak-out. A steel roller bearing cup of 200 mm (8 inch) diameter and 25 mm (one inch) thickness required a soakout time of 6 hours to reach a temperature of 0.1°C (0.2°F) above a 21°C (70°F) ambient starting from a soaked temperature of 40°C (100°F) in still air. When the same component was subjected to 600 m/min (2000 fpm) velocity air stream it reached ambient (as defined above) in 25 minutes. It took only 2 minutes to reach ambient when submerged in a still grinding coolant. A separate American standard governs soak-out requirements for granite surface plates. Because of the low thermal conductivity of granite it requires long soak-out times. Bryan lists an example of a granite surface plate with dimensions 2 m *x* 1 m *x* .5 m thick requiring 74 hours to soak out within ± 0.2 of 20°C from an original temperature of 40°C.

An example of temperature transients (soak-out time) is presented here. A 1 inch (cube) gage block is moved from a room in which it was used for a long time at 72°F to a metrology lab maintained at 68°F, Figure 5.19. How long must we wait before it comes to within 10 μ inch of its calibrated length. The coefficient of thermal expansion for steel gage blocks can be approximated as 6.2 (10^{-6})/°F.

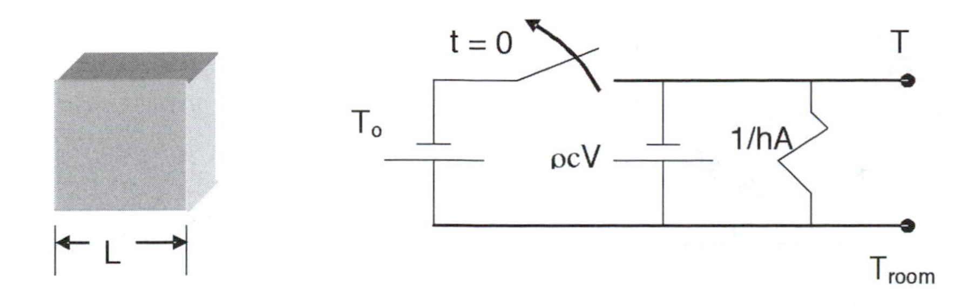

Figure 5.19. Gage block and "electrical" analogy for soak-out determination.

The heat transferred by convection is balanced by the decrease in internal energy of the block. This can be written as:

$$q = hA\,(T - T_{room}) = -\,c\,\rho\,V\,dT/dt \tag{5.6}$$

where

- q = heat flow (BTU/hr)
- T_{room} = environmental temperature (°F)
- h = film coefficient (0.3 BTU/(hr ft^2°F)
- A = area = 6/144 ft^2
- V = volume = 1/1728 ft^3
- c = specific heat capacity = 0.11 BTU/ (lb$_m$°F)
- ρ = density = 489 lb$_m$/ft^3 for steel
- L = length = 1/12 ft
- t = time (hr)

The solution to the first order differential equation is given in Figure 5.20 below. τ is known as the time constant of the system and indicates the speed of response. When time = τ, the temperature has dropped to e^{-1} or 36.8% of its total drop at $t = \infty$.

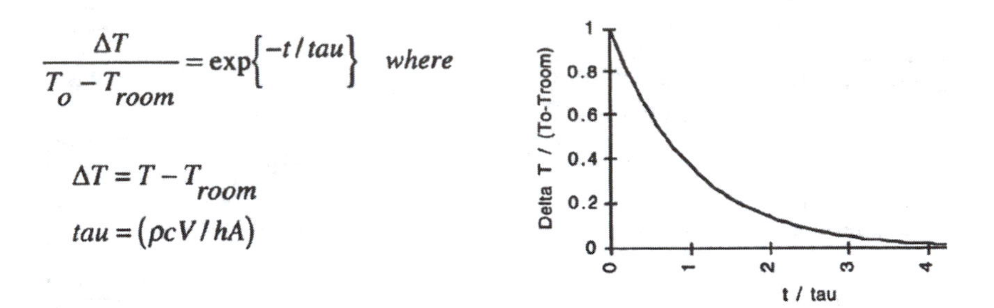

Figure 5.20. Transient response of first order system with time constant, τ.

For the gage block sitting in the 68°F room,

$$\tau = [(\rho c V)/(hA)]$$

$$= \frac{[(489\ lb_m/ft^3)\ x\ (0.11\ BTU/lb_m{}^\circ F)\ x\ (1/1728\ ft^3)]}{[(0.3\ BTU/(hr\ ft^{2\circ}F)\ x\ (5/144 ft^2)]}$$

$$= 3 \text{ hours}$$

To be within 10 μ inch of its calibrated length, we can calculate the allowable Δ T from the expression used before,

$$\Delta L = \alpha\, \Delta T; \text{ then } \Delta T = 1.6^\circ F \text{ for } \Delta L = 10\ \mu \text{ inch.}$$

We can then calculate the necessary time for this to occur from

$$\Delta T/ (T - T_{room}) = e^{(-t/\tau)} \text{ or } t = 2.74 \text{ hours}$$

To see the effect of the medium in which the part (the gage block in this example) soaks-out consider what would happen if the block was dropped into circulating oil (h =100) instead of still air. Note that the film coefficient is some 300 times larger.

$$\tau = [(\rho c V)/(hA)]$$

$$= .0089 \text{ hour (or .54 minutes!)}$$

and the required wait is only .49 minutes.

The above examples and discussions related to the soak-out of a piece in a constant temperature environment. On occasion the temperature to which a part or machine tool is exposed varies, sinusoidally for example with the temperature cycle in a building. The question then is, how does the temperature of the object respond to this dynamic environment? If we can determine that the time constant of the piece is much greater than the oscillation of the environmental temperature, the part may not be affected by this change for short processing times or short periods of residence in that environment.

The following example illustrates some of the main variables in the analysis of these varying thermal environments. A steel plate is exposed to an air environment which is itself sinusoidally varying. Both sides of the .25 inch steel plate are exposed to air, Figure 5.21. The forced response of the steel plate to a sinusoidal input is also a sinusoid of the same frequency, but of different amplitude and phase. The temperature of the plate is a function of the period and amplitude of the temperature of the environment. The temperature of the environment, T_e, is represented as

$$T_e = M_e \sin[2\pi t)/P]$$

where

M_e = amplitude of the sinusoidal variation in air temperature
P = period of environmental sinusoid
t = time

The temperature of the plate, T_s, can then be expressed as

$$T_s = M_s \sin[2\pi t)/P + \phi]$$

Where M_S/M_e the ratio of the amplitudes of temperature variation of the plate to the environment is represented as

$$\frac{M_s}{M_e} = \frac{1}{\sqrt{(2\pi t/P)^2 + 1}}$$

where

M_e = amplitude of the sinusoidal variation in air temperature
L = thickness = 1/48 ft
τ = time constant defined earlier
= $[(\rho cV)/(hA)]$
= $[(\rho cL/2)/(h)]$
= 1.87 hour

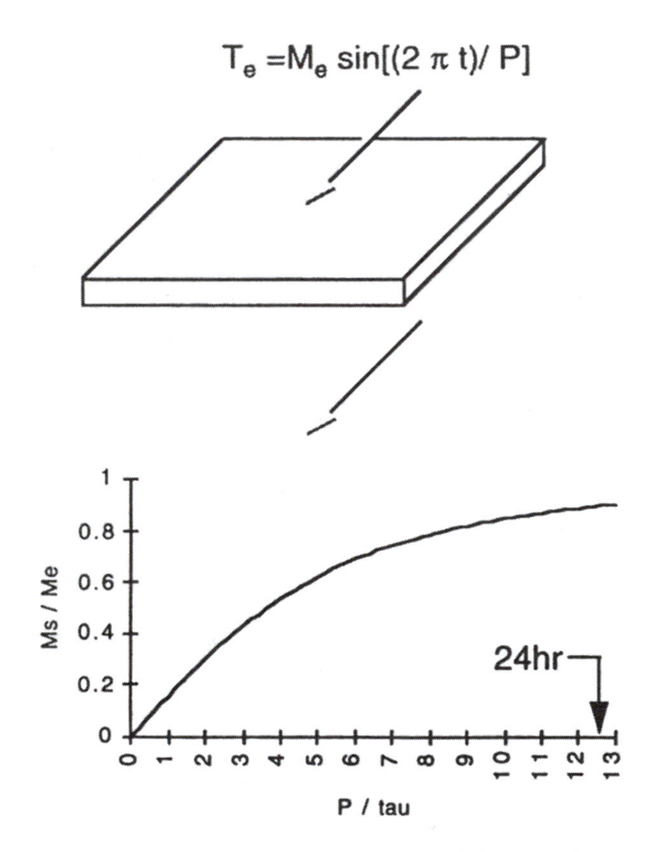

Figure 5.21. Steel plate exposed to sinusoidally varying temperature environment.

Figure 5.21 shows the variation of M_S/M_e as a function of P/τ. For P/τ much less than 1 the plate response is not much affected by the environmental variation.

5.4 Conductive, convective, and radiative heat transfer parameters

Clearly, the last example shown the previous section was for a condition in which the plate is thin and relatively small mass compared to the surface area. Other situations are apparent, coinciding with other regimes of heat transfer for more pieces that have more mass and less surface area. One way of representing the effect of the physical dimensions of the part (or machine component) as well as the thermal constraints is shown in Figure 5.22 which shows the regimes of cooling for a part with distinction between when the part "dominates" and the coolant "dominates." As can be seen, for thinner parts and lower values of h the coolant dominated and dictates the behavior of the system. Conversely, for thicker parts and higher values of h the part dominates. Typical values of h for materials as diverse as plastic and metal are given. And, natural and forced convection are illustrated for both air and water.

As utilized above, Newton's law of cooling governs convection, and q, the rate of heat transfer is defined by the difference between fluid and surface temperature, surface area and convective heat transfer coefficient, as $q = hA(T_{surface} - T_{fluid})$. The film coefficient, h, depends on the viscosity, density, thermal conductivity, specific heat and, specially, the velocity of the fluid past the surface. The table below reviews the film coefficient of some common fluids (1 BTU/(hr ft^2°F) = 5.678 W/(m^2°K)).

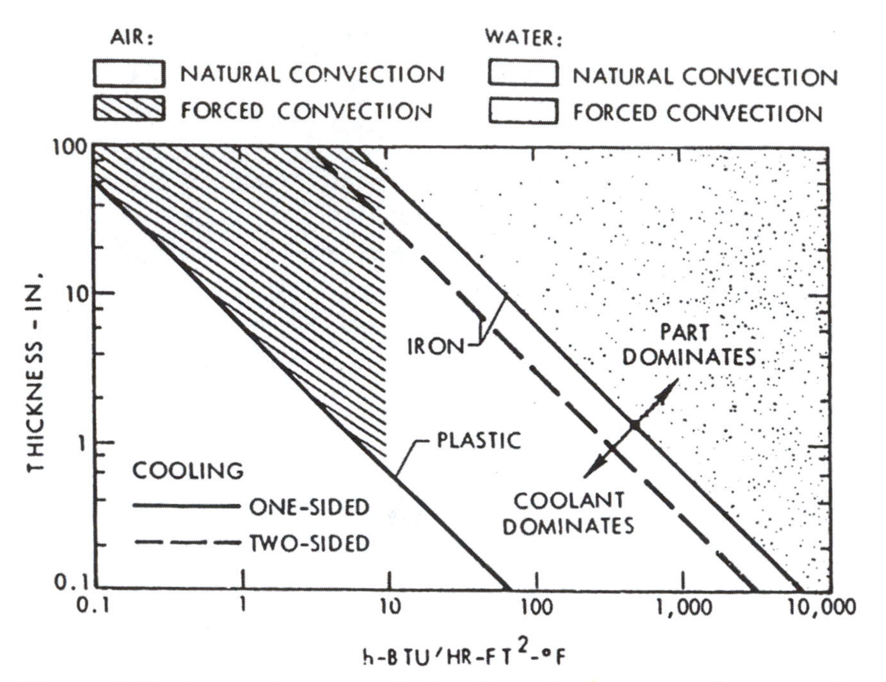

Figure 5.22. Lumped mass analysis of transient response for convection.

Table 5.2. Film coefficient of several fluids.

Fluid and Mode	h (BTU/Hr Ft ^{2} °F)	h (W/(m ^{2} K)
Air, free conv	0.1 to 0.5	0.5 to 2.5
Water, free conv	0.5 to 20	20 to 100
Air, forced conv	0.1 to 5	0.5 to 30
Water, forced conv	50 to 2k	300 to 10k
Oil, forced conv	10 to 300	50 to 2k
Boiling Water	500 to 10k	3k to 60k

Similarly for conduction, Fourier's law of heat conduction gives rate of heat transfer as a function of thermal conductivity, k, area, A, temperature along the path and location along the path x, as q = -kA (dT/dx). The thermal conductivity, k, is a physical property of the material and varies greatly for otherwise similar materials. Table 5.3 below shows k values for a number of different materials (1 BTU/(hr ft °F) = 1.731 W/(m°K)).

Table 5.3. Thermal conductivity for a variety of materials.

Material	k (BTU/Hr Ft °F)	k (W/(m K)
Al 1100-H14	128	221
Al 2024 T3	70	121
Cu 110	227	391
Steel 1020	27.0	46.7
Steel 4340	21.7	37.6
CRES 304	8.50	14.7
Acrylic ("Lucite")	0.10	0.17
Phenolic ("Bakelite")	0.05	0.09
Diamond	1156	2000
Ti (6Al-6V-2Sn)	4.1	7.1
Ag	241	417

Finally, another type of heat transfer that must be considered is radiation. This is governed by the Stephan-Boltzman law

$$E_b\,(T) = \sigma\,T^4$$

where

$E_b(T)$ = total blackbody emissive power (perfect radiator)
σ = 5.56 (10^{-6}) W/(m^2 K^{-4})
T = absolute temperature

Two specialized cases are often considered:

1. solar radiation — the sun (5672°K) can impart about 1 kW/m^2 to a "black" horizontal surface around noon, at mid-latitudes on a clear day.
2. radiative coupling between an object and its enclosure, both near room temperature.

We will look more at the second case as it often occurs in machine tool systems. Consider the case where body 1 is enclosed by body 2 with surface areas A_1 and A_2, reflectivity ρ_1 and ρ_2, emissivity ε_1 and ε_2, and body temperatures are T_1 and T_2. The heat transfer between 1 and 2 per area is

$$\frac{q_{12}}{A_1} = \frac{E_{b1} - E_{b2}}{1 + (\rho_1 / \varepsilon_1) + [(A_1 / A_2)(\rho_2 / \varepsilon_2)]}$$

For low temperatures, we can linearize this and treat it like a convection problem as

$$q_{12}/A_1 = h_r (T_1 - T_2)$$

where

$$h_r = \frac{\sigma(T_1^2 + T_2^2)(T_1 + T_2)}{1 + (\rho_1/\varepsilon_1) + [(A_1/A_2)(\rho_2/\varepsilon_2)}$$

The parameters associated with radiative heat transfer for many surfaces are listed in Table 5.4 below.

Table 5.4. Parameters for determining radiative heat transfer.

Surface	Absorpty @ 20 °C	h (BTU/Hr Ft 2 °F)	h (W/(m 2 K)
Polished Metal	0.01 to 0.15	0.01 to 0.05	0.06 to 0.85
Machined Metal	0.20 to 0.50	0.20 to 0.51	1.14 to 2.90
White Paper	0.95	0.96	5.45
White Paint	0.90 to 0.95	0.91 to 0.96	5.17 to 5.45
Flat Black Lacquer	0.95	0.96	5.45
Cast Iron	0.21	0.21	1.19

5.5 Specific heat sources and examples of thermal problems

We have already seen three examples of thermal energy induced variation in the performance of a machine. Figure 5.3 showed a schematic of potential distortion due to thermal inputs from a variety of sources, Figure 5.7 illustrated the "growth" in a precision gage exposed to variations in the room temperature and Figure 5.9 showed how well the length of a mechanical component tracks room temperature variation. A few additional examples are presented here of thermal growth in other machine components.

Most of the interest in thermally induced errors in machine tool systems is directed to the behavior of spindles of machine tools.

These generally operate at high speeds for various time periods and under various loads depending on the nature of the part being machined. The objective in most studies of spindle thermal errors is to either develop and validate systems of cooling that will nullify the influence of thermal effects on the spindle performance (generally growth in some direction) or to accurately model and predict the spindle growth behavior so that other compensation can be done to offset the effects of this error on the workpiece (recall the discussion around Figure 5.2 earlier).

The first example shows the effect of a forced oil cooling systems on a spindle design, Figure 5.23, from Bryan[38]. The concern is to minimize the axial growth of the spindle due to frictional heating of the rolling element bearings. Frictional heat generation in the spindle unit is the largest source of thermal distortion in machine tools such as lathes and mills, External surface temperatures can reach 40 to 50°F above ambient after lengthy full speed operation. Variation with spindle speed is seen of course. The largest resulting thermal error is axial spindle growth, which influences face flatness and shoulder positions in lathe work and the overall part geometry in end milling. Typical values of axial spindle growth are in the range of 0.003-0.006 inches for lathes and up to 0.015 inches for mills. This data is from some thirty years or so ago referring to primarily mechanical manual machine tools. But it is reasonably close to the size of effects seen on present day machines — especially lower end machines with roller bearing spindles. The re-designed bearing had, as can be seen in the figure, a flow through lubrication system with a temperature controlled oil supply was implemented for each of the spindle bearings[38].

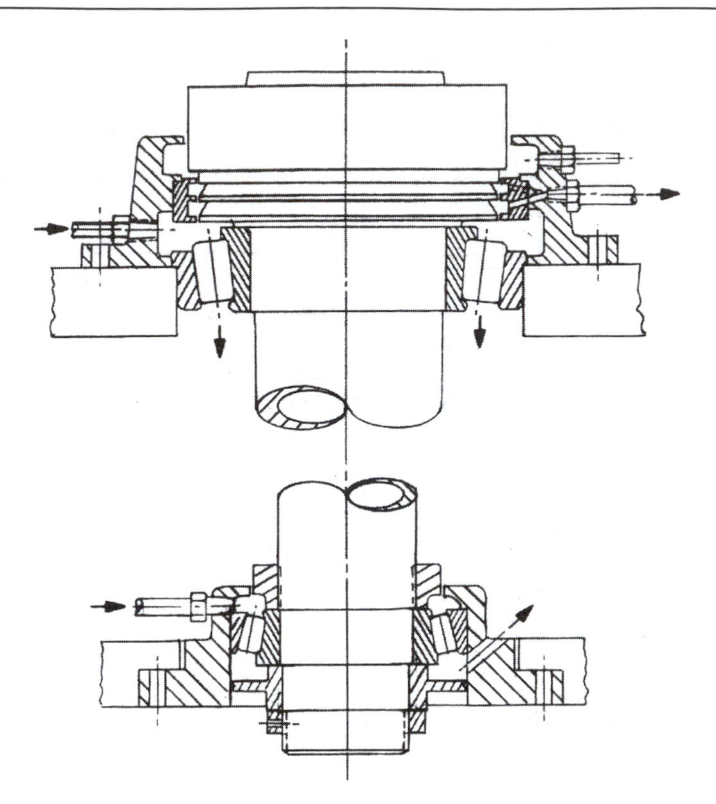

Figure 5.23. Spindle design for test of enhanced oil flow cooling system, from Bryan[38].

The results of the modification were quite remarkable, Figure 5.24. The two curves in Figure 5.24 show the growth axially of the original and modified spindle lubrication design, A and B, respectively. The reason for this remarkable improvement (over say simply flowing cooling oil over the bearing assembly rather than flowing it through the bearings themselves) is due to the nature of the action of the bearing under the temperature increase during operation, The temperature distribution within the bearing is quite complex. One component of spindle error growth is temperature rise and expansion of the spindle shaft causing a reduced preload of the end bearings. But a second, important mechanism is based on the temperature distribution within the tapered roller bearings themselves. A significant portion of the frictional heat is produced by the rubbing if the large ends of the rollers against the thrust lip of the cone. The majority of frictional heat is removed through the cup to the housing because of the larger surface area. The cone and rollers run hotter than the cup and housing, resulting in a differential expansion of the diameters. A wedging effect caused by the conical geometry tends to cause a much larger axial displacement, proportional to the cotangent of the angle from the axis, which tends to increase the preload. Both the elastic elongation of the shaft between bearings and compression of the smaller and more compliant rear bearing contribute to the forward movement of the spindle nose. To minimize this second effect, it is necessary to place the cooling oil as close to the point of friction generation as possible — hence forcing the temperature controlled oil directly through the bearing (i.e. in the spaces between the cup, cone, cage and rollers). Some consideration was given to the potential for heat generation in the fluid in the bearing as well[38].

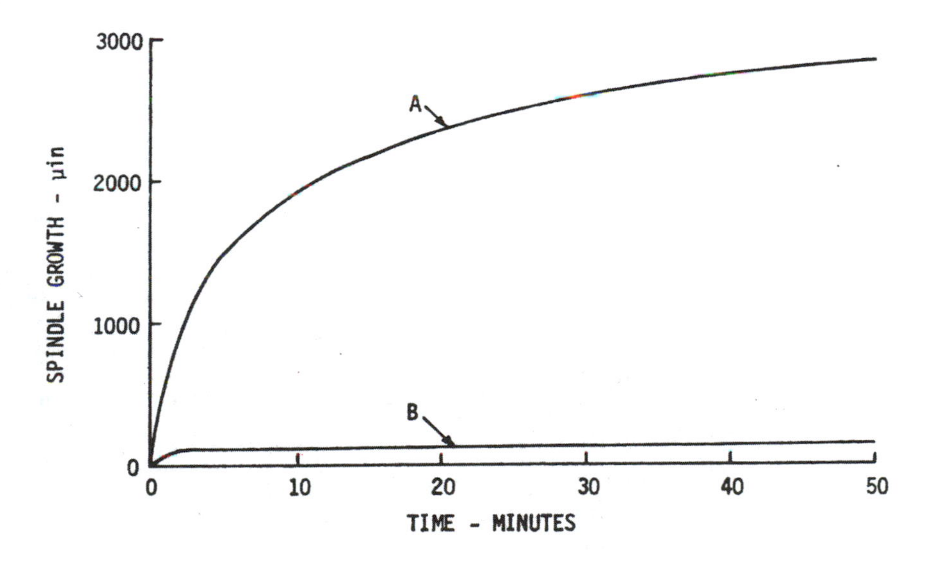

Figure 5.24. Experimental results: Curve A – original system; Curve B – forced flow system, from Bryan[38].

Others have measured the performance of spindles in thermally active environments in order to model and then predict the behavior. One excellent example is Krulewich[39]. This study used experimental design to develop a linear model coupling discrete temperature measurements on a machine spindle to deflection. The biggest challenge to the successful model development is choosing the location of the temperature sensors. The assumption used was that adequate knowledge of the temperature distribution completely determines the deflection of a structure at any instant in time. This work developed a methodology for this. Machine deformation is caused by a varying temperature field throughout the machine component due to changing operating conditions, Figure 5.25.

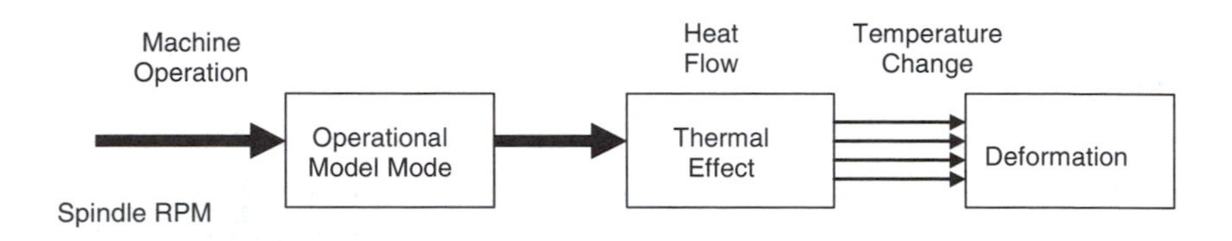

Figure 5.25. Causal chain of thermal distortion in machine spindle, after Krulewich[39].

The behavior of the spindle was measured during a "typical" spindle cycle of varying spindle speeds. The resulting thermal behavior induced spindle growth for simply accelerating the spindle to a steady state 5000 RPM is shown in Figure 5.26. The spindle growth is shown along the vertical axis. This is similar to the behavior seen in Figure 5.24. Both the experimental data (points) and model prediction is shown. When spindle speed variation is used, the growth behavior of the spindle is more interesting, Figure 5.27. These figures are similar to the one in Figure 5.26 except show the variation due to spindle RPM. The spindle speed profile has the speed increasing from 1000 RPM to 6000 RPM in steps with some dwell time at each spindle speed. The "time constant" of the spindle is clearly seen in the increasing spindle growth on acceleration and decreasing growth on deceleration.

Finally, thermal effects have a dramatic effect on machine performance over time, especially as the machine "warms up." This is often the case in small lot manufacture where much of the cycle of production is composed of setup and tear-down time and, hence, the machine may not operate long enough to establish a steady state temperature. Figure 5.28, from Donmez[40], shows the change in behavior of Z axis displacement error measured as the machine (a "Superslant" lathe) warms during continuous operation of the Z slide. One can also see from the figure that there is an initial 200 μ inch of backlash in the ballscrew mechanism comprising the axis. The growth in amplitude and variation in error at a location along the axis is apparent.

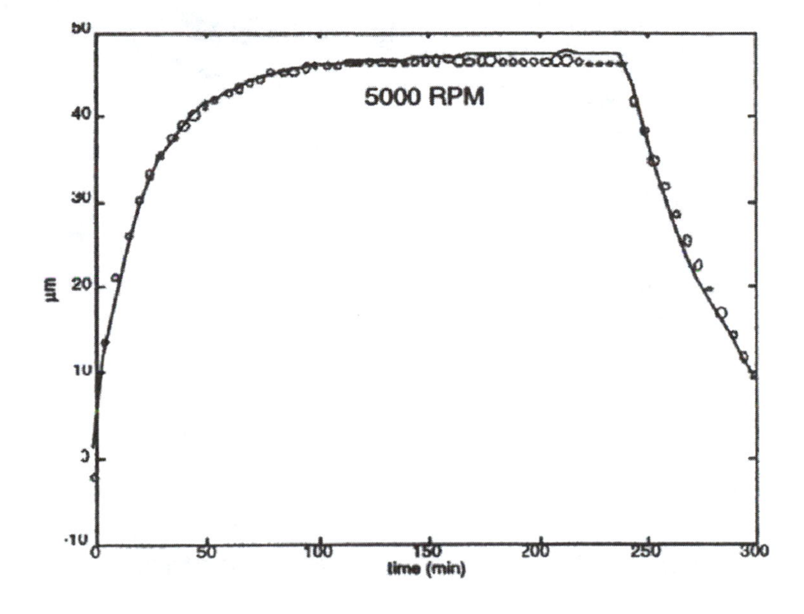

Figure 5.26. Model prediction and measured axial growth at constant 5000 RPM, Krulewich[39].

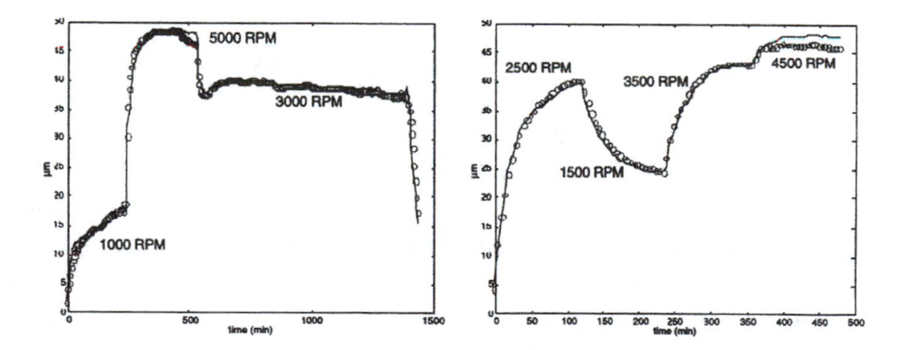

Figure 5.27. Model prediction and measured axial growth for variable RPM, Krulewich[39].

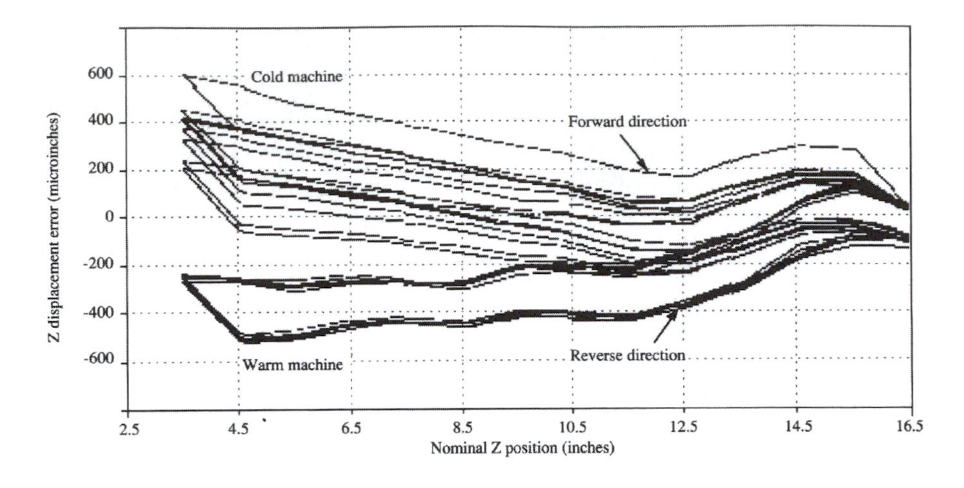

Figure 5.28. Z displacement error measurement taken as machine warms, from Donmez[40].

We will discuss some of the other examples of thermally induced errors relative to metrology equipment in section 5.9 below.

5.6 Environmental control of precision machinery

It is common to use temperature controlled environments for purposes of metrology and construction of precision equipment. The environment can consist of an entire room for which the temperature is controlled to some specification, a "box" on the machine in which the manufacturing process takes place (machining for example) or some other type of environmental control. One temperature-controlled environment developed and perfected at the Lawrence Livermore National Laboratory is a system based on flows of temperature controlled oil over the machine tool and workpiece before, during and after machining. In fact, it is often easier (and cheaper and more effective) to control the part on the machine through such a oil flow setup than to attempt to control the whole room. In some cases, it is most effective to both control room environment and machine temperature using circulating fluids that are temperature controlled. We will review some of these in this section.

5.6.1 Machine enclosures

One common means to control the environment the workpiece sees is to utilize a machine enclosure as part of the machine tool, Figure 5.29. These are also needed for safety (prevent access or contain debris and fluid) as well. Figure 5.29 shows a Mori Seiki milling machine with an enclosure that must be closed during machining for safety and environmental control.

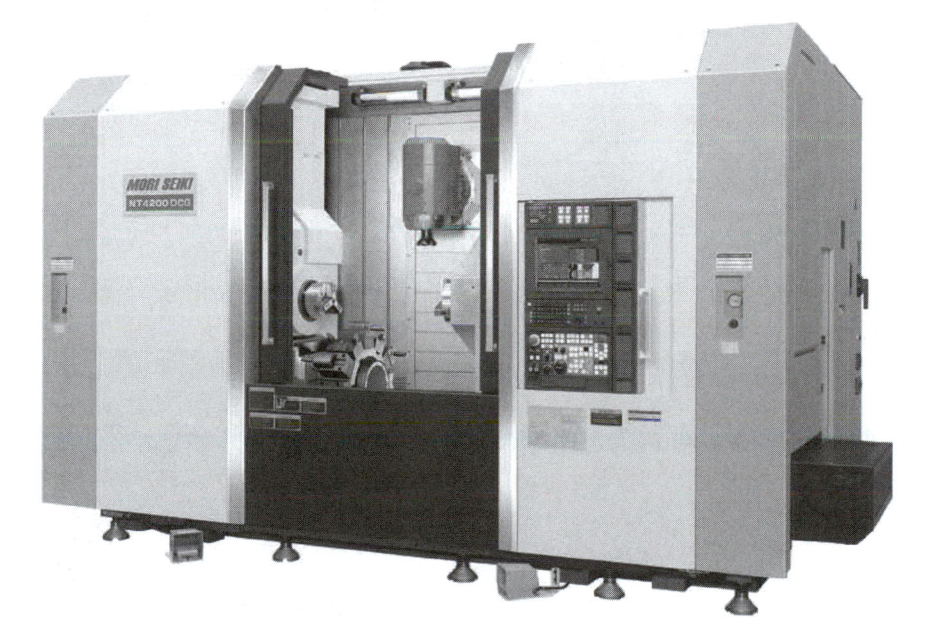

Figure 5.29. Mori Seiki "mill-turn" machine operating area enclosure open for access, courtesy of Mori Seiki.

Figure 5.30, also from Mori Seiki, shows a smaller milling machine with the enclosure closed, and close up showing machining area open to allow access to the work platform and tool.

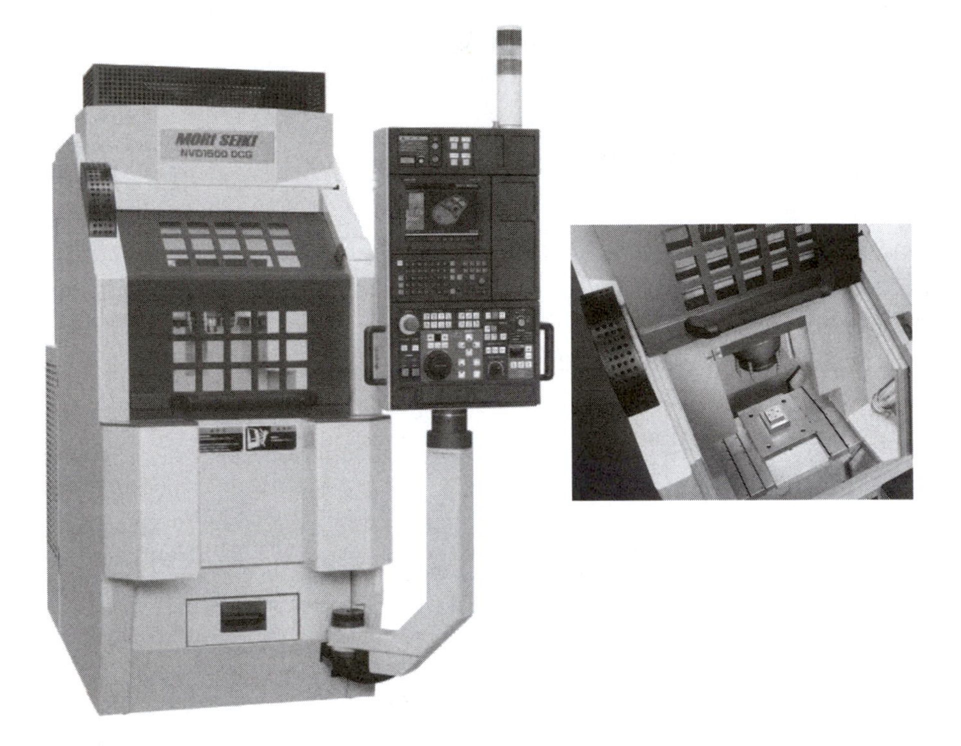

Figure 5.30. Mori Seiki milling machine, left, and with "box" open for access, right, courtesy of Mori Seiki.

5.6.2 Factory and room enclosures

The idea of "thermally optimized factory" as mentioned by Bryan[37] suggests rooms or whole assembly areas maintained at a 68°F. How this is accomplished is a challenge. The engineers at Lawrence Livermore National Laboratory have built several such enclosures for housing precision diamond turning machines (the capability of the Large Optics Diamond Turning Machine (LODTM) — one of those machines — will be discussed more in Chapter 6). These rooms are temperature controlled by carefully treating the incoming air to maintain a temperature of 68°F ± 0.1°F (or the equivalent in°C). Many of these techniques are now in common use in other production environments requiring close temperature control (as in semiconductor fabrication.)

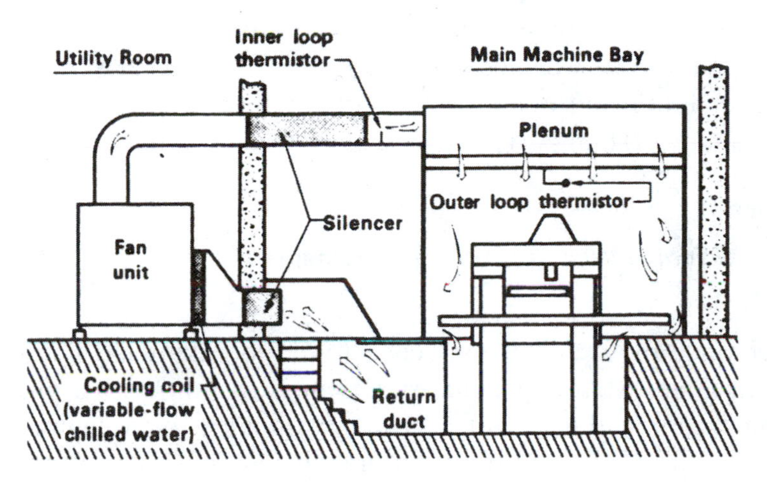

Schematic of room layout with air handling equipment

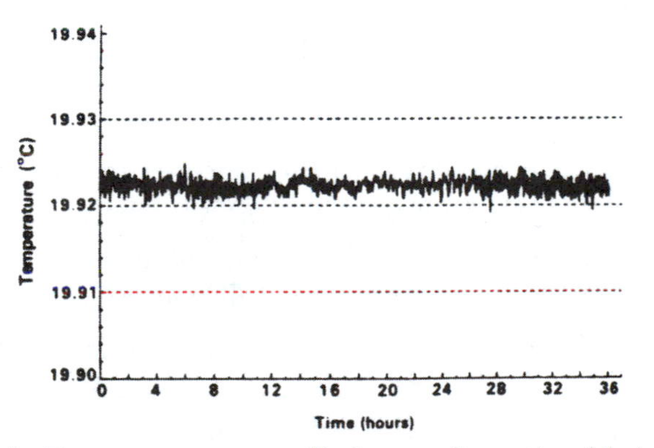

Figure 5.31. Temperature controlled room for optics fabrication, top, and trace of air temperature control achieved, from Roblee[41].

A schematic of a controlled room, and the temperature control achieved is shown in Figure 5.31 above, from Roblee[41]. The figure shows, top, a cross sectional schematic with the necessary air handling equipment and the plenum arrangement for providing a laminar flow of conditioned air over the machine tool exhausting it through the floor. This is the same approach used in other "high tech" applications for both particulate and temperature control. The lower trace in Figure 5.31 shows the quality of temperature control achieved in this design. Temperature control over a period of 36 hours, as measured by thermisters in the room, shows a stability of ± 0.002 °C (!). This was accomplished by using a heat exchanger for air temperature control and water with a temperature stability of approximately ± 0.00025 °C.

5.6.3 Machine treatment without enclosures

A technique pioneered at Lawrence Livermore National Laboratory by then Chief Metrologist James Bryan is to "shower" a precision machine tool and the part being fabricated with a temperature-controlled oil. This is useful both during soak-out to bring a part to the temperature of the machine as well as during production to maintain the part temperature in the face of thermal energy from the machining process and other sources in the machine. This was applied to measuring machines as well. A detailed schematic of the application of an oil shower to a measuring machine is shown in Figure 5.32, from Bryan[42]. The primary advantage of a liquid shower is its greater heat removal capability and the fact that it is easily directed to the critical areas of the machine and workpiece surfaces. Although air showers, as discussed above, is effective to an extent, one of its major drawbacks is the operator chill and noise. Bryan points out that 68°F air at a velocity of 800 ft/min. is roughly equivalent to still air at 55°F — not suitable for humans for any length of time. Add to this the cost of the enclosure and air handling equipment, it makes the liquid shower approach cost effective and efficient. An order of magnitude improvement was realized in thermal stability with the use of the oil shower in this example, Figure 5.33.

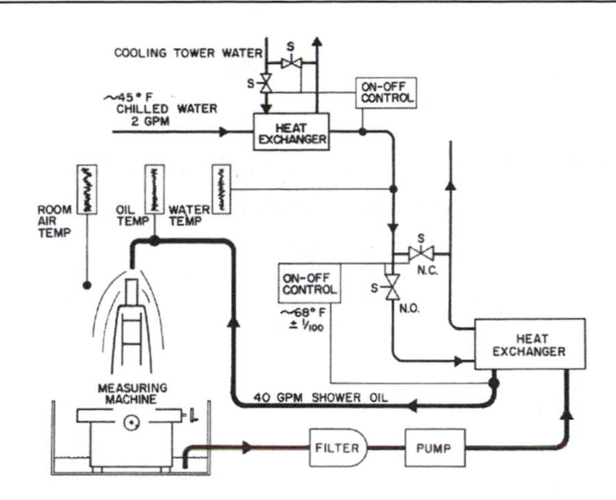

Figure 5.32. Schematic of oil showered measuring machine with shower oil at a nominal 68°F; the maximum temperature variation averaged over 30 seconds is 0.01°F, from Bryan[42].

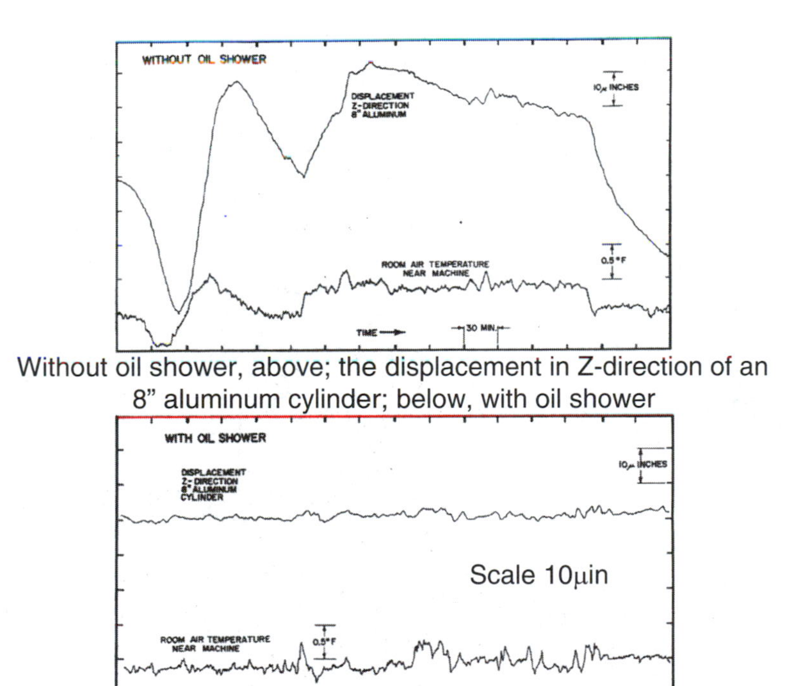

Figure 5.33. Improvement in thermal stability with the use of an oil shower, from Bryan[42].

5.7. Thermal effects and metrology

Finally, we will look into the thermal effects in dimensional metrology in more detail. The concern about temperature and the deformation resulting from differential temperatures or the uncertainty error in coefficients introduced can be analyzed using metrology principles. First, we need to review some basics.

There are three elements in a length measuring system, Figure 5.34. These are the part (piece to be measured; the physical object for which a linear dimension is to be determined — the unknown or desired length), the master (the known length to which the measuring instrument is calibrated), and the comparator (the device used to perform the comparison of the part and master)[36].

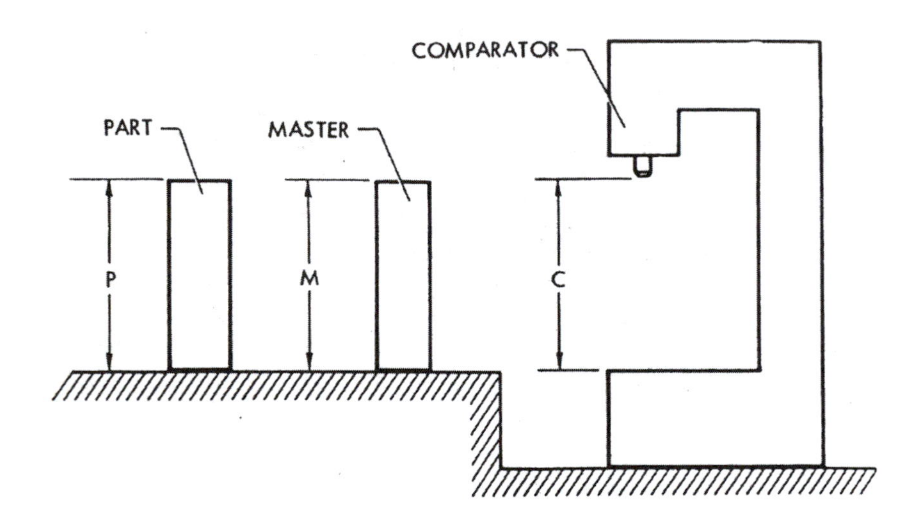

Figure 5.34. Sketch of part, master and comparator, similar to ANSI B89.6.2-1973 Standard for temperature and humidity environment for dimensional measurement.

How these are realized in a metrology instrument, a micrometer, for example, can differ, Figure 5.35. Here, we see two images of a "conventional" micrometer showing calibration with a master and measurement of a thin walled tube. The various elements in Figure 5.34 are embodied differently depending upon how the system is configured.

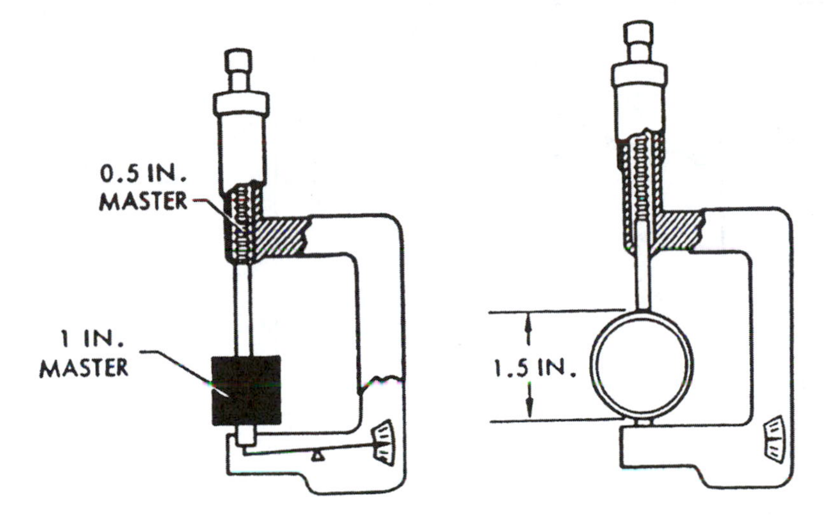

Figure 5.35. Views of two conventional calipers (one with internal "master" scale), after ANSI B89.6.2-1973[36].

Importantly, the thermal environment that the comparator, part and master are in will have a major impact on the precision of the measurement made. The temperature of the part, master and comparator will lag behind the temperature of the air, Figure 5.36. It is unlikely that all three will have the same temperature except after all have had a chance to adjust to the ambient (assuming it is stable). If ambient air temperature is varying (cyclically, for example) the temperatures (and thus size) of the three elements will vary as well according to their thermal time constants. The graph in Figure 5.36 shows a steady state dimensional response of 3-element system to sinusoidal ambient-temperature variation.

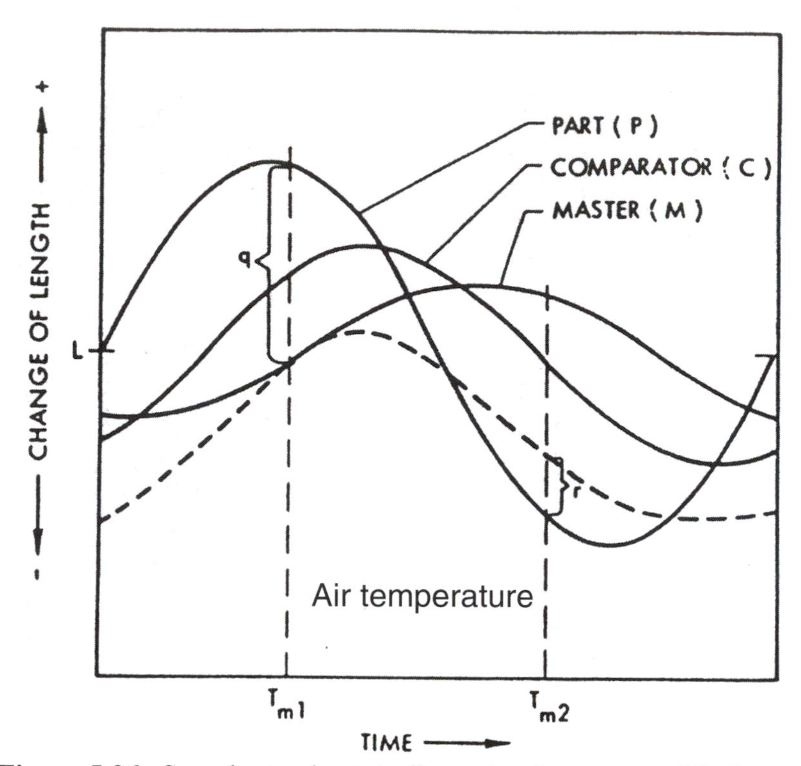

Figure 5.36. Sample steady state dimensional response of 3-element system to sinusoidal ambient-temperature variation, from ANSI B89. 6.2-1973[36].

Bryan and his co-authors[37] review some of the fundamentals of these systems starting first with a two-element system representation of a "C" frame comparator, Figure 5.37. This is similar to those shown in earlier figures. Here the tubular part has a much lower mass (and stiffness) than the comparator. All length measuring apparatus can be represented as consisting of a number of individual elements arranged to form a "C". In fact, it does not take too much imagination to represent most machine tools in this fashion as well. In Figure 5.37, the comparator and part form the two elements. Bryan[37] describes the behavior of both of these elements with respect to changes in environment. That description is repeated here *verbatum.*

> "If the coefficient of expansion of the comparator is exactly the same as the part, the gage head will read

zero after soak-out at any uniform temperature that we might select. If we induce a change in temperature, however, the relatively thin section of the tubing will react sooner than the thick section of the comparator frame and the gage head will show a temporary deviation. The amount of the deviation will depend on the rate of change of temperature. If the rate is slow enough to allow both parts to keep up with the temperature changes, there will be a small change in gage-head reading. If the rate is so fast that even the thin tubing cannot respond, there will again be a small change in reading. Somewhere in between these extremes there will be a frequency of temperature change that results in a maximum change in reading. "

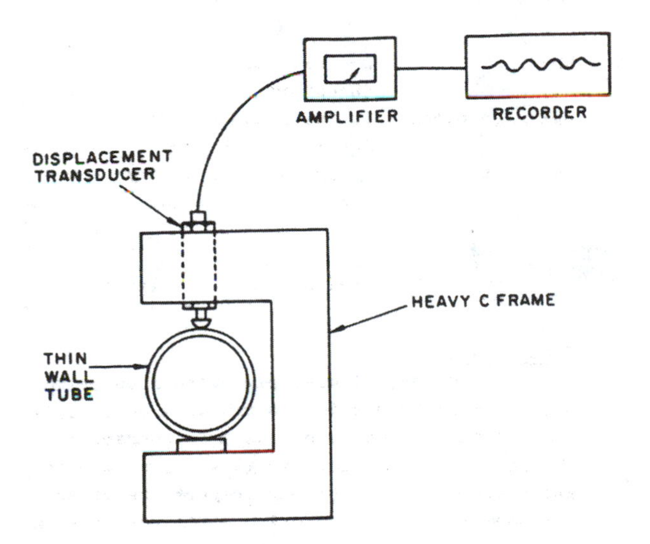

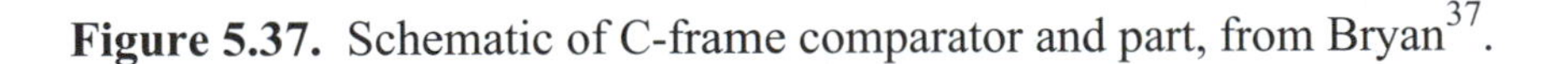
Figure 5.37. Schematic of C-frame comparator and part, from Bryan[37].

The comparator and work can be represented schematically as in Figure 5.38 with a cylinder representing each element. The size of each cylinder is adjusted to resemble the comparable masses of the two elements, Bryan[37] made sample heat-transfer calculations for each cylinder. The cylinder (in Figure 5.38) with the transducer placed on it can be considered the comparator. The two cylinders were made of steel and are 4 inches long. Cylinder A (comparator) is 2 inches in diameter and cylinder B (part) is 0.5 inch in diameter.

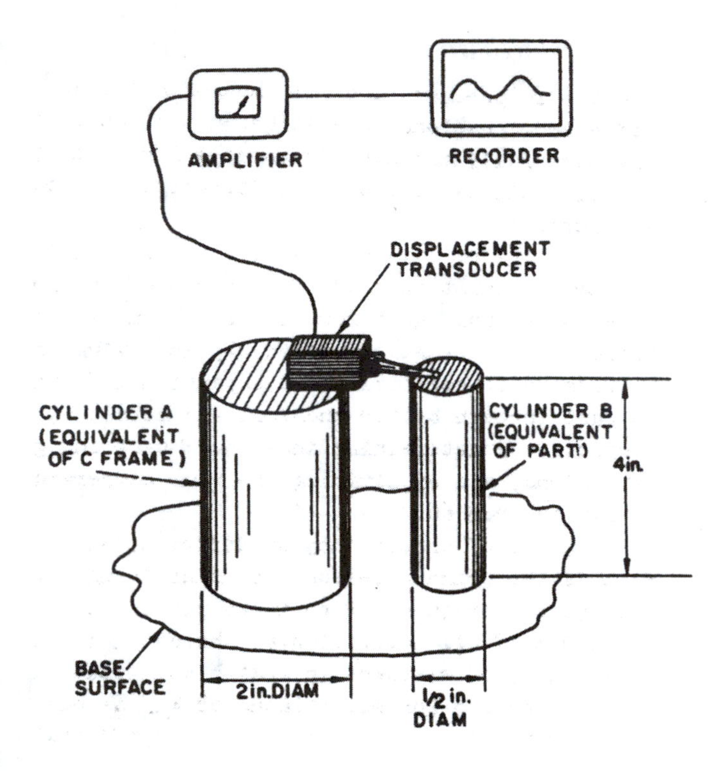

Figure 5.38. Model of C-frame comparator and part, from Bryan[37].

Figure 5.39 shows the results of the simulation and the predicted changes in the length of the two cylinders as a result of a plus/minus 1°F sinusoidal variation in air temperature with a frequency of 1 cycle per hour. The comparator cylinder (A) shows less than one third the temperature change of the thinner "part" cylinder (B) and its temperature lags the thin cylinder by about 3 or 4 minutes. The dotted line in the upper figure shows the predicted gage-head reading which is the same as the instantaneous difference in lengths of the two cylinders. Bryan calls this the "thermal drift" of the system.

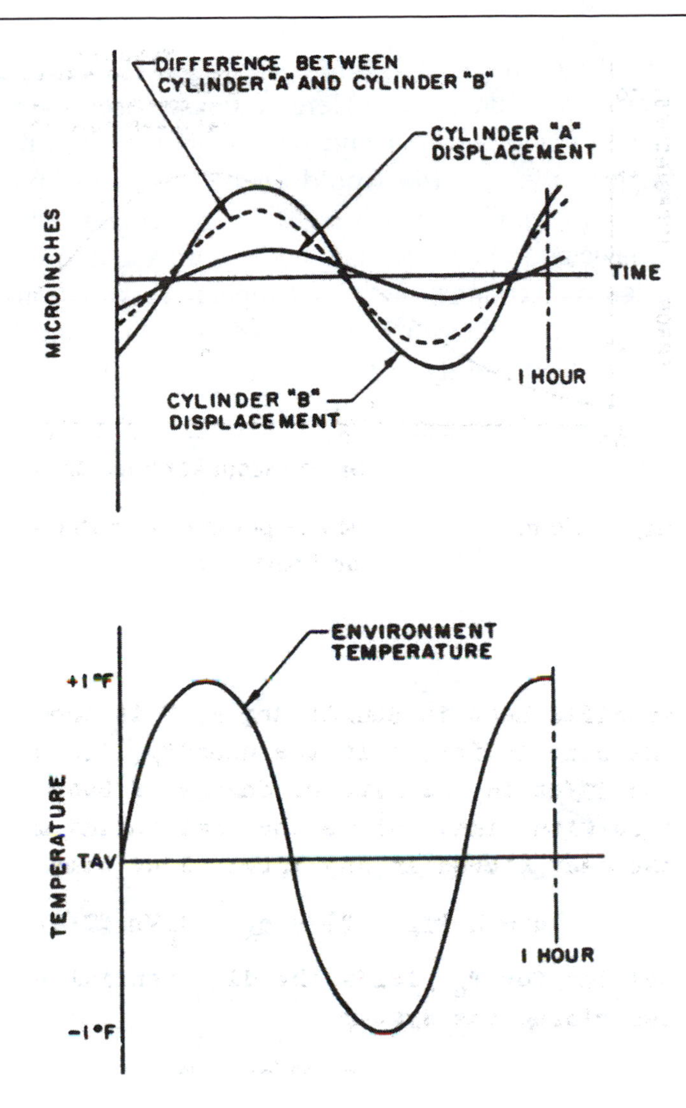

Figure 5.39. Individual and net thermal response of the two – element system, from Bryan[37].

The effect of varying thermal variation frequency on this two element system is shown in Figure 5.40. The drift is small for very high or low frequencies of temperature variation and reaches a maximum amplitude at a point in between which Bryan calls "resonance" in this thermal system. This Figure 5.40 effectively represents the frequency response of the system. Here, resonance occurs for a variation at one-half cycle per hour and results in an error value of 15 μinches. Importantly, this error would result even if the time average temperature of the environment was 68°F. In actuality, "real" machines are expected to be worse as they do not have uniform behavior (expansion and contraction with temperature variation) due to structural non-uniformities. It is not unusual to have error magnitudes of 150 μinches/°F or more in real systems at resonant frequency.

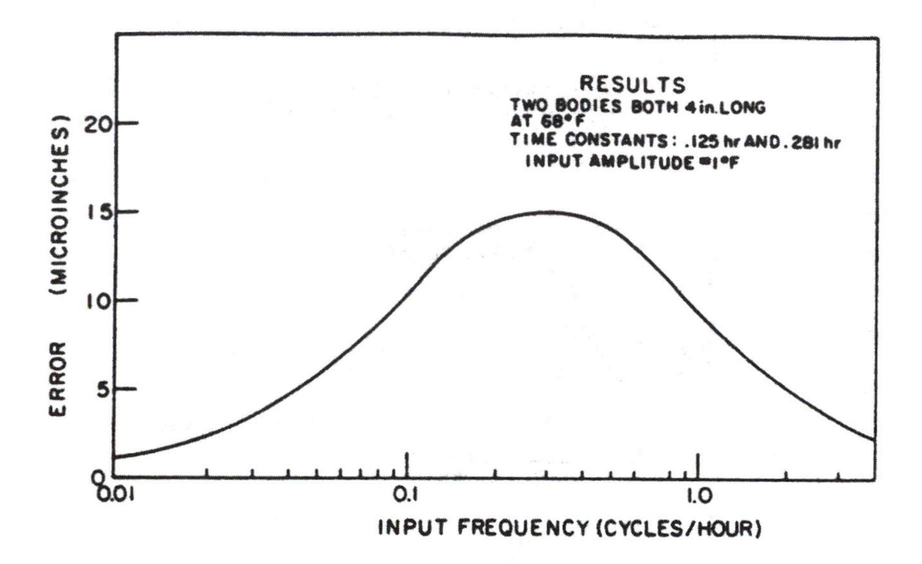

Figure 5.40. Simulated frequency response of two-element C-frame model, from Bryan[37].

There is much more that could be said about the importance of and design for minimizing the effects of thermal errors. This chapter presented a review of major issues and considerations. The references cited offer much more detail and the reader is encouraged to investigate them.

5.8. Observations

Thermal effects in manufacturing and machine tools have been noticed for centuries both from the point of view of the impact on the machine and the process. Early manufacturing derived knowledge supported the first law of thermodynamics. Benjamin Thompson, Count Rumford, describes a set of innovative calorimetric experiments completed while he was a superintendent in a factory for the manufacture of brass cannons in Munich in the late 1700's[43]. He was observing the boring of a cannon and Rumford was taken with the considerable temperature which a brass gun acquires, in a short time, during the boring operation and intrigued as well with the intense heat of the metallic chips created during the machining process. Rumford showed that the heat generated by the action of the steel boring tool, friction primarily, during machining of the hollow metallic cylinder was greater than that produced equally by "nine wax candles, each of an inch in diameter, all burning together, or at the same time, with bright clear flames." The machine was driven by horses, as seen in Figure 1.16 (although Rumford's machine was horizontal) so Rumsford was able to derive the work input for this temperature rise measured. Davies, et al[44], point out that this work was clearly ahead of its time as reference to the temperature of the machining process was not again cited in the literature until Taylor's work in 1901, cited previously[11].

VI ERROR MAPPING AND ERROR BUDGETS

6.1 Introduction

Assuming that all of the sources of error in a precision machine can be identified and quantified, it is helpful to be able to determine the degree to which they impact the quality of the part, i.e. dimension, form, surface roughness, principally. In Chapter 2 we discussed generally error sources and their interactions, for example, Figure 2.4. We will discuss in greater detail the determination of the "sensitive direction" of a machine tool which is, in most cases of machining, normal to the machined surface. The degree to which the elements of the machine tool, structure, spindle, tooling, and fixture interact with the error sources (thermal, mechanical, etc.) to cause displacement or other undesirable movement in a sensitive direction with respect to the workpiece, indicates the inaccuracy that will be imparted to the workpiece. There are several determinations to this — machine components and error sources mapping onto directions affecting the work. This is, in short, error mapping. And, the formulation of a list of these errors and magnitudes, as they affect the workpiece (or desired motion, or any other output of the system) is called an error budget.

The ultimate goal is to derive an error budget. The error budget relies on two sets of rules — connectivity and combinational. Connectivity rules define the behavior of machine components and interfaces in the presence of errors. The combinational rules define how the errors are to be combined to determine the impact on the accuracy of the workpiece. The procedure is comprised of three steps as follows:

Step 1 — determination of a kinematic model of the machine and its principal components in the form of a series of homogeneous transformation matrices (HTM).

Step 2 — analyze systematically each type of error in the system and use the HTM to determine the relative tool -work errors.

Step 3 — combine the errors to yield upper and lower bound estimates of the total error of the machine.

Step 1 we will refer to as "error mapping" and discuss this in the next section.

6.2 Error mapping

With respect to a typical machine tool and its behavior in use, we identified six degrees of freedom relative to the three axes of motion, viz. translation along x, y and z as well as rotation about these three axes, Figure 4.1. These three rotational degrees of freedom are traditionally referred, reference to one of the axes, as "roll" (rotation about x, for example), "pitch" (rotation about y affecting the x axis by causing up and down motion in the direction of the axis) and "yaw" (rotation about z affecting the x axis by causing left and right rotation with respect to the motion along the axis.) An interferometry setup to measure this roll, pitch and yaw motion in a three axis system is illustrated in Figure 6.1. Altogether there are 21 geometric causes of error on a three axis machine tool — the twelve listed above plus straightness in two included planes for each axis (that adds six more) and orthogonality in the three planes defined by the axes for a total of 21. Figure 6.2 from McKeown[27] gives them all and the direction of the error. This will be important as we develop the HTM for a machine.

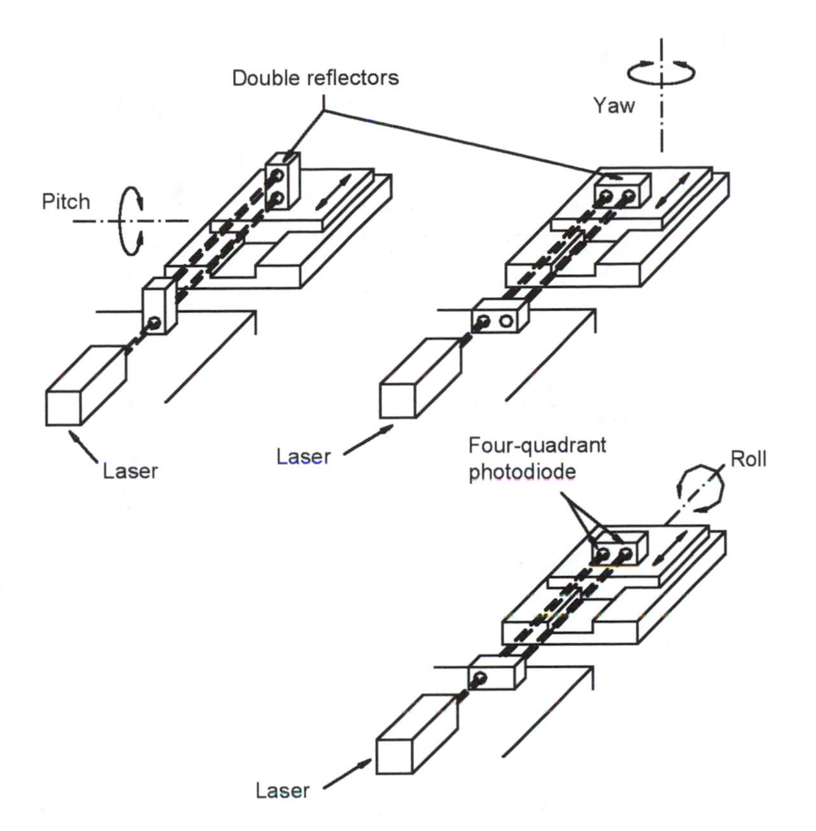

Figure 6.1. Measurement of roll, pitch and yaw movements, from MTTF[25].

Axis of motion	Degree of Freedom/ Error Motion	Direction of Error		
		X	Y	Z
	pitch	*		
	yaw	*		
X	roll		*	*
	straightness-in XY plane		*	
	straightness-in XZ plane			*
	pitch		*	
	yaw		*	
Y	roll	*		*
	straightness-in YX plane	*		
	straightness-in YZ plane			*
	pitch			*
	yaw			*
Z	roll	*	*	
	straightness-in ZX plane			
	straightness-in ZY plane			
	orthogonality X to Y	*		
	orthogonality X to Z	*		
	orthogonality Y to Z		*	

Figure 6.2. Twenty-one geometric causes of systematic error in a three linear axis machine, from McKeown[27].

In a machine tool for material removal, the HTM is designed to allow us to assess the error in the surface generated (location and form) based on the cumulative errors affecting it. This error is comprised of four major elements, the error in the position of the tool itself from its ideal location (as caused, for example, by temperature, direction, position, load, speed); the error in the position of the tool tip from ideal (as caused by, for example, toolsetting errors, tool wear, vibration or tool tip deformation); error in the workpiece position from ideal (as caused by, for example, part thermal growth, residual stress distortion, work setting, vibration, deformation); and error in the measurement of the process error (as caused by, for example, probe lobing, probe accuracy, kinematics).

The HTM defines the spatial relationship between the tool tip and a corresponding position on the surface of the work. Any errors will cause deviation from this ideal relationship. To represent the relative position of a three dimensional body in space with respect to a reference coordinate frame requires a 4x4 matrix. We will review some of the basics of these transformations here to clearly define the terminology. If we represent the reference frame by R and the coordinates system of the body as N, then the notation ${}^{R}T_{N}$ is a HTM describing the position and orientation of a coordinate system N in a coordinate system R. Then, the full HTM can be represented as

$$^{R}T_{N} = \begin{bmatrix} O_{ix} & O_{iy} & O_{iz} & d_{x} \\ O_{jx} & O_{jy} & O_{jz} & d_{y} \\ O_{kx} & O_{ky} & O_{kz} & d_{z} \\ 0 & 0 & 0 & d_{s} \end{bmatrix} \tag{6.2.1}$$

where the terms O_{ix} etc. are the direction cosines (unit vectors *i, j, k*) representing the orientation of the body's coordinate system relative to the reference frame and d_x, d_y, d_z and d_s displacements of the frame along the x, y and z axes and a scale factor, respectively. The scale factors on the unit vectors are zero.

Thus, the equivalent coordinates for (x_N, y_N, z_N) in the reference frame can be represented as

$$\begin{bmatrix} x_R \\ y_R \\ z_R \\ 1 \end{bmatrix} = {}^{R}T_{N} \begin{bmatrix} x_N \\ y_N \\ z_N \\ 1 \end{bmatrix} \tag{6.2.2}$$

A pure translation of the body along the y axis by a distance d_y (that is, from position x_0, y_0, z_0 to position x_1, y_1, z_1) would yield a HTM of the form

$$^{x_0 y_0 z_0}T_{x_1 y_1 z_1} = \begin{bmatrix} 1 & 0 & 0 & 0 \\ 0 & 1 & 0 & d_{y} \\ 0 & 0 & 1 & 0 \\ 0 & 0 & 0 & 1 \end{bmatrix} \tag{6.2.3}$$

And a similar pattern is followed for translation along the other two axes, i.e. pure translations only affect the last column of the matrix.

Rotations about any axis are treated similarly but are a bit more complicated due to the nature of perceived motion (roll, pitch or yaw) from the axis referred to (the one we are directed along.) If the coordinate system of the body (x_N, y_N, z_N) is rotated by an amount θ_x about the x-axis, the HTM representing the transformation of a point in that coordinate frame (x_a, y_a, z_a) relative to the original frame (x_0, y_0, z_0) is

$$ {}^{x_0 y_0 z_0}T_{x_a y_a z_a} = \begin{bmatrix} 1 & 0 & 0 & 0 \\ 0 & \cos\theta_x & -\sin\theta_x & 0 \\ 0 & \sin\theta_x & \cos\theta_x & 0 \\ 0 & 0 & 0 & 1 \end{bmatrix} \tag{6.2.4} $$

And a similar pattern is followed for rotation about the other two axes.

Once the HTM is determined that reflects the displacement and/or rotation of the body relative to a reference, it can be linked to other HTMs representing other bodies. Machine structures are composed of many sub-parts N all linked to another through intermediate coordinate frames ultimately referenced to the frame of the machine tool, R. Then, the position of the Nth part relative to the frame of reference is

$$ {}^{R}T_N = \prod_{m=1}^{n} {}^{m-1}T_m = {}^{0}T_1 \, {}^{1}T_2 \, {}^{2}T_3 \, {}^{3}T_4 \bullet\bullet\bullet {}^{n-1}T_n \tag{6.2.5} $$

The sub-parts can represent either axes of the machine tool or rigid bodies comprising the machine. For example, if four machine elements, A, B, C and D are “stacked” to build a machine, and the machine base reference system is R, then, using the terminology developed here, the position of A relative to B is ${}^{B}T_A$, B relative to C is ${}^{C}T_{B}$, C relative to D is ${}^{D}T_{C}$, and D relative to the base is ${}^{R}T_{D}$. So, finally, A relative to the machine base can be represented as

$$ {}^{R}T_{A} = {}^{R}T_{D}\,{}^{D}T_{C}\,{}^{C}T_{B}\,{}^{B}T_{A} \tag{6.2.6} $$

To aid in the determination of the nomenclature for rotations refer to Figure 6.3, from Slocum[18]. The figure identifies the displacement and rotation errors relative to the axis of motion, the x-axis in this case. Thus, rotation about the y-axis is seen as yaw error, ε_y, rotation about the x-axis is seen as roll error, ε_x, and rotation about the z-axis is seen as pitch error, ε_z — all relative to the x-axis.

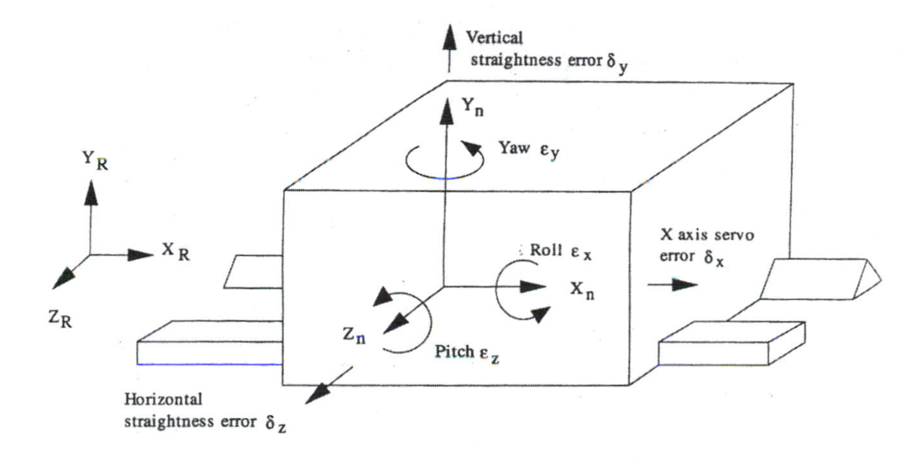

Figure 6.3. Motions and errors indicated for a single axis linear mechanism, from Slocum[18].

Consider first the linear displacement of an ideal machine axis carrying a table, Figure 6.3. The three rotational (ε_x, ε_y, and ε_z) and three translational (δ_x, δ_y, and δ_z.) error components of a rigid body as applied in this case are shown in the figure. These errors are occurring along (displacement, due to d_x, d_y and d_z) and about (rotation, due to θ_x,θ_y and θ_z) the machine axis. Clearly, rotational errors will contribute to the error in position of a point on the table relative to the reference frame. The machine table has offsets (i.e. linear errors relative to a parallel cartesian frame of reference) in all three directions, d_x, d_y, and d_z and its HTM can be written from (6.2.1) as

$$ {}^{R}T_{N} = \begin{bmatrix} 1 & 0 & 0 & d_x \\ 0 & 1 & 0 & d_y \\ 0 & 0 & 1 & d_z \\ 0 & 0 & 0 & 1 \end{bmatrix} \tag{6.2.7} $$

We can consider the effect of the error about each axis on the relative error by writing similar HTM's for each rotational error motion and combine them as in (6.2.6). Since the result follows a usual pattern we can often do this by observation and some imagination. In eq. 6.2.1, the terms represent displacements and rotations according to the right hand rule. A rotation ε_x about the *X* axis (which we refer to as *roll*) will cause the tip of the *Y* axis to rotate in the positive *Z* direction by an amount proportional to sin ε_x and in the negative *Y* direction by an amount proportional to 1-cos ε_x. Here, and in most cases for precision engineering, we assume the angles involved in these rotations to be sufficiently small so that small angle approximations are valid. So, with respect to (6.2.1), the term representing the interaction between the *Z* axis and the *Y* axis, $O_{ky} = \varepsilon_x$. The roll action similarly causes the tip of the *Z* axis to move in the negative *Y* direction by an amount $-\varepsilon_x$ so term $O_{jz} = -\varepsilon_x$.

A rotation about the *Y* axis (that we call *yaw*) of ε_y will cause the tip of the *X* axis to move in a negative *Z* direction so that $O_{kz} = -\varepsilon_y$. This also causes thetip of the *Z* axis to move in a positive *X* direction so $O_{iz} = \varepsilon_y$. Similar analysis of rotation in a positive direction ε_z about the *Z* axis will result in terms of (6.2.1) of $O_{jx} = \varepsilon_z$ representing motion of *X* axis relative to *Y* and $O_{iy} = -\varepsilon_z$ representing the motion of the *Y* axis relative to the *X* axis. The terms for rotational error impact on the displacement along the axes are assumed to be zero as previously explained and we have ignored the second order terms for the rotational errors.

We can summarize the above discussions, with respect to the X axis, as follows:

Errors in	are represented as	and are due to
X	δ_x	displacement in X
Y	δ_y	displacement in Y
Z	δ_z	displacement in Z
θ_x	ε_x	roll of X
θ_y	ε_y	yaw of X (roll of Y)
θ_z	ε_z	pitch of X (roll of Z)

And, for the rotational motions, the directions of the errors for the X axis relative to (6.2.1), we can similarly summarize these discussions as follows:

ERRORS	Vector			
		X	Y	Z
Direction	X	—	pitch (negative)	yaw (positive)
	Y	pitch (positive)	—	roll (negative)
	Z	yaw (negative)	roll (positive)	—

For the complete story we must add the displacements in X, Y, and Z. to this.

This yields a resultant HTM describing the error in position of the table with respect to its ideal location as

$$E_N = \begin{bmatrix} 1 & -\varepsilon_z & \varepsilon_y & \delta_x \\ \varepsilon_z & 1 & -\varepsilon_x & \delta_y \\ -\varepsilon_y & \varepsilon_x & 1 & \delta_z \\ 0 & 0 & 0 & 1 \end{bmatrix} \tag{6.2.8}$$

If we include the displacement of the axis as in (6.2.7), the actual HTM for the linear motion with errors of the table on the axis is then ${}^{R}T_{N_{error}} = {}^{R}T_N E_N$

$$
{}^{R}T_{N_{error}} = \begin{bmatrix} 1 & -\varepsilon_z & \varepsilon_y & a+\delta_x \\ \varepsilon_z & 1 & -\varepsilon_x & b+\delta_y \\ -\varepsilon_y & \varepsilon_x & 1 & c+\delta_z \\ 0 & 0 & 0 & 1 \end{bmatrix} \tag{6.2.9}
$$

We can derive a similar HTM for rotational errors occurring in, for example, a spindle or other revolute joint. First, it is necessary to define the problem and notation, Figure 6.4, from Slocum[18]. Radial displacements (linear) are made relative to the *X* axis. The *Z* axis is the axis of symmetry of the joint or spindle by definition and displacement can occur along the *Z* axis.

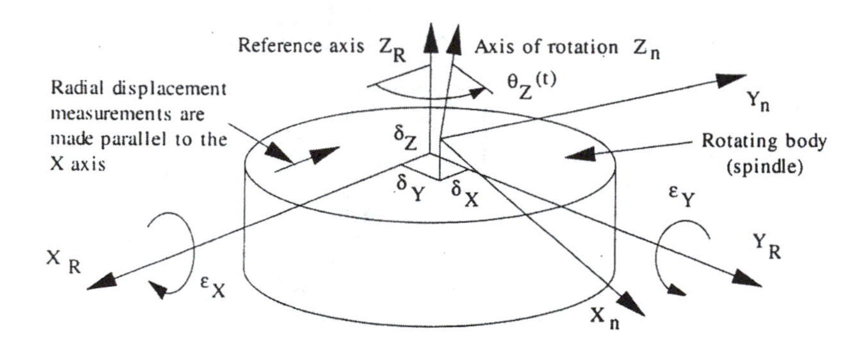

Figure 6.4. Motions and errors indicated for an axis of rotation, from Slocum[18].

Hence, the HTM representing the error relative to ideal of the spindle is comprised of two translational and three rotational components accumulated as in (6.2.5) and can be written as

$$
{}^{R}T_{N_{error}} = \begin{bmatrix} \cos\varepsilon_y \cos\theta_z & -\cos\varepsilon_y \sin\theta_z & \sin\varepsilon_y & \delta_x \\ \sin\varepsilon_x \sin\varepsilon_y \cos\theta_z + \cos\varepsilon_x \sin\theta_z & \cos\varepsilon_x \cos\theta_z - \sin\varepsilon_x \sin\varepsilon_y \sin\theta_z & -\sin\varepsilon_x \cos\varepsilon_y & \delta_y \\ -\cos\varepsilon_x \sin\varepsilon_y \cos\theta_z + \sin\varepsilon_x \sin\theta_z & \sin\varepsilon_x \cos\theta_z + \cos\varepsilon_x \sin\varepsilon_y \sin\theta_z & \cos\varepsilon_x \cos\varepsilon_y & \delta_z \\ 0 & 0 & 0 & 1 \end{bmatrix} \tag{6.2.10}
$$

In many cases the second-order terms are neglected (such as $\varepsilon_x \varepsilon_y$) or small angle approximations (such as $\cos\varepsilon = 1$ and $\sin\varepsilon = \varepsilon$) an be used. In these cases equation 6.2.10 simplifies to

$$^{R}T_{N_{error}} = \begin{bmatrix} \cos\theta_z & -\sin\theta_z & \varepsilon_y & \delta_x \\ \sin\theta_z & \cos\theta_z & -\varepsilon_x & \delta_y \\ \varepsilon_x \sin\theta_z - \varepsilon_y \cos\theta_z & \varepsilon_x \cos\theta_z + \varepsilon_y \sin\theta_z & 1 & \delta_z \\ 0 & 0 & 0 & 1 \end{bmatrix} \quad (6.2.11)$$

Slocum[18] points out that there are a set of specific standard terms and definitions for evaluating (that is measuring and discussing) the errors of rotating bodies in the ANSI B89.3.4M and these standard terms should be referred to in cases of confusion about terminology or in setting up an evaluation.

With these definitions and representations in mind, we are now ready to assess the relative error between the tool and workpiece in a machine tool. We should first consider the placement of the reference frame on the machine tool. To the extent possible it is desirable to locate the reference at the "roll center" of the machine or portion of machine of interest. This will reduce the likelihood of un-necessary rotational translations. Often the coordinate system (reference or component) is located at the origin of motion that will result in angular errors. If not, the pitch, yaw and roll errors can cause Abbé errors that must be incorporated into the model. It is often convenient, also, to place the coordinate system on the surface of the part so that measurements can be made more easily. Recall that for a reasonable error model (HTMs) it is necessary to be able to physically determine the location of the machine elements relative to the coordinate frames, for exampleone frame on the roll center and one on the surface of the part.

The procedure for determining the relative error between the tool (on the machine) and the workpiece surface is relatively straightforward and breaks in to a series of steps:

i. the kinematic model of the machine is determined based upon the individual HTMs of each of the critical machine elements (axes and major components, for example). Starting at the tool tip, each HTM is developed for the elements leading back to the reference frame of the machine located at a convenient point. Again, starting now at the correlative position on the work surface, each HTM is developed for the elements leading back to the reference frame of the machine. The first sequence of HTM's defines the relative position of the tool to the frame, ${}^{R}T_{tool}$. The second defines the relative position of the work surface to the frame, ${}^{R}T_{work}$.

ii. The difference between the two sequences of HTM's defines the error of the tool relative to the ideal work surface. The elements of the HTM's must be measured or estimated and will include, for example, encoder resolution, straightness data or machine specifications, cutting load induced tool deflection, typical tool setup error, work fixturing error, etc. The relative error HTM E_{rel} is determined from

$$
{}^{R}T_{work} = {}^{R}T_{tool}\ E_{rel} \text{ or}
$$

$$
E_{rel} = {}^{R}T_{work}\ {}^{R}T_{tool}{}^{-1} \tag{6.2.12}
$$

The vector **D** representing the translational component of E_{rel} (recall the displacements d_X, d_Y, d_Z from equation 6.2.1) represents the translation of the tool point's reference frame required so that the tool tip is on the work surface. If one is interested in compensating for the relative error between the tool tip and the workpiece, the machine controller would have to implement a correction vector based on the displacement vector **D** relative to the machine frame as follows:

$$
{}^{R}\begin{bmatrix} d_x \\ d_y \\ d_z \end{bmatrix}_{correction} = {}^{R}\begin{bmatrix} d_x \\ d_y \\ d_z \end{bmatrix}_{work} - {}^{R}\begin{bmatrix} d_x \\ d_y \\ d_z \end{bmatrix}_{tool} \tag{6.2.13}
$$

${}^{R}D_{correction}$ represents the incremental motions of the X, Y, and Z axes on a Cartesian machine necessary to compensate for machine errors in tool tip location. This may not, however, represent the position vector **D** if angular errors or Abbé offsets are present. Similarly, for additional degrees of freedom (for tool orientation relative to the work surface) **D** will not represent that orientation error.

iii. Finally, the error gain or sensitivity of each of the six error types (three rotational (ε_x, ε_y, and ε_z) and three translational (δ_x, δ_y, and δ_z.)) of a particular axis can be calculated. These can then be used as a multiplier for the error sources that contribute to the error component of the axis.

We will present here an example of the formulation of a two-axis Cartesian machine's error gain matrix from Slocum[18]. Figure 6.5 shows the location and orientation of the coordinate frames for the machine. One coordinate frame has been associated with each axis and the reference systems have their axes in parallel. So, we will determine the HTM describing the motion and errors of axis 2 (on top) relative to axis 1 (below) and then the HTM describing the motion and error of axis 1 relative to the reference coordinate frame. The locations of the axis coordinate frames relative to each other and the reference frame are detailed in the figure.

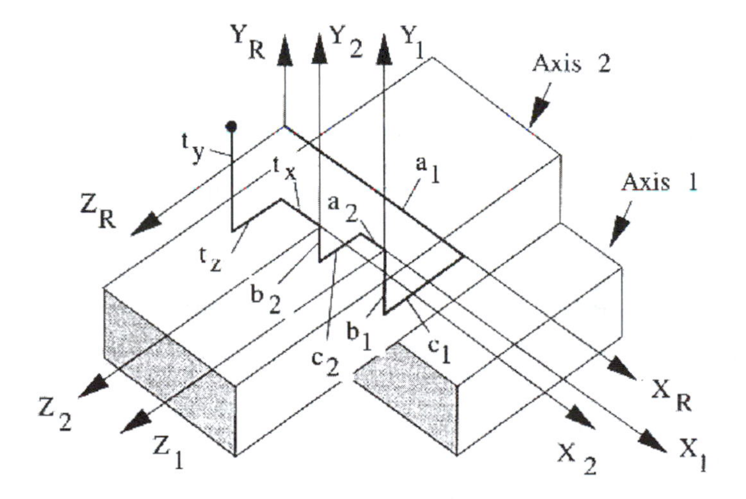

Figure 6.5. Coordinate frame definition for a two-axis Cartesian machine tool, from Slocum[18].

The errors in each axis are identical to those shown in Figure 6.3. Since the axes of the coordinate frames for the axes are parallel, the HTM given by equation 6.2.9 as follows:

$$^{R}T_1 = \begin{bmatrix} 1 & -\varepsilon_{z1} & \varepsilon_{y1} & a_1 + \delta_{x1} \\ \varepsilon_{z1} & 1 & -\varepsilon_{x1} & b_1 + \delta_{y1} \\ -\varepsilon_{y1} & \varepsilon_{x1} & 1 & c_1 + \delta_{z1} \\ 0 & 0 & 0 & 1 \end{bmatrix} \tag{6.2.14}$$

for axis 1 and

$$^{1}T_2 = \begin{bmatrix} 1 & -\varepsilon_{z2} & \varepsilon_{y2} & a_2 + \delta_{x2} \\ \varepsilon_{z2} & 1 & -\varepsilon_{x2} & b_2 + \delta_{y2} \\ -\varepsilon_{y2} & \varepsilon_{x2} & 1 & c_2 + \delta_{z2} \\ 0 & 0 & 0 & 1 \end{bmatrix} \tag{6.2.15}$$

for axis 2 relative to axis 1.

The actual coordinates of the tool point in the reference coordinate system are given by

$$\begin{bmatrix} X_t \\ Y_t \\ Z_t \\ 1 \end{bmatrix}_{actual} = {}^{R}T_1 {}^{1}T_2 \begin{bmatrix} t_x \\ t_y \\ t_z \\ 1 \end{bmatrix} \tag{6.2.16}$$

and the ideal coordinates of the tool point would be the sum of all of the individual components along their respective axes:

$$\begin{bmatrix} X_t \\ Y_t \\ Z_t \\ 1 \end{bmatrix}_{ideal} = \begin{bmatrix} a_1 + a_2 + t_x \\ b_1 + b_2 + t_y \\ c_1 + c_2 + t_z \\ 1 \end{bmatrix} \tag{6.2.17}$$

The translational errors in the tool point position are given by, then,

$$\begin{bmatrix} \delta_{xt} \\ \delta_{yt} \\ \delta_{zt} \end{bmatrix} = \begin{bmatrix} X_t \\ Y_t \\ Z_t \end{bmatrix}_{actual} - \begin{bmatrix} X_t \\ Y_t \\ Z_t \end{bmatrix}_{ideal} \tag{6.2.18}$$

If we evaluate the terms of equation 6.2.18 while ignoring the second order terms we get the following error gains:

$$\begin{bmatrix} \delta_{xt} \\ \delta_{yt} \\ \delta_{zt} \end{bmatrix}_{actual} = \begin{bmatrix} -t_y(\varepsilon_{z1} + \varepsilon_{z2}) + t_z(\varepsilon_{y1} + \varepsilon_{y2}) - b_2\varepsilon_{z1} + c_2\varepsilon_{y1} + \delta_{x1} + \delta_{x2} \\ t_x(\varepsilon_{z1} + \varepsilon_{z2}) - t_z(\varepsilon_{x1} + \varepsilon_{x2}) + a_2\varepsilon_{z1} - c_2\varepsilon_{x1} + \delta_{y1} + \delta_{y2} \\ -t_x(\varepsilon_{y1} + \varepsilon_{y2}) + t_y(\varepsilon_{x1} + \varepsilon_{x2}) - a_2\varepsilon_{y1} + b_2\varepsilon_{x1} + \delta_{z1} + \delta_{z2} \end{bmatrix} \tag{6.2.19}$$

These gains are the coefficients of $\delta_{x1}\,\delta_{y1}, \delta_{z1}, \varepsilon_{x1}, \varepsilon_{y1}, \varepsilon_{z1}, \delta_{x2}, \delta_{y2}, \delta_{z2}, \varepsilon_{x2}, \varepsilon_{y2}, \varepsilon_{z2}$, the errors. In a machine with a rotary axis, an angular error about one axis may have components about other axes. However, in these cases the gain associated with these errors will not represent an amplification by distance and, thus, would not appear in a table of error gains. The error gain matrix of this two-axis machine tool is, thus

	δ_{xt}	δ_{yt}	δ_{zt}
Axis 1 errors			
ε_{x1}	0	$-t_z-c_2$	t_y+b_2
ε_{y1}	t_z+c_2	0	t_x-a_2
ε_{z1}	t_y-b_2	t_x+a_2	0
Axis 2 errors			
ε_{x2}	0	$-t_z$	t_y
ε_{y2}	t_z	0	$-t_x$
ε_{z2}	$-t_y$	t_x	0

6.3 Error budget

6.3.1 Definition of error budget

An error budget can be defined simply as a list of all error sources and their effect on part accuracy. The error budget is a powerful tool for the precision machine designer as it can provide a rank order of sources of machine tool error for design optimization. It is similarly useful as an evaluation tool by placing upper bounds on the accuracy of existing machine tools. We consider here two categories of errors, random, so called, and systematic including hysteresis (recall the Taniguchi chart in Figure 1.2b). The basic assumptions on which an error budget is constructed are that

i. The instantaneous value of total error in a specified direction is the sum of all individual error components, i.e. linear superposition of errors.

ii. Individual error components have physical causes that can be isolated (identified), controlled, and measured to allow reduction and prediction.

This usually gives an upper bound estimate on the error.

In this discussion we rely on the following definitions for the error budget analysis:

- displacement error — the distance between the actual and ideal feature on the workpiece; we have used the term "bias" for this earlier. Displacement errors are always measured in a sensitive direction.
- error source — the physical cause of a displacement error.
- coupling mechanism — physical "mechanism" such as stiffness, thermal coefficient of expansion, damping which may have a filter effect on amplitude or frequency.

6.3.2 Error budget flow chart

Donaldson[45] detailed a flow chart for generating an error budget for a two-axis lathe as seen in Figure 6.6. This is similar to Figure 2.4 except that there are weights applied to the interconnections and the error directions are present representing distinct axes of errors with respect to the workpiece features. In Figure 6.6 the error sources, S_i, correspond to the physical causes of the error (thermal error, for example). The coupling mechanism, C_i, is equivalent to a directional factor (to be detailed in Chapter 7 following) which determines the magnitude of the error in a "sensitive direction" — and here referred to as a displacement error, E_i. We know that the displacement of the tip of a cutting tool, lathe tool for example, perpendicular to the work surface in turning (the sensitive direction) will generate an error in the diameter of the part (the workpiece geometry). Workpiece error categories are size (diameter, length), form (roundness, straightness and flatness, for example) and surface finish (with or across the lay direction) and are represented by E_{iA} for the i-th component of the A category. Finally, the error directions shown are either parallel or perpendicular to the machine axis, X and Z for the two axis lathe illustrated here, resulting in components E_{iAX} and E_{iAZ} for the X and Z axis elements of component A. Then, the application of a combinational rule will result in resultant displacement errors, E_{AX} and E_{AZ}. These appear as workpiece errors E_{WA} through E_{WC} for the three workpiece error categories illustrated here, dimension, form and surface roughness, for example.

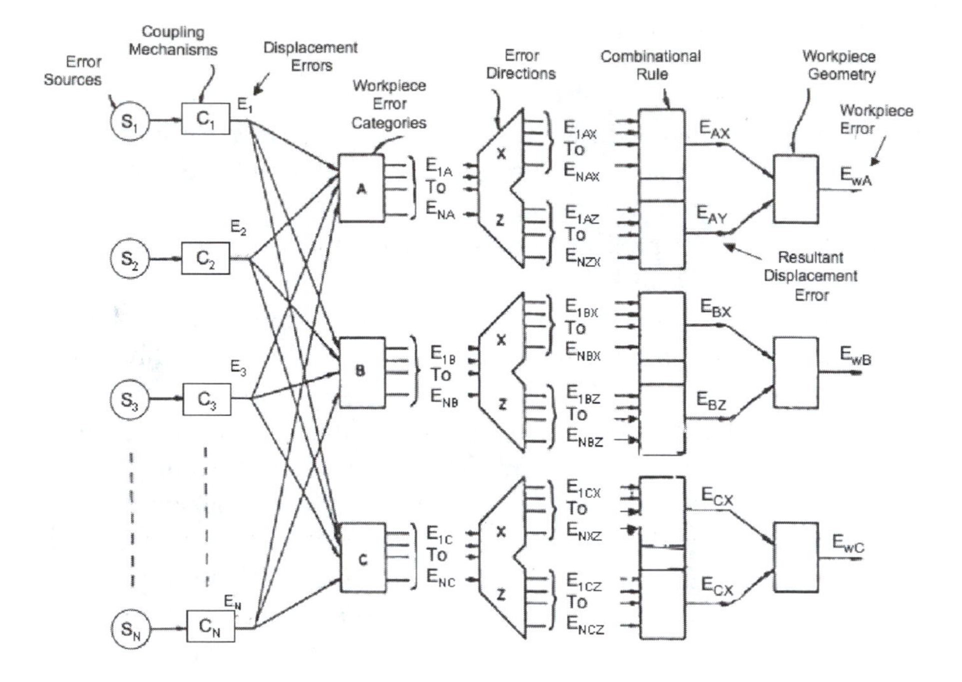

Figure 6.6. Flow chart for generating an error budget for a two-axis lathe, from Donaldson in MTTF[45].

6.3.3 Combinational rules for errors

As stated earlier we consider three types of errors defined as follows:

- random — under equal conditions at a given position the error doesn't always have the same value.
- systemmatic — error is always the same value and sign at a given position under given circumstances.
- hysteresis — systematic error, highly reproducible, but with sign equal to the direction of approach.

In the application of combinational rules, i.e. methods for combining and determining the amplitude, for individual error values (magnitude only — they are assessed for each axis of concern) we must chose between the largest isolated value (peak-to-valley) or a root mean square (RMS) amplitude. We have seen both of these examples in the representation of surface roughness. Then, having either, we can make the combination.

We consider the example or RMS here. The RMS_{total} can be determined as follows:

$$RMS_{total} = \left[\sum_{i=1}^{N} (RMS_i)^2 \right]^{1/2} \tag{6.3.1}$$

where RMS_i is the value of the displacement error *for each error source*, for each axis for example. The peak to valley error value for a source is related to the RMS_i value by a constant depending on the probability distribution of the error. Hence,

$$PV_i = K \bullet RMS_i \tag{6.3.2}$$

where K equals 2.83 for a sinusoidal distribution, 3.46 for a uniform probability density and 4 for $\pm 2\sigma$ Gaussian distribution. If the peak-to-valley information is known then the RMS_{total} value can be determined as

$$RMS_{total} = \frac{1}{C} \left[\sum_{i=1}^{N} (PV_i)^2 \right]^{1/2} \tag{6.3.3}$$

where C depends upon the distribution.

These combinational rules are applied to the data to render the resultant displacement errors seen in Figure 6.6. That is,

$$E_{1Ax} \bullet\bullet\bullet E_{NAx} \Rightarrow E_{Ax}$$

$$\bullet$$

$$\bullet$$

$$\bullet$$

$$E_{1Cz} \bullet\bullet\bullet E_{NCz} \Rightarrow E_{Cz}$$

This exercise is done for all major axes and sources of error as detailed in Figure 6.6. The errors are then summarized to determine a "worst case" (a.k.a. upper bound) and "best case" (a.k.a. lower bound) error. In both cases the error is comprised of the contributions from the systematic, hysteresis and random sources. Worst case is

$$\varepsilon_{worstcase} = \sum \varepsilon_{systemmatic} + \sum \varepsilon_{hysteresis} + 4\sum \varepsilon_{random} \tag{6.3.4}$$

The first two components, systematic and hysteresis are evaluated at the one standard deviation, 1σ, level while the random error is evaluated at the four standard deviation, 4σ, meaning that the chance that an error will not exceed the 4σ limit is better than 99.9937% likelihood.

And the best case error is represented as

$$\varepsilon_{bestcase} = \sum \varepsilon_{systemmatic} + \sum \varepsilon_{hysteresis} + 4\left(\sum \varepsilon^{2}_{random}\right)^{1/2} \tag{6.3.5}$$

In many cases it is sufficient to average the two. Table 6.1 shows the calculation of the surface finish error budget for the large optics diamond turning machine (LODTM) at Lawrence Livermore National Laboratory. The 11 error sources identified and their magnitudes *affecting the surface finish* are listed. The RMS value of the effect is 4.2 nanometers (0.17 μin.) or 42 $\overset{o}{A}$ RMS.

Table 6.1. Surface finish error budget of the LODTM, from Donaldson in MTTF[45].

Error source	peak-to-valley amplitude, nm (μin)
external mechanical disturbances	5 (0.20)
airborne noise	2.5 (0.10)
hydraulic noise	2.5 (0.10)
spindle drive	2.5 (0.10)
spindle air pressure	2.5 (0.10)
theoretical finish	2.5 (0.10)
cutting mechanics/tool edge	4 (0.16)
MCU resolution	5 (0.20)
servo tool mount	10 (0.40)
metrology phase distortion	2.5 (0.10)
refractive index diff.	3.5 (0.14)
sum of squares	215.75 (0.3452)
	RMS = 4.2 nm (0.17 μin)

VII ERROR DUE TO COMPLIANCE AND VIBRATION

7.1 Introduction

We have, in past chapters, looked into various sources of error and in the last chapter developed a methodology to allow the estimation of the magnitude of an error source and its contribution to dimension, form or surface characteristics of a workpiece. In general, the error terms considered were "macroscopic" in nature. That is, they dealt with the kinematic errors in machine axes and components, thermal errors as well as system error, for example, the resolution of an encoder feeding back positional information. In this chapter we will delve a little deeper into other sources of error in manufacturing – those due to structural compliance and vibration (or chatter).

This is a complicated topic and, as many of the other topics covered here, can take many different directions depending on the focus of the discussion. Basically, we are addressing the static and dynamic stiffness of machines. That is, static deflection under the load of machining (and the sensitivity of the position of the load in the machine tool loop) and dynamic deflection under the varying periodic as well as non-periodic influence of the manufacturing process and other machine components. Since all machines are elastic structures at some level of observation, the effect of these static and dynamic forces and influences will be seen in the dimension, form or surface characteristics of a workpiece. For example, Figure 7.1, from Tönshoff[46], shows the chatter marks in a workpiece processed on a cylindrical grinder. The wavy surface marks, as indicated in the roundness profilometry readout, show the results of vibration during machining creating undesirable surface features.

During the design process for a machine, consideration must be given to such things as required working space, number of axes and travel, power and speed (spindle as well as translation) and structure. It is the structure that concerns us here.

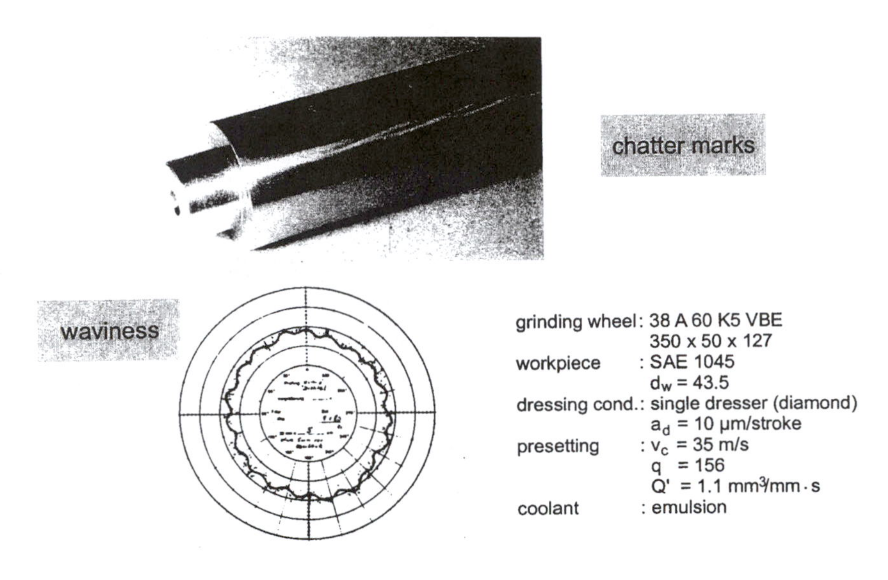

Figure 7.1. Chatter marks in grinding, from Tönshoff[46].

The design of machine structures (from machine tools to lithography machines) is influenced by performance criteria that can be summarized in the following four categories, from Tlusty[47]:

i. weight deformation – changes in the form or shape of the structure as a result in movement of workpieces or machine components (headstocks, tables, etc.)
ii. cutting force deformation – since the machine tool (in cutting) acts as a "structural loop," the forces applied to this loop will cause deformation. Hence, when the cutting process generates forces during chip formation, those forces will cause deformation of the loop structure. Similarly, variation in cutting forces, as

with changes in depth of cut of width of cut along the tool path, will cause variation in deformation. The force variation can be separated into two different classes:

a. Type A – cutting force varying with depth of cut, and
b. Type B – relative stiffness between tool and workpiece varying

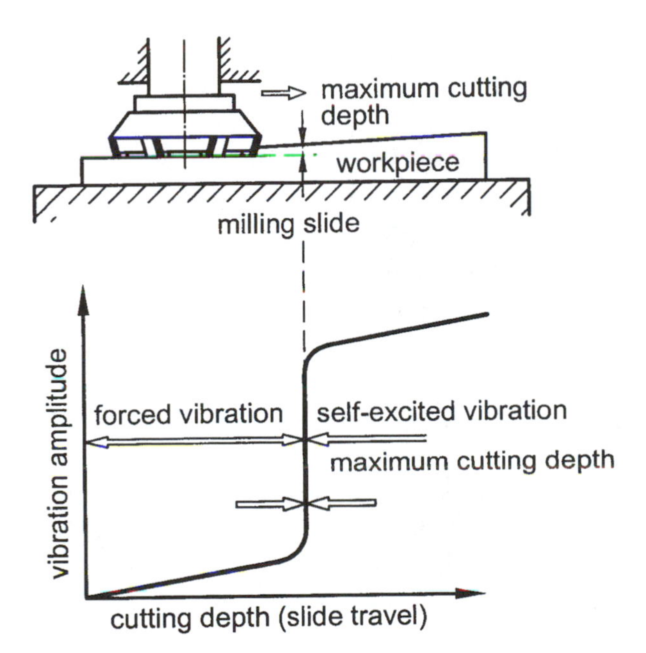

Figure 7.2. Illustration of forced and self-excited vibration sensitivity to machining conditions, from Tönshoff[46].

iii. forced vibrations – periodic forces due to machine action (rotation of slightly eccentric component, for example) or from external environmental sources

iv. self-excited vibrations (chatter) – conditions within the machining process can generate vibrations. This is usually more prominent at heavier depths of cut, Figure 7.2. The classic example is the vibration due to regeneration of waviness on a machined surface (so-called wave removal). Cutting force and self-excited vibration are not common outside of machining processes and machinery whereas the other two deformations are found in other machine structures during operation.

Lest we focus too much on machine tools in our discussion, Figure 7.3 below shows a schematic of a commercial wafer stepper lithography tool[48]. This figure shows a machine structure similar to many machine tools except that the "spindle" is replaced with an optical column of lens and reticle masking carrier, the table is a "wafer stage" for handling the wafer during exposure and the load/unload system typical of many automated machine tools is a wafer handler. One significant difference is that there is no contact between the column and the wafer. Note the vibration damping elements attempting to isolate the machine from environmental vibration sources (floor vibration, for example) and the "active motion compensation" which is attempting to offset vibration induced by motion of the wafer stage between exposures (a type of forced vibration).

We cannot attempt here to cover all the subtleties of dynamic behavior of machines and their components. We will attempt to illustrate some of the major principles underlying the performance of machines and how their performance is affected by the process. In Chapter 11 when we discuss some applications, we will introduce more detail about machines such as those used in lithography processes.

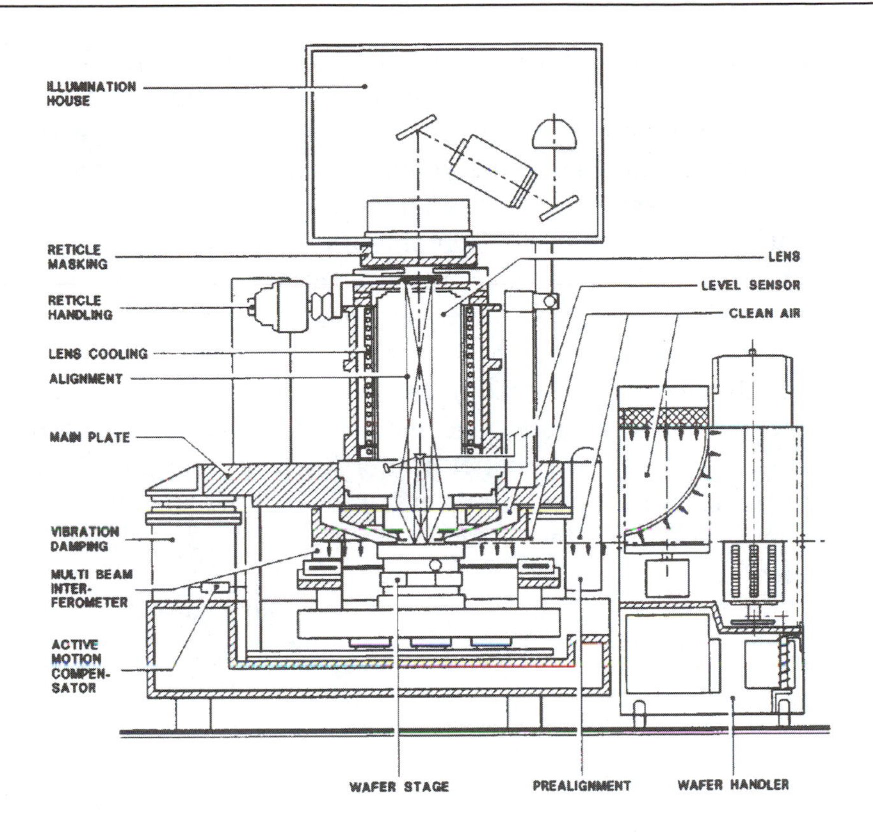

Figure 7.3. Schematic of wafer stepper lithography tool, van den Brink [48].

7.2 Excitations in machine tools

Even though we may try to design structures so that they have the widest possible operating ranges, it is not possible to prevent some excitation due to machine or process operation. But, we can attempt to restrict the undesirable behavior to conditions that are transitory or not normally encountered in operation. This technique is employed in the design of most machines and structures from aircraft engines to bridges. What are some of the sources of excitation in machines that we are concerned about? Figure 7.4, from Tönshoff [46] shows some of the typical excitation types in a conventional machine tool and Figure 7.5, also from Tönshoff[110], gives more specifics as to the sources of the excitation. Basically, any rotating part that is likely to have any imbalance (by poor design, construction, or operation) as well as components that induce or experience periodic loads or impacts in operation are excellent sources of vibration. We restrict our discussion here to machining processes but one can imagine similar situations in other manufacturing machinery such as presses (with tremendous "once per stroke" vibration), injection molding machinery (with either pressure pulses (if hydraulic) or rotational vibrations if screw fed injection), all the way to lithography machines where short, fast stage motions can induce inertia loads and vibration (watch your inkjet printer lurch back and forth along with movement of the print head next time you print something!)

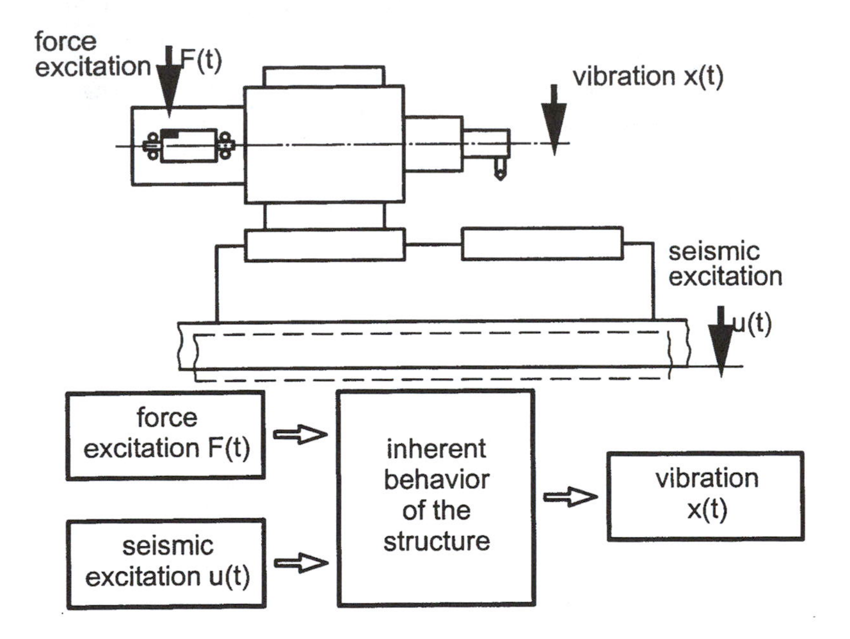

Figure 7.4. Types of excitation in machines, from Tönshoff[46].

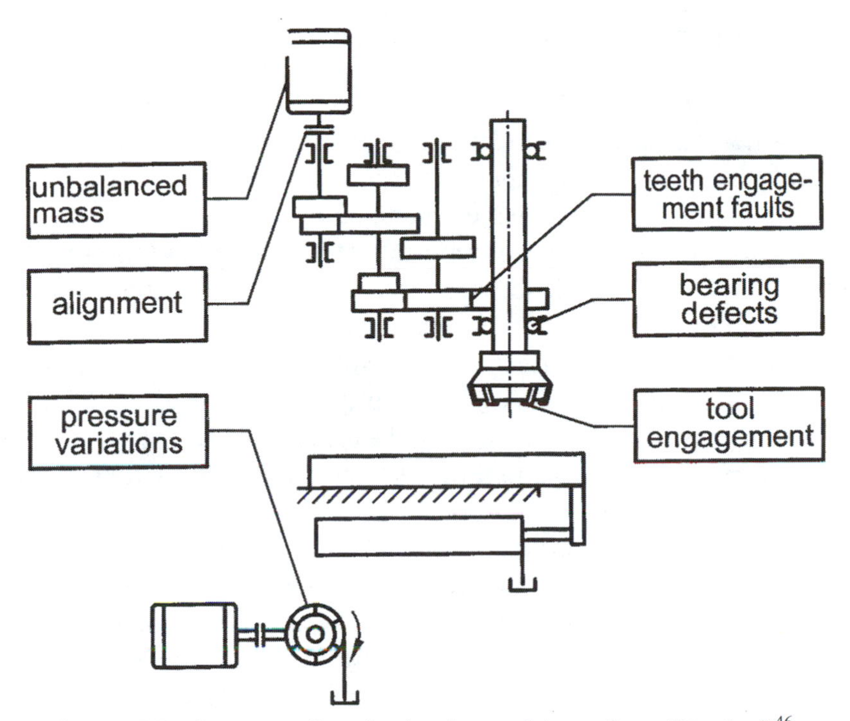

Figure 7.5. Sources of excitation in machines, from Tönshoff[46].

Solutions to the problems of excitation rely on either altering the machine to inhibit vibration generation (for example, by improving pump performance to minimize pressure variations or insuring that rotational components (like shafts) are sufficiently balanced and aligned), altering the process (by changing process parameters (spindle rotational speed, or tooth engagement, for example) or adding compensation to offset the potential vibration source (hydraulically damped boring bars or "adaptively" tuned grinding wheels.) These can often increase the range of performance so that production specifications can be met.

Our focus here is to discuss some of the more common sources of compliance and vibration induced errors in machines. The following sections cover the four basic categories of deformation (that is static errors) and vibration (or dynamic errors) introduced above.

7.3 Weight deformation

Our earlier discussion in Section 4.4.2 on macro scale compliance has covered most of the important issues with respect to deformation due to the weight of machine elements and, if part of a moving table or spindle carrier, the variation on the deformation with that motion. While Section 4.4.2 discussed foundation displacement due to weight of components and compliance of joints, ways and bearings, we add here only the effect on the workpiece of weight deformations. Figure 7.6, from Tlusty[47], illustrates, for several machine tool configurations, the effect of weight deformations. In Figure 7.6a, the cross support rail of a plane mill (as used for milling spar structures in aerospace products) both bends and twists due to the manner in which the head is mounted on the rail. Pay special attention to the effect on the plane of the surface of the workpiece of these deformations. The component deformation can result in non-planar surface creation (from bending) as well as more localized surface finish effects due to the non-perpendicular cutting tool axis orientation in Figure 7.6d, for example, due to torsion. The distinction here is between more global effects and local surface roughness effects due to the "scalloping of the surface" by a misaligned cutting tool.

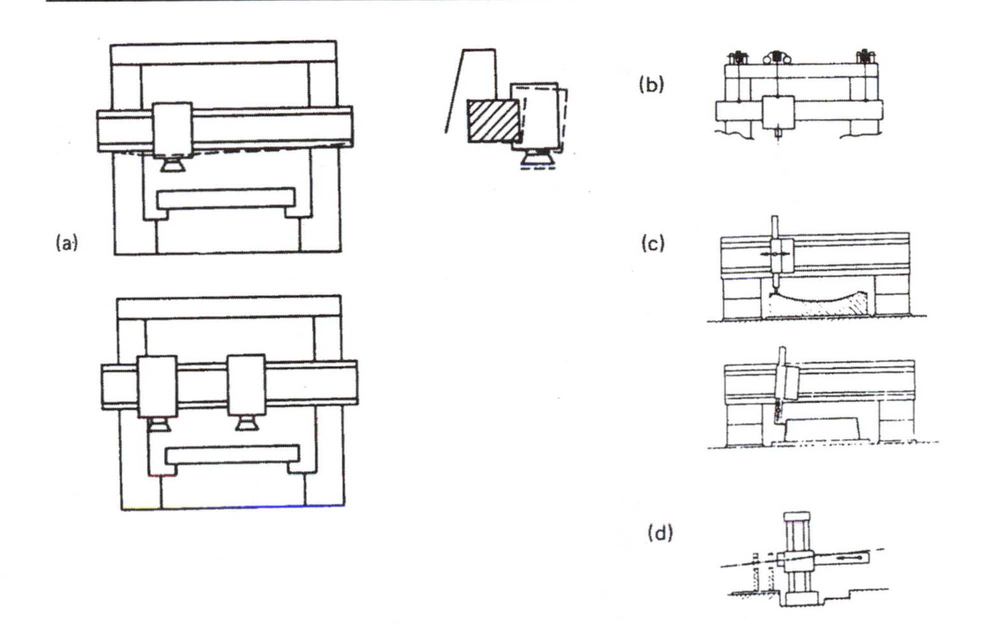

Figure 7.6. Weight deformation effects on a number of machine tools, a) plane milling machine, b) jig boring machine, c) vertical boring machine, and d) horizontal milling machine, from Tlusty in MTTF[47].

An extreme case is seen in Figure 7.6d with a machine tool with a horizontal "cantilevered" spindle support arm. As expected, movement of the spindle head towards and away from the machine column will cause varying deviation in the position of the spindle head.

There can be an interaction between the shifting weight of a machine during operation, its foundation and resulting structural errors due to deformation, Figure 7.7, from Tlusty[47].

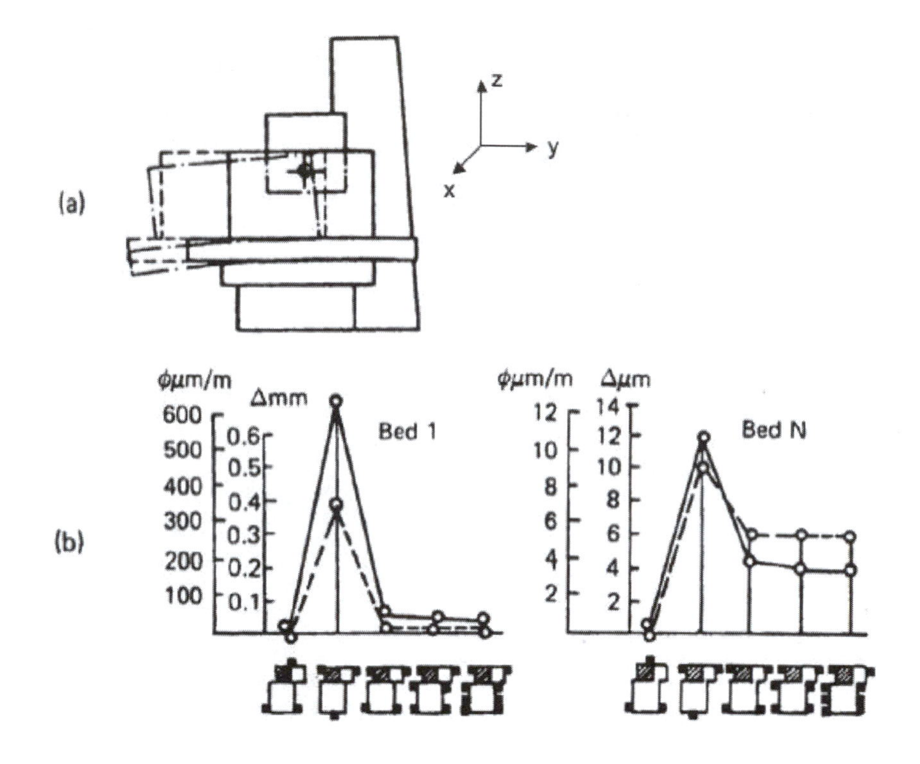

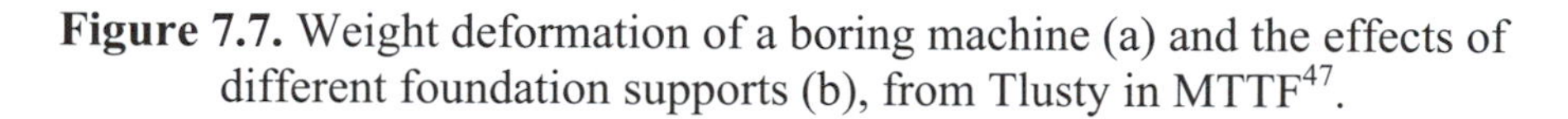

Figure 7.7. Weight deformation of a boring machine (a) and the effects of different foundation supports (b), from Tlusty in MTTF[47].

The table deformation due to motion along the y axis will result in varying distances and orientations of the tool relative to the workpiece in x-y (usually a cylindrical feature being produced). The machine saddle deforms and the bed twists. As seen in Figure 7.7b, depending on the number and location of support pads for the machine base, the machine will experience deviations between the spindle and workpiece (changing with the z-location). The deviations plotted in the figure as ϕ μm/meter were measured between the spindle and workpiece at a height of 1 meter above the table surface. These deviations were induced by mounting a heavy workpiece on the table and moving the table to the extreme +y and –y positions.

Ideally, only three pads should be used to maintain kinematic support. But, some of the most dramatic deviations are seen with only three pads depending on where they are placed under the machine base. For example, in “bed 1”, we notice that for the first two configurations (2 in front/1 in rear and 1 in front/2 in rear) there is a tremendous difference in deviation. Tlusty points out that in the first case, the moment and twist due to the eccentric weight are transferred through the bed to the column with essentially no deformation between spindle and workpiece. In the second case, the moment is counteracted at the column and the bed has a substantial twist. When more supporting pads are used (non-kinematic), specially for larger machines indicated as “bed N”, there is always some twist depending on the stiffness of the bed, the foundation and how they are connected.

This illustrates the importance of the stiffness of the relative components of the machine tool relative to the supports on the foundation, If the behavior of the machine under such shifting eccentric loads is repeatable, this information (as in Figure 7.7) can be used to determine error budgets for the machine and product specifications.

7.4 Cutting force deformation

This section is substantially the material given by Tlusty[47] in the MTTF study report cited earlier for deformations under cutting forces. It has been edited and additional materials added. The theoretical basis for this criteria was developed significantly from *Structure* of Machine Tools (Kongsberger and Tlusty, Pergamon Press 1971). The theoretical background is presented here as the significance of the impact on performance assessed. Tlusty’s original presentation is masterfully done and very instructive.

The cutting force produces deformation between tool and workpiece. If this deformation remains constant throughout the machining operation a workpiece of accurate form could be obtained. However, if it varies, as it usually does, this variation causes an error of form of the workpiece.

The deformation X depends on the cutting force F and on the stiffness k_a between the tool and the workpiece as in a simple elastic spring equation:

$$X = \frac{F}{k_a} \tag{7.1}$$

The definition of the stiffness k_a will be given later. First, it can be seen that a variation of X is caused by either:

A. a variation in the cutting force F due to the varying depth of cut, or
B. a variation of stiffness k_a along the tool path.

Often both variations A and B occur simultaneously. However, we will consider them separately in this discussion as the effects can always be superimposed.

7.4.1 Type A deformation: Deformation due to the variation of the cutting force

7.4.1.1 Introduction and background

The raw material of a workpiece entering a machining operation does not have an exact geometric form due to prior processing. In the case of a cylinder, its sections will not necessarily be exactly round; longitudinally it maybe "barrel" shaped or conical, and, due to inaccurate clamping, it will be rotating with a "run-out," e, as shown in Figure 7.8.

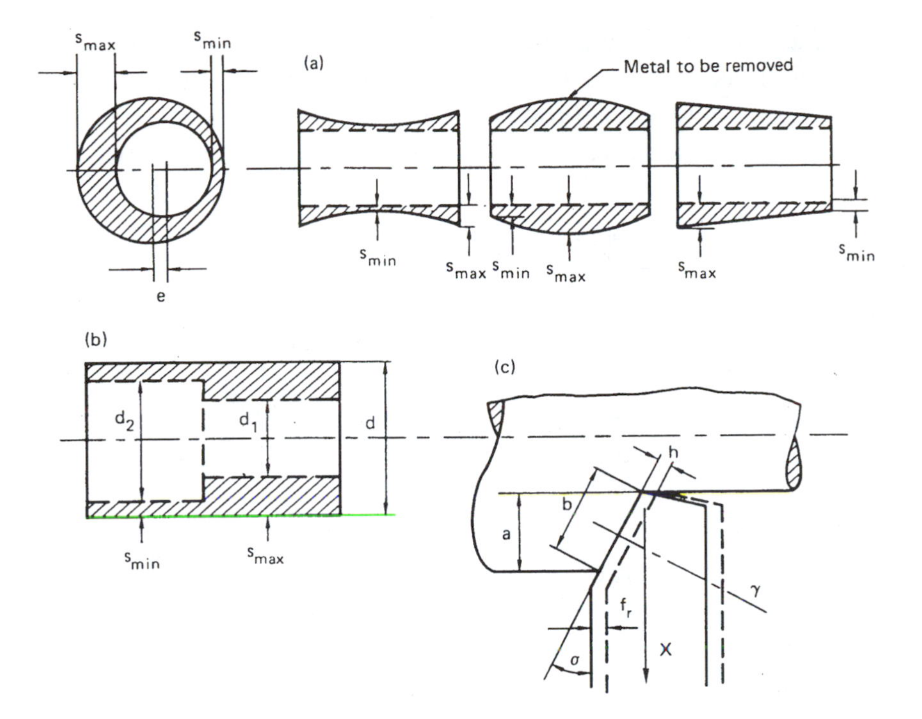

Figure 7.8. Illustration of imperfect work raw material and variations in amount of work material to be removed (depth of cut), from Tlusty in MTTF[47].

Similarly, a surface to be face milled will not initially be perfectly flat. Or, it may be that in one turning cut the tool path is so programmed that the initial diameter d will be reduced to two different diameters d_1 and d_2 as shown in Figure 7.8b.

In all the above examples, the cutting force varies during the cut. Consequently, deformation between the tool and workpiece also varies and initial form errors will be "copied" onto the machined surface.

In Figure 7.8c the cutting force, *F*, is a function of the chip width, *b*, and chip thickness, *h*. This relationship can be expressed empirically as

$$F = Cbh^m \tag{7.2}$$

where C is a material constant and m is an exponent smaller than 1 (we can assume m = 0.85 here) and is basically a "geometry" constant. The chip width b is proportional to the depth of cut a and h is proportional to feed per revolution f_r (as shown in Figure 7.8c)

$$b = \frac{a}{\cos\sigma}, \quad h = f_r \cos\sigma \tag{7.3}$$

where σ is the side cutting edge angle (SCEA). Thus the cutting force is proportional to the depth of cut a as

$$F = \cos^{m-1}\sigma f_r^m a \tag{7.4}$$

The coefficient of proportionality r_a between a and F is called the "cutting stiffness" with respect to the depth of cut and will vary with different materials (proportional to hardness, usually). We can determine r_a as

$$r_a = C\cos^{m-1}\sigma f_r^m \tag{7.5}$$

or

$$F = r_a a \tag{7.6}$$

Now, to determine the effect of this force on deformation of the machine or workpiece, we need to consider the directions in which the force acts as well as directions of likely movement (distortion) of the machine. The cutting force has a direction that depends on the angle σ and lies in a plane γ that is approximately perpendicular to the cutting edge, and, therefore, has a larger tangential component (in the direction of the cutting speed) than normal component. This force causes a deflection of the system. The component X of this deflection is normal to the surface behind the cut. It is this deflection component, which affects the dimension, and hence the geometric form, of the machined workpiece. We've previously defined this as the "sensitive" direction.

Therefore, a particular stiffness k_a related to the Type A effect discussed here is defined as a ratio between a force F acting in the direction of the cutting force and a deflection in the direction X, as expressed in equation 7.1.

The ratio of cutting stiffness r_a to the machine stiffness k_a has a special significance and we can denote it μ:

$$\mu = \frac{r_a}{k_a} \tag{7.7}$$

Referring now to Figure 7.8, note that the actual depth of cut a is the difference between the stock s to be removed and the actual deflections at each point of the cut, or

$$a = s - x \tag{7.8}$$

If we combine equations (7.1), (7.4), (7.6) and (7.7) we have

$$X = \frac{\mu}{1+\mu} s \tag{7.9}$$

The maximum form error, Δ, of the part before the machining operation starts is (see Figure 7.8a)

$$\Delta = s_{\max} - s_{\min} \tag{7.10}$$

and the maximum error after the operation is

$$\delta = \frac{\mu}{1+\mu} \Delta \tag{7.11}$$

The ratio

$$i = \frac{\delta}{\Delta} = \frac{\mu}{1+\mu} \tag{7.12}$$

expresses the "rate" of improvement of the form error of the workpiece in one pass. This ratio depends on the feed used (see equation 7.4). If several passes, say m, are taken as part of the machining process, the total improvement is then

$$i = i_1 \text{ x } i_2 \text{ x } \ldots \text{ x } i_m \quad (7.13)$$

If in all the passes the same feed i_1 were taken, then the total improvement would be

$$i = i_1^m \quad (7.14)$$

Thus, if the improvement in one pass is $i = 0.01$, after three passes an error is left which is only $\delta = 0.000001\Delta$.

We will review some examples of the practical implementation of this analysis for three typical machining processes.

7.4.1.2 Examples for single edge cutting

We start out considering the cutting of steel (common in conventional manufacturing; not at all common in precision manufacturing) for three processes: milling, grinding and turning. First for turning, an average value of C will be assumed as 1500 N/mm^2. The equations above are suitable for turning as shown in Figure 7.8c. Consider first a SCEA, σ, of 30° and three different feed values, giving cutting stiffness as shown:

Finishing cut: $f_r = 0.1$ mm, $r_a = 217$ N/mm
Averaging cut: $f_r = 0.25$ mm, $r_a = 472$ N/mm
Roughing cut: $f_r = 0.5$ mm, $r_a = 850$ N/mm

In milling, essentially equation 7.2 can be used by considering the "average" chip thickness, h_{av} (to be discussed in more detail later, Chapter 11) during the cut of each tooth and taking the feed per tooth f_t equal to the above feeds per revolution f_r, and multiplying by the number of teeth cutting simultaneously. This last determination can be a bit tricky for milling as often a number of teeth are engaged with the workpiece but not fully cutting (that is chip load different than "average") due to helix angle and engagement. For the same $\sigma = 30°$, a face milling cutter 150 mm in diameter with 12 teeth and width of cut (axial engagement" 80 mm has a corresponding cutting stiffness 3.27 times higher.

In grinding, the situation is different from turning and milling. The cutting efficiency of the grinding wheel is much less than in cutting with defined edges; specific power (per mm^3/min removal rate) is about 10 times higher. Also, this width of cut is rather large, see Figure 7.9a, from Tlusty[47], for plunge grinding and, Figure 7.9b, for grinding with tangential (transversal, longitudinal) feed. The depth of cut a usually very small, on the order of 10 μm, and a small

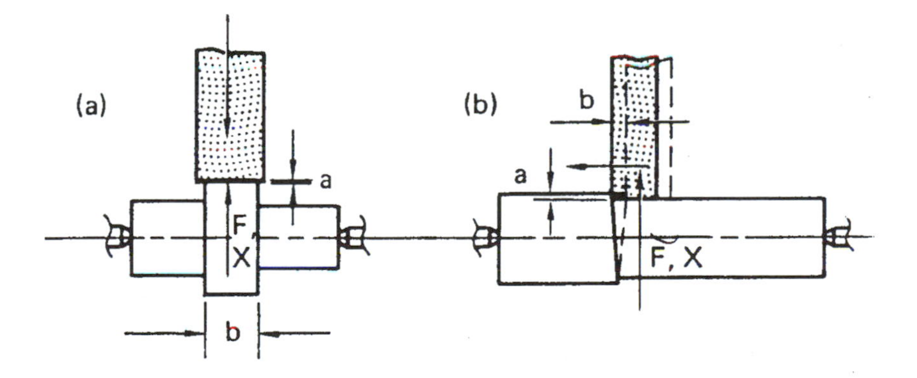

Figure 7.9. Definitions of chip dimensions in plunge grinding (a) and grinding with tangential feed (b), from Tlusty in MTTF[47].

change of a represents a large relative change of force. For example, displacement in the direction X by 10 μm represents a 100% change of force.

The cutting stiffness $r_a = {}^{F}\!/_{a}$ for a 10 μm wide plunge grinding operation is typically:

for external grinding, $r_a = 5{,}000$ N/mm
for internal grinding, $r_a = 50{,}000$ N/mm

Note that these values are 10 to 100 times higher than for turning. Therefore, the effect of cutting force deformations on accuracy is very significant in grinding.

Although specific machine tool components have specific, measurable stiffness, we will start by considering a typical average value of this stiffness to be 10 N/μm, or 10,000 N/μm. Then, we may calculate average values of the stiffness ratio μ to be:

for turning, $\mu = 0.05$
for milling, $\mu = 0.16$
for grinding, $\mu = 2$

Given the small values of μ for turning and milling, we can simplify with reasonable approximation equation 7.11 for these processes as

$$\frac{\delta}{\Delta} = \mu \tag{7.15}$$

7.4.1.3 Machine stiffness and directional orientation

We will now analyze the character and typical values of the machine stiffness k_a and, specially, of the directional orientation involved. As explained earlier, the particular stiffness denoted as k_a relates the cutting force to the *X* component of the relative displacement between tool and workpiece. The direction *X* is normal to the surface behind the tool – the sensitive direction.

Machine stiffness depends upon the configuration of the machine tool an on the position of the cutting process in the machine tool structure. As this position changes during the tool motion, stiffness k_a varies during the cutting operation. The effect of variation in k_a will be considered in the next section. For a periodically varying force, the stiffness k_a can be considered to be a "dynamic" stiffness ratio between force and deflection. Usually, however, the variation of cutting force is slow and k_a can be considered a "static" stiffness. In this section we will consider the stiffness k_a for a position of the tool where, most interestingly, k_a is at a minimum.

The directional orientation of the cutting process in the machine is involved in a double sense:

i. the structure is usually most flexible in a certain direction; dynamically, the system only deforms in certain "modes." Depending on the direction of cutting, the projection of the force in this sensitive direction many be large.
ii. The deflection of the system has to be projected into the sensitive direction X to be effective.

We can illustrate this by using the example of a lathe, Figure 7.10, from Tlusty[47]. The cutting force F acts between the tool and the workpiece, Figure 7.10b, in a plane perpendicular to the cutting edge, which itself has a SCEA σ as seen in the figure. The

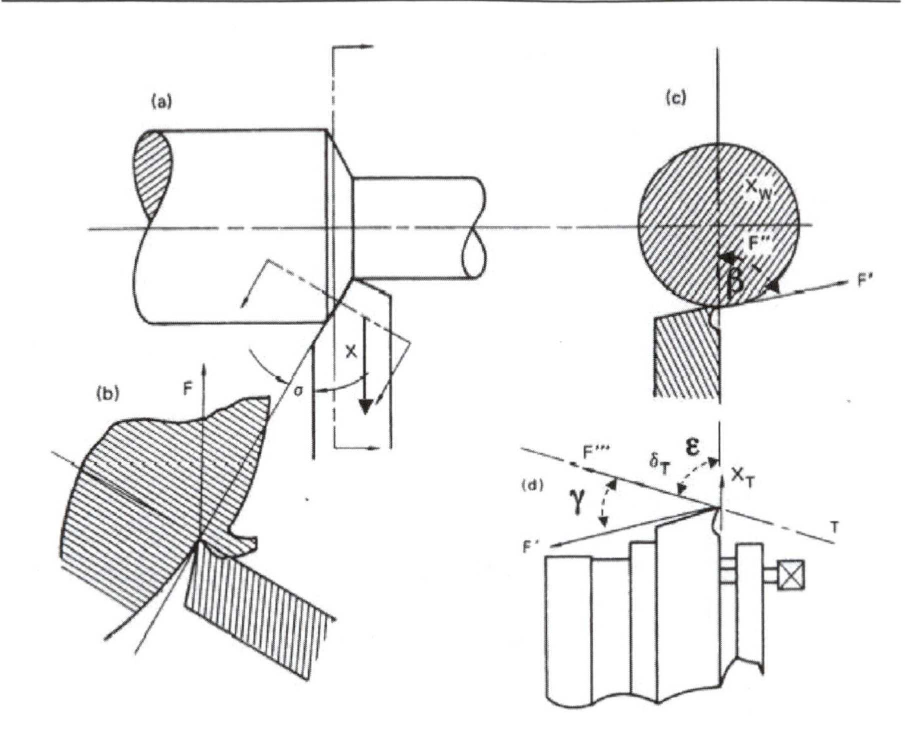

Figure 7.10. Direction of forces and resulting deformation on a lathe, from Tlusty in MTTF[47].

decisive direction X of the deformation (Figure 7.10a) affects the dimension and form of the surface behind the tool. The sensitive direction of the displacement of the workpiece is any radial direction. Force F has a component F' in this plane ($F' = F\sin\sigma$), Figure 7.10c, and another component F'' in the direction (and at an angle β with F') X, which produces the displacement X_w. The action of the force F' on the tool is depicted in Figure 7.10d. Depending on the design of the slides and of the toolholder, the sensitive direction of the deformation of this part may be the direction T. Force F' is projected into F''', which produces the displacement δ_T with a component X_T in the direction X. In the diagram this displacement is shown to be "negative" – into the workpiece. Most often it will be positive – out of the workpiece. The resulting displacement will be the sum

$$X = X_w + X_T \tag{7.16}$$

If the "direct stiffness" on the workpiece and tool are:

$$k_w = \frac{F''}{X_w} \tag{7.17}$$

(force acting in direction X, displacement in direction X)

$$k_T = \frac{F'''}{\delta_T} \tag{7.18}$$

(force and displacement in direction T)

then the decisive stiffness k_a would be expressed by forces and displacements in terms of its geometric angular components as follows

$$F'' = F \sin\sigma \cos\beta \tag{7.19}$$

$$X_w = \frac{F''}{k_w} \tag{7.20}$$

$$F''' = F \sin\sigma \cos\gamma \tag{7.21}$$

$$\delta_T = \frac{F'''}{k_T} \tag{7.22}$$

$$X_T = \delta_T \cos\varepsilon \tag{7.23}$$

and, finally

$$X = F \sin\sigma \left(\frac{\cos\beta}{k_w} + \frac{\cos\gamma \cos\varepsilon}{k_T} \right) \tag{7.24}$$

Obviously, the SCEA angle σ plays an important role and for σ = 0 there will be almost deflection X. In reality, there are also some smaller flexibilities in directions other than those indicated, and even with σ = 0 the force has an X component due to the nose radius of the tool and therefore a small X-deflection. Note, however, that the tool deflection does not depend on only the *X* component of the force.

The lengthy example of the lathe should indicate the importance of the directional orientation. We will not discuss in detail the directional orientation and the corresponding value of machine stiffness k_a for all types of machine tools, but rather survey a few other examples. In Figure 7.11 an example of a vertical milling machine, used for face milling, is given.

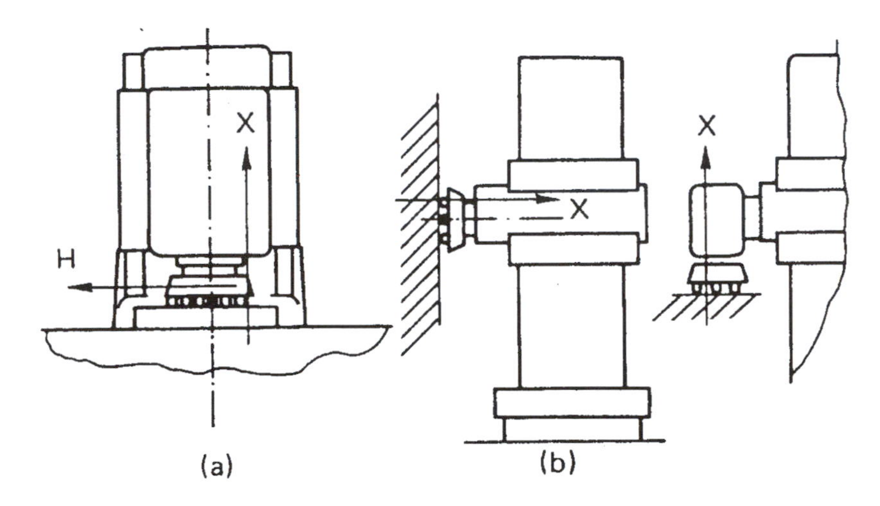

Figure 7.11. Orientation of flexible and decisive directions a) on a vertical milling machine and b) on a floor type milling and boring machine, from Tlusty in MTTF[47].

The horizontal direction *H* is the flexible direction most sensitive to the bending of the column, the twist of the attachment between headstock and column, and the bending of the spindle. Thus, the flexible direction is almost perpendicular to the decisive (sensitive) direction *X*. In Figure 7.11b face milling by a floor type machine is shown. In this case, due to the flexibilities of the ram of the machine, of the column (at its base), and of the vertical head attachment, there may be significant deflection in the decisive direction *X*. In Figure 7.12a the cases of a boring bar and boring spindle are shown. The effect of the side-cutting edge angle here is similar to the case of the lather. The main concern here is the inevitably low stiffness of the boring bar. In addition, this is a case where dynamic flexibility is often involved. If a boring spindle produces a periodic force – say a variation of force once per revolution – and the periodic frequency n is close to the natural frequency f_n of the bar, then the dynamic flexibility h_{dyn} may be several times higher than the static flexibility h_st and must be considered, see Figure 7.13.

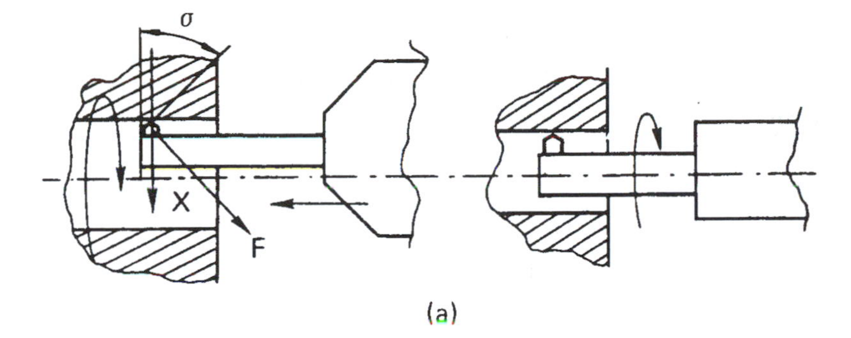

Figure 7.12. Directional orientation of boring tools, from Tlusty in MTTF[47].

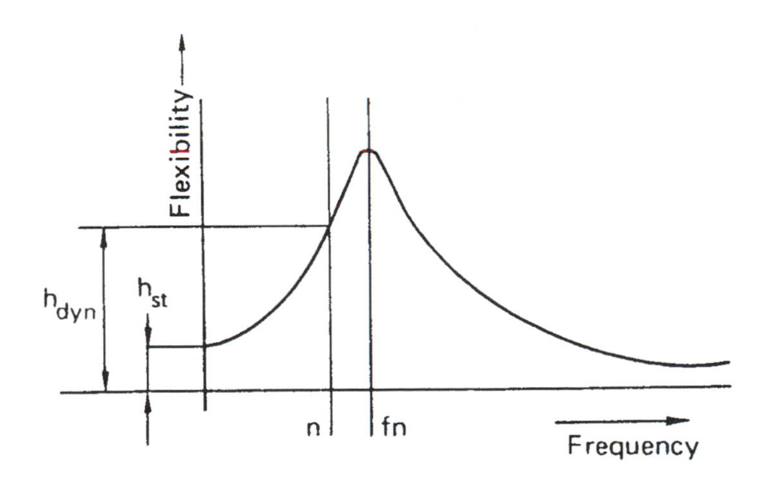

Figure 7.13. Dynamic effect in boring, from Tlusty in MTTF[47].

Finally, we will consider a grinding example. Figure 7.14 shows external and internal cylindrical grinding. In each case the force *F* lies in the same plane as the direction of maximum flexibility, which coincides with the sensitive direction *X*. In addition, the case of internal grinding has the same problem as boring: a rather flexible spindle and the possible presence of dynamic stability. Because of these problems and the abovementioned high cutting stiffness, criteria relating accuracy of grinding machines to the effect of deformations under cutting force are of extreme significance. Often, internal grinding may require that special attention be given to increasing the stiffness of the spindle and its mounting as much as possible.

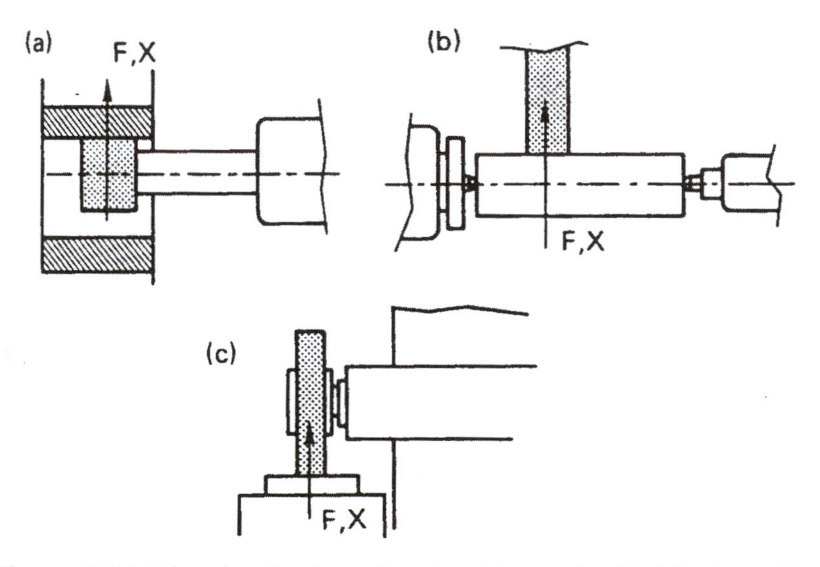

Figure 7.14. Directional orientations in a) internal cylindrical grinding, b) external cylindrical grinding, and c) surface grinding, from Tlusty in MTTF[47].

7.4.2 Type B deformation: Deformation due to the variation of the stiffness along the tool path

This situation can be first analyzed if we assume that the cutting force does not vary at all. The raw material work piece has a perfect form, and the stock to be removed is uniform. The constant value of the cutting force F depends on the depth of cut and feed as seen before in equations 7.2 and 7.3.

In many cases the stiffness k_a, as it was previously defined, varies during the cut. Let us take the ideal case of a completely rigid workpiece clamped between equally flexible centers in a lathe, Figure 7.15. If the tool and cutting force act in the position p the reactions at the centers will be

$$R_1 = F\frac{l-p}{l},\ \ R_2 = F\frac{p}{l} \tag{7.25}$$

with displacements $x_{1,2} = (R_{1,2})/k_c$, where k_c is the stiffness of a center. The displacement X is then

$$X = (X_1 + X_2)\frac{p}{l} = \frac{F}{k_c}\left(1 - \frac{p}{l} + \frac{p^2}{l^2}\right) \tag{7.26}$$

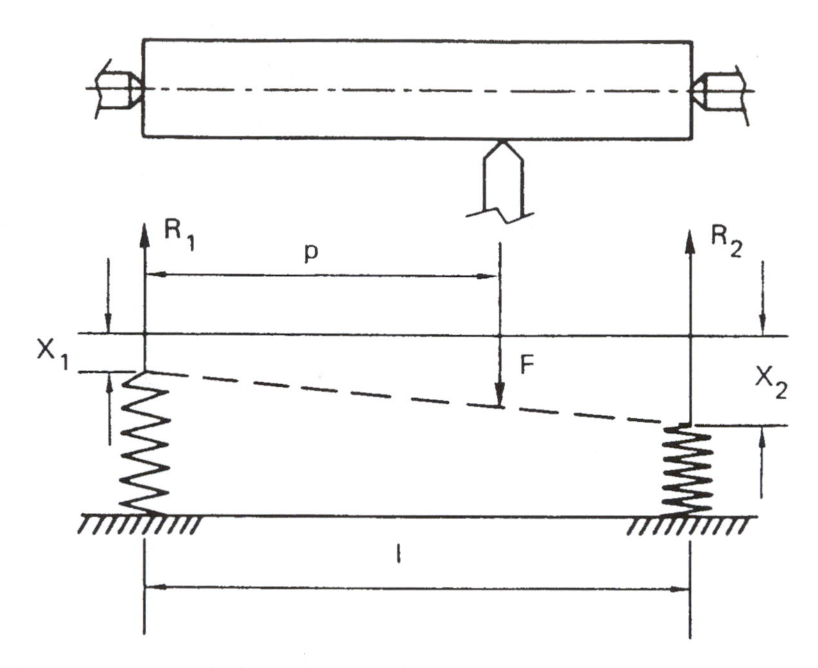

Figure 7.15. Variation of flexibility along the tool path for a lathe, from Tlusty in MTTF[47].

The displacement varies as a parabolic function of p. When the tool I sin the middle of the workpiece, deflection under the tool is $X = F/(2k_c)$. When the tool is at either end, the deflection under the tool is $X = F/k_c$. That is, the displacement at the middle is half that at the ends. In other words, stiffness is twice as large in the middle of the workpiece as at its ends.

The error due to the variation of the oriented stiffness k_a is

$$\delta = \frac{F}{(k_a)_{\min}} - \frac{F}{(k_a)_{\max}} = F(c_{a\max} - c_{a\min}) \tag{7.27}$$

where c_a is the oriented flexibility. Then

$$\delta = F(\Delta c_a) \tag{7.28}$$

where $\Delta c_a = (c_{a\max} - c_{a\min})$ is the variation in flexibility.

In reality, the stiffness is different at the two ends and the workpiece has a stiffness related to its diameter (for cylindrical turning). In Figure 7.16, also from Tlusty[47], flexibility along the workpiece length is plotted for a particular lathe and various workpieces. c_m represents compliance due to machine (flexible centers) alone (that is with an infinitely rigid workpiece), c_{wa} represents compliance for a cylindrical workpiece 600 mm long and 100 mm in diameter and c_{wb} for a workpiece 600 mm long with an 80 mm diameter.

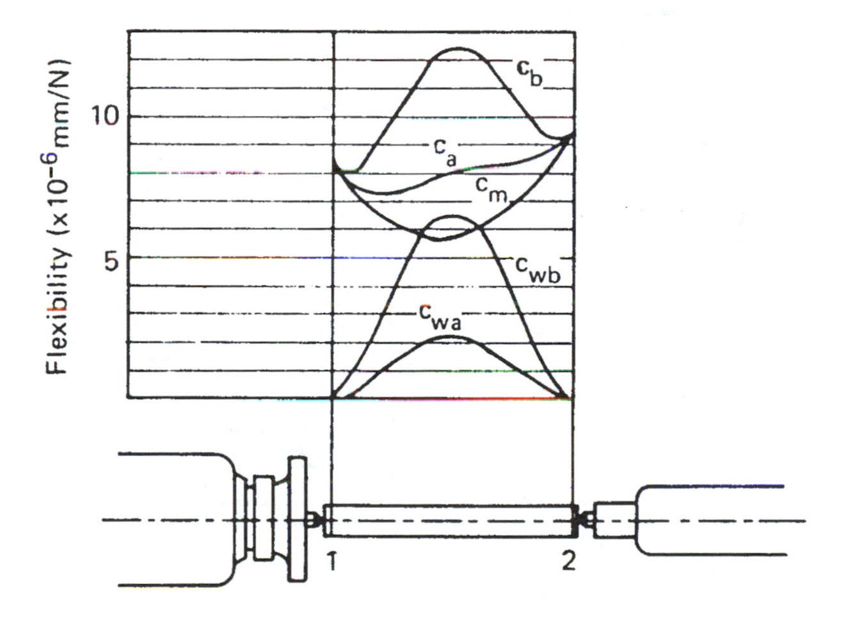

Figure 7.16. Separate and combined flexibility of lathe centers and work, from Tlusty in MTTF[47].

In Figure 7.17, the variations of flexibility between tool and workpiece are diagrammatically shown for various machine tool types. Figure 7.17a applies to an external cylindrical grinding machine. Two curves are shown. The flexibility varies periodically between these two curves due to the effect of a one-sided driving pin. Figure 7.17b, related to internal grinding, shows the variation of flexibility along a workpiece in overhang.

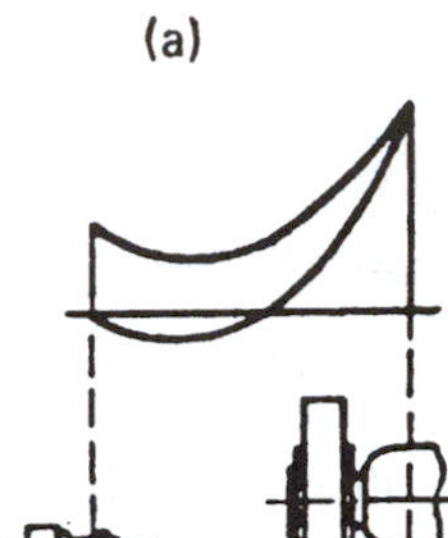

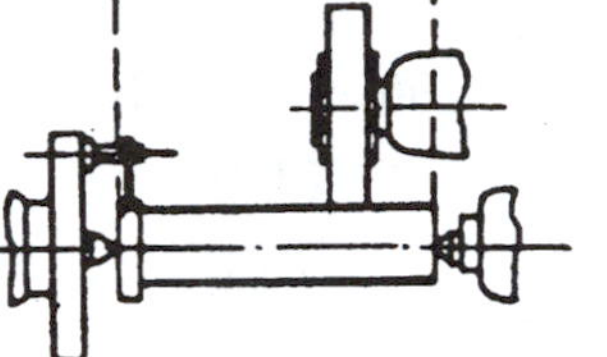

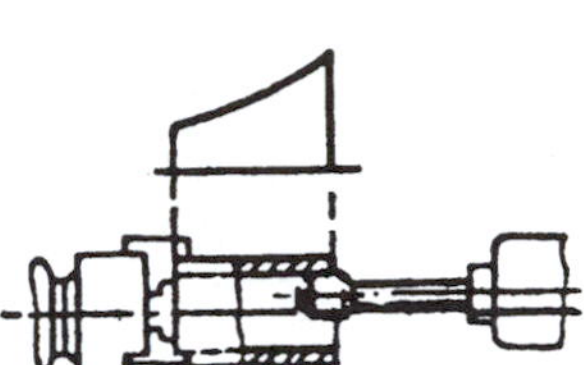

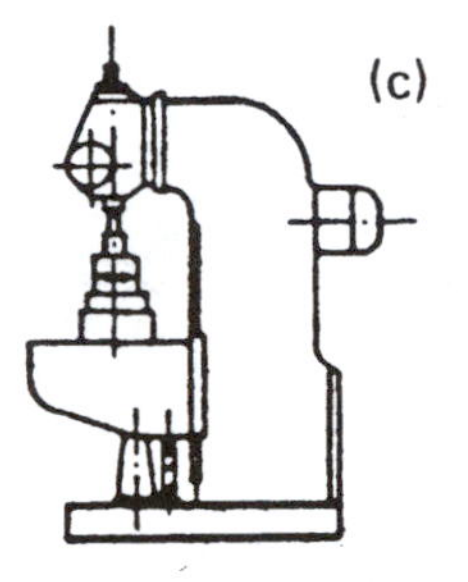

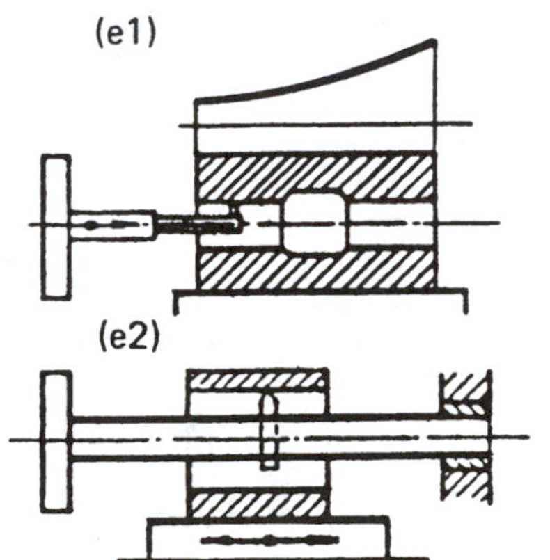

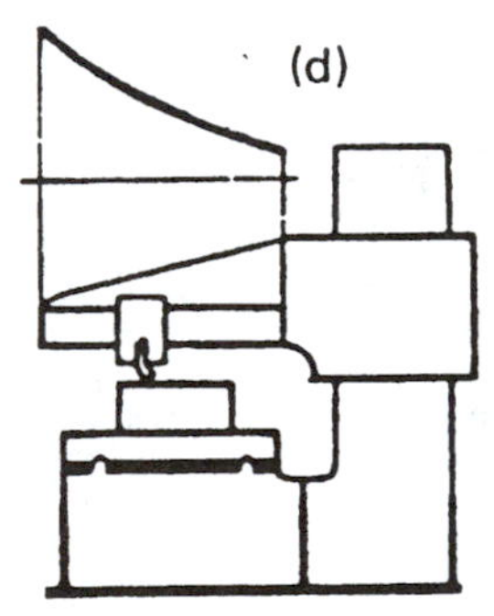

Figure 7.17. Variations in flexibility on various machines: a) external grinding machine, b) internal grinding machine, c) vertical milling machine, d) single column planning, turning or vertical milling machine, e1) boring spindle, and e2) supported boring bar, from Tlusty in MTTF[47].

The flexibility increases with the distance from the chuck. In a vertical milling machine of the knee type as shown in Figure 7.17c – and especially of the bed type – there is hardly any variation of flexibility. For a single column planning, milling, or vertical turning machine, flexibility increases with the distance from the column as shown in Figure 7.17d. In Figure 7.17e1, the variation in flexibility due to the extension of a boring spindle is shown. By contrast, with a supported boring bar and a table feed (used instead of extending the spindle), as shown in Figure 7.17e2, no flexibility variation occurs (recall Wilkinson's cylinder boring machine in Figure 1.18).

7.4.3 Comparison of the errors from deformation types A and B

In deformation type A, due to decreasing variation in stock being removed, the error diminishes with subsequent cuts. However, errors of type B depend, according to equation 7.27, almost entirely on the force used in the last cut only.

In order to illustrate this distinction let us take an example where both errors A and B are significant, such as boring in overhand, see Figure 7.18.

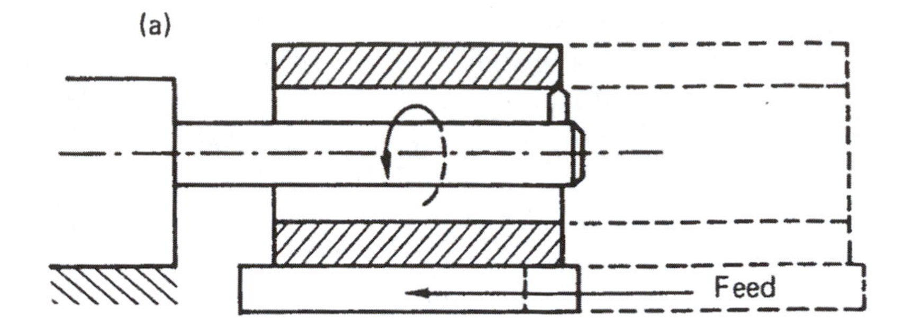

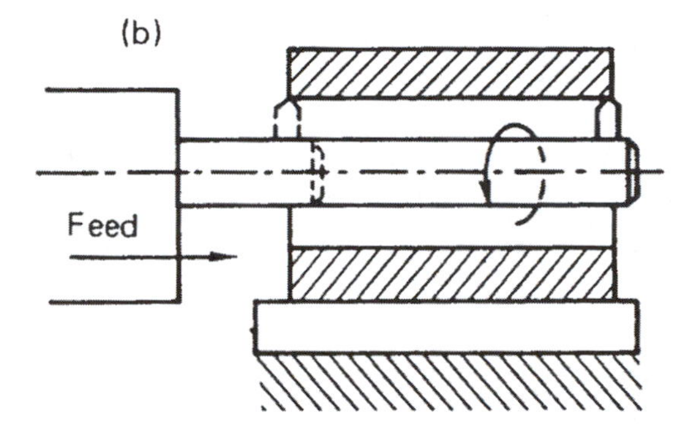

Figure 7.18. Boring with a) table feed and b) spindle feed (extension), from Tlusty in MTTF[47].

Assume that the only flexible part is the boring spindle with a diameter of 63 mm and that a bore 300 mm long is to be produced. Two methods will be investigated:

i. boring is done with table feed with the spindle extended 400 mm all the time, and
ii. boring is done with the table stationary and the spindle extension varies between 100 and 400 mm.

A tool with a SCEA $\sigma = 30°$ is used in the example. The oriented flexibility on the tool is

$$c_a = \frac{X}{F} = \frac{\sin\sigma\cos\beta}{k_{sp}} \tag{7.29}$$

where k_{sp} is the "direct" stiffness of the boring spindle of diameter d and length l.

$$k_{sp} = \frac{3E\pi d^4}{64l^3} = 3 \text{ x } 10^4 \frac{l^3}{d^4} \tag{7.30}$$

for a steel spindle. Using $\sigma = 30°$, $\beta = 60°$ gives

$$c_a = 8.3 \text{ x } 10^{-6} \frac{l^3}{d^4}$$

For the two extreme extensions,

$l = 100$ mm, $c_a = 5.3 \text{ x } 10^{-7}$ mm/N
$l = 400$ mm, $c_a = 3.3 \text{ x } 10^{-5}$ mm/N
$\Delta c_a = 3.25 \text{ x } 10^{-5}$ mm/N

Let's assume one roughing cut with an average depth of cut of 4 mm and with feed $f_r = 0.2$ mm followed by any number of finishing cuts with a depth of cut of $a = 0.75$ mm and feed $f_r = 0.05$ mm. For the error of type A let us assume an initial form error run-out of $\Delta = 2$ mm and require it to be brought down by a factor of 200 to 0.01 mm. In the worst case of type A (at the longest spindle extension), the cutting stiffnesses and stiffness ratios are, according to equations 7.4, 7.6 and 7.7 and using $C = 1500$ N/mm^2:

In roughing – $r_a = 390$ N/mm, $\mu_r = 0.013$
In finishing - $r_a = 120$ N/mm, $\mu_f = 0.004$

It may be seen that after the roughing cut an error of $\delta_r = 2\mu_r = 0.026$ mm remains, which is more that double the allowable tolerance. After one finishing cut the error decreases to $\delta_f = 0.001$ mm. This corresponds to the operation i. with the moving table.

In the operation ii. with feed accomplished by extending the spindle, one has to take in to account the cutting forces. Calculations give F_r = 1560 N and F_f= 90 N for roughing and finishing, respectively. The form error due to the variation of flexibility Δc will be, after the roughing cut, $\delta_r = 0.05$ mm and after the finishing cut $\delta_f = 0.003$ mm. A comparison of results using methods a) and b) shows that the type B error is larger than the type A error by a factor of 2 after the first cut and 30 after the second cut. In general, however, it may be stated that, with the exception of grinding, the errors due to the cutting force deformations are negligible in most metal cutting operations if one roughing and one finishing cut are taken. This is generally valid if errors of type B are avoided.

A special case is multi-tooth machining, end milling for example. In end milling the tool itself is rather flexible and cutting stiffness is rather high because of the usually large axial depth of cut. For example, for a 19.05 mm (3/4 inch) diameter cutter 63 mm (3.5 inch) long and an axial depth of cut 40 mm (or 1.6 inch), with tool stiffness of about 15,000 N/mm and cutting stiffness in steel of 40,000 N/mm, we find $\mu = 2.67$ and $i = 0.73$. The improvement of the error in one pass would be merely 0.73 times.

The "copying ratio" i (improvement factor of form error) is very high – as high as in internal grinding. After two passes the improvement is still only 0.53, after three passes 0.4, etc. Actually, the cutting force has a periodic component of frequency anywhere between 30 and 1000 Hz. Consequently, the "dynamic" value of μ may be as much as 5 times higher, resulting in an i value of up to 0.96.

However, in end milling the relationship between the cutting force, the deflection, and the error of the machined surface is very special and the rather straightforward deliberations based on the above values of i do not apply. Figure 7.19, from Tlusty[47], is a simplified picture of cutters shown with straight teeth. The end mill and its shaft and support are flexible. In up-milling, cases a) and c), the end mill deflects mostly towards the workpiece. Every tooth starts with a very small force at Point A and reaches the maximum force at point B. Every tooth generates the final surface while at point A. In the case of a two-fluted cutter as shown in a), there is practically no deflection when the tooth is at A. In b), when the tooth is at B, there is a large deflection but there is no tooth at A to imprint this deflection on the generated surface. However, a four-fluted cutter shown in c) will transfer the deflection on the generated surface. In down-milling, case d), the situation is similar but the deflection is away from the workpiece. In reality, the teeth are helical and the relationship between forces, deflections, and the form of the machine surface is more complicated.

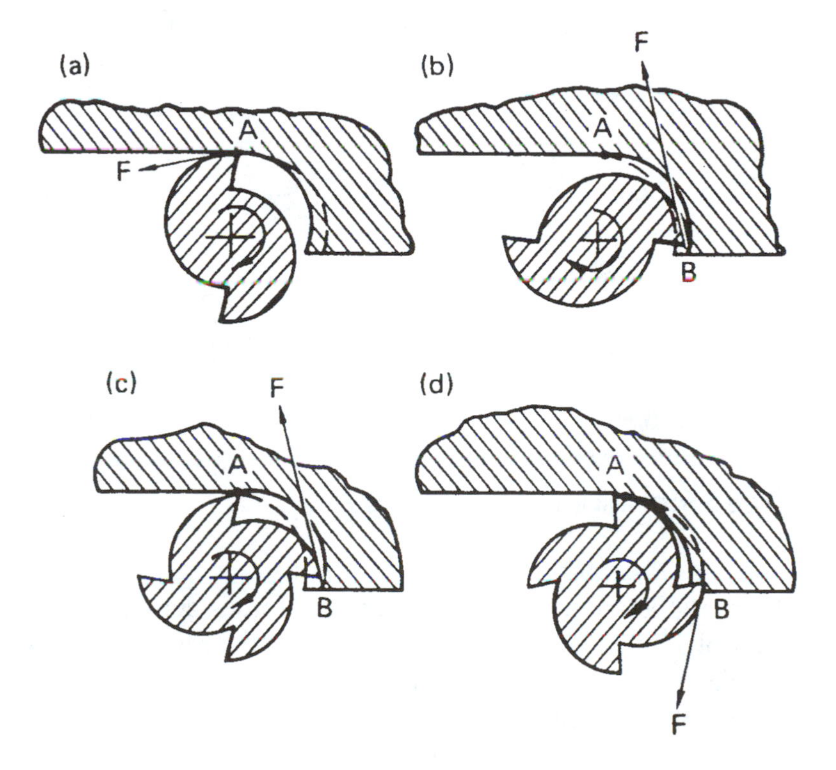

Figure 7.19. Deformations of end mills and their effect on accuracy of the milled surface, from Tlusty in MTTF[47].

It is possible to compute the forces, deflections, and the envelope representing the generated surface – which may be undersize (up milling), oversize (down milling), straight, inclined, convex, or concave depending on the number of teeth, helix angle, and radial and axial depth of cut. Some of the more comprehensive work on this has been done by the R. Devor and his colleagues at the University of Illinois under the name of EMSIM (for **E**nd **M**ill **SIM**ulation)[49]. Figure 7.20 illustrates a typical output of the simulation for an end mill machining a step showing the predicted result of end mill deflection and the resulting surface profile.

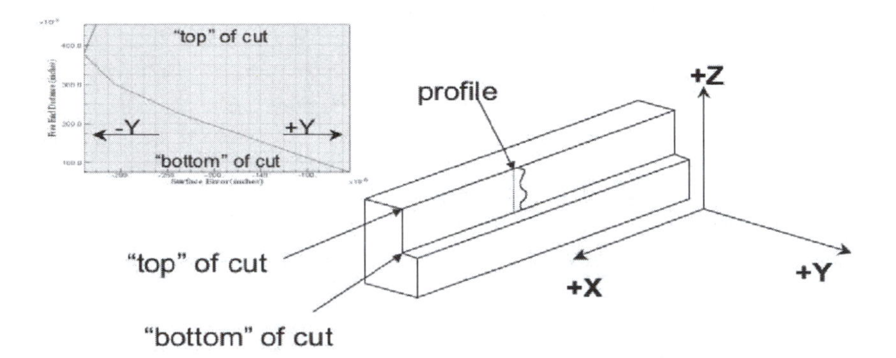

Figure 7.20. Output of EMSIM simulation for end mill deflection simulation.

7.5 Forced vibrations

Forced vibrations are produced by periodic forces acting within the machine tool as well as in the environment of the machine. If external periodic disturbances enter the machine via floor and foundations, it may be necessary to use isolation mountings (recall Figure 7.3). Mostly, their effect is of concern only if they produce waviness on the machined surface beyond an acceptable level. Concern is generally limited to fine machining: fine turning, fine boring, and grinding. Indeed, grinding is the most common case where it is necessary to deal with forced vibrations. Another special case of forced vibrations occurs when the frequency of milling cutter teeth coincides with natural frequencies of torsional vibrations of the spindle drive.

The periodic disturbing forces may be due to the following, and refer to Figures 7.4 and 7.5 for a complete illustration:

- mechanical unbalance of the grinding wheel, electric motors, or shafts running with higher speeds
- electromagnetic unbalance of electrical motors
- inaccuracies of ball bearings of electric motors
- pulsating pressure in a gear pump.

Serious problems arise if any of these disturbing forces is in resonance with one of the natural modes of vibration of the structure.

Thus, a statement of the criterion of forced vibrations is rather straightforward and may be formulated to specify that waviness due to forced vibrations between tool and workpiece must not exceed a specified level. Methods used to satisfy this criterion are identical with those of structural dynamics is discussed in the following section.

There are problems in the diagnosis of vibrations. In most cases it is rather easy to identify the source of a disturbing periodic force. However, there are cases where it is difficult to measure the waviness of the machined surface, identify the sources, and identify the combination of vibrations which produced it. This topic is not covered in detail in these notes. An excellent reference on the current analysis and prediction methodologies is Altintas[50].

7.6 Self-excited vibrations (chatter)

7.6.1 Introduction

Chatter is a self-excited type of vibration occurring in metal cutting. As distinguished from forced vibration, chatter is not caused by an external periodic force but rather generates its own periodic force to sustain itself. In further distinction to forced vibration, chatter either does not exist at all – stable cutting – or it exists and usually builds to large amplitudes which may cause poor surface finish, loosening of the tooling or workpiece, or tool and workpiece breakage. Thus, as a rule, chatter is unacceptable and if it occurs, cutting must immediately be changed – usually the depth of cut (width of chip) decreased – to eliminate the chatter. With chatter, the reason for analysis is usually not as concerned with the amplitude of the vibration, which may easily grow beyond acceptable values, as in the conditions of the limit of stability.

The limit of stability in a particular operation depends on three groups of factors:

i. on the dynamic characteristics of the structure of the machine tool.
ii. on the parameters of the cutting process: cutting speed, feed, depth of cut, tool geometry, and workpiece material.
iii. On the location of the cutting process in the structure and on directional orientation.

These factors will be discussed and their influences assessed. We will discuss the fundamentals first as illustrated in Figure 7.21. The figure shows a conventional turning and milling process with the key feature that there is some "overlap" between passes. This could be, at an extreme, a plunge cut (parting) operation where each subsequent pass covers that same surface area as the previous cut. Depending on the SCEA and the feed per revolution, as introduced in Figure 7.8c, the amount of overlap between cuts will vary. Under any circumstances, the transitional surface will retain the "surface imprint" of the previous pass. As shown in Figure 7.22, the tool used in turning or boring continuously removes chips from the surface generated during the previous revolution (previous pass). In milling, the tooth of the cutter removes chips from the surface generated by the preceding tooth (pass) as seen in Figure 7.23.

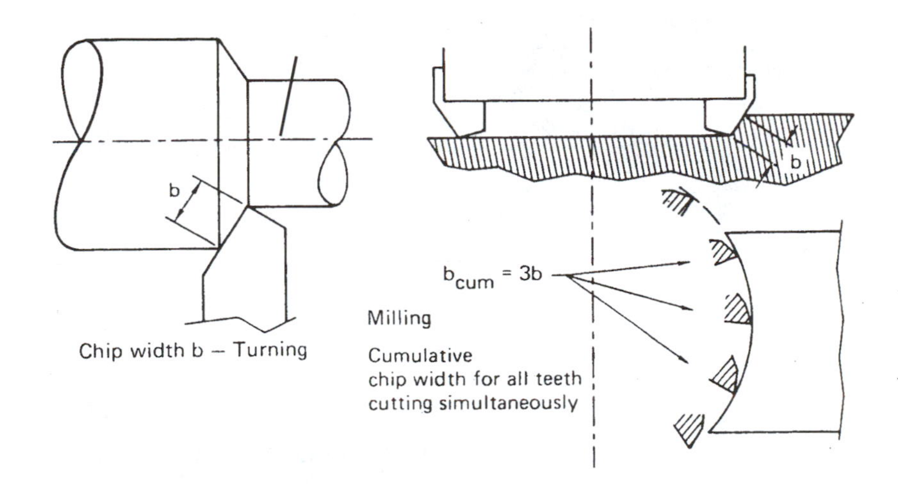

Figure 7.21. Illustration of basic turning and milling configurations, from Tlusty in MTTF[47].

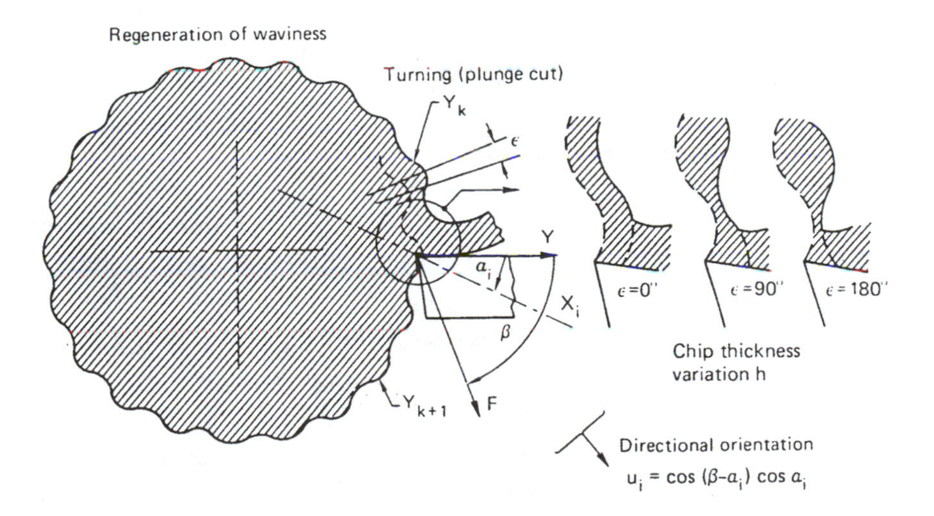

Figure 7.22. Illustration of wave regeneration in cutting that can lead to chatter depending on chip width *b* and directional orientation, from Tlusty in MTTF[47].

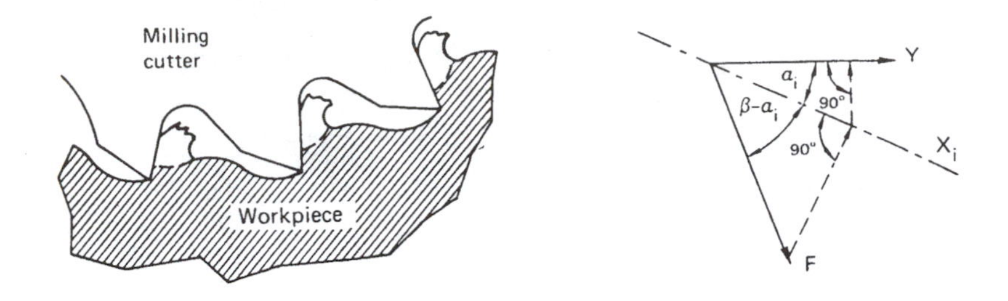

Figure 7.23. Detail of wave regeneration and directional orientation, from Tlusty in MTTF[47].

In a stable cut, any vibrations and resulting surface undulations die out during subsequent passes due to damping in the system. In an unstable cut, undulations will increase, at least initially. At the limit of stability any undulation will remain constant in magnitude. The width of chip b as defined in Figure 7.21 is the most crucial parameter. In any operation, for a sufficiently small b cutting will be stable and for a sufficiently large b chatter will occur. The limit chip width b_{lim} is the smallest b at which chatter might start, and below which chatter will not occur. Indeed, in tests of cutting stability, and with all other cutting conditions kept standard, the value of b_{lim} is established as a measure of stability.

At the limit of stability, the magnitude of undulations remains constant. Undulations created in subsequent passes have a phase shift ε with respect to each other. The difference between the two undulations γ_k, γ_{k+1}, that is radial distances to the two surfaces along the same radius, Figure 7.22, is the chip thickness variation h. Correspondingly, variable cutting force F is produced. If the structure is free to vibrate between tool and workpiece in a direction X_i, the component $F\cos(\alpha_i - \beta)$ of the force F produces vibrations in this direction. In a multi-degree of freedom structure, vibrations will be produced simultaneously in all degrees of freedom (in the various directions X_i). The component of vibration in direction X_i which falls into direction Y (that is, $Y = X_i \cos\alpha$) modulates the chip thickness h. The direction Y is normal to the cut surface (that is, to the surface being generated by the primary cutting edge.)

The directions of the normal Y and of the cutting force F constitute the directional orientation of the cutting process. Every "mode of vibration" of the structure having a direction of freedom X_i is associated with a directional factor $u_i = \cos(\alpha_i - \beta)\cos\alpha_i$. This is for a two dimensional case as seen in Figure 7.22.

In a general, three dimensional case, the directional factor will also involve projection of the cutting force into the direction of the mode and from there into the direction of the normal Y to the cut surface. The corresponding transfer function – defined as a response (in the cutting force F) – is called the Oriented Transfer Function (OTF).

The directional orientation of a mode is very important, for example, modes with $\alpha = 90°$ ($\cos\alpha$=0) will not induce chatter because they have a zero component in Y; modes with $(\alpha - \beta) = 90°$, which are perpendicular to the force F, do not excite chatter. Generally the value of u_i expresses the attenuation of each mode in the regenerative process.

The phase shift ε between undulations of the two successive passes has a significant effect on the amount of self-excitation. For $\varepsilon = 0$, there is no chip thickness variation, that is, $h = 0$. Phase shift angle varies with spindle speed and there is a certain value ε_m (close to 90°), of ε for which self-excitation is a maximum. For this value of ε, the limit chip width b_{lim} is minimum and represents the basic, sometimes called borderline, stability limit and is designated $b_{basic,\ lim}$ (or $b_{b,lim}$).

The phase shift ε depends on the frequency of vibration and no the spindle speed, n. With variations of n, the angle ε varies and affects the value of b_{lim}. This variation will be discussed later in some detail.

The cutting process may inject positive or negative damping in to the self-excitation and thereby affect b_{lim}. The cutting speed has a decisive effect on the damping in the cutting process. However, at higher cutting speeds like those used with carbide tools the damping effect is small. These aspects of damping will also be discussed in some detail later.

We first disregard the effect of spindle speed on the phase shift ε and assume that this phase shift is free to adjust itself to the value of ε_m. This is in reality what happens in planning and shaping and is often approximately so in other operations like turning and milling. We will also assume that there is no damping in the cutting process. The situation of basic stability and no damping will be called a *basic case* and will be used for assessing the effects of the characteristics of the structure.

7.6.2 Basic stability; effect of structural dynamics

The various types of machine tools in their various configurations contain various normal modes of vibrations. Each mode can be characterized by its "modal" parameters: frequency, damping ratio, modal stiffness, and mode shape. We will review some basic examples of this. There are a number of references that can provide additional details, for example from Tlusty[51].

In Figure 7.24, an example is given of the mode shapes of a workpiece on a conventional lathe. The series of sketches show the increasing frequencies of the first six modes of vibration and the shapes of the modes. Modes similar to the 3rd and 6th modes can be found in the vertical direction as well, with frequencies generally higher than those shown for the horizontal direction.

This figure illustrates the variety of modes found in machine tool structures. Although no example of a milling machine is shown here, similar modes are seen in the individual elements and, often, couple to supporting structure as well. Higher frequency modes are usually stiffer and the number of modes that are of significance, usually between two and six, almost never exceed ten. In some instances there is one mode so prominent that all the other modes may be neglected. A case with such a prominent mode is called a *very basic case*.

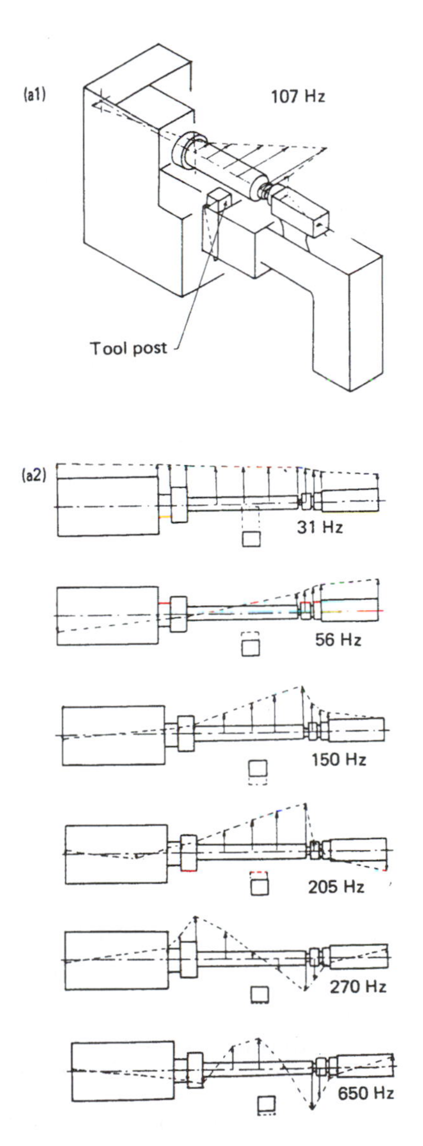

Figure 7.24. Mode shapes of a workpiece mounted in a conventional lathe, from Tlusty in MTTF[47].

We will first present the condition for the limit of stability for *basic* cases. Referring to equation 7.2 and starting from the transfer function of the cutting process:

$$F = r_b b(Y - Y_o) \tag{7.31}$$

where F is the variable component of the cutting force, r_b is the "cutting stiffness" in the direction normal to the cutting edge, b is the chip width, and $(Y - Y_o)$ represents the amplitude of chip thickness.

Actual chip thickness, h, consists of a steady state value h_m (the mean chip thickness), which is equal to infeed per pass, and the modifications to that value due to the variations Y_o at the top (the undulation of the surface from the previous pass), and the variations Y at the bottom due to the vibration in the current cut. This can be written as

$$h = h_m + (Y - Y_o)e^{j\omega t} \tag{7.32}$$

In accordance with the assumption of no damping in the cutting process, r_b, is a real number. Note that r_b applies per unit width of chip. Also,

$$Y = F\Phi(\omega) \tag{7.33}$$

where $\Phi(\omega)$ is the Oriented Transfer Function. That is, it is a ratio of the complex amplitude of the Y component of all the X vibrations over the complex amplitude of a force acting in the direction F, as a function of frequency ω; both the vibrations and the force are taken as relative between the tool and the workpiece.

The OTF is obtained as a sum of all the direct transfer functions of the modes Φ_i multiplied by the directional factors u_i as

$$u_i = \cos(\alpha_i - \beta)\cos\alpha_i$$

$$\Phi = \sum_{1}^{i} u_i \Phi_i \tag{7.34}$$

If we combine equations 7.31 and 7.33 and eliminate the force

$$Y = r_b b \Phi (Y - Y_o) \tag{7.35}$$

and this can be modified to

$$\frac{Y_o}{Y} = \frac{1/(r_b b) + \Phi}{\Phi} \tag{7.36}$$

The condition for the limit of stability can be formulated so that vibrations do not decay or increase from pass to pass as discussed earlier. This means that the amplitudes $|Y_o|$ and $|Y|$ are equal, or

$$\frac{|Y_o|}{|Y|} = 1 \tag{7.37}$$

If we combine equations 7.36 and 7.37 we obtain

$$\left| \frac{1}{r_b b} + \Phi \right| = |\Phi| \tag{7.38}$$

which expresses the equality of the absolute values of two complex numbers (the function $\Phi(\omega)$ is complex, while $r_b b$ is real). This condition then has two parts

$$\text{Im}(\Phi) = \text{Im}(\Phi)$$

which is obvious, and

$$\frac{1}{r_b b} + \text{Re}(\Phi) = \pm \text{Re}(\Phi)$$

Here, the + sign leads to $b = \infty$; the – sign gives

$$\frac{1}{r_b b} = -2\,\mathrm{Re}(\Phi)$$

which is the actual condition for the limit of stability. We can then express the limit value of the chip width as follows using the term $b_{\lim}$ previously introduced

$$b_{\lim} = \frac{-1}{2r_b\,\mathrm{Re}(\Phi)} \tag{7.39}$$

The chip width b is a positive number. Equation 7.39 can therefore only be satisfied for the negative part of the function $\mathrm{Re}[\Phi(\omega)]$. Further, of all the values b that satisfy equation 7.39, there is a minimum one, the smallest chip width at which chatter can occur. This is the actual critical limit of stability and it corresponds to the largest negative value of $\mathrm{Re}(\Phi)$, to the minimum $\mathrm{Re}(\Phi)_{\min}$

$$b_{\lim,cr} = \frac{-1}{2r_b\,\mathrm{Re}\,\Phi(\omega)_{\min}} \tag{7.40}$$

For chip widths $b < b_{\lim}$ cutting is stable – there is no self-excited vibration. For $b > b_{\lim}$ chatter will occur and grow. In practice, because of nonlinearities in the phenomenon, the amplitude of chatter will stabilize at a finite value.

For additional details and discussion on the development of the chatter stability limit, the reader is referred to Tlusty[47] from which portions of this above discussion were taken or Tlusty[51] for a more comprehensive discussion.

Let's look at a very basic example of a single degree of freedom system, the so-called *very basic case* as shown in Figure 7.25. The structure is oriented so that its direction of freedom X coincides with the direction of the normal Y. Correspondingly, its directional factor is

$$u_i = \cos 60° \cos 0° = 0.5$$

The system with stiffness *k*, mass *m*, and damping ratio ξ has a real part of the Direct Transfer Function *G** as shown in Figure 7.25b (and shown as compliance *vs* frequency) with its minimum being

$$G_{\min}^{*} = -\frac{1}{4k\xi(1+\xi)} \cong \frac{1}{4k\xi} \tag{7.41}$$

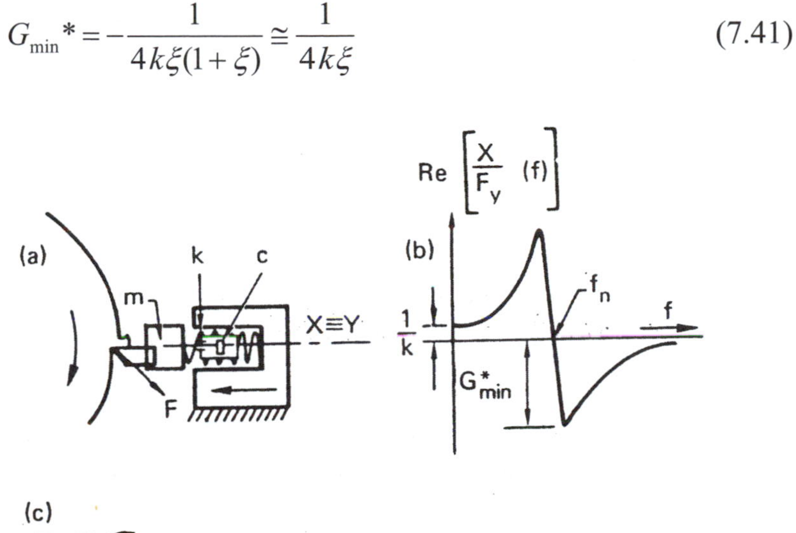

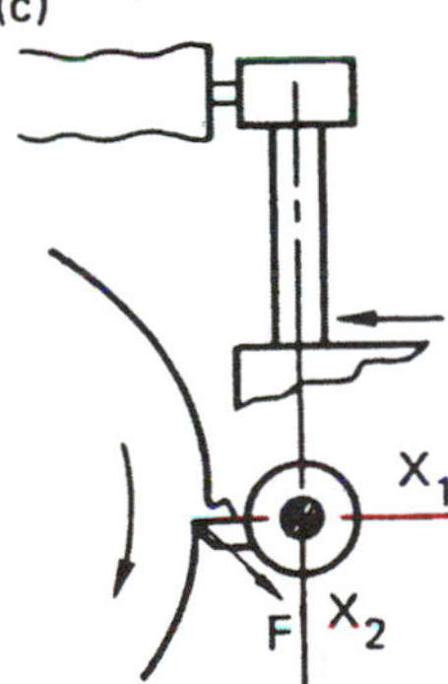

Figure 7.25. a) Model for a simple basic vibrating cutting system, b), real part of transfer function, and c) application to a round tool bar case, from Tlusty in MTTF[47].

For this simple case

$$b_{\lim} = -\frac{1}{2r_b u G^{*}_{\min}} = \frac{2k\xi}{r_b u}$$

where the OTF is obtained as $\Phi = uG^*$. A case which acts as the one in Figure 7.25a but with a round bar as in Figure 7.25c can be thought of as having two degrees of freedom, X_1 and X_2, with identical stiffness and identical frequencies. We can choose X_1 and X_2 in any direction, for example $\alpha_1 = 0°, \alpha_2 = 90°$. Then X_2 has a zero directional factor and the case is equivalent to the one in Figure 7.25a.

In reality, structures have more than one degree of freedom. There are numerous cases of two fairly prominent modes with frequencies close to one another. Boring is an example of a process where there are often two prominent modes due either to the nonsymmetry of the clamping of the boring bar, to a key slot in a boring spindle, or to different stiffnesses of the spindle housing in two perpendicular directions. The existence of a second degree of freedom may substantially influence the limit of stability. This can be illustrated with a simple example presented symbolically as a boring bar but which represents a general case of a system with two mutually perpendicular degrees of freedom.

Figure 7.26a shows a boring bar with a circular cross section, similar to that in Figure 7.25c. In Figure 7.26b, the bar with rectangular cross section is chosen (although many other rectangles could also be chosen) so that its moment of inertia I_a about the axis A is

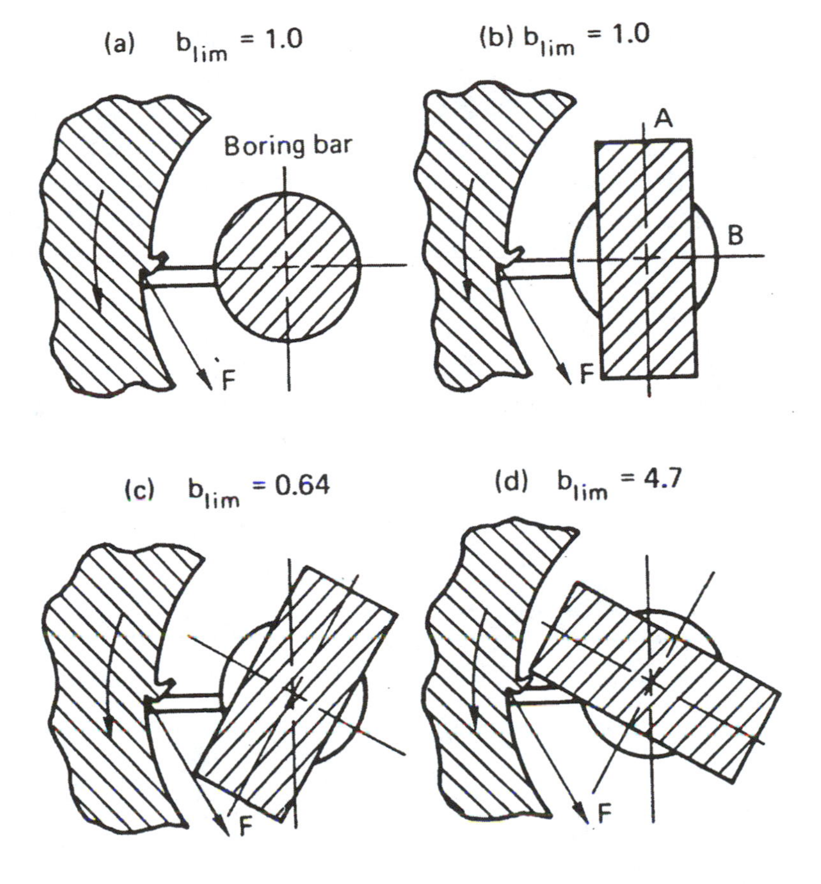

Figure 7.26. System with two mutually perpendicular degrees of freedom; the limit of stability is dependent on orientation, from Tlusty in MTTF[47].

is the same that of the round bar while I_B is increased by a factor of about six. The two cases a) and b) give the same limit of stability. Case c) and d) in Figure 7.26 show two other orientations of the same rectangular bar that result in limits of stability 0.64 and 4.7 times that of case a), respectively,

In multi-degree of freedom systems various combinations of modal participation in the resulting OTF arise. The real part of the OTF may have a shape like the one shown in Figure 7.27a. The OTF denoted by G is the sum of the Oriented Transfer Functions G_i corresponding to the individual modes. Some of them, like G_3, are "inverted" because the corresponding directional factor u_3 had a negative value. Quite often one mode is dominant, as in the stability diagram in Figure 7.27b. Such a case approximates the single mode case in Figure 7.25a except that it may have a different directional orientation. Also, cases are frequent where there are two modes with close frequencies and not very different stiffnesses. In such cases, depending on the orientation, the two modes can amplify each other as in Figure 7.27c or subtract from each other as in Figure 7.27d, leading to either very low or very high stability. These cases best express what is often referred to a "mode coupling." Cases of two dominant modes correspond to structures like a boring bar, ram on a vertical boring mill, overarm on a milling machine, etc.

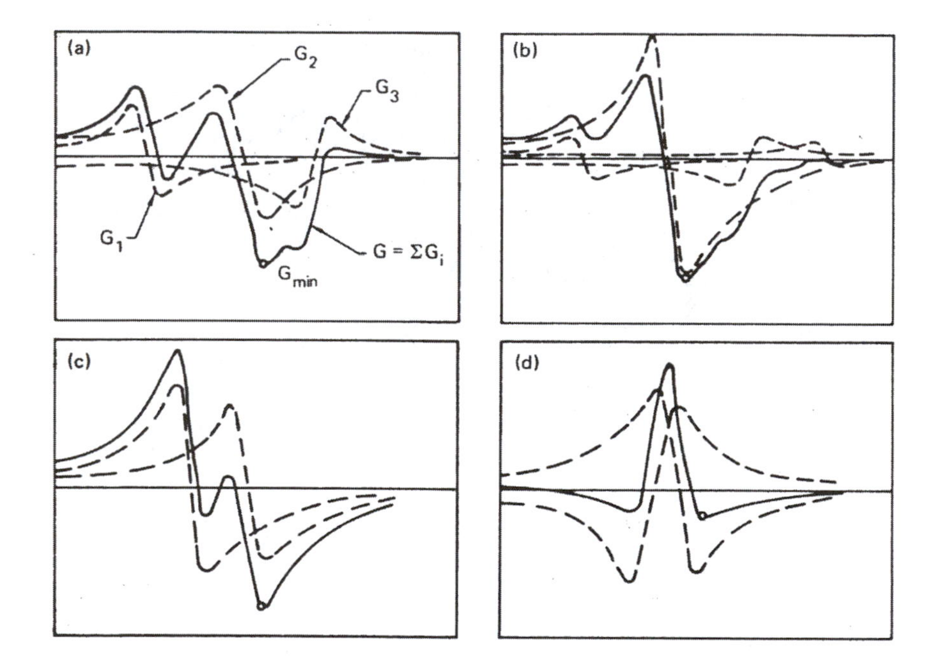

Figure 7.27. Oriented Transfer Functions arising from various combinations of modes; a) sums of several modes, b) one dominant mode, c) two dominant modes that are additive, and d) two dominant modes that are subtractive, from Tlusty in MTTF[47].

On some machine tools the directional orientation of the cutting process varies little, as on a lathe, Figure 7.28a, where the normal to the cut surface is always horizontal and the direction of the cutting force varies in a small range. On other machines a great variety of orientations occur. In face milling, as shown in Figure 7.28b, the average direction of the cutting force and of the normal *Y* vary with the mutual position of the cutter and the workpiece, and with the direction of feed.

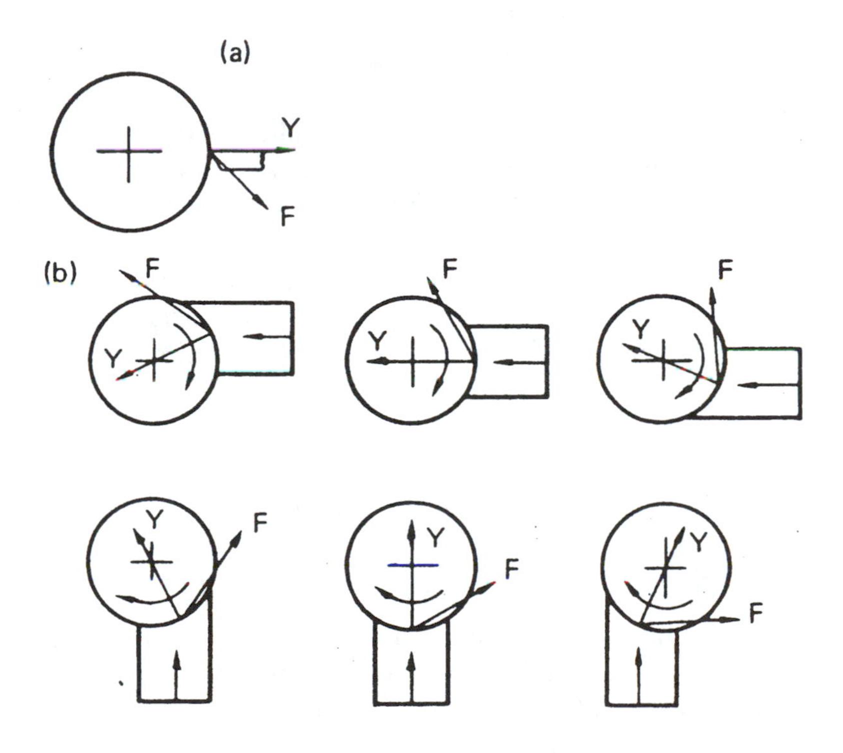

Figure 7.28. Directional orientations on a lathe (end view, a), and in face milling (top view, b) from Tlusty in MTTF[47].

From the preceding simple presentation and discussions the following conclusions may be drawn from the structure of a machine tool:

i. the "basic limit width of cut, $b_{b,\lim}$" without chatter (if no damping is generated in the cutting process and no interference of wave regeneration is introduced) is indirectly proportional to the "cutting stiffness" (specific for the given workpiece material) and to the maximum dynamic flexibility between tool and workpiece (the magnitude of the minimum of the real part of the OTF between tool and workpiece). For simple dynamics this means that $b_{b,\lim}$ varies with the stiffness and damping of the system. For actual, more complex dynamics, design of the machine tool may require increasing the stiffness of selected parts of the structure, even decreasing the stiffnesses of some other parts, selecting optimum mass distribution, and increasing the damping in selected parts of the structure.

ii. a particular machine tool is used in various configurations, in each of which the cutting process may have various "directional orientations." Limit of stability may substantially vary in these cases.

7.6.3 Variation of spindle speed and stability lobes

In the preceding section basic stability was derived with the assumption that the phase angle ε between undulations of subsequent cutting passes was free to adjust itself to the value of ε_m corresponding to maximum self-excitation. In practice this is true only for operations like planning. In turning or boring, the frequency of chatter and the phase angle ε are related (constrained) by the geometric condition of the finite length of the circumference of the workpiece. In Figures 7.22 and 7.23, the undulation has an average wavelength $w=v/f$, where v is the cutting speed and f is the frequency, and there area N integer plus a fraction undulations per revolution. Then

$$\left(N + \frac{\varepsilon}{360}\right)\frac{v}{f} = \pi d \tag{7.42}$$

or

$$\left(N + \frac{\varepsilon}{360}\right)\frac{\pi n d}{f} = \pi d$$

and, therefore

$$N + \frac{\varepsilon}{360} = \frac{f}{n} \tag{7.43}$$

where n is spindle speed, N is an integer and $\varepsilon < 360°$. However, for a large number N (high frequency, low spindle speed) a small change in frequency produces a large variation of ε. For example, if $N = 100$, a 0.25% change of frequency corresponds to a change of 90° in ε.

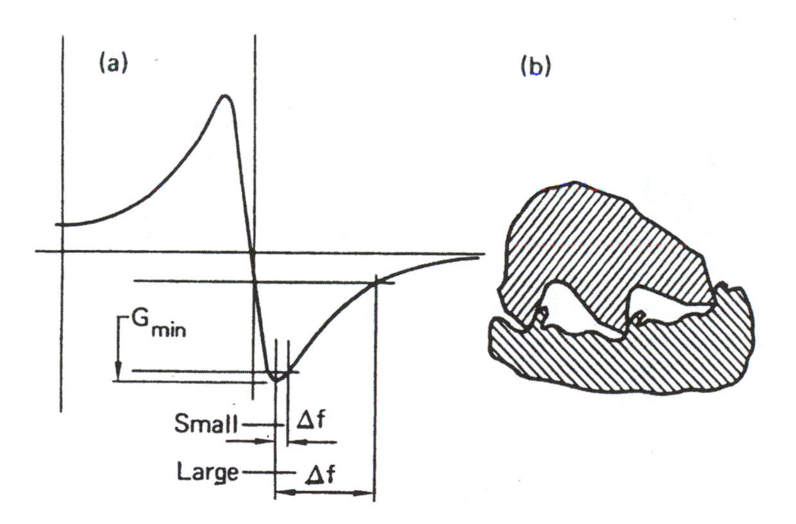

Figure 7.29. Effect of frequency change on stability limit, a) and phasing of subsequent chips in milling, b), from Tlusty in MTTF[47].

However, as shown in Figure 7.29a, a small change of frequency does not change the value G much from the G_{lim} corresponding to the basic case. Further, the "geometric condition" does not affect stability much in the cases where there is a large number of waves between subsequent cuts, as in turning or boring. In milling, however, where two subsequent cuts are produced by subsequent teeth, equation 7.42 must be modified into

$$N + \frac{\varepsilon}{360} = \frac{f}{nz} \tag{7.44}$$

where z is the number of teeth. In this case there are fewer waves between subsequent teeth. In Figure 7.29b, where $N = 2$, a 90° change in ε now gives a 12.5% change of frequency, and may cause a large instability.

The effects of the geometric condition on stability ε, known as "stability modes," are clearly shown in a graph like that seen in Figure 7.30, which applies to a system with a single degree of freedom (its transfer function shown), plotted with coordinates $b_{\text{lim}} / b_{b,\text{lim}}$ *vs* basically speed. Each lobe corresponds to a number (N plus a fraction) of waves between consecutive passes. It can be seen that for large N a variation of spindle speed does not produce appreciable variation of b_{lim} compared to the "basic" $b_{b,\text{lim}}$. But, for N=2 a speed may be found for which stability is increased by a factor of 6; for N=1 a speed exists to increase the ratio by more than 20 times. The spindle speed corresponding to N=1 is $n = f/z$. If spindle speed is increased well beyond this speed, f/z, no chatter can occur – the lobe diagram curve goes to infinity.

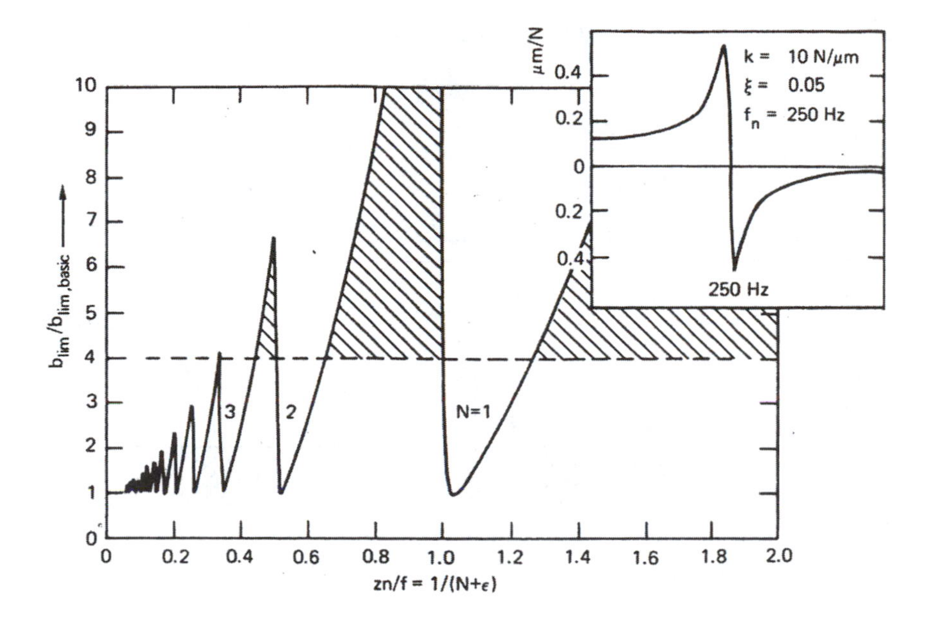

Figure 7.30. Stability "lobes" for a single degree of freedom system, From Tlusty in MTTF[47].

Setting the speed to achieve values in the shaded are of the diagram will avoid chatter. In those cases N is typically small, with low frequency and/or high cutting speed. Typical applications are i) carbide face milling of cast iron or steel (high cutting speed) on large machines, like floor type horizontal boring and milling machines (low frequency), or ii) a very high speed milling of aluminum (where undulations are long even with high frequencies.) It is essential that such machines be provided with stepless speed variation to achieve cutting conditions within stable regions.

It is also important to know that a system with more than one degree of freedom has more restrictions on the use of the "stability lobes" than a singular degree of freedom system, which has been most often discussed in research papers. Figure 7.31 shows a two degree of freedom system with OTF included. Note that this differs from Figure 7.30 by having one additional mode. Comparing the graph in Figure 7.31 with Figure 7.30, it can be seen that the speed range for which stability can be increased, say by a factor of four, is narrowed and the maximum increase of stability for N=1 is reduced to 9.

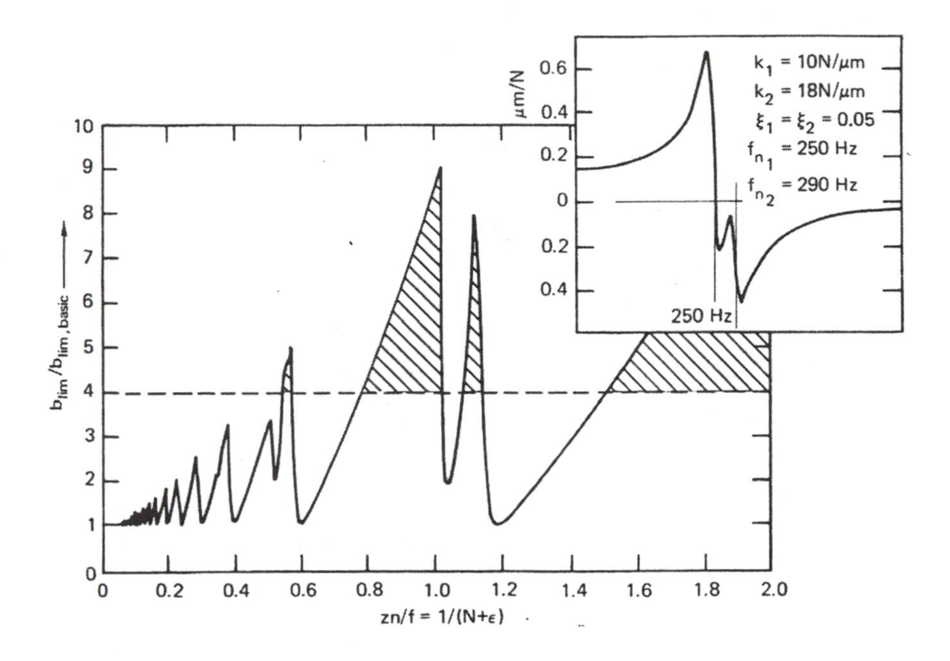

Figure 7.31. Stability "lobes" for a system with two degrees of freedom, from Tlusty[111].

The technique for using diagrams of "stability lobes" is important for securing higher cutting stability. On manually operated machines stability may be achieved by trial and error – not a very efficient process. This capability can often be incorporated in modern machine tool controllers if one is able to first determine the dynamics of the machine tool and availability of a "lobes" diagram for the particular machine, tool, and cutting process orientation.

7.7 Advanced analysis

The review of dynamics of machine tools presented here is based primarily on the work of Professor J. Tlusty. There has been substantial additional work done on the analysis of machine tool vibrations, machine tool design, and operating modes to insure stable machining under a variety of complex operations. Much of this is motivated by challenging machining operations required by the aerospace industry (airframes or engines). An excellent reference on these more comprehensive, and computationally based, techniques is the work of Professor Yusuf Altintas at the University of British Columbia in his book *Manufacturing Automation: Metal Cutting Mechanics, Machine Tool Vibrations, and CNC Design*[50].

VIII SENSORS FOR PRECISION MANUFACTURING

8.1. Introduction

8.1.1 The relevance of precision manufacturing and the need for in-process monitoring and control

New demands are being placed on monitoring systems in the manufacturing environment because of recent developments and trends in machining technology and machine tool design (high speed machining and hard turning, for example). In-process sensors play a significant role in assisting manufacturing systems in producing products at a cost affordable to the mass consumer market. In-process sensors are used to generate control signals to improve both the control and productivity of manufacturing systems. For example, acoustic emission sensors are used in many precision metal cutting processes to monitor the degree of tool wear, chip formation, surface features, etc., and in precision grinding, acoustic emission sensors are used to detect both the near approach of the grinding wheel to the work surface, and the initial wheel contact with the work[52]. In particular, in-process sensors are needed in precision manufacturing systems, because human oversight of the manufacturing process is inadequate to achieve the performance levels necessary in precision manufacturing processes. In the precision grinding example just mentioned, stock removal can be just a few tens of microns, and a human operator has a poor ability to monitor and control material removal in this regime.

Numerous different sensor types are available for monitoring aspects of the manufacturing and machining environments as illustrated by Moriwaki[53] in Figure 8.1. Of the range indicated, the most common sensors in the industrial machining environment are force, power and acoustic emission sensors. This chapter first reviews the state of sensor technology for manufacturing process monitoring in general. Then, details are given about the application of acoustic emission (AE) sensing to process characterization and monitoring. Special attention is paid to a review of the source of AE in metal cutting, signal processing of acoustic and other sensor signals and several examples are given. Finally, we review the basic design parameters and methodology for the development of acoustic emission sensors for particular applications.

A few themes are recognizable in all precision manufacturing operations. There is always a need to manufacture components at increasing levels of accuracy and finish. Often difficult-to-machine materials such as ceramics or composites are involved. Finally, increased productivity and price performance are required to bring the product to market at an affordable cost.

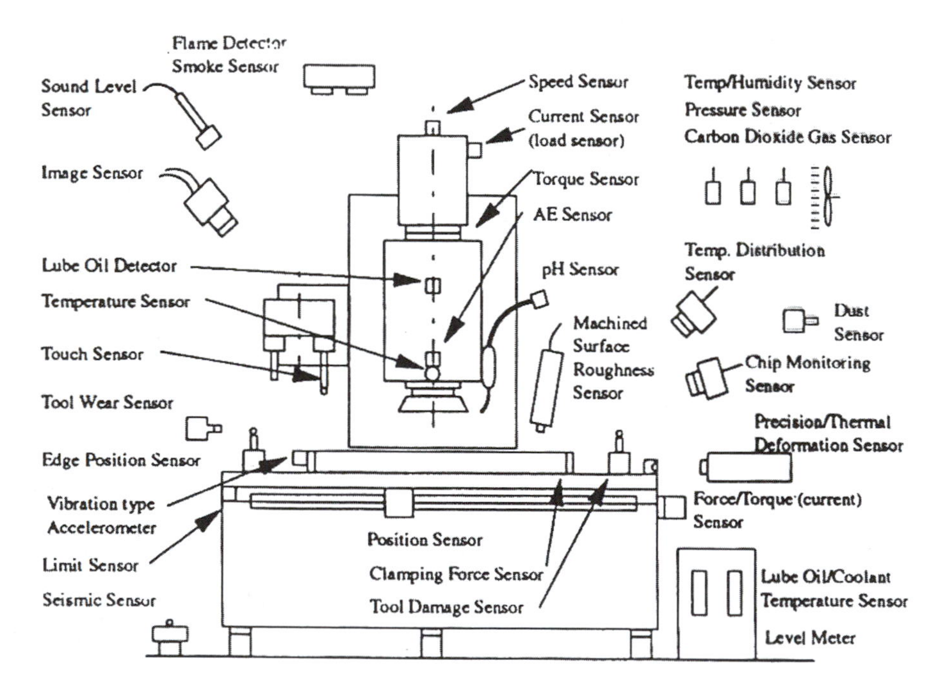

Figure 8.1. Abundance of sensors for machine tool monitoring, from Moriwaki[53].

In-process sensors are a significant technology helping manufacturers to meet the challenges inherent in manufacturing a new generation of precision components. In process sensors play different roles in manufacturing processes. First and foremost, they allow manufacturers to improve the control over critical process variables. This can result in the tightening of control limits of a process as well as improvements in process productivity. For example, the application of temperature sensors and appropriate control to traditional machine tools has been demonstrated to reduce thermal errors, the largest source of positioning errors in traditional and precision machine tools, and the work space errors they generate[119]. Secondly, they serve as useful productivity tools in monitoring the process. For example, as already stated, they improve productivity by detecting process failure as is the case with acoustic sensors detecting catastrophic tool failure in cutting processes. They also reduce dead time in the process cycle by detecting the degree of engagement between the tool and the work, allowing for a greater percentage of machining time in each part cycle.

Incorporation of an in-process sensor as a mission critical component of a manufacturing unit operation requires a high level of engineering confidence in the ability of the sensor to reliably detect the desired process characteristic. Without this confidence, manufacturers justifiably do not leverage in-process sensor technology to achieve the higher levels of process productivity they offer.

8.1.2 Requirements for sensor technology for precision manufacturing

Precision machining takes place at the sub-micron to nano scale dimensions (with respect to the uncut chip thickness, for example.) At these levels, the machining process, surface finish and chip formation are more intimately affected by the material properties such as ductile/brittle behavior or transitions in grinding or single point turning of brittle materials. These effects can adversely affect the surface quality or integrity of the machined component. Critical sensor information in precision machining is required mostly for assessing material removal at the sub-micron level, surface finish and subsurface damage. In addition, it is of interest to track for control purposes the variation in process parameters such as material removal rate (MRR), tool condition (e.g. wheel in grinding, abrasive in lapping, pad in chemical mechanical polishing) as well as process cycle related characteristics (e.g. contact or sparkout in grinding, air time in machining). These parameters are generally measured using sensors with very high sensitivity and with effective frequencies ranging to several MHz. In precision processes, sensor feedback information is critical for higher yields and process throughput.

Not surprisingly, different sensors have different applicability at different levels of precision, or displacement or MRR. Figure 8.2 below, see ref. 54 for details, shows a schematic diagram of different types of sensor applications for different precision levels and control parameters. The boundary represents the approximate range of usage with the shaded area emphasizing the core application range. Acoustic emission (AE) as illustrated here shows the greatest sensitivity (with the lowest noise level, i.e. highest signal to noise ratio) to the most critical process conditions in precision machining. Precision machining requires an attention to a number of work characteristics in addition to tolerance on dimension and, as the control parameters approach subsurface damage, the conventional sensing technologies from conventional manufacturing are less suitable. There are, of course, a host of other techniques for assessing roughness and subsurface damage but they are not usually considered to be useful "on line."

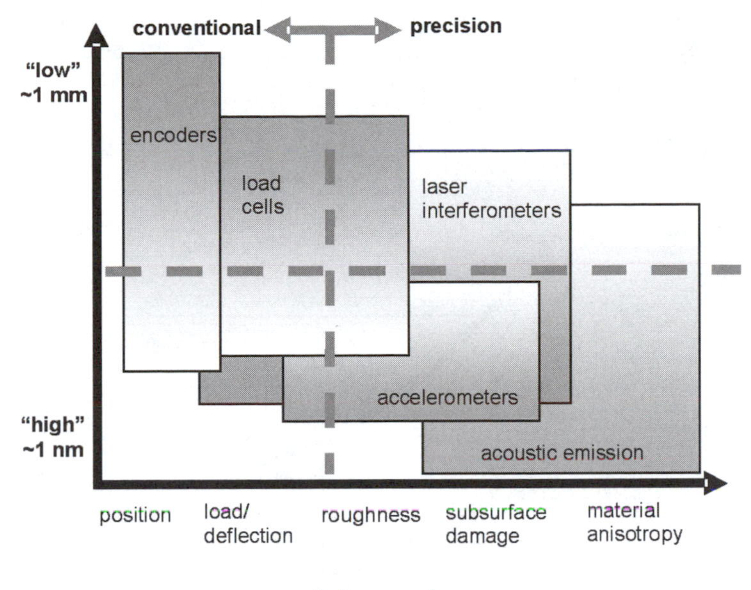

Figure 8.2. Sensor application vs. level of precision and error control parameters.

When material removal reaches the sub-micron level, essential signal features may be difficult to obtain. Conventional sensors such as force and vibration sensors suffer from inaccuracies due to the loss of sensitivity in the extremely high frequency range, where most of the micro cutting activities are sensed. However, sensors such as AE sensors exhibit improved response in the high frequency range, where much of the machine induced low frequency disturbance signals are diminished and the frequencies from sub-micron level precision machining activity becomes dominant (see Figure 8.3, details in ref. 54.) Therefore, by using sensors such as AE, noise from disturbance sources (bearings, spindle, slides, etc.) that generally contaminates the desired signal (for example, increases the variance compared to the mean) can be minimized and the micro cutting mechanism can be more effectively monitored.

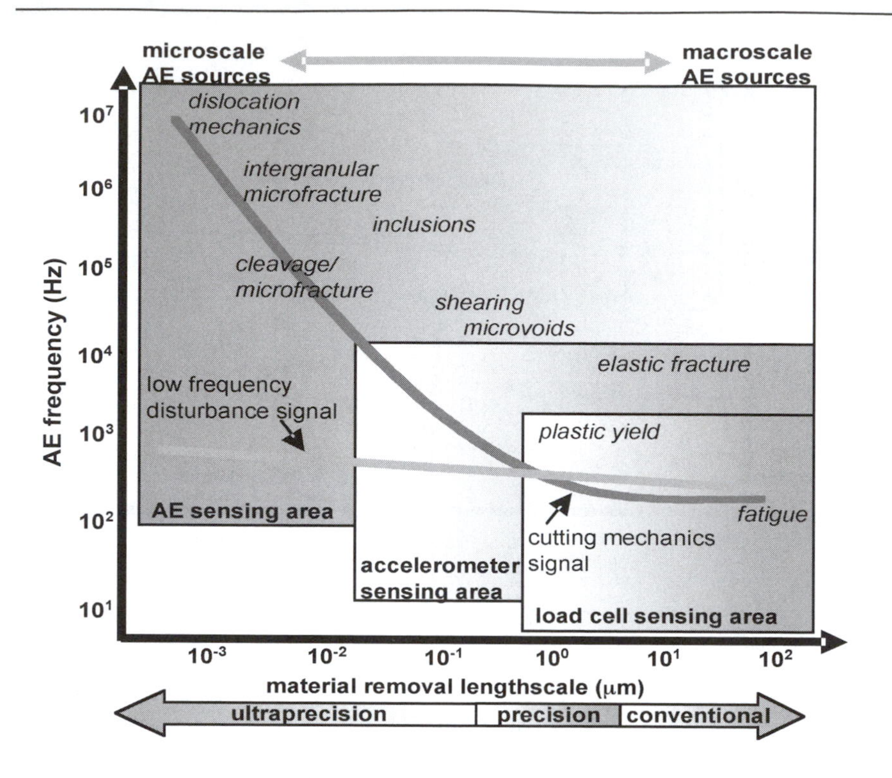

Figure 8.3. Signal/noise characteristics and sources of AE at different lengthscales of material removal.

8.2 Overview of sensors in manufacturing

8.2.1 Introduction

The transformation of stand-a-lone sensors used primarily as diagnostic devices in a machining process to sensors part of an intelligent system for tool and process monitoring and control has occurred most actively over the last decade. Kegg summarized the history of machine tool applications of sensors, and from the 1950's through the 1980's, these sensors were characterized by application of specific physical phenomena to sensing (thermocouples, piezoelectric crystals, accelerometers, strain gauges, acoustic emission, for example) a specific feature of the process (tool wear, spindle torque, tool vibration, for example)[55]. In the late 1980's and early 1990's, Byrne and Dornfeld discussed the influence of advanced signal processing techniques and artificial intelligence on the development and application of sensors and sensing systems[56, 57]. These so-called intelligent sensors included as part of their "packaging" abilities for self-calibration and self-diagnostics, signal conditioning, and, importantly, decision making. A new focus of much of the research on sensors was sensor fusion - the integration of information from several sensors to better and more robustly characterize a process or machine. This addresses Moriwaki's requirement for handling ambiguous or noisy inputs. It also lays the ground work for input to learning schemes, such as neural nets, to capture process knowledge when the process is sufficiently complex to defy clear mathematical modeling.

As mentioned earlier, monitoring of manufacturing systems is required now to insure that the optimum performance of these systems is obtained. The focus of monitoring is on either the machine (diagnostics and performance monitoring), the tools or tooling (state of wear, lubrication, alignment), the workpiece (geometry and dimensions, surface features and roughness, tolerances, metallurgical damage) or the process itself (chip formation, temperature, energy consumption). All four focus areas are subject to monitoring needs, often with competing requirements for time response or location of sensors. Thus, sensing systems for manufacturing processes must balance a number of options if they are to be effective.

There is a substantial amount of information in the literature on this topic area - mostly associated with elements of the intelligent machine tool such as control or monitoring. Comprehensive surveys have been published by Byrne and Dornfeld[56, 57], covering monitoring and control, Tönshoff, et al[58], and Tlusty[59] on sensors for unmanned machining. Prior to that one of the most complete reviews was done by Birla as part of the Machine Tool Task Force Study in 1980[60]. Other detailed reviews have been published on various aspects of machining and tool/workpiece monitoring. For example,

Shiraishi[61-63] reviewed, with numerous examples of applications and specifications on performance, sensors for machine, tool, workpiece and process monitoring in machining and Dornfeld[68] reviewed recent sensing techniques with respect to future requirements and intelligent sensors. Iwata[64] published the results of a survey of Japanese machine tool builders on their requirements and preferences on machine tool monitoring updating some of Birla's information on the same requirements. Finally, with a focus on drilling and tapping, Hoshi[65] reviewed techniques for automatic tool failure monitoring. The complete literature on machining and tool condition monitoring is extensive and is not reviewed here. The most recent compilation by Teti[66] lists over 500 citations in the last several years. Finally, Szafarczyk[67] has edited a volume of papers focusing on automatic supervision of manufacturing processes as part of an intelligent machine concept and includes, perhaps, the most recent comprehensive review of the subject from the perspective of sensors, signal processing, control, process modeling and integration with product design.

The sensors most frequently utilized for research on machine monitoring, and especially tool condition monitoring, are shown in Table 8.1 from Dornfeld[68]. Table 8.1 categorizes research activity in terms of the sensor(s) used but not, uniquely, the physical principle of operation of the sensor. That is, rather than piezoelectric elements for an accelerometer, ultrasonic sensor or acoustic emission sensor, the type of sensor is listed. In the cases just mentioned, distinction is made between "active" high frequency as in ultrasonics and "passive" high frequency as in acoustic emission. Audio sensors (i.e. in the range of 0-15 kHz) are listed as acoustic but low frequency. Although not a sensor, a category for "system design" is included because several very good contributions were reviewed. Applications listed include quite generic "machine diagnostics" as well as specific conditions monitored such as chatter or tool wear. This data was compiled from reviews of published literature and the *research activity level* indicated reflects the number of citations reviewed. In many of the references, a multi-sensor system or intelligent sensor is proposed as part of a monitoring strategy for machining. Some of these are described in more detail below.

Table 8.1. Sensor research for tool and process monitoring in machining.

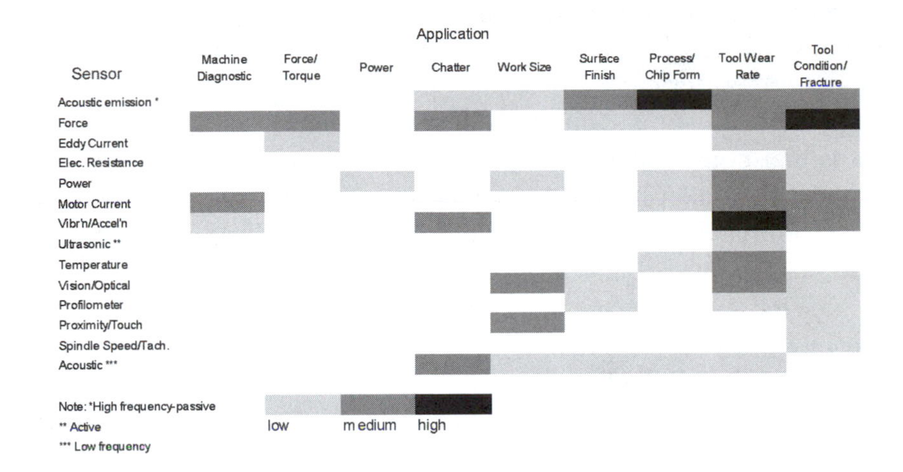

8.2.2 *Sensor systems for process monitoring*

With regard to sensor systems for manufacturing process monitoring, a distinction is to be made on the one hand between continuous and intermittent systems and, on the other hand, between direct and indirect measuring systems[69]. In the case of continuously measuring sensor systems, the measured variable is available throughout the machining process; intermittently measuring systems record the measured variable only during intervals in the machining process. Direct measuring systems employ the actual quantity of the measured variable, e.g. tool wear, while indirect measuring systems measure suitable auxiliary quantities, such as the cutting force components, and deduce the actual quantity via empirically determined correlations. Direct measuring processes possess a higher degree of accuracy, while indirect methods are less complex and more suitable for practical application. Continuous measurement enables the continuous detection of all changes to the measuring signal and ensures that sudden, unexpected process disturbances, such as tool breakage, are responded to in good time. Intermittent measurement is dependent on interruptions in the machining process or special measuring intervals, which generally entail time losses and, subsequently, high costs. Furthermore, tool breakage cannot be identified until after completion of the machining cycle when using these systems, which means that consequential damage cannot be prevented. Intermittent wear measurement nevertheless has its practical uses, provided that it does not result in additional idle time. It would be conceivable, for example, for measurement to be carried out in the magazine of the machine tool while the machining process is continued with a different tool. Intermittent wear-measuring methods can be implemented with mechanical[70-75], inductance-capacitance[76, 77], hydraulic-pneumatic[78] and opto-electronic[75, 79-87] probes or sensor systems. In view of the described disadvantages, this paper will not enter into further detail with regard to these systems or simple systems which are limited to detecting the correct position or the presence of a tool[88-97].

The direct sensor continuously measuring is the optimal combination with respect to accuracy and response time. For direct measurement of the wear land width, an opto-electronic system has been available, for example, whereby a wedge-shaped light gap below the cutting edge of the tool, which changes proportional to the wear land width, is evaluated[98]. The wear land width can also be measured directly by means of specially prepared cutting plates, the flanks of which are provided with strip conductors which act as electrical resistors[99-101]. A number of measuring processes have been developed for turning and milling, in order to detect the offset of the cutting edge[70, 76-78, 81, 82, 99, 102-107]. In each instance, the change in the distance from the sensor to the cutting edge or from the sensor to the workpiece is determined as the cutting-edge offset. As the cutting-edge offset alone is insufficient to describe tool wear, in addition to which the stated measuring methods - including those for measuring the width of wear land - remain unsuitable for practical deployment, these processes can only be regarded as laboratory methods for the time being. A new approach uses an image processing system based on a linear camera for on-line determination of the wear on a rotating inserted-tooth face mill. Non-productive time due to measurement is avoided and the system reacts quickly to tool breakage. There are, however, problems due to the short distance between the tool and the camera, which is mounted in the machine space to the side of the milling cutter, and due to chips and dirt on the inserts[108].

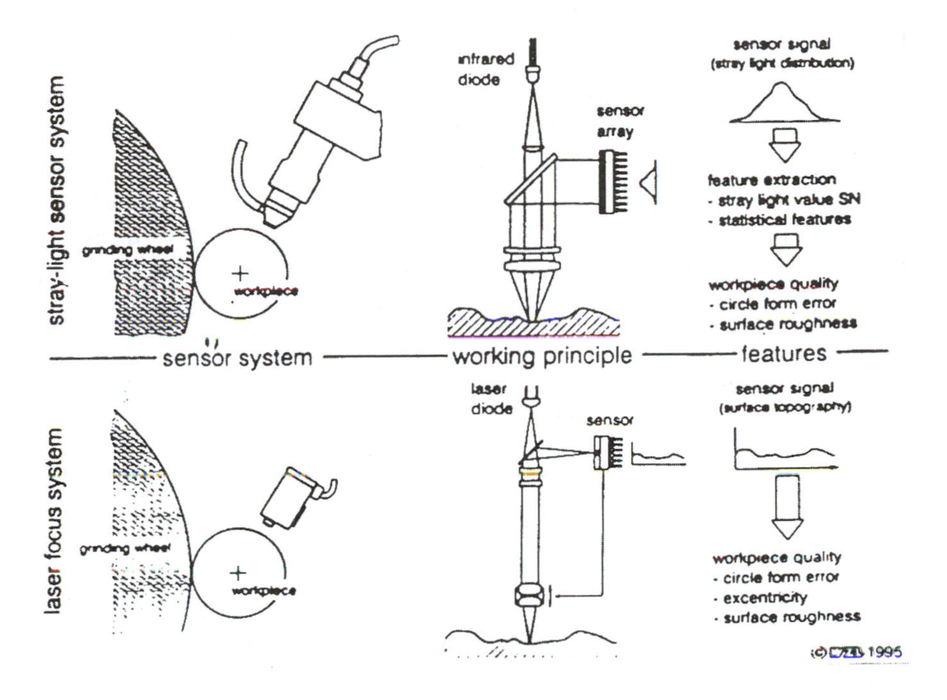

Figure 8.4. Optical sensor systems for surface quality measurement.

Two types of sensors are used to measure workpiece surface quality, which is relevant for the finished product in a finishing process and from which inferences can be made on tool wear. These are, first, ultrasonic sensors, for example coupled to the workpiece via the coolant[109]. At their present stage of development, they can only be used to measure stationary tools; direct tactile sensors also suffer from this drawback. The second possibility is to use optical methods[110-114] which can again be subdivided into two types: Stray light and laser beam sensors, Figure 8.4. With the stray light method, a diode, emits an infra-red beam which is focused on the workpiece, reflected from it and deflected onto a diode array. Various parameters can be deduced from the intensity distribution, indicating the waviness and concentricity of the workpiece. Because of the large light spot diameter, it is, however, difficult to achieve a good correlation between these parameters and workpiece roughness. This method is also confined to surfaces which have been ground. In the laser focus method, a sharply-focused laser beam is substituted for the infra-red light, giving much higher resolution compared to tactile measurements.

In grinding operations a laser triangulation sensor can be used to sense grinding wheel profiles at working speed. Light emitted from a laser diode is reflected from the grinding wheel and focused on a receiver diode by a lens. The reduced peak roughness R_{pk} and the reduced valley roughness R_{vk} can be detected from the signal, allowing the wheel topography to be determined, Figure 8.5.

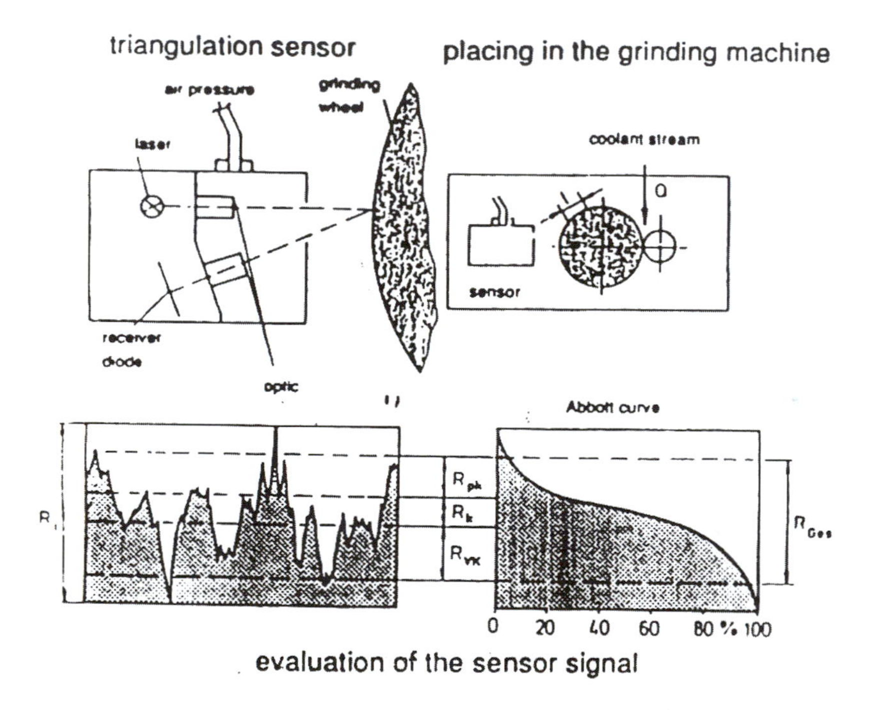

Figure 8.5. Optical system for monitoring of grinding wheel topography.

The problem of correlating signal values with tactile roughness values has meanwhile been solved through the use of appropriate peak filters. This method can also be used to determine the waviness and concentricity of the workpiece. It is suitable for use in cutting processes with a geometrically-defined cutting edge[112-114]. Because of the speed of measurement, the best that can be achieved with either method is quasi-on-line measurement, e.g. measurement during retraction of the grinding wheel. Generally, optical systems are sensitive to dirt and chips, and suitable countermeasures have to be taken. For these reasons, such measuring systems have been slow to find acceptance in industry.

The indirect continuous measuring processes, which are able to determine the relevant disturbance, e.g. tool wear, by measuring an auxiliary quantity and its changes, are generally less accurate than the direct methods. A valuable variable which can be measured for the purpose of indirect wear determination is the cutting temperature, which generally rises as the tool wear increases as a result of the increased friction and energy conversion. However, all the known measuring processes are pure laboratory methods for turning[115-120] which are furthermore unfeasible for milling and drilling, due to the rotating tools. Continuous measurement of the electrical resistance between tool and workpiece[121] is also unfeasible for practical applications, on account of the required measures, such as insulation of the workpiece and tool, and due to short circuits resulting from chips or cooling lubricant. Systems based on sound monitoring using microphones[122] also have not yet reached industrial application due to the problems caused by noises that are not generated by the machining process.

8.3 New developments in signal and information processing for tool condition monitoring

8.3.1 Introduction

The concern for the highest possible reliability, widest range of operating conditions (i.e. robust) and most sensitivity to the phenomena under observation (i.e. rich measurement vector) has driven the development of sensing methodologies for intelligent machining. The sensors most frequently utilized for research on machine monitoring, and especially tool condition monitoring (TCM), were shown in Table 8.1. Table 8.1 indicates a wide distribution in sensors and applications for machining monitoring. Figures 8.6 and 8.7 illustrate an array of force and acoustic emission technology applied to machine and tool monitoring. Over one quarter of the contributions can be classed as multi-sensor systems or intelligent sensors. This reflects the challenges of reliably detecting such phenomenon as tool wear over a reasonable range of industrial conditions using either only one sensor or straightforward signal processing. The bulk of the activity is in tool condition monitoring, fracture and wear, in turning and milling. The sensor technology showing the greatest "activity" is acoustic emission (AE), often employed with force sensors, for

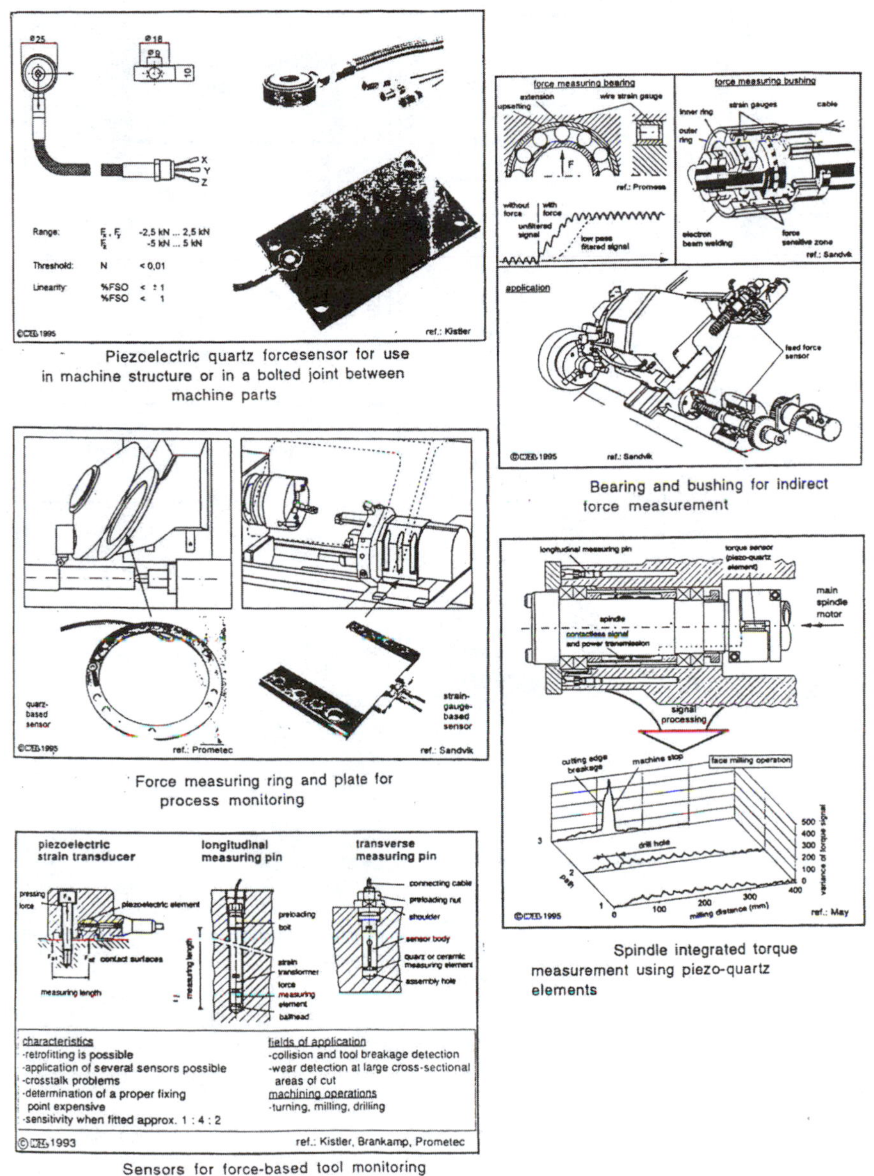

Figure 8.6. Sensors for force-based tool monitoring and their integration into the machine tool.

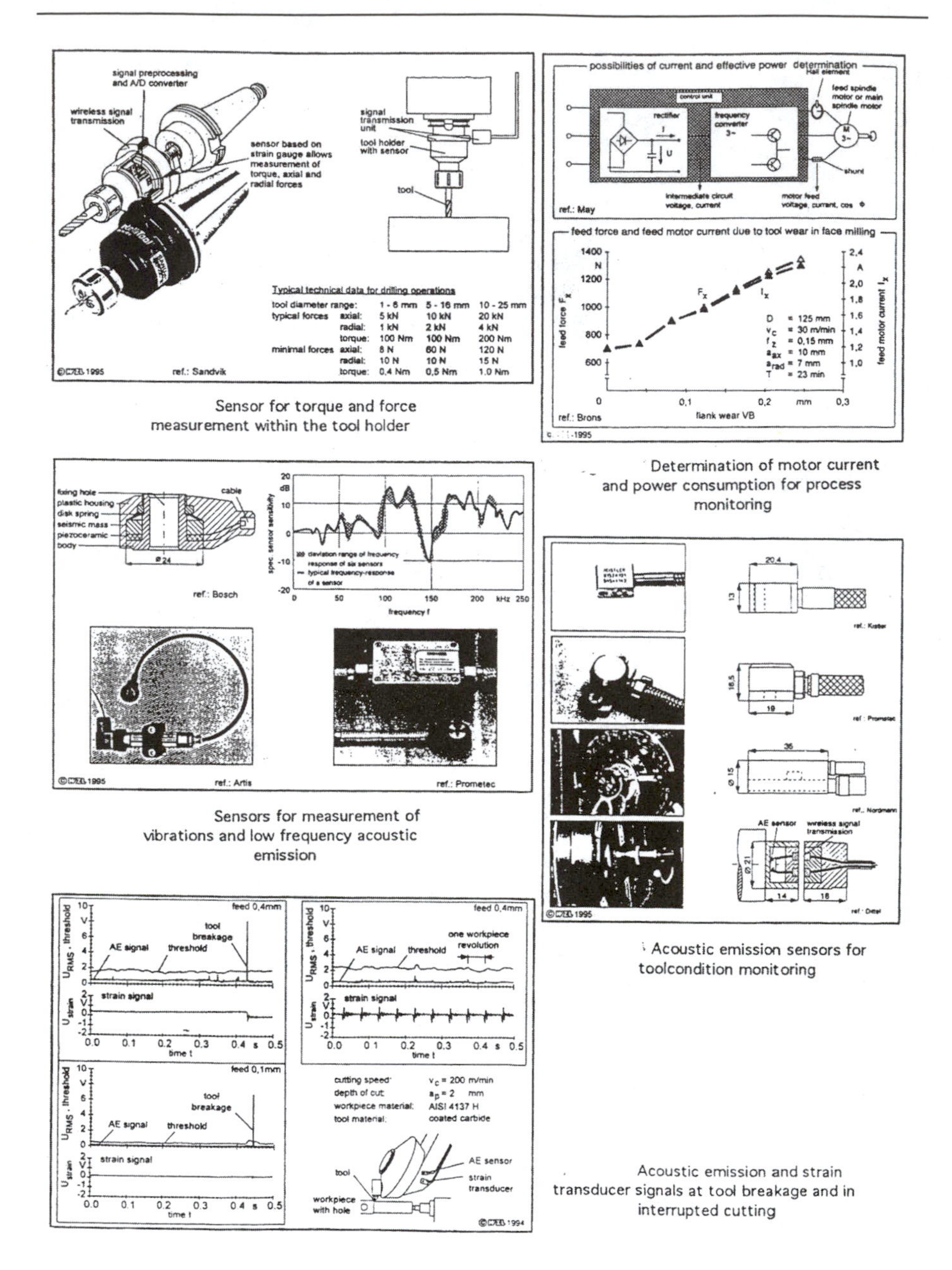

Figure 8.7. Vibration, acoustic emission, and motor current sensors and their application in tool and process monitoring.

tool wear monitoring. The additional sensitivity of acoustic emission to wear and fracture coupled with the high response rate of the signal has lead to increased utilization. However, this sensitivity has required coupling with more conventional sensor technology to reduce the heavy dependence of the AE signal with process parameters. As such, thus, it provides an excellent example of a sensor technology that requires careful signal processing/feature extraction and, often, integration with other sensor(s) to be most effective. A later section describes the source of acoustic emission in manufacturing processes and appropriate signal processing methodology.

8.3.2 Intelligent sensors

To a great extent, the multi-sensor systems referred to above exhibit many of the features of intelligent sensors. Generally, intelligent sensors have a much greater functionality than conventional sensors because they must respond to the special needs of the machine tool or process they are monitoring. Moriwaki refers to the capabilities of such sensors with respect to his discussion of the intelligent machine tool[123] when he distinguishes an intelligent machine tool from conventional machines as one driven based on "self decision making" as opposed to "predetermined commands." In addition to sensor feedback of the machining process (which Moriwaki classifies as part of adaptively controlled machines driven by predetermined commands), the intelligent machine is able to utilize experience and know how accumulated during past operation, accumulates knowledge through learning and can accommodate ambiguous inputs. Clearly, one of the critical elements in information accumulation, knowledge acquisition and accommodating "noisy" information, at least from the point of view of the process on the machine, is monitoring systems used in-process during machining, Figure 8.8[123].

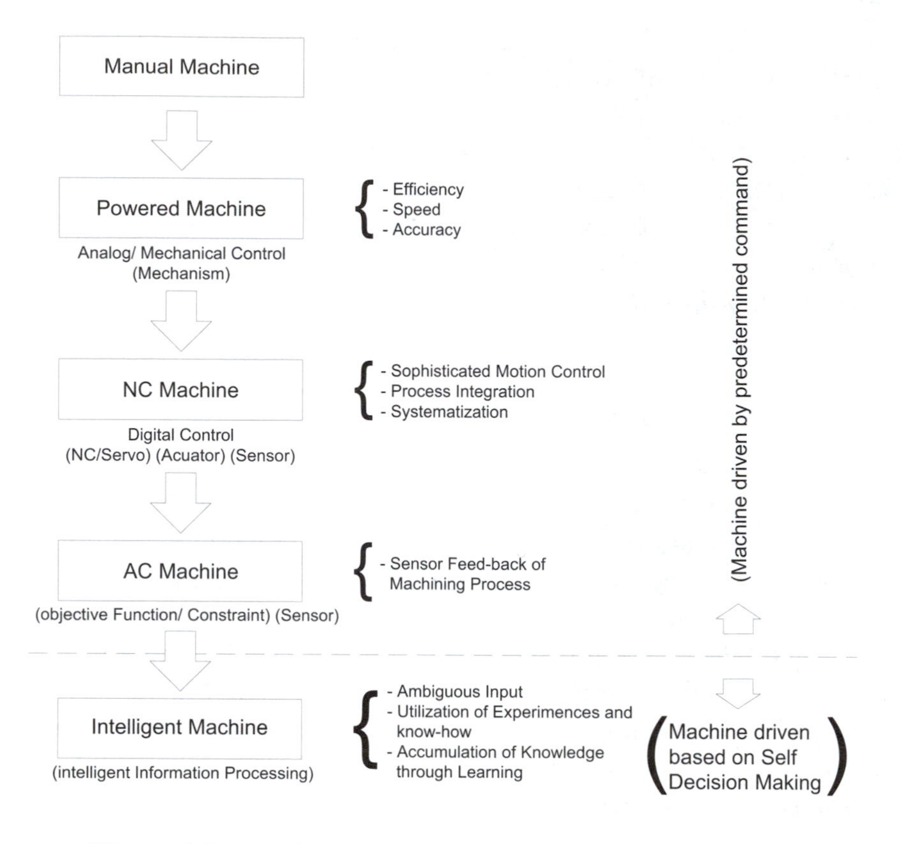

Figure 8.8. Development of the intelligent machine tool, from Moriwaki[123].

Intelligent sensors fill this requirement. They are able to do some or all of the following things:

- self-calibration
- signal processing
- decision making
- fusion ability
- learning capability

Signal processing in this case means that the sensor has the capability to do feature extraction from the measurement vector so that a data stream comes out of the sensor not just a signal. Decision making as part of the sensor system enables it to do such things itself not relying on the controller or other processors to do this. Sensor fusion describes the ability to combine or add the output of other sensors to provide a more robust decision on the process state. A very important aspect of the sensor is that it should be able to "learn" from past information using neural network or other knowledge representation scheme in order to continuously increase its reliability and robustness. An "intelligent sensor" is, thus, more or less a combination of conventional sensors, signal processing and feature extraction methods as well as implementation strategies that are integrated in the sensor or sensor system.

Developments in signal and information processing for TCM include the application of advanced sensor signal analysis (e.g. frequency content, cepstrum, time series and higher order spectrum analysis) in order to do feature extraction (determine and extract sensor signal features useful for conventional and artificial intelligence-based pattern recognition procedures). Using different approaches, researchers attempt to characterize a sensor signal (measurement vector) through a limited number of parameters (feature vector). Feature determination and extraction is particularly important in the case of acoustic emission (AE) signals as the AE sensor will generate between 10^6 and 10^7 data points per second during monitoring if one wishes to capture the full dynamic range of the signal. Analysis of information at this rate for real time applications cannot be carried out at any level of sophistication using currently available computers. The "curse of dimensionality", which plagues so many pattern recognition procedures, reaches here an exceptional intensity. A variety of methods for dimensionality reduction have been proposed.

Dornfeld conducted a study on some of the characteristics of research proposing multi-sensor or intelligent sensor systems[68]. The study indicates the sensor(s) used, the objective or condition(s) to be monitored, whether or not additional control or optimization is included in the study, nature of the signals processed (time or frequency domain), unique signal processing or fusion techniques employed or developed, any models used to relate the sensed parameters and extracted features to the process variables (usually spindle speed, feed or depth of cut), and the control variables used. The applications range from machine diagnostics to surface finish control although the categorization of objectives used is restricted to tool monitoring (TM), process control (PC), process monitoring (PM), and optimization (O). Several of the contributions evaluate the linkage of sensor(s) to the process control through ACO or ACC-like approaches and a few utilize advanced artificial intelligence-based techniques, neural networks, fuzzy logic or genetic algorithms for sensor fusion or sensor signal characterization and decision making. Obviously this is only a representation of the activity in laboratories around the world but it indicates the breadth of approaches. Most of the approaches are, at least, complementary to Moriwaki's adaptive control level, Figure 8.9, and many form the basis of the sensor portion of the intelligent machine.

8.3.3 Implementation strategies

The philosophy of implementation of any sensing methodology for diagnostics or process monitoring can be divided into two simple approaches. In one approach, one uses a sensing technique for which the output shows some relationship to the characteristics of the process. After determining the sensor output and behavior for "normal" machine operation or processing one observes the behavior of the signal until it deviates from the normal, thus indicating a problem. In the other approach, one attempts to determine a model linking the sensor output to the process mechanics and then, with sensor information, uses the model to predict the behavior of the process. Both methods are useful in differing circumstances. The first is, perhaps, the most straightforward but liable to misinterpretation if some change in the process occurs that was not foreseen (that is, "normal" is no longer normal.) Thus some signal processing strategy is required.

The signal that is delivered by the sensor must be processed to detect disturbances. The simplest method is the use of a rigid threshold, Figure 8.9. If the threshold is crossed by the signal due to some process change affecting the signal, collision or tool breakage can be detected. Since this method only works when all restrictions (depth of cut, workpiece material, etc.) remain constant, the use of a dynamic threshold is more appropriate in most cases[180]. The monitoring system calculates an upper threshold from the original signal. The upper threshold time-lags the original signal. Slow changes of the signal can occur without violating the threshold. At the instant of breakage, however, the upper threshold is crossed and, following a plausibility check (the signal must remain above the upper threshold for a certain time duration), a breakage is confirmed and signaled. Because of the high bandwidth of the AE signal fast response time to a breakage is insured. Of course, process changes not due to tool breakage (some interrupted cuts, for example) that affect the signal similarly to tool breakage, will cause a false reading.

Another method is based upon the comparison of the actual signal with a stored signal. The monitoring system calculates the upper and lower threshold value from the stored signal. In case of tool breakage, the upper threshold is violated. When the workpiece is missing the lower threshold is consequently crossed. The disadvantage of this type of monitoring strategy is that a "teach-in" cycle is necessary. Furthermore, the fact that the signals must be stored means that more system memory must be allocated. These methods have found applicability to both force and AE signal-based monitoring strategies.

These strategies work well for discrete events such as tool breakage but are often more difficult to employ for continuous process changes such as tool wear. The continuous variation of material properties, cutting conditions, etc. can mask wear related signal features or, at least, limit the range of applicability or require extensive system training.

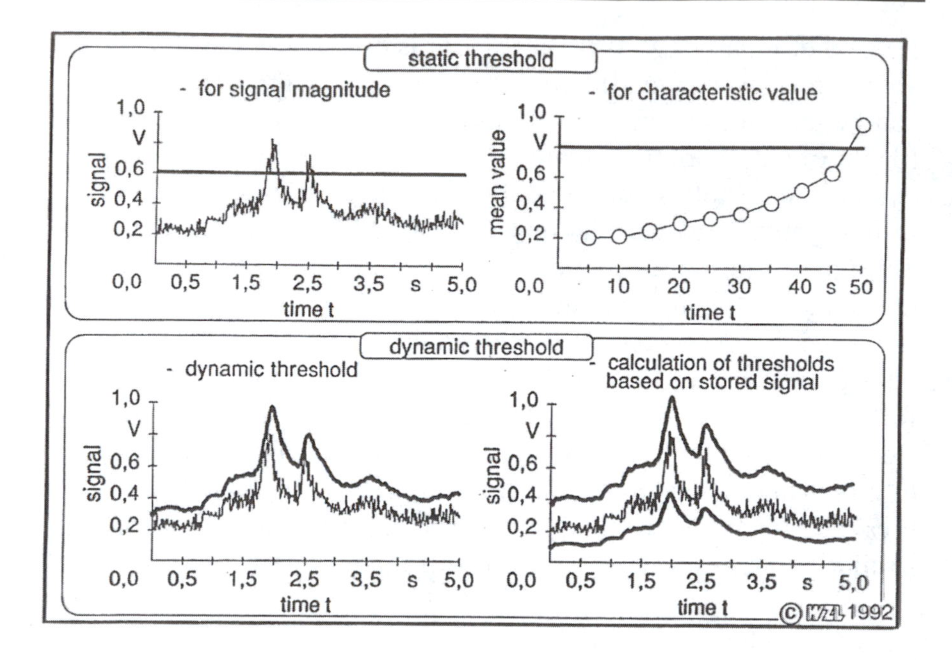

Figure 8.9. Implementation strategies with thresholds, from Byrne[56].

A more successful technique is based on the tracking of parameters that are extracted from signal features that have been filtered to remove process related variables (e.g. cutting speed). For example, using parameters of an auto-regressive model (filter) of the AE signal to track continuous wear[124]. The strategy works over a range of machining conditions.

8.3.4 Multisensor Approaches

The combination of different, inexpensive sensors today is ever increasing to overcome shortages of single sensor devices. There are two possible ways to achieve a multisensor approach. Either one sensor is used that allows measurement of different variables or different sensors are attached to the machine tool to gain different variables. The challenge in this is both electronic integration of the sensor as well as integration of the information and decision-making. Techniques for this information integration, often called sensor fusion, and decision-making will be described later in this paper.

8.3.5 Sensors for High Speed Machining

A number of process developments have occurred that impose additional challenges on the use of sensors in manufacturing processes. Specifically, these include high speed machining as well as hard material machining and the new initiatives at minimization of cutting fluids, so-called MQL or minimum quantity lubrication. Interestingly, it appears that the applications of high speed machining (HSM, of all materials, not just hard materials) and hard machining (at all speeds, not just HSM) intersect in a number of applications. This is driven by enhanced economics of manufacturing resulting from:

- reduced lead times and improved surface finishes (in the case of die manufacturing, for example),
- replacement of grinding by hard machining (for engine transmission components, for example),
- handling of more complex pieces and reduced part counts (for example, compared with sheet metal assemblies in the aerospace industry),
- reduced disposable costs (lubrication purchase, handling and disposal, for example).

The application interactions between these new technologies can be summarized in the table below.

Table 8.2. Application of new manufacturing technologies for various industries.

	Die/Molds	Electronics	Gear/transmission	Aerospace	Automotive
HSM	X	X	O	X	X
Hard M/cing	X	X	O		
MQL	X	X	X	X	X

Key: X in use O potential application

Future trends in these processes call for ever higher spindle speeds and smaller chip sizes. As a result, the "percentage change in *X* " (where *X* is the feature of the tool or workpiece under investigation …or the characteristic of the sensor signal being tracked) from "new" to "used" or "acceptable" to "unacceptable" which is the basic driver for monitoring is becoming increasingly smaller. As the percentage change decreases this reduces the effective signal for monitoring and, hence, reduces the signal to noise ratio (all else being equal). So the real challenge in sensing for high speed machining (with or without MQL, etc.) is the need for increased sensitivity and/or reduced noise. Demand for increased utilization of machine tools also requires improved reliability and availability for sensing systems.

These new developments pose some special challenges for sensing systems. These challenges include:

- Machining processes often have small, non-traditional uncut chip thickness, a_c, and shape; this is especially true for ball end mills; this yields low energy consumption in machining or unusual patterns of energy usage. Hence, some traditional sensing methods, based on cutting forces or torque, for example, may not work.
- Spindle speeds contemplated are at or above 100,000 rpm. The response time of the sensing system for reaction to events such as tool failure decreases in proportion to spindle speed. This, coupled with high rates of table feed (required to

- maintain material removal rates at high spindle speeds) which often approach 70-75 meters/minute and accelerations of 1.5 *g*, makes the required response times of sensor/signal processing/control system in the millisecond or sub-millisecond ranges to avoid damage to part or tooling.
- Tooling systems for these processes have become more complicated. For example, tooling must be carefully balanced to perform at high rotation speeds making the installation of the sensor more difficult.
- Failure of tooling may generate high energy particles or pieces sensor systems need to be more rigorously protected.
- Dynamic effects of tool entry and exit at high speeds can add additional noise to the sensor signal.
- MQL strategies can complicate the machining process and add noise to the sensor system. Chip clogging of tooling or adhesion to cutting edges are common problems.
- Dry cutting, the extreme of MQL, is often implemented with special tool coatings and treatments. This can complicate failure modes of tooling (wear patterns or chipping) and complicate sensing tasks.

However, there are also some benefits due to high speed machining that can be realized for sensing. These include the following:

- Better designed, stiffer machine tools and spindles mean lower levels of background "noise" and vibration for sensor systems.
- Tooling design for high speed use insures less geometric or size variability from tool to tool.
- Finish cutting means less chip thickness or cross section variability so "cleaner" sensor system.
- Many tools for high speed machining systems have integral chip disposal systems (for example, the Mitsubishi "Q" system) so there can be less interference from chips.
- Enhanced insert clamping techniques for solid failure-free clamps will insure less insert-to-insert geometry and size variability. This reduces sensor "noise".

- In some cases tooling has been designed to be more "flexible" in the sense of accommodating or adjusting to different diameter tools with the same shank. This means that shank mounted sensor systems can accommodate a wider variety of tool diameters with one built-in sensor.
- MQL and dry cutting, if used, reduce noise due to fluid splash and reduce the environmental protection needed for the sensor system.
- New tool coating materials reduce friction and reduce sensor noise due to secondary energy dissipation and chip stiction.

These advances in high speed machining will encourage some additional developments in sensor design, such as integrated sensors on tooling (thermocouple patterned inserts, for example) and will require non-contact high data rate signal transmission and advanced signal processing for detecting subtle (but important) changes in tooling conditions.

We review next the generation of sensor signals in manufacturing processes. The discussion centers around the acoustic emission sensing technology as it typifies the opportunities and challenges of process monitoring. It is important for the overall reliability of the system to understand the physics of the source of the sensor system.

8.4 Acoustic Emission in Manufacturing

8.4.1 Background

The philosophy of implementation of any sensing methodology for diagnostics or process monitoring can be divided into two simple approaches. In one approach, one uses a sensing technique for which the output shows some relationship to the characteristics of the process. After determining the sensor output and behavior for "normal" machine operation or processing one observes the behavior of the signal until if deviates from the normal thus indicating a problem. This was illustrated in the previous section. In the other approach, one attempts to determine a model linking the sensor output to the process mechanics and then, with sensor information, use the model to predict the behavior of the process. Both methods are useful in differing circumstances. The first is, perhaps, the most straightforward but liable to misinterpretation if some change in the process occurs that was not for seen (that is, "normal" is no longer normal.) To insure against this type of misinterpretation, intelligence has been added to the sensors to add sophistication to the feature extraction and decision-making process. This will be discussed later. We will first look at the potential of acoustic emission in machinery diagnostics. Although this discussion is focused on acoustic emission, much of the analysis and interpretation can be applied to other similar sensing technologies, especially with regard to process and machinery variations and their impact on sensor signals.

Machinists have long used audible feedback during machining as a means of monitoring the cutting process, with skilled machinists able to judge even minute variation in tool wear and surface finish simply by listening to the machining process. The unique sound given by metals during plastic deformation is nothing new, with the unique "tin cry" sound given by tin being plastically deformed as a common one. The above examples all share the common trait of the generation of elastic stress waves within a medium due to plastic deformation. The term *acoustic emission* typically refers to elastic wave propagation in the ultrasonic frequency range (~20-2000 kHz). Unlike ultrasonic non-destructive techniques (NDT), which are a means of active scanning (i.e. generation, transmission, and collection of signal), acoustic emission is mostly a passive means of scanning, much akin to holding a microphone or other sensor and "listening" for various phenomena.

8.4.2 Acoustic emission sources-diagnostics

The table below, Table 8.3[127] lists potential AE sources in machinery as well as variables that can affect the characteristics or amplitude of the AE signal for application in machinery diagnostics. One could apply either of the two abovementioned approaches to implementation of AE but the sources listed would be contributing to the signal detected. With this signal information cycle consistency, material consistency, machine component behavior, etc. could be determined. To any of the above signal amplitude or characteristics one could pose questions about the constancy of the effect. That is, will there be:

- constant or variable rate/occurrence
  - -variable in time/cycle
  - -variable in frequency or periodicity
  - -speed
- constant or variable transducer distance to source
- constant or variable volume of material deforming
- constant or variable clamping force
- constant or variable hydraulic fluid flow rate
- variation in lubrication (area covered, viscosity, temperature, quantity)

Normally, for most effective diagnostic application, the variation of only the source must be insured. The reliability of the AE-based diagnostic system is dependent on the designer's ability to consider all of these potential variables. In many cases, the major factors affecting the AE signal are sufficiently dominant as to render the "second order" effects inconsequential. The literature does not list extensive application of AE to machinery diagnostics except for specific applications such as bearing monitoring, and electrical transformer malfunction detection.

8.4.3 Acoustic emission sources-process monitoring

The use of acoustic emission-based sensing for manufacturing process monitoring is much better documented. Whether or not it has been much more successful is not clear as many applications are complex and ones in which competing technology either does not exist or has not been effective either.

Table 8.3. Acoustic Emission and Machinery Diagnostics.

AE Mechanism Sources	Signal Amplitude or Characteristics
mechanical impact	contact force between elements
rolling elements	relative velocity of element surfaces
translational elements	contact/impact velocity
electrical noise	pressure
hydraulic noise	periodicity/frequency of source*
hydraulic cycle variation	contact material(s)
gear engagement	contact material relative properties
mechanical contact	(hardness, density, mechanical
	strength)
clamping action	surface characteristics on contacting
	elements (impact and localized de
	formation)
	size of contacting areas
arcing/sparking	signal transmission distance (fixed
	or varying)
arc voltage/current variation	transmission media
friction/rubbing	transducer coupling efficiency
element fracture	voltage/current of discharge
element distortion	area of fracture
(plastic deformation)	rate of fracture propagation
	rate of deformation
	volume of deforming material
fluid pressure and flow rate	entrapped gas in fluid
	eccentricity/runout effects
	lubrication

*as distinguished from frequency of AE signal (typically in 100 kHz to 1 MHz range)

Hoshi[126] reviewed sensing applications in use in industry and compared the reported success of many differing sensing applications in machining processes. Included by Hoshi are reviews of the following measurements or sensors (the most common "competing technologies") for edge chipping, fracture, and wear (related to cutting tools), and poor hole quality (drilling): touch sensors, load amperage, vibration, torque/force limiter, and acoustic emission. The bulk of the processes monitored are drilling and the more straightforward approaches of monitoring, tool touch, and load amperage, are the most frequently used. Vibration sensing and acoustic emission sensing are the next most often applied but, according to Hoshi, only touch sensing was reported to be 100% successful. Load amperage, vibration and AE were reported to be 80%, 75% and 33% successful, respectively. This indicated that much additional research is needed to make these techniques useful.

The review of Hoshi listed applications for which some attempts at AE sensing have been made. In fact, the potential for AE sensing is much greater. Dornfeld[127] has listed manufacturing processes and their likely AE generating sources as well as potential process changes and product defects that could be detected using AE signals. All of the processes are material deformation-based manufacturing processes. They use either continuous or discontinuous application of energy to reform or remove material in one way or another. The process monitoring or product defect monitoring listed is based on either deformation (including friction and rubbing) or fracture derived AE. In one or two cases material metallurgical transformation is an AE source. The reference defines the likely zone or area of the process in which the AE is generated based upon the mechanics of the process. Thus process behavior likely to deform the material being "processed" will be a likely source of deformation-based emission. In deep drawing, for example, the deformation is non-steady (meaning variable over the process cycle *and* variable over the workpiece. Defects are due to undesirable changes in the formability of the material due to nonuniform lubrication, misalignment of punch and die or wear. These changes will alter the observed pattern of AE during a punching cycle and can be directly associated with an aspect of the process. Product defects, such as wrinkling or edge cracking, will also generate AE appropriate to the source. That is, wrinkling is instability in the deformation of the material due to insufficient hold-down pressure on the flange or the effects of peculiar part geometry in the presence of die wear, for example. This results in additional deformation-based AE. Edge cracking will generate distinct fracture-based AE. Thus, depending on the type of the AE source, and whether the source is expected to be steady or non-steady, on must use the appropriate signal processing methodology. Examples of AE in deformation process monitoring include upsetting[128], punch stretching[129] and wire drawing [130].

Sensor location and signal processing are not always straightforward considerations. As seen in Tables 8.2 and 8.3 many other considerations are present as well. Thus the choice of time or frequency-based techniques, energy (root mean square (RMS)) or count/count rate-based analysis, etc. are quite dependent on the specifics of the process being monitored and objective of the monitoring. There is insufficient room here to explore all of the interactions detailed in Table 8.3. Some examples of monitoring of machining processes (generally steady or continuous sources of AE except for distinct "events" such as tool fracture) will be presented in more detail. There is also great possibility for AE application in non-traditional (meaning here "non-deformation based") as well and some attempts have been made, especially in welding and casting. An excellent review of other process monitoring in the curing of composite materials, resistance spot weld monitoring, electron beam and laser spot welding and a wide variety of applications in the electronics industry (cracks in ceramics and silicon substrates, contact flaws, flux jump in superconducting magnets, particle contamination, etc.) is Vol. 5 of the *ASNT Nondestructive Testing Handbook,* Miller[131].

8.4.4 Acoustic emission in machining

Research over the past several years has established the effectiveness of AE based sensing methodologies for machine tool condition monitoring and process analysis. The problems of detecting tool wear and fracture of single point turning tools motivated much of this early work. In addition, the sensitivity of the AE signal to the various contact areas and deformation regions in the cutting and chip formation process has led to the analysis of AE signals as a basic tool for the analysis of the cutting process. Investigations of AE from metal cutting have often been limited to two-dimensional or orthogonal machining because of the simplicity of the geometry and chip flow. Principal areas of interest with respect to AE signal generation are in the primary generation zone ahead of the tool where the initial shearing occurs during chip formation, the secondary deformation zone along the chip-tool rake face interface where sliding and bulk deformation occurs, and the tertiary zone along the tool flank face-work surface interface. Finally, there is a fourth area of interest that is associated with the fracture of chips during the formation of discontinuous chips. If one is studying the milling process (or other interrupted cutting) an additional source of AE is the impact of the tools on the workpiece and the noise due to the swarf motion on the tool and work. Moriwaki[132] reviews other sources of AE from metal cutting.

A number of studies on developing models of AE generation in machining (Dornfeld[133], Dornfeld and Kannatey-Asibu[134, 135] and Rangwala and Dornfeld[136, 137]) have established the principle role of process parameters, especially cutting speed, in the determination of RMS energy of the signal. For conventional machining the friction and rubbing accompanying the cutting are, perhaps, the most significant sources of AE and are dependent on the cutting speed as well, Heiple, et al[138]. For precision machining, such as diamond turning, the model-based predictions for AE sources are much more accurate. Teti and Dornfeld[139] view both event-based (count-rate) and energy-based research on AE from metal cutting.

A basic model for the generation of AE during machining (in this case primary and secondary shear generated AE in orthogonal machining) was proposed by Dornfeld and Kannatey-Asibu[134, 135]. The formulation of the model is based on the simplified Ernst and Merchant model of orthogonal machining and builds a dependency of AE energy on material properties such as flow stress, volume of material undergoing deformation and the strain rate. Incidentally, almost every other effort at modeling acoustic emission from deformation-based manufacturing processes is built on this same approach.

The generation of AE was derived from the portion of the applied energy that results in plastic deformation. Hence, a strong dependency on cutting speed (influence strain rate), depth of cut and feed rate (both influencing volume of material undergoing plastic deformation) was seen. The relationship between the emission signal and the cutting parameters based on the Ernst and Merchant model can be written as:

$$V_{rms} = C\left\{\tau_k dU\left[\frac{\cos\alpha}{\sin\phi\cos(\phi-\alpha)}f + \frac{1}{3}(l+l_1)\frac{\sin\phi}{\cos(\phi-\alpha)}\right]\right\}^{0.5} \tag{8.1}$$

where

d = depth of cut (width of chip)
f = uncut chip thickness
τ_k = average material shear strength
ϕ = shear angle
α = rake angle
U = cutting velocity
l = contact length between the chip and the tool rake face
l_1 = length from the tool edge to the end of the sliding zone on the tool rake face
U_c = chip velocity - $\sin\phi/\cos(\phi-\alpha)\,U$

and C is a proportionality constant influenced by tool geometry, instrumentation gain, etc. The first term

$$df\tau_k\frac{\cos\alpha}{\sin\phi\cos(\phi-\alpha)}U$$

corresponds to the work rate in the primary shear zone. The second term

$$\frac{1}{3}\tau_k d(l+l_1)\frac{\sin\phi}{\cos(\phi-\alpha)}U$$

corresponds to the work rate in the secondary shear zone. This model was reasonably successful in predicting the energy (RMS) of the AE signal during orthogonal machining under conventional machining conditions (i.e. typical feedrates and depths of cut) due to speed variations but did not accurately predict the influence of feed and depth of cut variations.

Tests by Pan and Dornfeld[140] and Liu and Dornfeld[141] of diamond machining of a variety of materials indicated better correlation between the model predicted values and experimental results. This is due in some sense to the reduced contact area at the tool flank, the more sharply defined cutting edge, and reduced friction on the rake face with diamond machining. Lan and Dornfeld [90] modified the basic model to include the contribution due to flank contact and to include relative attentuation of the signal from the sources in the process. This work added the term $\tau_d Uw$ (where w is the average length of the tool flank-workpiece interface), to equation (8.1) to account for flank wear land generated AE. In addition, the coefficient of friction set as 0.5 in equation (1) was allowed to vary, which is more consistent with the results of the almost "ideal" diamond turning operations carried out in.[140, 141] Equation (8.1) was, thus, rewritten as:

$$V_{rms} = C_1 \left\{ \tau_k dU \left[C_2 \frac{\cos\alpha}{\sin\phi\cos(\phi-\alpha)} f + C_3 \frac{1}{3}(l+2l_1) \frac{\sin\phi}{\cos(\phi-\alpha)} + C_4 w \right] \right\}^m \tag{8.2}$$

where C_2, C_3, and C_4 are factors of signal attenuation, and m is material dependent. The factors of signal attenuation C_2, C_3, and C_4 correspond to signal transmission losses between the shear zone, chip interface zone, and wear zone and the transducer on the tool shank, respectively. From Lan[142], the attenuation factors were experimentally evaluated during turning tests. The magnitude of C_2 was found to be between 0.20 and 0.25, based on lead break calibration tests, while both C_3, and C_4 could be assumed to be 1. This latter assumption is reasonable since, for C_2, the contact between the primary shear zone and the tool is only a very small area, i.e. the cutting edge, so that significant signal attenuation is expected. The AE generated at the tool-chip interface and at the tool flank-workpiece is directly on the tool surface, so little signal attenuation results. Attenuation of the AE in the cutting tool was included in the factor C_1. These tests for signal attenuation were done with time domain signals under the assumption that the various interfaces involved had no significant affect on the frequency domain characteristics.

Additional studies have focused on AE generated during machining based on experimental observations. Lan and Naerheim[143, 144] have investigated the effect on the AE generated of changes in machining conditions and the influence of lubrication with differing materials. Schmenk[145] suggested the use of metal removal rate to normalize the AE energy measured during machining to reduce the sensitivity to process parameters. Finally, Diei and Dornfeld studied the AE generated during face milling operations with form and insert tooling.[146]

Rangwala and Dornfeld[136] further developed the dependence of acoustic emission signal energy and spectral characteristics on machining parameters in orthogonal machining. All previous AE generation mechanisms in machining proposed deal with the low strain rate case, where thermal activation is an important factor for unpinning dislocations from impurity points in the material. In this case, V_{rms} has been found to be proportional to the square root of the plastic strain rate.[147, 148] James and Carpenter have shown that in these strain rate regimes, AE is generated by the unpinning of dislocations from impurity points. Since dislocation unpinning is controlled by the thermal component of the shear stress (which is a logarithmic function of strain rate), V_{rms} is proportional to the square root of the strain rate.

At higher strain rates (such as in the damping region), a dislocation spends more time traveling between obstacles rather than waiting at pinning points. In this case therefore, it is possible that the actual motion of the dislocation between obstacles rather than unpinning is the major contributor to the detected AE. As shown below, the dislocation damping power is proportional to the square of the strain rate so that if it is assumed that AE signal power is proportional to the damping power, we expect that V_{rms} will be proportional to the plastic strain rate. Asibu and Dornfeld[135] had suggested that a possible mechanism for AE generation at the high strain rates in metal cutting could be the strain energy released in overcoming drag forces in the medium. Since the flow stress at high strain rates is a function of the strain rate, a power law relationship between V_{rms} and the strain rate was proposed as

$$V_{rms} \propto \dot{\varepsilon}^{m}$$

where m is related to the strain rate sensitivity of the material as in (8.2) above.

If dislocation damping is responsible for AE generation in metal cutting, the power in the AE signal should be proportional to the damping power P_d. Power in the AE signal is given as:

$$\text{AE Signal Power} = V_{rms}^{\;2} \tag{8.3}$$

The damping power, F, can be calculated by considering that the damping force on a unit length of dislocation segment is $F = B\,v$, where B is the damping coefficient and v is the dislocation velocity, so that power expended to move a unit length of dislocation is Bv^2. The total length of dislocations moving at any instant of time is given by $\rho_m V$, where ρ_m is the mobile dislocation density and V is the plastically deformed volume. Thus:

$$P_d = Bv^2 \rho_m V. \tag{8.4}$$

Using the relationship between strain rate and dislocation velocity

$$\dot{\varepsilon} = \rho_m b v$$

where b is the Burger's vector, and the macroscopic viscosity coefficient defined as

$$\mu = \frac{B}{\rho_m b^2}$$

equation (8.4) can be written as:

$$P_d = \mu \varepsilon \acute{Y}^2 V. \tag{8.5}$$

This shows that the RMS value of the AE signal in metal cutting is proportional to the plastic strain rate. Two experimental results indicate that damping may be responsible for AE generation in metal cutting. First, from the data presented by Asibu and Dornfeld[135], V_{rms} is seen to vary linearly with the cutting velocity which indicates a linear relationship with the plastic strain rate. Secondly, the viscosity coefficient for steel is approximately 1.8 kN-s/m^2 and for aluminum, it is 1.5 kN-s/m^2, so that according to equations (8.3) and (8.5), at a given strain rate, steel should generate AE of a higher energy level than aluminum[149]. Since the strain rate is essentially a function of U (some dependence of strain rate on material properties is present, however, the dependence on cutting velocity is much stronger), we expect that at a given cutting velocity, AE RMS would be greater for steel than for aluminum. This is verified by Asibu and Dornfeld[135] which shows the slope of the linear relationship between V_{rms} and cutting velocity U is higher for steel than for aluminum. The relationship is especially applicable at low chip-tool contact lengths where the primary source of AE generation is due to plastic deformation. At higher contact lengths, more of the contibution to the AE energy is due to sliding along the chip- tool interface.

Rangwala and Dornfeld[137] also attempted to develop basic models for the spectral content of acoustic emission generated from machining to complement the energy-based analysis. The power spectrum reveals the variation of the AE signal power at discrete frequencies, however, phase information is lost during transformation from time domain to power spectral representation. The AE signal detected at the sensor is an averaged effect of a large number of discrete dislocation sources operating in the medium. Rouby et al[151] have developed a model to explain the spectral characteristics of acoustic emission generated due to disclocation motion. They propose that the motion of a dislocation segment between obstacles produces a stress pulse which is transmitted as an elastic wave to the surface of the medium. This is consistent with the AE generation mechanism discussed above.

The connection between dislocation motion and spectral characteristics can be summarized as follows. The time of flight of the dislocation segment between obstacles is given by:

$$\tau = \frac{D}{v}$$

where τ is the mean time in seconds, and D is the mean dislocation distance between obstacles (also defined as the mean free path of the dislocation). This assumes that the time for acceleration and deceleration of the dislocation is negligible. The motion of the dislocation produces a surface displacement at the sensor. Most AE sensors are sensitive to the surface velocity, so that the spectrum of the velocity wave produced at the sensor (and hence the spectrum of the detected AE) is expected to reflect the average behavior of the dislocation velocity spectrum. Since the accelerative and decelerative periods of the dislocation motion are assumed to be negligible, the dislocation velocity *vs* time profile is essentially a rectangular pulse in the time interval $[0, \tau]$ with a fundamental frequency of

$$\omega = \frac{1}{2\tau} = \frac{v}{2D} \tag{8.6}$$

Equation (8.6) can only be interpreted in a qualitative manner because, in reality, the spectral characteristics will also be affected by transmission properties of the medium sensor-surface coupling characteristics, sensor response and the response of the recording media. These considerations will be discussed later in this paper. The general trend to be expected, however, is that as the dislocation velocity increases or the mean free path decreases a larger proportion of the AE signal will be concentrated at higher frequencies.

In the context of metal cutting, it is of interest to observe how the mechanical variables of the operation affect the spectral characteristics of the generated AE. Dislocation velocities are related to the plastic strain rate according to the following:

$$\dot{\varepsilon} = \rho_m b v \tag{8.7}$$

where ρ_m is the mobile dislocation density and b is the magnitude of the Burger's vector. From equations (8.6) and (8.7), it is expected that process parameters which increase the strain rate would tend to shift the signal power towards higher frequencies. Additionally, the high levels of strain in the primary and secondary shear zones affect the state of strain hardening in the material. At high levels of strain hardening, the mean free path of the dislocations is reduced because of the presence of a large number of dislocation forests and sessile dislocations which effectively act as pinning points and obstruct the motion of mobile dislocations. Therefore, plastic deformation of highly strain hardened material is expected to shift AE signal power towards higher frequencies. On the other hand, thermal softening effects in the machining zone increase the mean free path D, and shift the spectra toward lower frequencies.

Rangwala and Dornfeld[137] experimentally observed how the AE spectrum changes with metal cutting process parameters such as speed, feed rate, and chip-tool contact length. The mean frequency f_m and the standard deviation of the frequency were calculated from the measured signal. The mean frequency is defined here as:

$$f_m = \frac{\sum_{i=1}^{n} S_i f_i}{\sum_{i=1}^{n} S_i}$$

where S_i is the signal power at the frequency f_i and n is the number of frequencies over which the spectrum is averaged. The parameter f_m indicates the frequency which divides the total power spectrum into two equal parts. The results showed that the spectral content of the

AE signal shifts towards higher frequencies as the cutting velocity is increased. Eventually, however, a point is reached when a further increase in velocity causes the signal power to shift to lower frequencies. This is attributed to thermal softening effects associated with increasing velocity, which causes the secondary zone AE to shift toward lower frequencies. The effects of feedrate on frequency do not show a clear trend except at the highest velocity where an increase in feedrate causes a drop in the mean signal frequency, f_m. It was also observed that the mean frequency increases with the chip-tool contact length (beyond a certain minimum length) and this was attributed to higher amounts of chip sliding at the higher contact lengths. The higher man frequencies at lower contact lengths are attibuted to instabilities and the associated high strain rates in the primary shear zone.

8.5 Signal processing, feature extraction and sensor fusion

8.5.1 Introduction

Human monitoring of manufacturing processes can attribute its success to the ability of the human to distingiush, by nature of the physical senses and experience, the "significant" information in what is observed from the meaningless. In general, humans are very capable as process monitors because of the high degree of development of sensory abilities, essentially noise free data (unique memory triggers), parallel processing of information and the knowledge acquired through training and experience. Limitations are seen when one of the basic human sensor specifications is violated; something happening too fast to see or out of range of hearing or visual sensitivity due to frequency content. These limitations have always served as some of the justification for the use of sensors. Sensors, of course, are also limited in their ability to yield an output sensitive to an important input. Thus we need to consider the use of signal processing and along with that feature extraction. In most cases the utilization of any signal processing methodology has as its goal one or more of the following: the determination of a suitable "process" model from which the influence of certain process variables can be discerned; the generation of features from sensor data that can be used to determine process state; or generation of data features so that the change in the performance of the process can be "tracked".

An overview of signal processing and feature extraction is conveniently summarized in Figure 8.10 below, see Rangwala[152] for details. Here the measurement vector extracted from the signal representation from the sensor (basic signal conditioning) is the "feedstock" for the feature selection process (local conditioning) resulting in a feature vector. The characteristics of the feature vector include signal elements that are sensitive to the parameters of interest in the process. The "decision-making" process is characterized in the lower portion of the figure. Based on a suitable "learning" scheme which maps a teaching pattern (i.e. process characteristics we desire to recognize) onto the feature vector, a pattern association is generated. The "pattern association" contains a matrix of associations between the desired characteristics and features of the sensor information. In application, the pattern association matrix operates on the feature vector and extracts correlations between features and characteristics - these

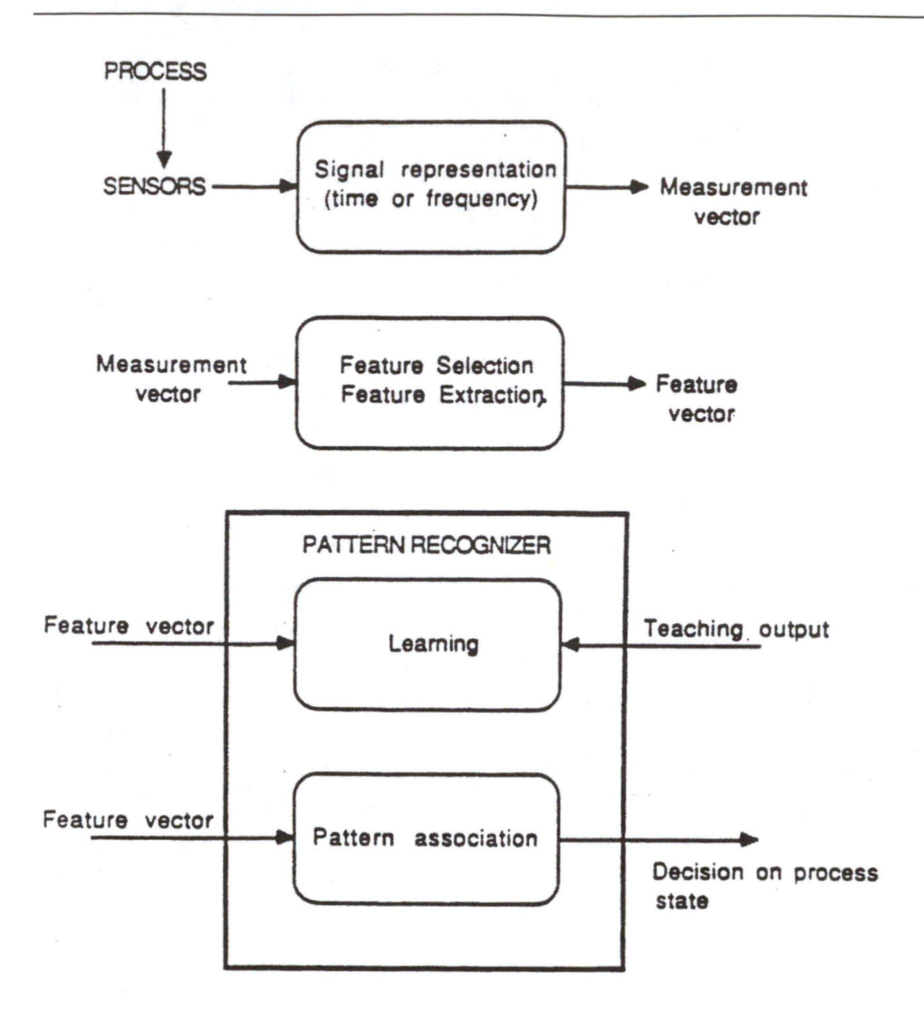

Figure 8.10. An overview of signal processing and feature extraction.

are taken to be "decisions" on the state of the process if the process characteristics are suitably structured (e.g. tool worn, weld penetration incomplete, material flawed, etc.).

There is a close relationship between sensor fusion methodologies and signal processing/feature extraction so some overlap in methodologies will exist. Advanced information processing and fusion methodologies are most often embodied in what could be referred to as an "intelligent sensor." Hence, we will first discuss, generally, the concept of an intelligent sensor and then describe the use of neural networks as a sensor fusion strategy with multi-sensor inputs including acoustic emission. An example of the application of recursive autoregressive models for feature extraction from acoustic emission signals for tool wear monitoring will also be presented.

8.5.2 Intelligent sensor defined

The philosophy of implementation of any sensing methodology for diagnostics or process monitoring was introduced earlier and was divided into two simple approaches. In the first approach, a sensing technique for which the output shows some relationship to the characteristics of the process was used regardless of our understanding of the source of the signal. In the second approach, one attempts to determine a model linking the sensor output to the process mechanics and then, with sensor information, use the model to predict the behavior of the process. Both methods are useful in differing circumstances. The first is, perhaps, the most straightforward but liable to misinterpretation if some change in the process occurs that was not foreseen (that is, "normal" is no longer normal.) To insure against this type of misinterpretation, intelligence has been added to the sensors to add sophistication to the feature extraction and decision making process. Intelligent sensing systems have been most commonly associated with robot systems operating in unstructured environments. In these applications, information from only one sensor is generally insufficient to allow complete specification of the environment for task planning and execution. Multiple sensors are often employed for object location and recognition, for example, and employ cameras, infrared, ultrasonic and tactile sensing devices. The integration of the data from all of these sensors operating simultaneously is the major challenge for sensor fusion methodologies.

The development of an intelligent sensor for monitoring a manufacturing operation generally requires the following three hierarchical stages[152]:

1. Determining the sensitivity of a sensor signal to the process parameter or parameters to be monitored,

2. Developing an appropriate in-process real time signal processing method for extracting signal parameters that are rich in information regarding the process parameters being monitored but relatively insensitive to other parameters, and

3. Developing a decision making scheme that can make a decision on the process state based upon the data obtained from all previous experience as well as current sensor information.

This strategy yields both hardware and software elements. The hardware elements of an intelligent sensor system include the usual sensor element (transducer element and electronics such as the piezoelectric element and charge amplifier in a load cell or bimetallic and reference junctions with a Wheatstone bridge in a thermocouple) and associated signal amplification and transmission electronics found in all basic sensors as well as additional elements.[218] These include signal conditioning (conversion of sensor output to information relative to a process parameter), local conditioning and decision making (feature extraction and state estimation) as well as self calibration and diagnostic functions (to insure the sensor system is functioning properly and that any faults can be easily found). The sensor system provides a high level signal to the process controller or for further fusion with the output of other sensors.

8.5.3 Sensor fusion defined

With a specific focus for the monitoring in mind, researchers have developed over the years a wide variety of sensors and sensing strategies each attempting to predict or detect a specific phenomenon during the operation of the process and in the presence of noise and other environmental contaminants. A good number of these sensing techniques applicable to manufacturing have been reviewed in the early section of this paper. Although able to accomplish the task for a narrow set of conditions, these specific techniques have almost uniformly failed to be reliable enough to work over the range of operating conditions and environments commonly available in manufacturing facilities. Thus, researchers have begun to look at ways to collect the maximum amount of information about the state of a process from a number of different sensors (each of which is able to provide an output related to the phenomenon of interest although at varying reliability). The strategy of integrating the information from a variety of sensors with the expectation that this will "increase the accuracy and ... resolve ambiguities in the knowledge about the environment", Chiu et al[154], is called sensor fusion. Sensor fusion is able to provide data for the decision-making process that has a low uncertainty due to the inherent randomness or noise in the sensor signals, includes significant features covering a broader range of operating conditions, and accommodates changes in the operating characteristics of the individual sensors (due to calibration, drift, etc.) because of redundancy. In fact, perhaps the most advantageous aspect of sensor fusion is the richness of information available to the signal processing/feature extraction and decision-making methodology employed as part of the sensor system. Sensor fusion is best defined in terms of the "intelligent sensor" as introduced above and in Dornfeld[155] since that sensor system is structured to utilize many of the same elements needed for sensor fusion.

8.5.4 Fusion methodologies

This section reviews, briefly, techniques for integrating information from several sources. These are discussed in light of the requirements for monitoring manufacturing processes. The objective of sensor fusion is to increase the reliability of the information so that a decision on the state of the process is reached. This tends to make fusion techniques closely coupled with feature extraction methodologies and pattern recognition techniques. The problem here is to establish the relationship between the measured parameter and the process parameter. There are two principal ways to encode this relationship, Rangwala[152]:

- theoretical - the relationship between a phenomenon and the measured parameters of the process (say tool wear and the process), and
- empirical - experimental data is used to tune parameters of a proposed model.

As mentioned earlier, reliable theoretical models relating sensor output and process characteristics are often difficult to develop because of the complexity and variability of the process and the problems associated with incorporating large numbers of variables in the model. As a result empirical methods which can use sensor data to tune unknown parameters of a proposed relation are very attractive. These types of approaches can be implemented by either a. proposing a relationship between a particular process characteristic and sensor outputs and then using experimental data to tune unknown parameters of a model, or b. associating patterns of sensor data with an appropriate decision on the process state without consideration of any model relating sensor data to the state. The second approach is generally referred to as pattern recognition and involves three critical stages, Ahmed and Rao[156]:

- sampling of input signal to acquire the measurement vector
- feature selection and extraction
- classification in the feature space to permit a decision on the process state

These were previously illustrated in Figure 8.10. The pattern recognition approach provides a framework for machine learning and knowledge synthesis in a manufacturing environment by observation of sensor data and with minimal human intervention. More importantly, such an approach allows for integration of information from multiple sources (such as different sensors) which is our principal interest here.

Sata, et al[157, 158] were among the first researchers to propose the application of pattern recognition techniques to machine process monitoring. They attempted to recognize chip breakage, formation of built up edge and the presence of chatter in a turning operation using the features of the spectrum of the cutting force in the 0-150 Hz range. Dornfeld and Pan[159] used the event rate of the RMS energy of an acoustic emission signal along with feedrate and cutting velocity in order to provide a decision on the chip formation produced during a turning operation. Emel and Asibu[160] used spectral features of the acoustic emission signal in order to classify fresh and worn cutting tools. Balakrishnan, et al[161] use a linear discriminant function technique to combine cutting force and acoustic emission information for cutting tool monitoring.

The manufacturing process may be monitored by a variety of sensors and, typically, the sensor output is a digitized time domain waveform. The signal can then be either processed in the time domain (for example, extract the time series parameters of the signal) or in the frequency domain (power spectrum representation). The effect of this is to convert the original time domain record into a measurement vector. In most cases, this mapping does not preserve information in the original signal. Usually, the dimension of the measurement vector is very high and it becomes necessary to reduce this dimension due to computational considerations. There are two prevalent approaches at this stage: select only those components of the measurement vector which maximize the signal/noise ratio or map the measurement vector into a lower dimensional space through a suitable transformation (feature extraction). The outcome of the feature selection/extraction stage is a lower dimensional feature vector. These features are used in pattern recognition techniques and as inputs to sensor fusion methodologies.

8.5.5 Neural Networks

It is preferable if the feature extraction and learning activities occur at the same time since this would allow the extraction of optimal information from the features along with noise rejection. Such an approach is possible using neural network pattern recognizers. A neural network is a collection of simple, interconnected processors which operate in parallel and store knowledge in the strength of the connections between the individual processors. Such parallel networks of computing elements crudely resemble processing activity in the brain and have been successfully applied to intelligent tasks such as learning and pattern recognition.

Artificial neural networks are an attempt to mimic the computational architecture of the human brain in electronic hardware, the objective being to incorporate intelligent functions such as learning and pattern recognition in computers. The human brain consists of a large number of interconnected neurons, each possessing very simple computational abilities. However, the interactions between the neurons allows for parallel processing of information, which greatly enhances the speed of computation and causes a large amount of knowledge to be brought to bear in processing this information, Hinton and Fahlman.[162] A typical neuron consists of three components: the cell body, the input lines into the neurons (dendrites) and output lines emerging from the cell body (axons). The axons of a neuron are connected to the dendrites of other neurons at points called synapses. This forms a highly interconnected system of neurons which communicate with each other via synapses. The synapses determine the strength of the connection between two neurons. A typical neuron receives inputs from various other neurons via the dendrites. The time averaged sum of these inputs causes biochemical reactions inside the cell body, which results in pulses of electrochemical activity being transmitted over the axon lines of the neuron. The pulse rate depends on the magnitude of the input excitation to the neuron, and is usually assumed to be a sigmoid function of the input. This is because the output saturates at extreme values of the input (the output pulse rate lies between 0 and 500 Hz for a typical neuron). The synapse between two neurons may be excitory (in which case, a high activity in one neuron causes a high activity in the other neuron) or inhibitory (in which case, a high activity in one neuron causes activity in the other neuron to be suppressed).

Artificial neural networks can be implemented by using amplifiers with a sigmoid input-output relationship as the “neuron” element and resistive connections between the amplifiers representing the synapses between the neurons. The conductance values of the resistive elements represent the strength of the connection between the individual processors and the sign of the conductance determines whether the excitory (positive conductance) or inhibitory (negative conductance). The voltage output of the amplifier represents the activity level of a given processor whereas the input to the processor is simply a current summing operation on the outputs of the other processors. A schematic of two such interconnected processors is shown in Figure 8.11. The connection strength between the jth and the ith neuron is represented as w_{ij} and is simply the conductance value of the connection between these processors. The input to the ith processor is given by

$$INPUT_i = \sum_{j=1}^{N} w_{ij} OUT_j \tag{8.8}$$

where N is the total number of processors connected to the ith processor and OUT_j is the output of the jth processor. The output of a processor is usually assumed to be a sigmoid function of the input or in some cases, a threshold function may also be used. The conductance values or the "weights" are the learning parameters of the system and encode the knowledge in the system. These learning parameters are acquired through learning. Physically, this implies that the resistors should be able to change their resistance values in response to the learning process, i.e. adaptive or programmable resistors are required.

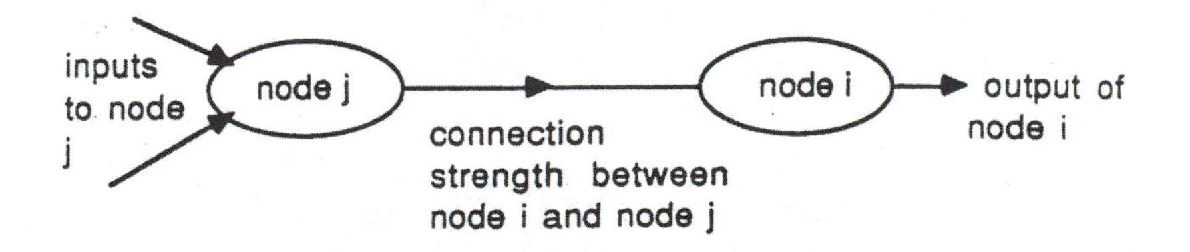

Figure 8.11. Schematic of two interconnected processors.

Computation in neural networks occurs by propagation of signals through the connections between the processors. Each processor is capable of very simple functions such as current summing and elementary arithmetic operations. In a general network, all the processors may be connected to each other, with the knowledge of the system encoded in the "strength" of connections between the individual processors. Some of the processors may receive inputs from the external world (input nodes) whereas others may transmit their outputs to the external world (output nodes). The remaining nodes which are not directly connected to the external world are called "hidden" nodes. These nodes are important since they are responsible for feature extraction and internal representation of the knowledge acquired through the learning process. One important feature of neural networks is that the knowledge in the system directly determines how the processors interact in contrast to being stored in a separate knowledge base, waiting to be accessed sequentially by a CPU. Also, the knowledge is distributed over a large number of connections which results in a fail soft operation: a failure of some of the processors will cause graceful degradation in performance rather than a complete loss of the knowledge base.

Pattern recognition tasks typically involve associating sets of patterns. In a neural network, the knowledge required to associate the correct sets of patterns is encoded in the values of the weights or connection strengths between the processors. Neural architectures proposed by Hopfield[163, 164] have been demonstrated to successfully implement content addressable memories, in which a system of interconnected processors can retrieve a complete memory, given only a partial knowledge of the memory. Hopfield networks consist of a system of N processors, each processor connected to all other processors. The operation of such a network can be described as a dynamic system to which the network state converges. The state of the network is represented by the activation levels of the individual processors at a given instant of time, and is an N dimensional vector. The important point here is that the location of the attraction basins (or equilibrium points) can be controlled by suitable selection of the connection strengths between the processor states (which represents an incomplete or incorrect memory), the network will converge to one of the attraction basins of the system. If the initial state is close enough to the desired state, the correct memory (represented by the converged state) will be retrieved.

One class of neural networks which have been shown to be successful at pattern recognition tasks are multilayered, feedforward networks of the type shown in Figure 8.12. There are three kinds of processing units in such networks: input layer nodes which accept patterns from the external world, output layer nodes which generate outputs to the external world and hidden nodes which do not directly interact with the external world. The role of the hidden nodes is to form internal representations of the patterns presented at the input layer. These networks perform pattern association tasks in which a

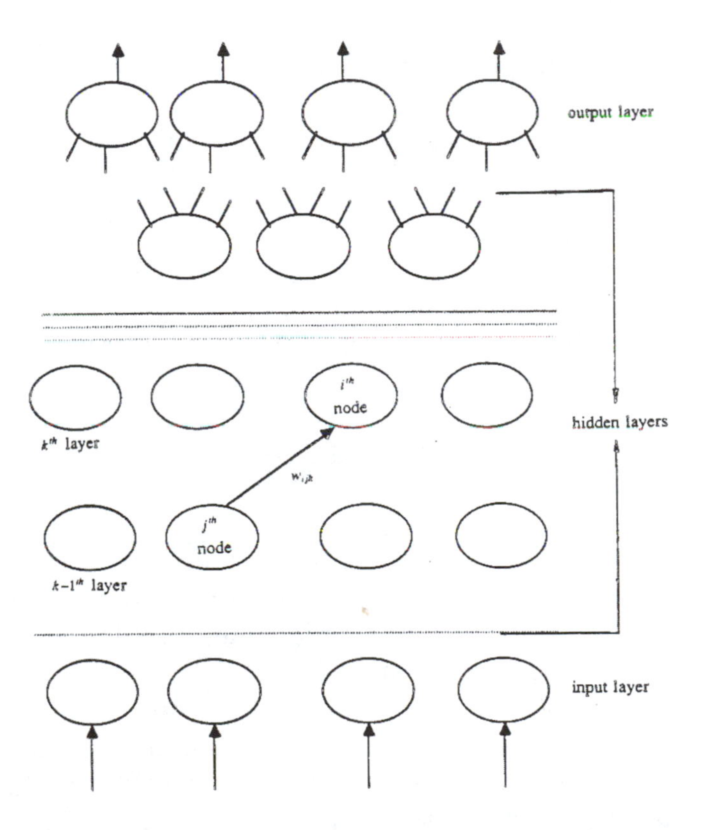

Figure 8.12. Structure of a feed-forward neural network.

pattern presented at the input layer of the network is associated with a pattern at the output layer. In these networks, information propagates from the bottom to the top layer, with connections existing only between processors in adjacent layers.

Let:

$w_{i,j,k}$ = weight between j^{th} processor in $(ki1)^{st}$ layer and i^{th} processor in the kth layer.

$net_{i,k}$ = input to i^{th} node in the kth layer.

$out_{i,k}$ = output of ith node in the kth layer.

$t_{i,k}$ = threshold value associated with the ith node in the kth layer.

The input to a processor is given as:

$$net_{i,k} = \left[\sum_j w_{i,j,k} out_{j,k-1} \right] + t_{i,k} \tag{8.9}$$

The output of a given processor is a sigmoid function of the input and can be expressed as:

$$out_{i,k} = F(net_{i,k}) = \frac{1}{e^{-net_{i,k}}} \tag{8.10}$$

Although recognized that multilayered neural networks possess many attractive properties, one of the obstacles to their development was the absence of an efficient learning algorithm for training such networks. The generalized delta rule developed independently by Rumelhart and McClelland[165] and Le Cun[166] fills this gap and has been shown to work efficiently on pattern association tasks. This is a supervised learning procedure in which examples of input and output patterns (representing the patterns to be associated) are used to train the network. The rule consists of presenting an input pattern to the network, propagating activity among the various processors according to Eqs. (8.8) and (8.9) and computing the pattern at the output nodes with the current set of learning parameters (thresholds and weights of the network). The actual output pattern is then compared to the desired output pattern and the error is calculated as follows:

$$E = \frac{1}{2}\sum_{i=1}^{q}(d_i - a_i)^2 \tag{8.11}$$

where d_i is the desired output at the ith output layer node, a_i is the actual output and q is the total number of nodes in the output layer. The procedure of calculating the error is repeated for all sets of training input-output patterns and the individual errors are added to compute the total error. This constitutes the forward pass through the network.

Next, the error is propagated from the top to bottom layer through the network to modify the weights and thresholds in such a way that the error term is minimized. This involves minimization of a non-linear error function with respect to the threshold and weight values, using gradient descent. Computation of the error term with respect to each weight and threshold is accomplished using local information at each node, so that gradient calculations at each layer can be accomplished in parallel. Rumelhart[165] uses the gradient information to adjust the weights and thresholds as follows:

$$\Delta w_{i,j,k} = -\eta \frac{\partial E}{\partial w_{i,j,k}} \tag{8.12}$$

$$\Delta t_{i,k} = -\beta \frac{\partial E}{\partial t_{i,k}} \tag{8.13}$$

where η and β are the step sizes in the minimization process.

A linear network is one in which the output of a processor is a linear function of its input, with the input to a processor defined as in Eq. (8.9). For a linear system, the error surface is bowl shaped and has only one minimum point so that convergence is guaranteed. In the present case, however, the error function is a non-linear function of the learning parameters, so that any gradient descent scheme for error minimization is prone to terminating in a local minimum. There is no guarantee that a gradient descent procedure will find a set of thresholds and weights so that the error term is zero. However, this does not seem to present difficulties in practical implementations, since the number of hidden layers and number of nodes in each hidden layer can be so chosen that a set of weights and thresholds which drive the error to zero can usually be found, for a given implementation.[165] The final values of the learning parameters are randomized and usually lie between -1 and 1.

Once such a network has been trained using a set of training patterns, it can be used to associate patterns presented at the input layer with appropriate patterns at the output layer. The advantages of using these networks for pattern recognition tasks is that learning can be accomplished purely by observation of sensor data, learning and pattern recognition is accomplished using parallel computation, and such networks can form internal representations of the raw sensory information independently. Internal representations are necessary because the raw information may be noisy and redundant and not suited for making pattern classification decisions. Another advantage of using such networks is that knowledge is stored in the connection strengths between the processors and directly determines how the network operates, rather than being stored in a data base, waiting to be accessed by the CPU.

There are similarities between the perceptron type (linear discriminant function) network discussed earlier and the multilayered networks discussed here. The perceptron performs pattern classification by computing a weighted sum of sensor inputs and comparing it to a threshold value. In case of a perceptron, there are only two layers of units with the outer layer unit implementing a threshold type function. The perceptron implements a hyperplane in feature space, with the hyperplane surface representing the decision surface. No processing of raw information is carried out. With multilayered neural nets, however, the raw input patterns undergo further processing to filter out noise in these patterns. The training procedure forces the hidden nodes to perform feature extraction on the raw features so that as information propagates through the network, the noise is filtered out. The feature extraction capability of the hidden nodes is developed during the training procedure so that the extracted features are better suited for the classification task. The last two layers of the neural network essentially implement a perceptron, however, the features used in this case are the internal features which are relatively noise free. This leads to better generalization abilities of the classifier. Further, another difference between the perceptron and multilayered neural network is that the perceptron can perform only a linear separation of the sensor features whereas the feature extraction abilities of the hidden nodes in a multilayered neural network allow the network to perform arbitrary mappings between input and output patterns.

We now review examples of signal processing methodologies applied to acoustic emission, and other, sensors for manufacturing process monitoring.

8.6 Applications of signal processing and sensor fusion

8.6.1 Introduction

Two examples of applications of intelligent sensor signal processing and sensor fusion are presented here to illustrate the practical implementation of the methodologies discussed in earlier sections. The examples are drawn from research work related to process monitoring for manufacturing - specifically metal cutting. However, the principles behind the applications are appropriate for a wide range of processes and problems. The methodologies presented include adaptive autoregressive models for feature extraction for tool wear monitoring and neural network sensor fusion for tool condition monitoring. Both examples rely on acoustic emission generated by the machining process as one of the sensor outputs observed during process monitoring. Acoustic emission from metal cutting was described earlier.

8.6.2 Tool wear detection using time series analysis of acoustic emission

Acoustic emission is primarily generated by the deformation in the primary shear zone and the sliding friction in the secondary shear zone and tool flank/work surface contact area. As the cutting tool wears, additional frictional action between the tool flank and the workpiece creates additional acoustic emission. The portion of AE that is attributed to friction on the wear land becomes more important when the flank/workpiece contact area increases as a result of tool wear. Acoustic emission generated from shearing and friction exhibit different signal characteristics since the mechanisms by which AE is produced in these occasions are fundamentally different. As a result, the signal characteristics of acoustic emission are expected to change when the tool wear-land progresses.

The detection of tool wear using acoustic emission has been attempted by several researchers as discussed briefly in the previous section. The signal analysis methodologies reported include the RMS (Root-Mean-Square) measurement, the event count analysis, and the frequency analysis. Using the first method, the RMS voltage level of AE is shown to increase in proportion to the amount of the flank wear. In the second method, correlation is found between the amount of the flank wear and the number of accumulated counts of the AE signal whose amplitude exceeds some pre-selected threshold value. RMS voltage level and event count based on a properly selected threshold are both representative of the signal power content which increases with wear land in general. However, the crater wear, due to its influence on the tool effective rake angle, tends to reduce the sensitivity of AE energy content to progressive flank wear. Tool coatings may also influence AE energy. Moreover, cutting parameters will change the AE energy content independently since the AE power released is proportional to the strain rate in the cutting process. As a result of using a single parameter to characterize the signal, power level analysis often fails to distinguish between the change of AE source mechanisms (such as the flank wear and the growth of crater wear on tool rake) and the change of cutting parameters (such as feed rate, cutting speed and depth of cut). Therefore, calibrations of AE signal power at all combinations of cutting parameters will be necessary to actually implement the monitoring technique. In the third method, spectral characteristics are found to be a function of the tool wear condition, Emel and Kannatey-Asibu.[160] However, difficulties usually encountered in the frequency analysis are the "signal coloring" effects caused by the propagation media, sensor frequency response and instrumentation system function. Furthermore, the discrete Fourier transform introduces a large number of orthogonal parameters to characterize the AE signal, therefore, a parameter selection process has to be implemented before the spectral analysis can be practically carried out.

In this example, a time-series analysis technique is used to characterize the acoustic emission RMS signal with an autoregressive model. The model parameters are constantly updated according to the RMS signal dynamics which are strongly dependent on the AE source mechanism. As a result, these time varying parameters are expected to contain information about the condition of cutting tool wear. A detailed description of the background and implementation of time-series modeling is given below. The detection scheme illustrated monitors the time series model parameters of the acoustic emission RMS signal generated during metal cutting. When the parameter pattern exceeds the preset allowable range, a severely worn tool condition is concluded and a tool change should take place. The merits of this technique are that the number of parameters to be monitored is much less than in spectral analysis and it provides a large degree of freedom to describe the acoustic emission signal than by just studying its DC power level.

8.6.2.1 Time series analysis

It is sometimes possible to derive the mathematical model for a dynamic system based on physical laws which then allows us to calculate the value of some time-varying quantity at any particular time. This type of model would be entirely deterministic. However, very few dynamic systems are totally deterministic because changes due to unknown or unquantified effects may take place during the process. Thus, it is convenient to construct stochastic models that can describe the dynamics from a probabilistic point of view. In this way the underlying system physics or system characteristics can be studied or ascertained from experimental data.

A time series is generally defined as a sequence of observed data ordered in time (or other variables such as space). In manufacturing process monitoring we often encounter such ordered sets of data corresponding to the output of a force transducer during a machining process, temperature of a sheet during forming or vibration amplitude of a grinding wheel. The statistical methodology associated with the analysis of these sequences of data is referred to as time series analysis, Wu and Pandit[167], and this approach has been applied to the analysis of a wide range of manufacturing processes. Any statistical dependence between the data is seen in the correlation or autocorrelation between successive observations. It is impossible to discuss here in detail time series analysis. We will address one simple form, the autoregressive model using a stochastic gradient algorithm. This method does not require any prior knowledge of the signal statistical properties and the modeling algorithm is adaptive in the sense that the parameters are updated at every time step as opposed to batch algorithms which estimate the parameters only after a whole data set has been collected. In the application of the time-series modeling approach to in-process characterization of systems or processes that have fast time-varying dynamics or features, the modeling technique has to be adaptive such that the information contained in the measured data can promptly be used to reflect instantaneous system dynamics or features. Not only is the adaptability needed for real-time analysis but it also reduces the memory size required since the oldest data point is discarded at each time step.

In an N-th order autoregressive (AR) model for a time series $y(k)$, where k is the discrete time index, $y(k)$ can be predicted (called $\hat{y}(k)$) based upon the current N previous values as

$$\hat{y}(k) = n(k) + \sum_{i=1}^{N} a_i(k) y(k-i) \tag{8.14}$$

where *n(k)* designates a white noise. In the present case, *y(k)* is the measured signal at the transducer site and a_i's are the model parameters.

If the measured data value of *y(k)* is different from the value predicted by using equation (8.14), some error will occur. The error signal, *e(k)*, is defined to be the difference between the model-predicted value $\hat{y}(k)$ and the actual sampled value *y(k)*:

$$e(k) = y(k) - \hat{y}(k) \tag{8.15}$$

The behavior of output signal *y(k)* is closely related to the model parameters a_i's as well. During a cutting operation, when the acoustic emission signal changes as a result of tool wear progression, the model parameters become time varying and are utilized to track the tool wear. The model parameters are calculated from the measurement of the acoustic emission sensor signal, y(k), using the stochastic gradient algorithm. With the stochastic gradient algorithm, the model parameters are adjusted every time a new data point is sampled. Each adjustment is an effort to minimize the square of the error signal at that instant, e^2 (k). The selection of the adaptation gain is critical since it governs both the stability of adaptation and the speed of convergence.

If some of the model parameters do not vary significantly with process state or if the model order is so high that the real-time implementation of signal analysis is difficult, it becomes necessary to ignore some less important model parameters. The importance of a model parameter is evaluated based on its ability to discriminate different process states. A discrimination index associated with each parameter is formulated here to quantitatively describe the relative importance of that model parameter. For any process state specified by "A," an "i-th parameter mean" can be defined for the i-th model parameter as its mean value with respect to time:

$$a_{i,A}^{o} = \frac{1}{M}\sum_{k=1}^{M} a_{i,A}(k) \tag{8.16}$$

where M is the number of total adaptation time steps. One natural way to evaluate the capability of a parameter a_i in separating two different states "A" and "B" is through the observation of its between-class variation $O_i\,[A,B]$ defined as:

$$Q_i[A,B] = \left| a_{i,A}^{o} - a_{i,B}^{o} \right| \tag{8.17}$$

When the variation of $a_i(k)$ between process states A and B is on the same order as its parameter mean, $Q_i\,[A,B]$ will decrease with increasing i. That is

$$Q_i[A,B] \geq Q_j[A,B] \quad i,j = 1-N \quad and \quad i \leq j \tag{8.18}$$

Therefore, $Q_i[A,B]$ alone cannot represent the relative importance of a_i in distinguishing process states. To formulate a better index for the relative importance of model parameters, a within-class variation in the i-th parameter for state A is defined as:

$$S_{i,A} = \left[\frac{1}{M}\sum_{k=1}^{M} (a_{i,A}(k) - a_{i,A}^{o})^2 \right]^{1/2} \tag{8.19}$$

Similarly, $S_{i,A}$ decreases with respect to i as the variation of $a_i\,(k)$ between conditions A and B is on the same order as its parameter mean:

$$S_{i,A} \geq S_{j,A} \quad i,j = 1-N \quad and \quad i \leq j \tag{8.20}$$

A discrimination index $J_i[A,B]$ between the two conditions A and B based on the i-th parameter can then be obtained by the normalization of $Q_i[A,B]$ with $S_{i,A}$ and $S_{i,B}$. That is

$$J_i[A,B] = \frac{Q_i[A,B]}{\left[S_{i,A} S_{i,B}\right]^{1/2}} \tag{8.21}$$

The discrimination index defined in this way does not necessarily decrease with i. A greater discrimination index implies that the difference in this specific parameter for two conditions is more pronounced, and that the parameter varies less within either of the conditions. Therefore, the discrimination index, $J_i[A,B]$, is a suitable indication of how successful the two conditions, A and B, can be separated through the observation of i-th model parameter. The most important parameter is the one that maximizes the discrimination index. Parameter reduction can then be achieved by ignoring the parameters with small discrimination indices.

The RMS signal is a function of the AE low frequency components and the total power contained in the individual bursts of AE generated during machining. To make the technique sensitive only to changes in the AE source rather than to the change of cutting parameters (RMS is proportional to cutting speed), the DC component of RMS voltage is filtered out. Thus the autoregressive model tracks only the dynamic properties but not the mean energy level of the AE signal. In this way, an off-line parameter calibration procedure is needed only once under any cutting condition to map out the allowable range for the parameters, since the parameters will be affected by tool wear but not by the change of cutting velocity, depth of cut, or feed rate.

8.6.2.2 Experimental Evaluation

A series of experiments were conducted to test the performance of the proposed technique, Liang and Dornfeld.[168] A Kennametal K68 tungsten carbide insert tool was mounted on a tool holder with 5 degree rake. The workpieces used were low carbon steel bars with 2 inch (5.08 cm) diameter. Experiments were performed under five different cutting conditions. Conditions included both fresh tools and those with different amounts of tool flank wear. No chatter, built-up-edge, or rake face crater was observed during cutting.

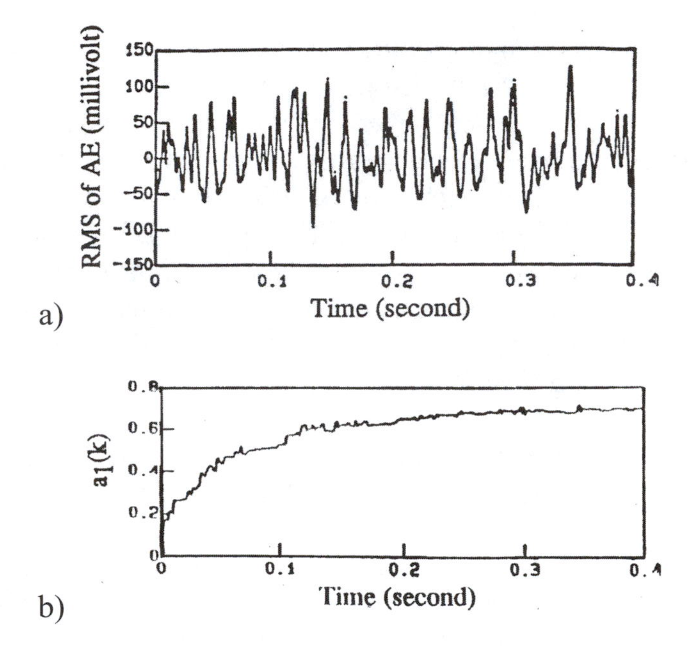

Figure 8.13. a) Typical time series of acoustic emission RMS signal recorded over 0.4 seconds, b) variation of first model parameter.

A typical time series of acoustic emission RMS signal recorded over 0.4 second is shown in Figure 8.13a. The DC level of the signal, normally about 300 mv, has been removed. Also plotted on this diagram is a dotted line which shows the predicted values of the AE RMS from a 6-th order autoregressive model with an adaptation gain of 1.0e-5.

Some signal dynamics at the beginning of cutting were lost for just a short period while the adaptation began. However, the predicted values agree with the original signal extremely well as seen from the figure. Figure 8.13b shows the time variation of the first model parameter. The decreasing trend in error signal with time is clearly seen. After 400 ms of adaptation, the error signal is confined within ± -5 percent of the measured signal. The first model parameter takes about 300 ms to achieve its optimal value. The time then required to track the time-varying optimal value will be much shorter than 300 ms since the initial condition will always be the current value instead of zero.

If an autoregressive signal is modeled by an autoregressive model with sufficiently high order, the error signal should be white since all the parameters are optimized in the sense of minimizing the mean-square error. The error signal e(k) shows no major autocorrelation. The existence of some minor correlated components indicates that:

1. The acoustic emission RMS signal generated during metal cutting operation is not a purely autoregressive process,
2. the model order, 6 in this case, is not sufficiently high so that the truncated higher order terms are incorporated into the error signal, and
3. based on the formulation of the stochastic gradient algorithm, the model parameters are not the exact solutions to the problem of minimizing the mean-square error but rather only approach the solution all the time.

However, these limitations are not significant here.

The modeling algorithm works on a pre-selected model order and adaptation gain, which are kept constant throughout the whole process. The guideline to the selection of the model order and gain are based on the minimization of the sum of square errors, defined as

$$S.S.E = \sum_{k=1}^{M} e^2(k) \tag{8.22}$$

where M is a large number, 2048 in this study. The correlation between model order, adaptation gain, and S.S.E. suggests that a larger adaptation gain in the region of low model order or a higher model order in the region of small adaptation gain will result in a smaller sum of square errors. However, a large adaptation gain associated with a high model order will cause unstable adaptation. An optimal order of 6 and an adaptation gain of 0.6e-5 were selected since they minimized the S.S.E.

The parameter vectors $[a_1, a_2, a_3, a_4, a_5, a_6]$ exhibit the distinction between parameters for various tool wear conditions and the agreement of parameters under various cutting speeds, feed rates, and depths of cut. The discrimination indices between classes 1 and 2, 1 and 3, 1 and 4, and 1 and 5 were calculated for each model parameter. The results of these tests show that the separation attributed to changes in cutting parameters are very low, whereas the separation resulting from various states of tool wear is comparatively high. All data collected under conditions 1, 2, and 3 are then combined into a single class, 1+2+3, and the discrimination index calculated. The resulting discrimination indices are lower than they are in the cases when fresh tool conditions are not combined. This is expected from the fact that the combined class has a higher within-class variation. Based on the discrimination analysis, parameters 1 and 4 are shown to be the two most important features in terms of detecting and tracking the amount of tool wear while maintaining insensitivity to machining tool condition.

A 2-D parameter plane is constructed by plotting the fourth parameter against the first one, as shown in Figure 8.14. Owing to the non-stationarity of the acoustic emission signal, the model parameters are time-varying and form clusters in the plane. However, these clusters, each representing a different tool wear condition, are clearly separable. As a result, the condition of tool wear can be effectively detected by monitoring the time trajectories of the model parameters. The parameter plane in Figure 8.14 suggests that a tool change should be called for when $[a_1, a_4]$ goes beyond [0.46, 0.07] since a 0.0197 inch wear-land tool is considered damaged for these studies.

8.7 Sensor integration using neural networks for intelligent tool condition monitoring

In this example, a technique for intelligent tool condition monitoring which employs information from multiple sensors is presented. This information is integrated via a neural network, a parallel computing architecture which can learn to recognize patterns of sensor information and associate them with decisions on the tool wear state. Initial efforts by Rangwala[152] and Rangwala and Dornfeld[169] demonstrated the feasibility of using neural networks for sensor integration in tool wear monitoring tasks. The networks were used as learning and pattern recognition devices, and were able to successfully associate sensor signal patterns with the appropriate decision on tool wear. Chryssolouris and Domroese[170] performed simulations in order to study the learning capabilities of these networks. Based on the simulation results, they proposed the use of neural networks as the decision-making component in an intelligent tool condition monitoring system. As shown in this example, neural networks are able to filter out noise in the sensor data and this enhances their ability for successful pattern association tasks. These aspects are experimentally evaluated for tool wear monitoring in a turning operation, under a range of machining conditions, to determine both discrete as well as continuous tool wear states.

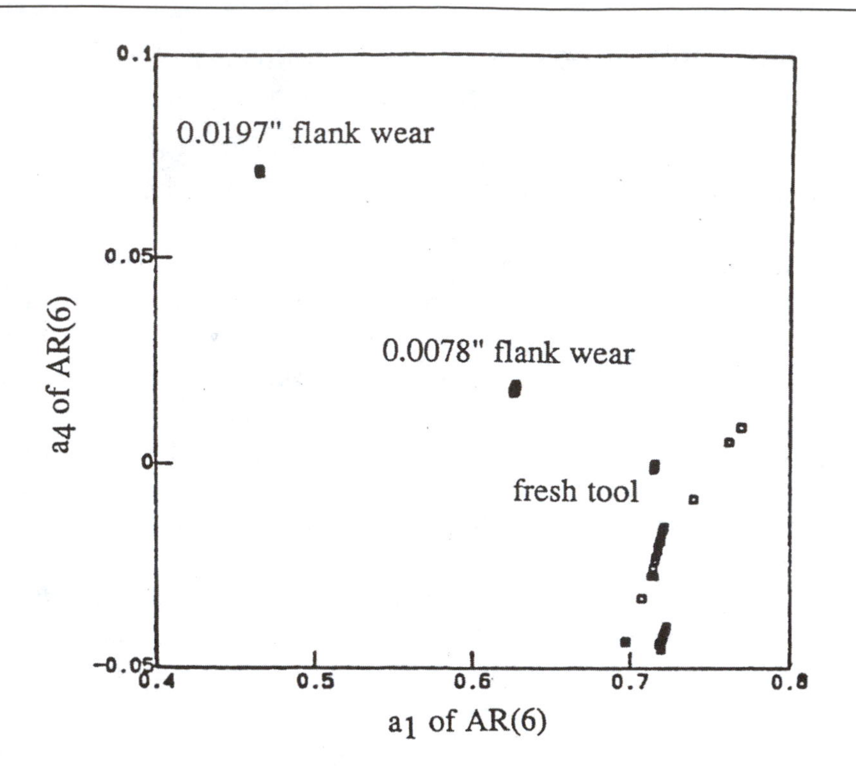

Figure 8.14. 2D parameter plane plotting the fourth parameter against the first.

8.7.1 Use of multiple sensors

In the application discussed here, it was decided to use AE and cutting force information in order to develop an intelligent tool condition monitoring system. The primary and secondary shear zones are important sources of AE when cutting with a fresh tool as discussed in the first portion of the paper. In the presence of flank wear, the tool-work interface becomes an additional zone of significant AE generation due to intense friction between the tool and workpiece surfaces which move past one another at high relative velocities. The effects of tool wear on AE generation in the primary and secondary zones must also be considered. Kobayashi et al.[171] conducted experiments with artificially ground worn tools and concluded that the presence of a flank land did not have an observable effect on the shear angle. This implies that flank wear does not affect the AE characteristics in the primary and secondary shear zones. However, the presence of crater wear affects the effective rake angle of the tool, and this could affect the generation of AE from the primary and secondary shear zones.

The root mean square (RMS) level of the AE signal (V_{rms}) measures the total power level of the signal and has been found to be sensitive to the degree of flank wear in a turning operation. Experiments conducted by Lan[172] for machining of SAE 4340 steel with carbide tools indicate that V_{rms} increases with machining time due to increased flank wear. However, in cases where the crater wear is significant, V_{rms} tends to decrease or remains constant. Since the presence of flank wear is expected to increase V_{rms}, Lan concluded that the effect of crater wear is to cause a drop in V_{rms}. The fact that V_{rms} remains constant with increased tool wear due to opposing effects of flank and crater wear makes it difficult to design an AE-based tool wear monitoring system which uses only information on the RMS level of the signal.

Emel and Kannatey-Asibu[160] present experimental data which shows that the power spectrum is sensitive to tool wear and process conditions. Results for machining of AISI 1060 with carbide inserts[152] also indicated that the AE power spectrum was sensitive to the level of flank wear and process parameters such as the cutting velocity. Table V , summarizes the qualitative effects of tool wear and cutting velocity on the AE power spectrum.[152] The mean frequency divides the total power of the spectrum into two equal parts, whereas standard deviation of frequency indicates the spread in power content around the mean frequency. An increase in flank wear and cutting velocity causes an increase in the low frequency (100 - 300 kHz) power of the AE signal. Other effects such as feed rate and depth of cut changes as well as chip tangling and chip breakage processes are also expected to affect the AE spectral characteristics. An important consideration for tool wear monitoring is that appropriate schemes should be used in order to identify spectral regions which show maximum sensitivity to tool wear under a range of process conditions.

The performance of an AE-based tool wear monitoring system can be enhanced by complementing the AE information with information from other sensors mounted on the machine tool (for example, force or power sensors). The magnitude of the cutting force is sensitive to the occurrence of tool wear in a turning operation (Andrews and Tlusty[59]). According to Wright[173] however, cutting force information by itself is inadequate for tool wear detection because its magnitude is also dependent on the cutting velocity. Another problem is that although flank wear tends to increase the cutting force, the accompanying crater wear tends to reduce it, so that the magnitude of the cutting force may not show any sensitivity to tool wear. Cook[174] and Martin et al.[175] have shown, along with others, that the cutting force spectrum (which reflects the dynamic characteristics of the cutting force) is sensitive to tool flank wear. Vibrations in the direction of the cutting force are induced due to flank wear in high frequency regions (> 5 kHz) and lower frequency regions (< 300 Hz). The former is due to vibrations of the tool holder whereas the latter is attributed to workpiece vibrations. The force spectrum is also dependent on process variables such as cutting velocity and feed rate and oscillations in the shear angle during chip formation.

The AE and cutting force information relate to different effects of tool wear. Acoustic emission is sensitive to the microscopic activities (and the resulting stress waves) related to plastic deformation and friction in the cutting zone. The cutting force spectrum is sensitive to the vibrations induced in the tool and workpiece due to the effects of flank wear. The advantage of using AE and cutting force sensors is that they provide information relating to microscopic (stress waves) and macroscopic (vibrations) effects of tool wear. This helps provide better signal features to the pattern classifier, allowing a greater reliability in making decisions on the state of tool wear.

8.7.2 Experimental evaluation

To apply the neural network-based machine learning approach discussed earlier, a series of machining tests were conducted on a precision lathe. The work material was case hardened AISI 1060 bars (hardened workpieces were used in order to induce faster tool wear). A Kennametal TPGF-322 insert of grade K68 was used. The bars were of nominal diameter 2" (50.8 mm) and 12" (305 mm) in length. An acoustic emission transducer (type D9201) was mounted on the tool shank. The tool shank was mounted in a fixture instrumented with a Kistler force dynamometer (type 9251A). The fixture was mounted in the tool turret. The AE sensor output was passed through a preamplifier (with a fixed gain of 40 dB) which high pass filters the incoming signal above 50 kHz. The preamplified signal was passed through an amplifier (5 dB gain) and recorded on the video channel of a modified Sony recorder. The cutting force signal was passed through a charge amplifier and recorded on the audio channel of the Sony recorder. The process variables were varied in the following range:

Feed rate: 0.002 ipr - 0.008 ipr (0.05 mm/rev - 0.20 mm/rev)
Depth : 0.01" - 0.03" (0.25 mm - 0.75 mm)
Velocity : 278 sfpm - 556 sfpm (85 m/min - 170 m/min)

No signals were collected while machining the hardened layer (approximately 1.5-2 mm thick) of the workpiece. Signals were collected only when the workpiece diameter was 45 mm (1.75") diameter or less. The tool flank land was measured using an optical comparator. The procedure used was to measure the flank land width after every two passes through the soft section of the bar and after every pass through the hardened layer. The tool wear was recorded at flank wear levels of of 0.1 mm (0.004"), 0.125 mm (0.005"), 0.25 mm (0.01"), 0.5 mm (0.02") and 0.75 mm (0.03"). Signals collected between these wear levels were ascribed to the wear value at the lower end of the interval. For example, signals collected between 0 and 0.1 mm flank wear were assumed to be generated due to cutting with a fresh tool. Between wear levels of 0.25 mm and 0.5 mm, no signals were recorded, although cutting proceeded. Signals collected during cutting below 0.25 mm flank wear were assumed to belong to fresh tool cutting, whereas signals associated with a flank wear level of 0.5 mm were assumed to belong to worn tool category.

During post-processing, the signals recorded on video tape were played back, filtered and digitized on a HP waveform recorder. The digitized AE signals had a record length of 1024 points, sampled at 5 Mhz, and the digitized force signals were of record length 512 points, sampled at 1 Khz. The sampled AE and force records were synchronized as closely as possible using the tape counter number as a reference. A total of 65 samples of fresh tool cutting and 58 samples of worn tool were collected for purposes of training and testing.The force time domain record is of length 512 (sampled at 1 kHz) and the AE time domain record length is 1024 (sampled at 5 Mhz). Using a Fast Fourier Transform (FFT) program yields the power spectrum representations of the time domain records. Consider the power spectrum as a vector whose components are the signal power at various discrete frequencies. The cutting force spectrum is of dimension 256 (256 discrete frequencies with a resolution of 2 Hz) and the AE spectrum is of dimension 512 (512 discrete frequencies with a 5 kHz resolution). Combining the AE and cutting force spectra yields a vector of dimension 768, each component of the vector representing the signal power at a discrete frequency in either the cutting force or the AE signal. This vector is referred to as the measurement vector.

Although valuable information may be contained in the entire measurement vector, from practical considerations, only a few of these components can be used for training and pattern association purposes. This is because in training a pattern classifier such as used here, the minimum number of training samples to be used is

$$N = 2(d+1)$$

where N is the number of training samples and d is the number of features used. This constrains the training procedure so that generalization behavior of the classifier is acceptable.

The approach for reducing the dimension of the measurement vector is to retain only those components of the spectra which show a high sensitivity to the process characteristics of interest and low sensitivity to noise or process parameters. Considering that the measurement vector is D dimensional, the objective is to select d features which maximize a criterion representing the signal to noise ratio of the features. The selected d features are the components of a d dimensional feature vector. The criterion function used in this case uses the concept of interclass Euclidean distance measures. A typical criterion is

$$J = trace(S_w^{-1} S_b)$$

where S_w is the within class scatter matrix and S_b is the between class scatter matrix of the d dimensional feature vector. S_w measures the scatter of data points within a class representing a process state and S_b measures the distance between clusters representing data points in the d dimensional feature space of different states. Intuitively, the value of J represents the signal/noise ratio of the feature vectors. Adding new features increases the J value since additional features cause the distance between mean values of the clusters representing states to increase. A high value of J indicates that the clusters corresponding to two different process states are far apart and that the scatter within the cluster is small.

In this example, 30 measurement vectors (equally divided between fresh and worn tool states) corresponding to various machining conditions were used to estimate S_w and S_b The final d features were selected using the Sequential Forward Search (SFS) algorithm.[242] It should be pointed out that the SFS algorithm is suboptimal in the sense that it does not guarantee that the best feature set is selected. However, it is computationally viable and yields feature sets whose signal/noise ratio is reasonably close to the optimal case.[243] It was decided that 30 samples (equally divided between fresh and worn tool cutting) would be used for purposes of training. The dimension of the feature vector was chosen to be 6. Three feature sets were selected using the procedure discussed above. Set 1 features were selected using a combined measurement vector of the

AE and force spectra. Application of the SFS algorithm yielded 4 AE and 2 force features in this case. Set 2 features were selected by considering the AE and force spectra as separate measurement vectors and selecting three features from each. In this case, the feature vector consists of three AE and three force features. Set 3 features were selected considering only the AE spectrum as the measurement vector. Adding new features increases J, since additional features cause the distance between the mean values of fresh and worn tool clusters to increase.

Since measurement vectors corresponding to different process conditions are used, the selected features should show a low sensitivity to changes in process variables. However, some sensitivity may still be present, so that it makes sense to use the process conditions as additional features. Information such as the feed rate and cutting velocity is easily available from the machine controller and can be used as additional features. Depth of cut information is difficult to obtain on-line, and is not used as a feature. Thus a change in sensor feature values due to a change in the depth of cut has the effect of noise corrupting the sensor feature value. Therefore a total of 8 features in the feature vector (input to the network) were used: 6 features from force and AE (2 force and 4 AE), 1 feature corresponding to cutting velocity and 1 feature corresponding to feedrate.

Various design parameters affect the performance of the tool wear monitoring system. These include factors such as the number of training samples, the number of sensors and sensor features used and the structure of the neural network. The effect of these factors on the performance of the tool wear monitoring system is evaluated next and a design which yields the best performance is shown. Although the exact design will change for different situations, the methodology presented here will yield practical design strategies for implementing on-line process monitoring systems.

The perceptron training algorithm was used to train a linear classifier, using set 1 features. In order to see the effects of sensor fusion, perceptrons were also trained using set 2 and set 3 features. Unless otherwise specified, all training sets contained 30 samples, equally divided between fresh and worn tool cutting. The trained classifiers were then tested on the remaining 93 samples (of which 50 correspond to fresh tool cutting and the remaining to worn tool cutting). Sets 1 and 2 yield comparable performance (88 % and 87% classification success rates respectively) whereas the performance of set 3 is lower (80 % success rate). This indicates that feature sets composed of multiple sensor information provide better classification performance. Simply looking at the AE feature would cause increases in depth of cut to be mistaken as a "tool worn" condition, and hence lead to classification errors. The sensitivity of the force feature (10 Hz) to tool wear and was seen to be reasonably high, regardless of changes in depth of cut. Including the force feature would, in this case, reduce classification errors. Of course, additional AE features may also provide sensitivity to tool wear under these operational conditions (in fact, this is the motivation for using a large number of features). However, as larger number of features from one sensor are used, the information provided by them becomes highly correlated, so that a loss of sensitivity to tool wear in one feature is accompanied by a loss of sensitivity in other features of the same sensor signal.

Feature sets 1, 2 and 3 were used to train and test multilayered neural networks. Sensor feature values were normalized in order to prevent saturation of the sigmoid function. This was done by dividing the feature value by its maximum value in the training set. Neural networks with a single hidden layer and three nodes in the hidden layer were used. The number of nodes in the input layer is equal to the number of input features, which in the current case is 8 (six sensor features and two process features). The output layer contains a single node, whose output level associates the current input pattern with a decision on tool wear. This yields a network with a 8-3-1 structure. During the training phase, the target state of the output node was fixed at 0.01 for fresh tool patterns and 0.99 for worn tool patterns. The minimization of the error was achieved by using conjugate gradient optimization, which adjusts the weights and thresholds in a direction which minimizes the error. The weights and thresholds were initialized to uniformly distributed random values lying between -1 and 1. During the testing stage, a pattern presented at the input layer was associated with a "fresh tool" decision if the output node activity was between 0 - 0.5, else the pattern was associated with a worn tool state.

A threshold value is associated with all nodes in the input, hidden and output layers. The role of the threshold is to compare the weighted sum of inputs to the node and generate an output which depends on the difference between this sum and the node threshold. The threshold value thus acts as a filter for incoming signals. Theoretically, the learning procedure maps worn tool samples to an output node activity of 1 whereas fresh tool samples are associated with zero activity of the output node, so that the signal to noise ratio (measured by the value of the discriminant index, *J*) of the output node feature approaches infinity. In practice, this does not occur because the output node error does not converge exactly to zero, however, since it is sufficiently close to zero, each filtering step in the network is expected to suppress noise and increase the signal/noise ratio as the signal propagates through the network. In order to observe the noise suppression behavior discussed above, the trained 8-3-1 network (set 1 features) was presented with all 123 samples and the variation of the J value of the features at each layer were calculated. The value of J increases at every layer, implying higher separability of the fresh and worn tool patterns at the decision layer.

The classification success rate of the 8-3-1 network used above, based on 93 test samples was found to be 94%. For comparison purposes, a perceptron network using the same normalized input features as the neural network was also trained and tested. The classification success rate in this case was found to be 88% (similar to that obtained with non-normalized features). The superior performance of the neural network is attributed to their noise suppression abilities. It is of interest to see how the classification performance is affected when the number of features presented to the input layer is varied. To observe this, the least significant feature in the input feature vector was dropped sequentially. Process features were always included as part of the feature vector. The modified vectors were used to train and test the performance of a perceptron and a neural network. An increase in the number of features used at the input layer generally improves classification performance. For a given number of features, the performance of the neural network is seen to be superior to that of the perceptron. The effect of not using the process features is that the performance is adversely affected when the process features are not included. In this case, it is possible that changes in process conditions are confused as being due to changes in tool wear state, so that the classification error rate increases. In this case also, the neural network performs better than the perceptron network. One aspect that is not fully explained is that although increase in the number of sensor features generally improves performance, in some cases, the use of an additional feature causes a deterioration in the performance level. A possible reason may be that the training and test data statistics for that particular feature may be very different, so that the trained network may not be able to respond correctly to test data. Training anomalies may also contribute to this sort of behavior.

8.8. The need for engineering models to design and predict the performance of in-process sensors

As we have seen earlier in the chapter, incorporation of an in-process sensor as a critical component of a manufacturing unit operation requires a high level of engineering confidence in the ability of the sensor to reliably detect the desired process characteristic. Without this confidence, manufacturers justifiably cannot leverage in-process sensor technology to achieve the higher levels of process productivity they offer. Sensor technology can often offer some assistance in signal interpretation, but it cannot compensate for a sensor whose behavior or relevance to the process mechanics are poorly understood.

To develop the engineering confidence in the performance of an in-process sensor, one of two methodologies may be pursued. First, as shown by the solid path in Figure 8.15, quantitative engineering models describing the process and the role of the sensor in the process are developed. This is the quantitative path. These models must be stable and verifiable in a variety of relevant experimental and production environments. The basis for a quantitative model for the generation of acoustic emission in manufacturing processes was detailed earlier. This development did not, however, address all of the elements in a monitoring system, for example, the transducer characteristics. In the case of the acoustic emission sensor, the transducer behavior can have a significant impact on the ability to monitor the process.

Often, engineering estimates of the all the parameters in the quantitative path are not obtainable. Then, testing can lead to an understanding of the signal characteristics relevant to different process parameters, and control can be implemented. This is shown in Figure 8.15 as the heuristic path. The burden of proof on the engineering researcher is somewhat higher when the heuristic path is taken, because the robustness of the various methodologies as they are ported from one system to another is always a question. Once the manufacturer is satisfied that a heuristic algorithm is sufficiently robust, it can be designed into a manufacturing process. At this point, an in-process sensor can be offered to a manufacturing environment with a quantitative level of confidence. The examples given for sensor applications earlier typify this approach.

Development and validation of models for the sensor and manufacturing process model in the first case, or the heuristic methodologies in the second case, is thus a critical task for improving manufacturing productivity. Fortunately, this task can be broken into sub-components, each of which is addressable through a variety of tools available to the manufacturing researcher. Figure 8.15 shows a diagram of these components.

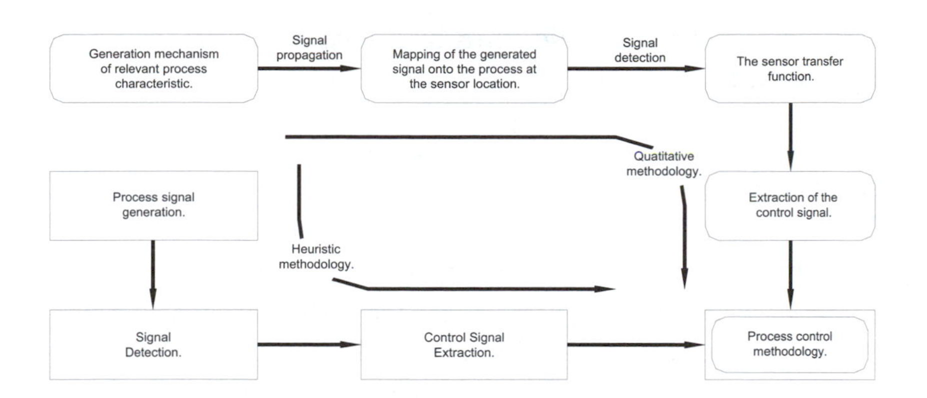

Figure 8.15. Quantitative and heuristic paths for the development of in-process monitoring and control methodologies.

The different paths have obvious advantages and disadvantages. The heuristic path is easier to implement, especially in the absence of existing models for the process physics. The challenge lies in developing the confidence that the transfer function for the process signal from the process to the sensor is invariant enough to utilize chosen methodology across platforms. With the quantitative path, the modeling is more intensive, but the issue of developing a viable model across platforms is not as critical. Which methodology is chosen is entirely situation dependent. Where one-of-a-kind processes exist with poorly characterized process physics, the heuristic approach is usually the most economic route to process optimization. Where the economic leverage of improved process control is significant, and the process control methodology can be spread across many machines, the increased investment associated with more quantitative models may be warranted.

8.9 Basic sensor classification and new sensing technologies

8.9.1 Introduction

We now review a basic classification of sensors based upon the principle of operation. Several excellent texts exist that offer detailed descriptions of a range of sensors and these have been summarized in the material below[176-178, 180]. We distinguish here between a *transducer* and a *sensor* even though the terms are often used interchangeably.

A transducer is generally defined as a device that transmits energy from one system to another often with a change in form of the energy. A good example is a piezoelectric crystal which will output a current or charge when mechanically actuated. A sensor, on the other hand, is a device which is "sensitive" to (meaning responsive to or otherwise affected by) a physical stimulus (light for example) and then transmits a resulting impulse for interpretation or control.[179] Clearly there is some overlap as in the case of a piezoelectric actuator (responding to a charge and outputting a motion or force) and a piezoelectric sensor (outputting a charge for a given force or motion input). In one case, the former, the piezoelectric device acts as a transducer and in the other, the latter, as a sensor. The terms can often be used interchangeably without problem in most cases.

A sensor, according to Webster's dictionary[179] "is a device that responds to a physical (or chemical) stimulus (such as heat, light, sound, pressure, magnetism, or a particular motion) and transmits a resulting impulse (as for measurement or operating control)." Sensors are, in this way, devices which first perceive an input signal and then converts that input signal or energy to another output signal or energy for further use. We generally classify signal outputs into six types as:

1. mechanical,
2. thermal (i.e. kinetic energy of atoms and molecules),
3. electrical,
4. magnetic,
5. radiant (including electromagnetic radio waves, micro waves, etc.) and
6. chemical.

Sensors now exist, and are in common use, that can be classified as either "sensors on silicon" as well as "sensors in silicon".[176] We will discuss the basic characteristics of both types of silicon "micro-sensors" but introduce some of the unique features of the latter which are becoming more and more utilized in manufacturing. The small size, multi-signal capability and ease of integration into signal processing and control systems make them extremely practical. In addition, as a result of their relative low cost, these are expected to be the "sensor of choice" in the future.

The six types of signal outputs listed above reflect the ten basic forms of energy that sensors convert from one form to another. These are listed in Table 8.4, collected from various sources.[178, 181, 186] Practically, these ten forms of energy are condensed into the six signal types listed as we can consider atomic and molecular energy as part of chemical energy, gravitational and mechanical as one - mechanical, and we can ignore nuclear and mass energy. The six signal types (hence basic sensor types for our discussion) represent "measurands" extracted from manufacturing processes that give us insight into the operation of the process. These measurands represent measurable elements of the process and, as well, derive from the basic information conversion technique of the sensor. That is, depending on the sensor, we will likely have differing measurands from the process. But, the range of measurands available is obviously closely linked to the type of (operating principle) of the sensor employed. Table 8.5, defines the relevant measurands from a range of sensing technologies.[182] The "mapping" of these measurand/sensing pairs on to a manufacturing process is the basis of developing a sensing strategy for a process or system. The measurands give us important information on the:

1. process (the electrical stability of the process - in electrical discharge machining, for example),
2. effects of/outputs of the process (surface finish, dimension, for example), and
3. state of associated consumables (cutting fluid contamination, lubricants, tooling, for example).

Table 8.4. Forms of energy converted by sensors.

Energy Form	Definition
Atomic	Related to the force between nuclei and electrons
Electrical	Electric fields, current, voltage, etc.
Gravitational	Related to the gravitation attraction between a mass and the earth
Magnetic	Magnetic fields and related effects
Mass	Following Relativity Theory ($E = mc^2$)
Mechanical	Pertaining to motion, displacement/velocity, force, etc.
Molecular	Binding energy in molecules
Nuclear	Binding energy in electrons
Radiant	Related to electromagnetic radiowaves, microwaves, infrared, visible light, ultraviolet, x-rays and gamma rays
Thermal	Related to the kinetic energy of atoms and molecules

Table 8.5. Process measurands associated with sensor signal types, after White.[182]

Signal Output Type	Associated Process Measurands
Mechanical (includes acoustical)	• position (linear, angular), velocity, acceleration • force, moment, torque • stress, pressure • strain • mass, density • flow velocity, rate of transport • shape, roughness, orientation • stiffness, compliance • viscosity • crystallinity, structural integrity • wave amplitude, phase, polarization, spectrum • wave velocity
Electrical	• charge, current • potential, potential difference • electric field (amplitude, phase, polarization, spectrum) • conductivity • permittivity
Magnetic	• magnetic field (amplitude, phase, polarization, spectrum) • magnetic flux • permeability
Chemical (includes biological)	• components (identities, concentrations, states) • biomass (identities, concentrations, states)
Radiation	• type • energy • intensity • emissivity • reflectivity • transmissivity • wave amplitude, phase, polarization, spectrum • wave velocity
Thermal	• temperature • flux • specific heat • thermal conductivity

Finally, there are a number of technical specifications of sensors that must be addressed in assessing the ability of a particular sensor/output combination to robustly measure the state of the process. These specifications relate to the operating characteristics of the sensors and are usually the basis for selecting a particular sensor from a specific vendor, for example:[182]

- ambient operating conditions
- full scale output
- hysteresis
- linearity
- measuring range
- offset
- operating life
- output format
- overload characteristics
- repeatability
- resolution
- selectivity
- sensitivity
- response speed (time constant)
- stability/drift

It is impossible to detail the associated specifications for the six sensing technologies under discussion here. A number of references have done this for specific sensors for manufacturing applications as we have discussed earlier in the chapter.

8.9.2 Basic sensor types

8.9.2.1 Mechanical sensors

Mechanical sensors are perhaps the largest and most diverse type of sensors because, as seen in Table 8.5, they have the largest set of potential measurands. Force, motion, vibration, torque, flow, pressure, etc. are basic elements of most manufacturing processes and of great interest to measure as an indication of process state or for control. Force is a push or pull on a body that results in motion/displacement or deformation. Force transducers, a basic mechanical sensor, are designed to measure the applied force relative to another part of the machine structure, tooling, or work piece as a result of the behavior of the process. A number of mechanisms convert this applied force (or torque) into a signal including piezoelectric crystals, strain gages, and potentiometers (as a linear variable differential transformer-LVDT). Displacement, as in the motion of an axis of a machine, is measurable by mechanical sensors (again the LVDT or potentiometer) as well as by a host of other sensor types to be discussed. Accelerometer outputs, differentiated twice, can yield a measure of displacement of a mechanism. Shiraishi[61-63], relies on a number of mechanical sensing elements to measure the dimensions of a workpiece. Flow is commonly measured by "flowmeters", mechanical devices with rotameters (mechanical drag on a float in the fluid stream), as well as venturi meters (relying on differential pressure measurement - using another mechanical sensor) to determine flow of fluids. An excellent review of other mechanical sensing (and transducing) devices is in Bray[178].

Mechanical sensors have seen the most advances due to the developments in semiconductor fabrication technology. Piezo-restive and capacitance-based devices, basic building blocks of silicon micro-sensors, are now routinely applied to pressure, acceleration and flow measurements in machinery. Figure 8.16a shows schematics of a capacitive sensor with applications in pressure sensing (the silicon diaphragm deflects under the pressure of the gas/fluid and modifies the capacitance between the diaphragm and another electrode in the device. Using a beam with a mass on the end as one plate of the capacitor and a second electrode, Figure 8.16b, an accelerometer is constructed and the oscillation of the mass/beam alters the capacitance in a measurable pattern allowing the determination of the acceleration. These sensor chips can be provided as basic OEM sensor elements or can be integrated into a next-level packaging schemes. These devices are constructed using conventional semiconductor fabrication technologies based on the semiconducting materials and miniaturization of VLSI patterning techniques (see, for example, Sze[176], as an excellent reference on semiconductor sensors). The development of microelectromechanical sensing systems (so-called MEMS) techniques has opened a wide field of design and application of special micro-sensors (mechanical and others) for sophisticated sensing tasks, for example, for use in positioning control of shop floor robotic devices. In fact, most of the six basic sensor types can be accommodated by this technology. Accelerometers are built on these chips as already discussed. Whatever affects the frequency of oscillation of the silicon beam of the sensor can be considered a measurand. Coating the accelerometer beam with a material that absorbs certain chemical elements, hence changing the mass of the beam and its resonant frequency, changes this in to a chemical sensor. Similar modifications yield other sensor types.

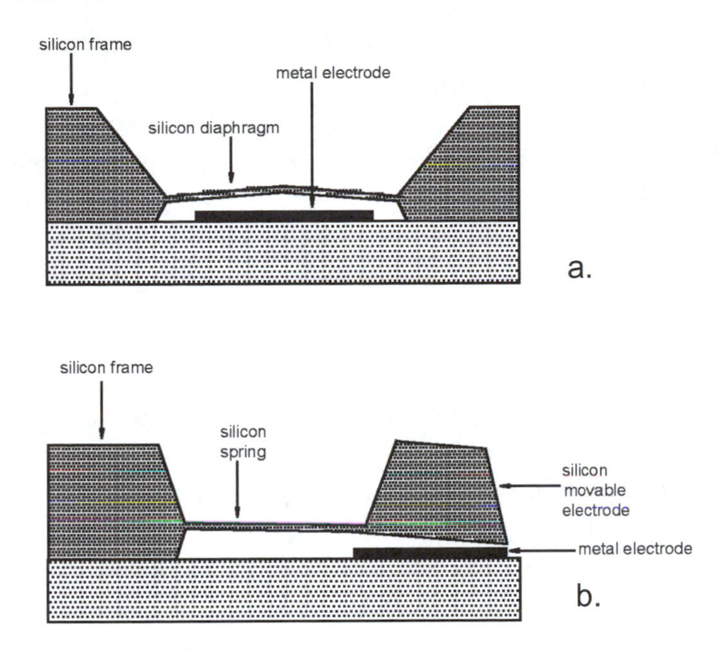

Figure 8.16. Schematic of a capacitance sensor for a) pressure and b) acceleration.

One particularly interesting type of micro-sensor for pressure applications, not based on the capacitance principles discussed above is silicon-on-sapphire (SOS). This is specially applicable to pressure sensing technology. Manufacturing an SOS transducer begins with a sapphire wafer on which silicon is epitaxially grown onto the smooth, hard, glass-like surface of the sapphire. Since the crystal structure of the silicon film is similar to sapphire's, the SOS structure appears to be one crystal with a strong molecular bond between the two materials. The silicon is then etched into a Wheatstone Bridge pattern using conventional photolithography techniques. Due to its excellent chemical resistance and mechanical properties, the sapphire wafer itself may be used as the sensing diaphragm. An appropriate diaphragm profile is generated into the wafer to create the desired flexure of the diaphragm and to convey the proper levels of strain to the silicon Wheatstone bridge. The diaphragm may be epoxied or brazed to a sensor package. A more reliable method of utilizing the SOS technology involves placing an

SOS wafer on a machined titanium diaphragm. In this configuration titanium becomes the primary load bearing element and a thin (thickness under .01") SOS wafer is used as the sensing element. The SOS wafer is bonded to titanium using a process similar to brazing, performed under high mechanical pressure and temperature conditions in vacuum to ensure a solid, stable bond between the SOS wafer and the titanium diaphragm. The superb corrosion resistance of titanium allows compatibility with a wide range of chemicals that may attack epoxies, elastomers, and even certain stainless steels. Since the titanium diaphragm is machined using conventional machining techniques and the SOS wafer is produced using conventional semiconductor processing techniques. SOS based pressure sensors with operating pressures ranging from 104 kPa to over 414 MPa are available.

Acoustic sensors have benefited from the developments in micro-sensor technology. Semiconductor acoustic sensors employ elastic waves at frequencies in the range of megahertz to the low gigahertz to measure physical and chemical (including biological) quantities. There are a number of basic types of these sensors based upon the mode of flexure of an elastic membrane or bulk material in the sensor is employed. Early sensors of this type used vibrating piezoelectric crystal plates referred to as a quartz crystal microbalance (QCM). It is also called a thickness shear-mode sensor (TSM) after the mode of particle motion employed. Other modes of acoustic wave motion employed in these devices (with appropriate design) include surface acoustic wave (SAW) for waves travelling on the surface of a solid, and elastic flexural plate wave (FPW) for waves travelling in a thin membrane. The cantilever devices described earlier are also in this class.

8.9.2.2 Thermal sensors

Thermal sensors generally function by transforming thermal energy (or the effects of thermal energy) into a corresponding electrical quantity that can be further processed or transmitted. Other techniques for sensing thermal energy (in the infrared range) are discussed under radiant sensors below. Typically, a non-thermal signal is first transduced in to a heat flow, the heat flow is converted into a change in temperature/temperature difference, and, finally, this temperature difference is converted into an electrical signal using a temperature sensor. Micro-sensors employ thin membranes (floating membrane cantilever beam, for example). There is a large thermal resistance between the tip of the beam and the base of the beam where it is attached to the device rim. Heat dissipated at the tip of the beam will induce a temperature difference in the beam. Thermocouples (based on the thermoelectric Seebeck effect whereby a temperature difference at the junction of two metals creates an electrical voltage) or transistors are employed to sense the temperature difference in the device outputting an electrical signal proportional to the difference. Recent advances in thermal sensor application to the "near surface zone" of materials for assessing structural damage (referred to as photo-thermal inspection) are reported by Goch et al.[183] This review covers other measurement techniques as well such as micromagnetic.

Thermal sensors are also employed in flow measurement following the well known principle of cooling of hot objects by the flow of a fluid (boundary layer flow measurement anemometers). They can also be applied in thermal tracing and heat capacity measurements in fluids. All three application areas are suitable for silicon micro-sensor integration.

Thermal sensors have also found applicability traditionally in "true-rms converters." Root mean square (RMS) converters are used to convert the effective value of an alternating current (AC) voltage or current to its equivalent direct current (DC) value. This is accomplished quite simply by converting the electrical signal into heat with the assistance of a resistor and measuring the temperature generated.

8.9.2.3 Electrical sensors

Electrical sensors are intended to determine charge, current, potential, potential difference, electric field (amplitude, phase, polarization, spectrum), conductivity and permittivity and, as such, have some overlap with magnetic sensors. Power measurement, an important measure of the behavior of many manufacturing processes, is also included here. An example of the application of thermal sensors for true root-mean-square power measurement was included with the discussion on thermal sensors. The use of current sensors (perhaps employing principles of magnetic sensing technology) is commonplace in machine tool monitoring. Electrical resistance measurement has also been widely employed in tool wear monitoring applications. Most of the discussion on magnetic sensors below is applicable here in consideration of the mechanisms of operation of electrical sensors.

8.9.2.4 Magnetic sensors

A magnetic sensor converts a magnetic field into an electrical signal. Magnetic sensors are applied directly as magnetometers (measuring magnetic fields) and data reading (as in heads for magnetic data storage devices). They are applied indirectly as a means for detecting non-magnetic signals (for example, in contactless linear/angular motion or velocity measurement) or as proximity sensors. Most magnetic sensors utilize the Lorenz force producing a current component perpendicular to the magnetic induction vector and original current direction (or a variation in the current proportional to a variation in these elements.) There are also Hall effect sensors. The Hall effect is a voltage induced in a semiconductor material as it passes through a magnetic field. Magnetic sensors are useful in nondestructive inspection applications where they can be employed to detect cracks or other flaws in magnetic materials due to the perturbation of the magnetic flux lines by the anomaly. Semiconductor-based magnetic sensors include thin film magnetic sensors (relying on the magnetoresistance of NiFe thin films), semiconductor magnetic sensors (Hall effect), optoelectronic magnetic sensors which use light as an intermediate signal carrier (based on Faraday rotation of the polarization plane of linearly polarized light due to the Lorenz force on bound electrons in insulators) and superconductor magnetic sensors (a special class)[176].

8.9.2.5 Radiant sensors

Radiation sensors convert the incident radiant signal energy (measurand) into electrical output signals. The radiant signals are either electromagnetic, neutrons, fast neutrons, fast electrons, or heavy-charge particles[176]. The range of electromagnetic frequencies is immense spanning from cosmic rays on the high end with frequencies in the 10^{23} rd Hz range to radio waves in the low tens of thousands of Hz. In manufacturing applications we are most familiar with infrared radiation (10^{11} Hz to 10^{14} Hz) as a basis for temperature measurement or flaw/problem detection. Silicon-on-insulator photodiodes as well as phototransistors based on transistor action are typical micro-sensor radiant devices for use in these ranges[176].

8.9.2.6 Chemical sensors

These sensors are becoming particularly more important in manufacturing process monitoring and control. It is important to Measure the identities of gases and liquids, concentrations, and states, chemical sensors for worker safety (to insure no exposure to hazardous materials or gases), process control (to monitor, for example, the quality of fluids or gases used in production; this is specially critical in the semiconductor industry which relies on complex process "recipes" for successful production), and process state (presence or absence of a material, gas or fluid, for example.) Chemical sensors have been successfully produced as micro-sensors using semiconductor technologies primarily for the detection of gaseous species. Most of these devices rely on the interaction of chemical species at semiconductor surfaces (adsorption onto a layer of material, for example) and then the change caused by the additional mass affecting the performance of the device. This was discussed under mechanical sensors where the change in mass altered the frequency of vibration of a silicon cantilever beam providing a means for measuring the presence or absence of the chemical as well as some indication of the concentration. Other chemical effects are employed as well such as resistance change caused by the chemical presence, the semiconducting oxide powder pressed pellet (so called Taguchi sensors) and the use of Field Effect Transistors (FET) as sensitive detectors for some gases and ions. Sze[176] gives a quite complete review of chemical microsensors and the reader is referred to this for details of this complex sensing technology.

8.10 Applications of sensors in precision manufacturing

The purpose of this section is to outline several case studies of the use of sensors for the monitoring of manufacturing processes at the three different manufacturing regimes outlined in the Taniguchi curve; the normal/conventional, precision, and ultraprecision scales.

8.10.1 AE-based monitoring of grinding wheel dressing

Currently, a conventional grinding process requires a complex sequence of tasks in order to start production on a workpiece, including equipment setup, machining of initial test samples, calibration of the tool, and frequent control and correction of the process parameters. AE can serve as a means for the complete automation of all these steps[184]. However, in grinding, the two main limitations of AE-based monitoring solutions are the oscillation of the RMS level and signal saturation. Despite these limitations, AE can be very effective for contact detection between moving surfaces. The following section will focus on a Fast AE RMS analysis and mapping technique for wheel condition monitoring in grinding.

8.10.1.1 Fast AE RMS analysis for wheel condition monitoring

A new method for process monitoring based on short-term analysis of AE RMS patterns is proposed. Since AE RMS changes occur after a considerable period of time, the short time evaluation could be a solution for a reliable AE application. Figure 8.17 shows a proposed system where the acoustic emission obtained from the contact between diamond tool and grinding wheel (or grinding wheel and workpiece in grinding monitoring) is converted to an RMS level and acquired by the computer by using an A/D conversion board.[184] The sampling rate range varies from 60-500 kHz, depending on the chosen resolution in the circumferential direction. A sampling rate of 2 kHz was used in this particular case to map each rotation of the grinding wheel, which corresponds to a resolution of about 0.5 mm/ sample. To be able to measure the contact of the diamond tool with each abrasive grain, the RMS calculation must be performed using a very fast time constant. The time constant was calculated as the average time spent for two consecutive hits between abrasive grains and the diamond tool. This calculation was done for a 60 mesh, L structure aluminum oxide wheel. The average distance between grains measured was about 0.38 mm with a grinding speed of 45 m/s, so a time constant of 10 microseconds was found.

In AE signal processing, the RMS calculation is normally done after having the raw signal filtered. Due to the very small time constant used in the RMS calculation, it was established that a high pass filter with a cut-off frequency of 100 kHz be used. A specific AE signal processing unit has been developed for this system. The data acquisition is made in data groups where each 1-dimention array of data corresponds to a full rotation of the grinding wheel. The AE signal acquisition starts in each rotation triggered by a magnetic sensor positioned in the wheel hub where a reference pin is installed. A graphical mapping of the AE signal with a color intensity scale is then created. During the wheel dressing operation, the mapping is constructed in real time by adding columns in the array as the dresser travels along the wheel surface.

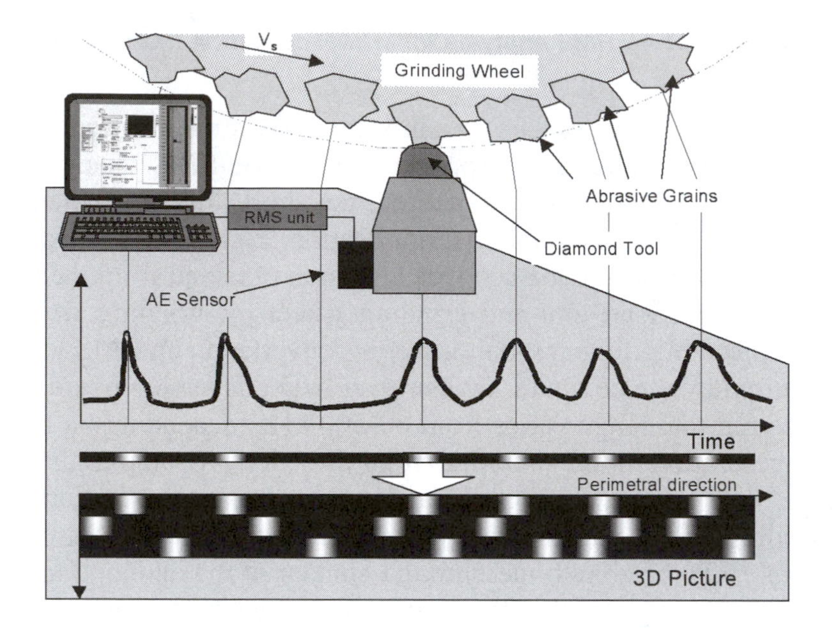

Figure 8.17. Procedure for AE signal graphical mapping construction.

The system can be used for different evaluation procedures, including:

- **Dressing evaluation:** During the dressing operation the interaction between dresser and grinding wheel can be acoustically mapped. Lack of contact between dressing tool and grinding wheel will appear as dark areas in the map.
- **Topographic mapping:** In this case the map is similar to that obtained for the dressing operation but using the dressing depth of cut nearly zero or with a value close to the undeformed chip thickness for the operation. In this case the map shows the active surface of the grinding wheel, which means the surface that will actually be in contact with the workpiece during grinding.
- **Grinding evaluation:** During a plunge grinding operation the interaction between the grinding wheel and the workpiece can be evaluated. In this case a different map is obtained

- where one axis is the grinding time and the other shows the average acoustic energy in the whole wheel length along its perimeter.

8.10.1.2 Grinding wheel topographical mapping

Figure 8.18 shows an output from the acoustic mapping system when used during a topographic mapping procedure. The vertical and horizontal directions are the wheel perimeter and width respectively, with a spatial resolution of 2 samples/mm. The depth of interaction between diamond tool and grinding wheel used was 1 micron (in the range of elastic contact). The color intensity shows the acoustic emission RMS value measured from the interaction between dressing tool and the abrasive grains. Darker areas correspond to lower acoustic energy detected by the sensor. The L shaped mark was created in the wheel surface in order to evaluate the system performance. Dark areas show worn regions of the wheel where the dressing tool had lower interaction with the abrasive grains. The main vertical worn band on the left side was a result from a grinding operation made with the workpiece shown in the photo of Figure 8.18a. In this figure, the "L" mark produced on the wheel surface and a magnified view of its representation in the AE graphical mapping is shown.

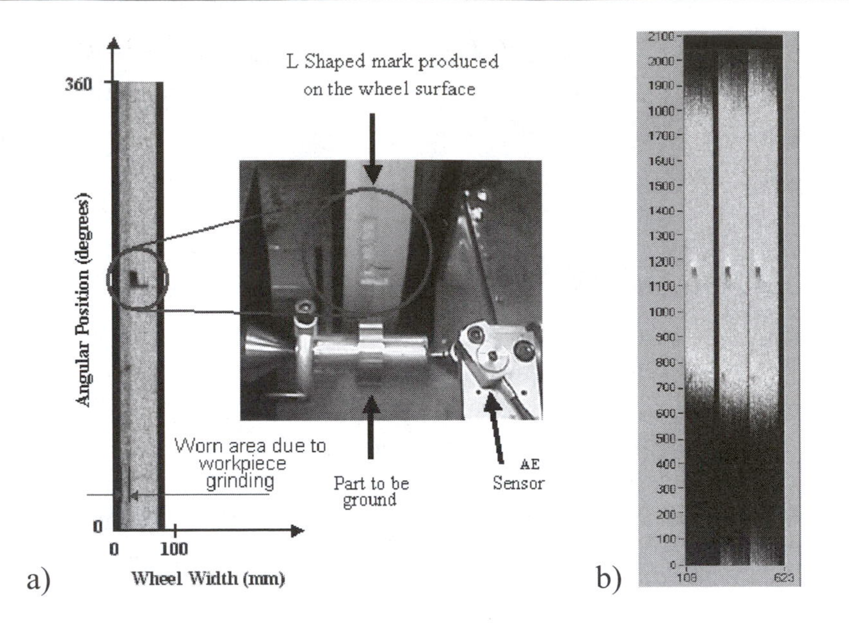

Figure 8.18. a) L-shaped mark produced on the wheel surface with corresponding AE map during dressing, b) Dressing of unbalanced grinding wheel (first 3 strokes, 2 microns per pass).

It was observed that the AE mapping system could generate an image similar to the surface topography present on the grinding wheel surface, even using a very small depth of interaction with the diamond tool. A spark-out dressing was also tested (depth of interaction = 0) and the result was nearly the same. The use of equal depth of interactions may lead to reading error in the case of any thermal deformation in the machine structure.

The same grinding wheel was dressed in order to remove the damage produced by the grinding operation. The dressing depth of cut was 2 microns per pass. This small value was chosen in order to evaluate the depth of the damage produced by the grinding operation. The maps of 3 consecutive dressing strokes made in an unbalanced grinding wheel are shown in Figure 8.18b. The wheel was dressed and then unbalanced using an automatic balancing system in manual mode. Each dressing operation was performed with a dressing in-feed of 1 micron. After six dressing strokes, the grinding wheel eccentricity disappeared, and the real unbalancing displacement level was calculated to be about 6 microns. The unbalancing device installed in the machine demonstrated a vibrational amplitude of only 1 micron (corresponding to the "green" zone for grinding). This strong difference is caused by the fact that the vibration sensor measures the vibration amplitude in the machine bearings while the acoustic mapping system measures the actual wheel eccentricity caused by the lack of balancing. The sequence shows exactly the position of the heavier side of the grinding wheel and the reduction in the eccentricity after dressing. This is another important piece of information provided by the mapping system, since most balancing devices for grinding machine use vibration sensors as feedback. It was observed that the minimum vibration level does not lead to the best concentricity of the grinding wheel. Therefore, the information provided by the mapping system could be used as a feedback signal when balancing CBN wheels during touch dressing operations where this problem is more critical[184].

8.10.1.3 Wheel wear mechanism

Figure 8.19 presents two experiments to demonstrate the influence of the wheel wear behavior in the AE map. Each stripe in the graph represents a single grinding cycle or a single workpiece in a production line of automotive components. The workpiece material was Inconel and the grinding wheel is a very hard grade, low friability aluminum oxide specification (DA 80 R V). The second experiment was a plunge grinding operation of hardened AISI 4340 steel using a soft white aluminum oxide grinding wheel (AA 60 G V).

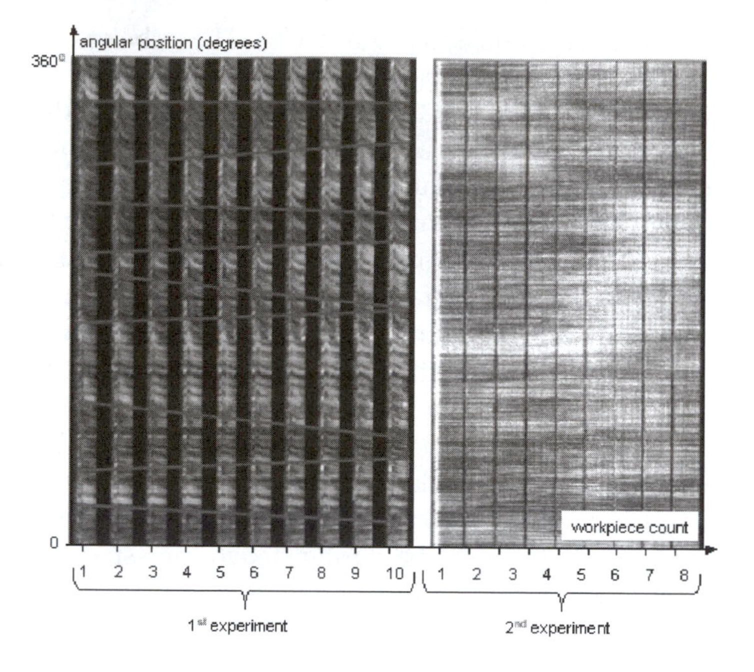

Figure 8.19. Mapping of AE from a grinding process showing two types of wear mechanisms.

The image composition of the several mappings in Figure 8.19 shows two distinct types of wheel wear behavior. In the first experiment the consistent pattern around the wheel shows that it is not losing grains. In the second experiment the transformation indicates that the wheel is indeed losing grains. These results were confirmed by checking the workpiece size plot (stable for the first case and an increasing tendency in the second case), workpiece temperature (grows for the first experiment and constant for the second) and power plots (power grows in the 1st experiment and remains constant for the second one).[184]

8.10.1.4 AE-based monitoring of face milling

The fast AE RMS graphic mapping system used for the AE monitoring work for grinding was also used to monitor a face milling process.

Previous work by Diei and Dornfeld established a high degree of sensitivity of the AE signal to the individual chip formation mechanism in face milling, with a significant variation observed in the AE signal (processed through time-difference signal processing techniques) with respect to different stages of the chip formation process, with significant increases in AE signal observed upon the initiation of chip formation (due to initial impact of the insert with the workpiece) and exit of the insert from the workpiece (see Figure 8.20a)[185, 186].

In the face milling monitoring work, a ROMI Discovery 3-axis CNC milling center was used with an AE sensor attached directly to the workpiece[187]. The AE RMS signal was collected during a typical machining operation (6 insert cutter, 200 RPM, 0.1mm/insert feed, 0.2 micron axial DOC) in a continuous fashion similar to the techniques and DAQ system parameters used in the grinding monitoring technique. Each vertical data trace in the AE RMS intensity mapping in Figure 8.20b corresponds to one spindle rotation, with a clear spike in AE signal occurring due to the initial impact of the insert with the workpiece (similar to that seen in Figure 8.20a). Successive spindle rotations are shown along the horizontal axis, for a total of approximately 200 total revolutions in the intensity mapping. The AE RMS signal for each cutting insert can be clearly seen (numbered 1-6) with a separation of 60 degrees between inserts. After several tens of spindle revolutions, AE signal due to the rubbing of inserts on the trailing edge of the cutter can be observed, although the AE signal due to rubbing of inserts 2 and 6 can barely be seen (see Figure 8.20b). This AE mapping technique serves as a means for tool condition monitoring in face milling, particularly for tool contact, breakage, and insert positional precision within the cutter[187].

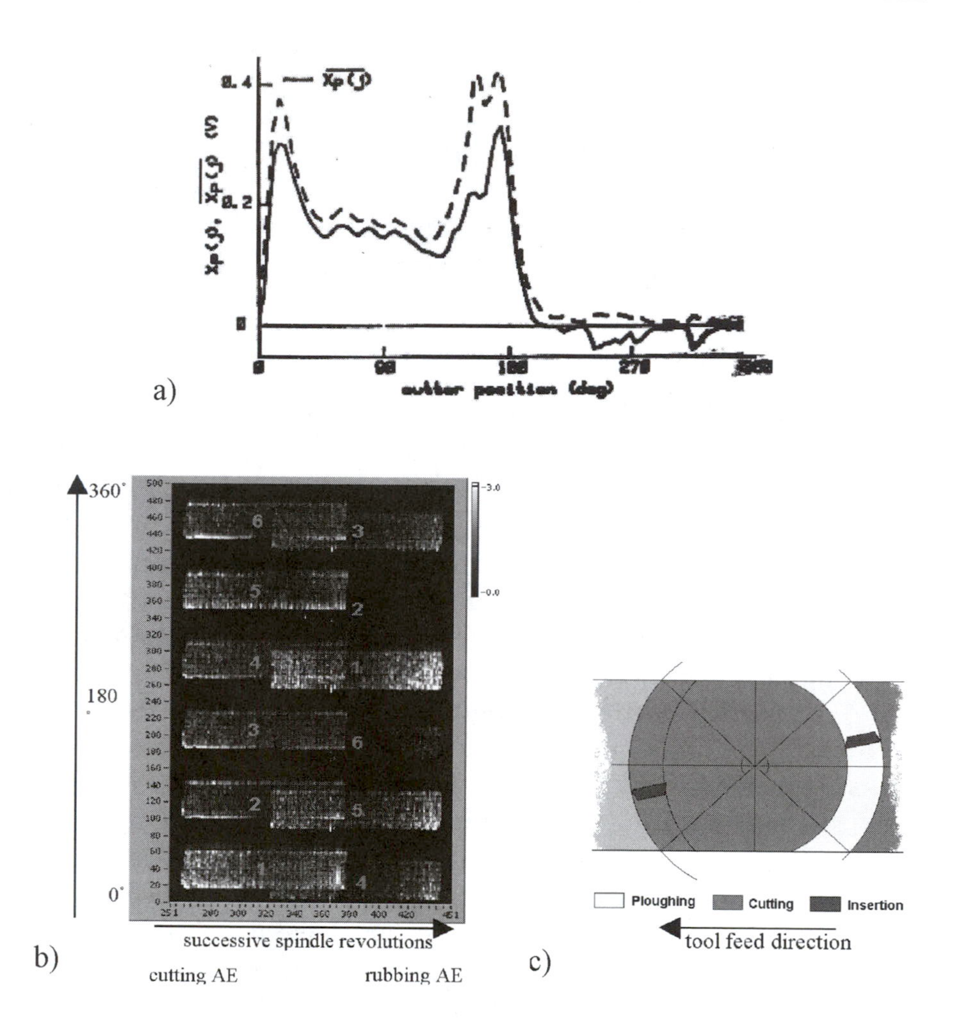

Figure 8.20. a) AE response for single chip formation in face milling, b) AE map for face milling operation, c) Tool motion relative to workpiece.

8.10.2 AE-based monitoring of chemical mechanical planarization

Chemical mechanical planarization (CMP) is one of the key enabling technologies in the semiconductor manufacturing industry today for the fabrication of extremely smooth and flat surfaces on a variety of semiconductor substrate materials, and processs details are discussed in Chapter 10. In order to meet the requirements of current lithography tools which require extremely stringent tolerances for flatness and planarity, CMP is capable of planarizing a 300 mm (current industry standard) diameter wafer achieving surface roughness on the order of 1–2 nm R_a and global planarity well below 0.5 μm. However, CMP has also become one of the key bottleneck or roadblock issues in semiconductor manufacturing today[188]. The decreasing line widths of semiconductor devices require new materials, such as copper and the so-called low-*k* dielectrics, which further challenge the process. Preferential polishing rates of adjacent materials, or surface features resulting from previous manufacturing steps, often lead to defects such as dishing which frustrate efforts to obtain planarity. The abrasive slurry can also induce defects such as surface contamination, scratches, slurry residue, etc., hence predicating the need for a reliable means of monitoring the CMP process.

Sources of AE at the pad asperity/surface interface are believed to dominate the measured AE signal, and are diagrammed schematically in Figure 8.21. AE generation due to plastic deformation induced by abrasive particle interaction with the wafer surface is believed to be a primary component of the total AE signal, with AE generation via elastic contact (stick-slip mechanisms, etc) between the wafer and pad asperities also contributing. At the macroscale, friction and rubbing between two surfaces (such as at the wafer/pad/retaining ring level) are potential sources; surface asperities come into contact and are elastically and/or plastically deformed, and possibly even welded together. As the surfaces slide over one another, these asperities are deformed further and possibly even fractured. A schematic of the AE generation mechanism at the atomic scale is shown in Figure 8.22, where individual atoms are first in contact in a relaxed state with a balance between nominal contact forces and repulsive van der Waals forces. As the two surfaces slide over one another with a relative velocity V, atomic bonds are either stretched or compressed. After a certain amount of relative motion, the two atoms then snap back into place (see Figure 8.22c). This mechanism for atomic-level friction has been used in tribology research to explain the stick-slip phenomena so commonly found in surface interaction[189-191]. The motion of atoms due to this stick-slip mechanism in sliding contact generates elastic waves (i.e. phonon generation) that propagate through the solid medium (such as the wafer in CMP) and can be detected as AE.

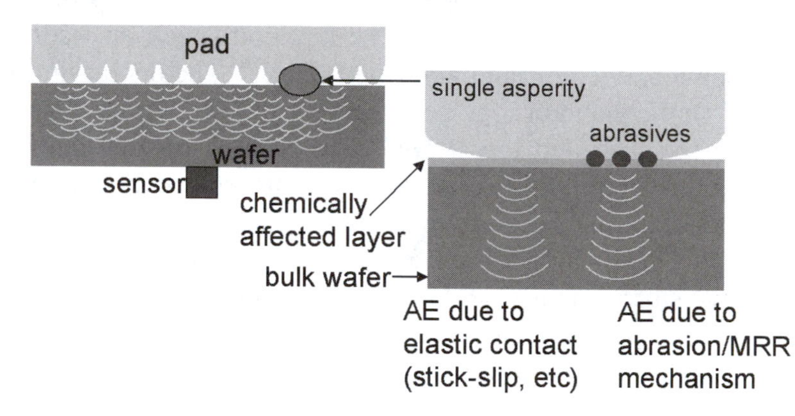

Figure 8.21. AE generation at the asperity scale

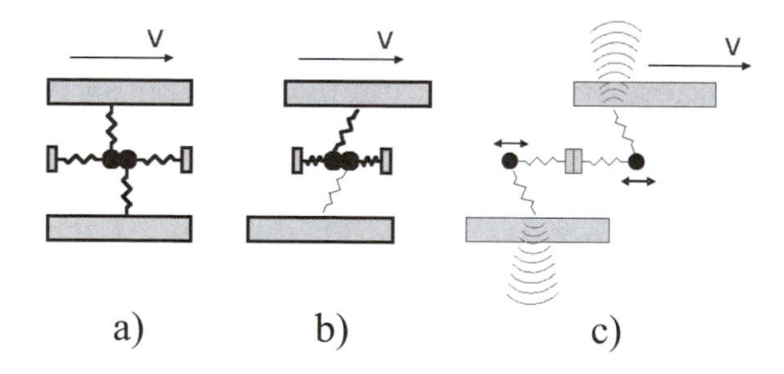

Figure 8.22. a) Atoms in relaxed state, b) Atoms pushed together, c) Phonon generation mechanism for AE.

The extent of AE generation due to solid/fluid interaction or internal fluid turbulence is still unknown as of yet. It is also believed that there is a significant difference between the time and frequency response of AE generated by the two sources shown in Figure 8.21. While a significant variation of AE in the time domain has been demonstrated (via RMS signal processing to demonstrate AE energy content) throughout past research, future work still needs to be done to examine any frequency dependence of the AE signal on the sources specifically at the CMP level. Nevertheless, it has been confirmed that AE energy and other signal features are very sensitive indicators of the degree and nature of contact between surfaces and will be the basis for the monitoring of the CMP process.

8.10.2.1 Monitoring of abrasive process parameters

Some of the earliest initial correlations between AE and abrasive process parameters such as MRR were made by Chang and Dornfeld[192], who found a direct linear relationship between MRR and AE RMS signal in a lapping operation on glass with Al_2O_3 abrasive slurry (see Figure 8.23).

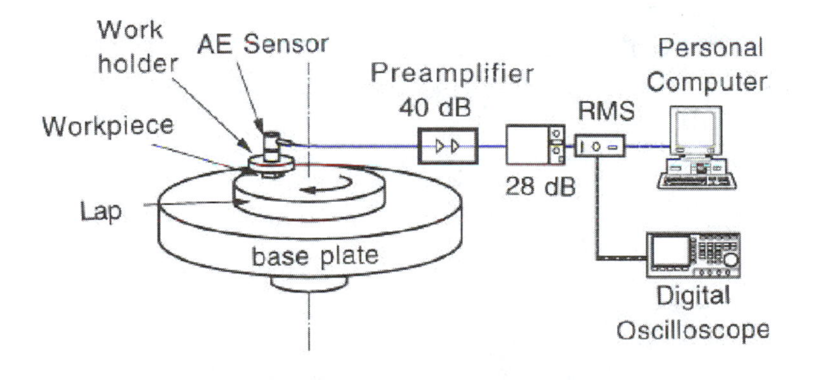

Figure 8.23. AE monitoring setup for lapping.

A direct (and nearly linear) correlation between MRR and AE_{RMS} signal was found (see Figure 8.24). The increase in AE_{RMS}

was believed to be attributed to a transition from a ductile to brittle material removal mechanism at higher removal rates[192].

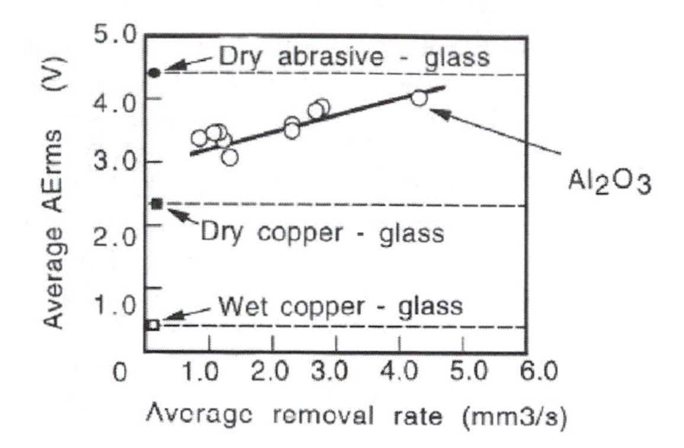

Figure 8.24. Correlation between AE_{RMS} and MRR.

Other early work includes the research of Tang and Dornfeld which demonstrated a significant variation in the AE_{RMS} signal during three distinct stages of CMP with a 200 mm bare silicon wafer polished with SC112 slurry, an IC1000 pad and a Strasbaugh CMP machine (see Figure 8.25)[193]. The process instability due to wafer set-down in the early stage of polishing (about 15 s) can be identified clearly from the raw data of Figure 8.25. Other tests have also shown the sensitivity of AE to different materials polished (see Figure 8.26).

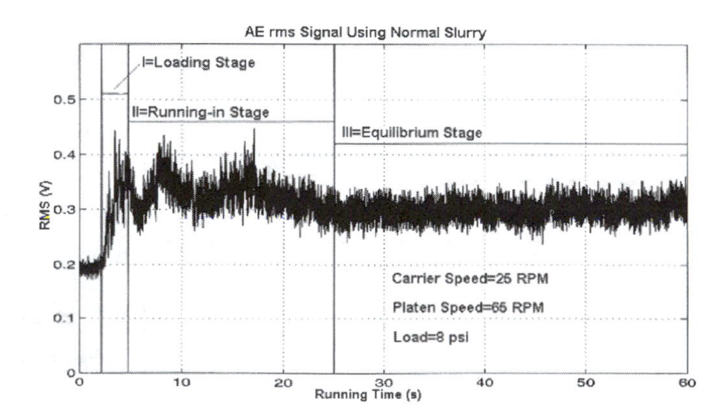

Figure 8.25. Typical AE_{RMS} signal in conventional CMP.

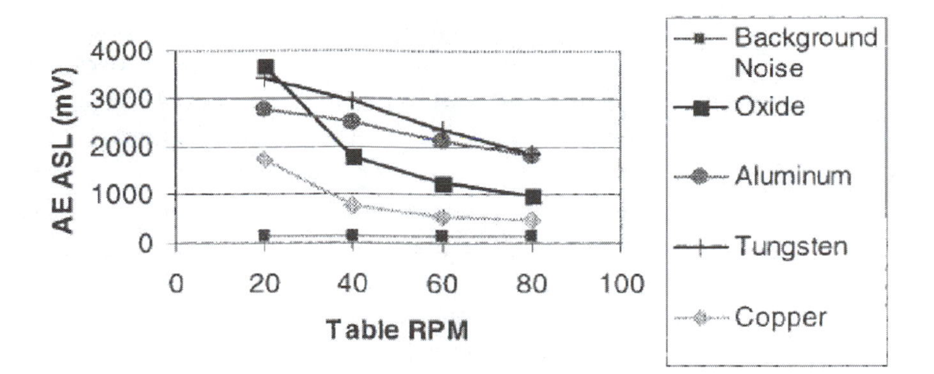

Figure 8.26. AE average signal level of polishing from various material layers.

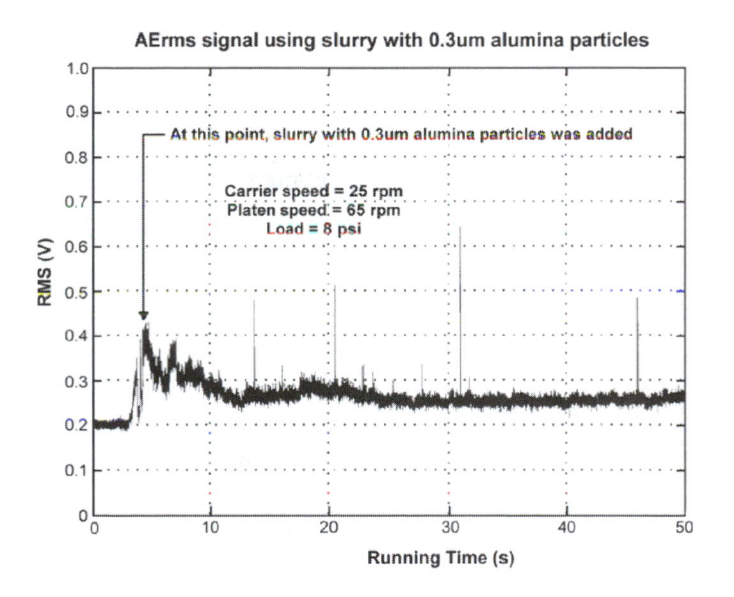

Figure 8.27. The AE_{RMS} signal with a silica-based slurry and 0.3 μm Al_2O_3 particles to induce scratches.

By microelectronic fabrication standards, CMP is an inherently "dirty" process and leaves micro defects, such as residual slurry, particles, pits and micro-scratches on the polished wafer surface. Some of the defects can be removed by post CMP cleaning, but defects such as micro-scratches cannot be recovered by simply cleaning the wafer, and therefore should be specifically addressed for the purpose of increasing chip yields. Tests were carried out to verify the feasibility of using AE for micro-scratch detection in the CMP process. In one set of tests, large diamond grits (here 1 μm) were artificially added to the slurry during CMP and the corresponding AE signals collected. In this case, a large number of mixed micro-scratches were observed in the polished wafer. Correspondingly, several spikes appeared in the AE_{RMS} signals in each wafer carrier rotation period (typically, about 9 s in these tests) especially as the pad/wafer gap decreases at lower speeds. Any scratches in the wafer surface generated owing to abrasive action during the process are visible as spikes of AE activity on top of the basic signal during steady-state polishing (see Figure 8.27). This information will also be useful in the development process models including the fluid and abrasive interactions.

8.10.2.2 Precision scribing of CMP-treated wafers

As a means of investigating the process physics of CMP, and to further evaluate the sensitivity of AE to the material removal physics taking place, a series of scribing tests on a CMP-processed oxide wafer were conducted[194]. In the CMP process for oxide planarization, the bulk of material removal takes place on a "chemically weakened" layer consisting of a highly hydrated and loosely bound network of silica on the order of a few nanometers in thickness (see Figure 8.28). The second layer is a "plastically compressed layer" around 20 nm deep from the chemically weakened layer, depending on process conditions. Unlike the chemically weakened layer, this layer is represented by a plastically compressed network of silica that has higher density. A bulk layer also exists below the plastically compressed layer that is not affected by the CMP process. Because of the variation in the material properties of each layer, it was initially postulated that the AE RMS signal obtained during material removal of these distinct layers would differ.

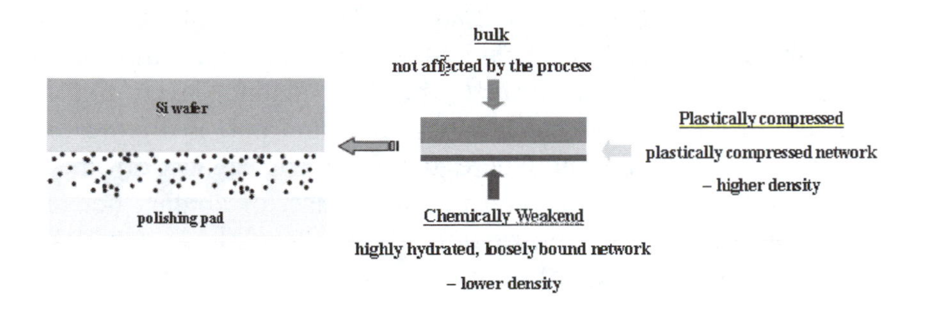

Figure 8.28. Three Distinct Layers Exhibited in Oxide CMP.

In order to assess the mechanical properties of the oxide layer, a scribing operation using a single point diamond turning machine was used (see Figure 8.29). The setup involved mounting a polished oxide wafer at a slight angle of ~2° and scribing across surface, with the tool engaging with an increasing depth-of-cut (DOC) as the tool traversed over the surface. An AE sensor was attached to the front side of the wafer and used to monitor these scratch tests.

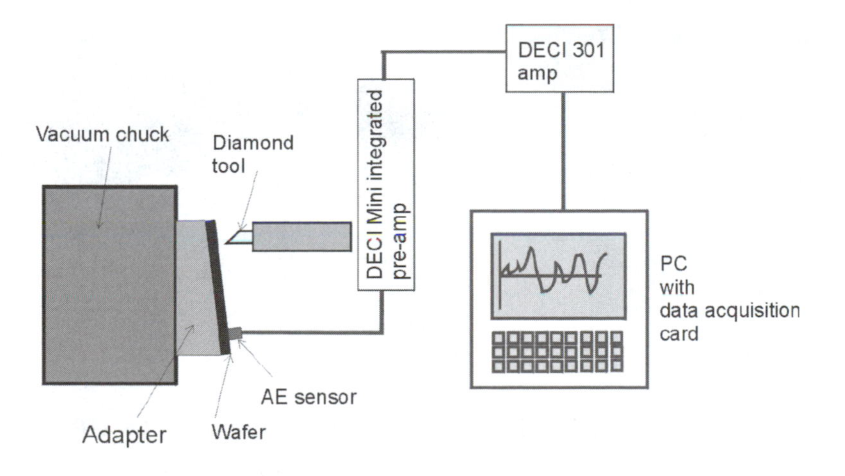

Figure 8.29. Experimental setup for CMP oxide wafer scribing.

During the scribing operation, two transitions (one from the chemically weakened layer to the plastically compressed layer, and the other from the plastically compressed layer to the bulk) appear as distinct features in the AE RMS signal. Figure 8.30 shows the AE signals from the scratch test of post-CMP wafers. As postulated, the variation in the AE signal reflects the three distinct layers and the transitions from one layer to another. The first part in Figure 8.30 shows an "air-cut" region where the tip is disengaged from the chemically weakened layer, and the very onset of contact between the tool and wafer. Since the chemically weakened layer consists of a loosely-bound network of silica and is only a few nm deep, the AE RMS signal from this layer differs only slightly from that of the air-cut region. As the tip starts to touch the wafer, within 10 milliseconds, the AE signals burst and increase over time for about 70 milliseconds, which corresponds to a DOC of 30 nm, and confirms that the plastically compressed layer was engaged at this time. Beyond this part, the AE signals monotonically increase, without significant deviation or burst signal, meaning that the tool is cutting in the bulk layer of the oxide.

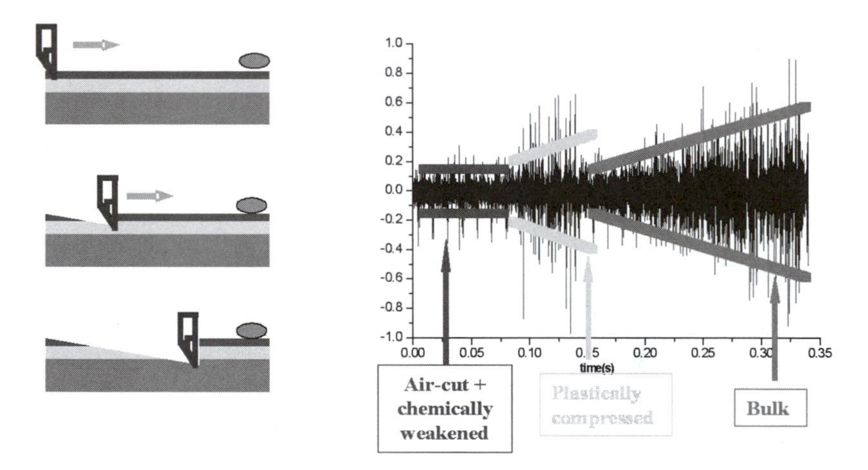

Figure 8.30. Variation in AE RMS signal during scribing of CMP-treated oxide wafer.

8.10.2.3 AE-based endpoint detection for CMP

In semiconductor manufacturing, the use of CMP in thin-film polishing for a fixed time is commonly used. However, due to fluctuations in the process, such as material-removal-rate (MRR) inconsistency, pad degradation, and non-uniformity issues, over- or under-polishing can occur. An endpoint detection system is required to insure that only the desired thickness of material is polished during the CMP process, thus offering many manufacturing advantages such as improved process yields, closer conformance to target requirements, and higher throughput.

In order to evaluate the feasibility of AE as an in-situ endpoint detection technique, two different sets of wafers were polished with a desktop CMP machine. The first set of wafers consisted of stacked films of oxide (5,000°A) at the bottom, tantalum (5,000°A) in the middle, and copper (1,500°A) at the top as an example of the copper damascene process. The wafer was polished with a conventional IC 1000 polyurethane pad and alumina-based slurry with 2.5% H_2O_2. The second set of wafers consisted of film stacks ranging from oxide (2,000°A) at the bottom to nitride (1,000°A) at the top, representing a shallow trench isolation process, and was polished with fixed abrasives and ph-adjusted deionized water (ph=11.5). During the CMP process, both AE signals and frictional force data were collected with a DAQ system, and the experimental data for the copper damascene test wafers are plotted vs. time in Figure 8.31a.

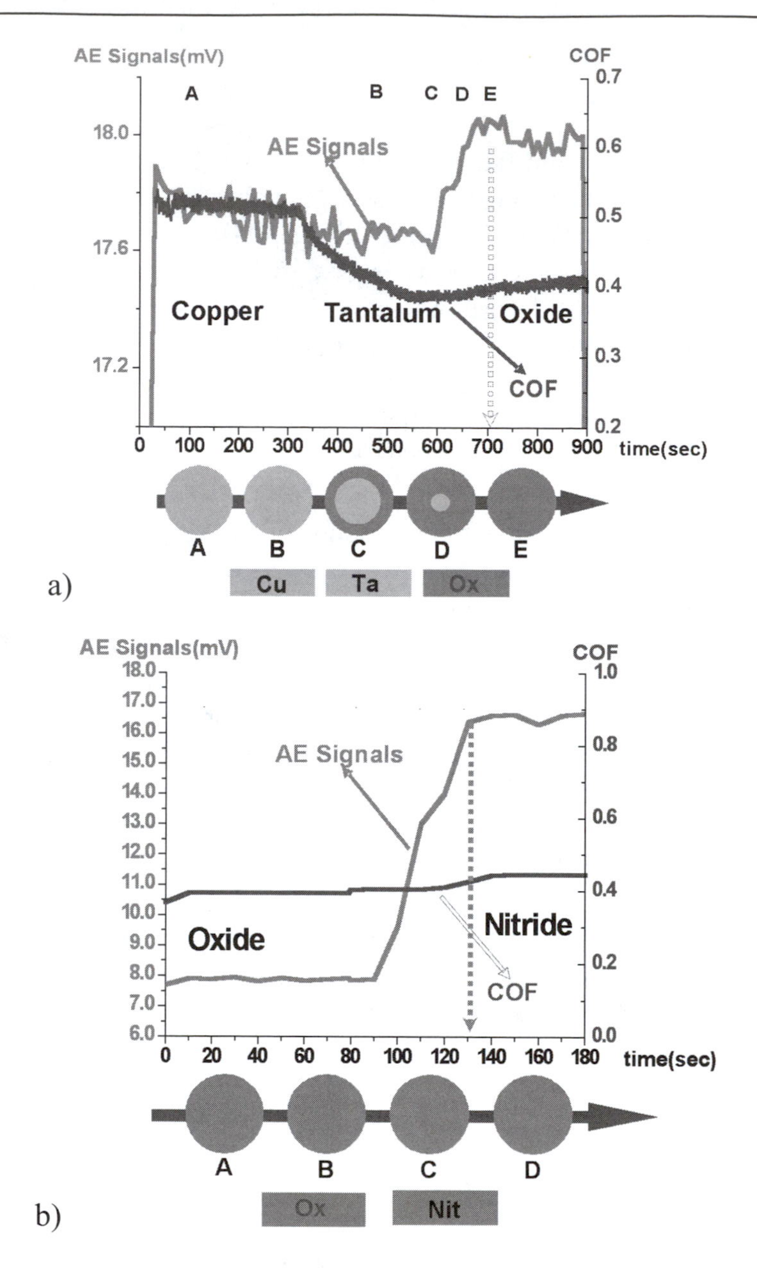

Figure 8.31. AE RMS endpoint detection for a) Cu damascene CMP process, b) STI CMP process.

The end-point was triggered at the edge of the wafer where copper was first cleared because of higher material removal at the edge. The phenomenon of the edge of the wafer clearing first also occurred in the second polishing step that removes the remaining copper and barrier layer. As shown in Figure 8.31a, AE RMS signals clearly show the transition from tantalum to oxide, indicating the ideal end-point whereas frictional force signals constantly increase as polishing time continues, making it difficult to detect the desired end-point with frictional force alone. A similar trend was also observed for the STI test wafers during polishing, with a very sharp transition observed in the AE RMS signal when the oxide completely cleared (see Figure 8.31b). These transitions are believed to be directly related to the variation in material properties of each of the films, with harder materials (such as nitride) demonstrating an increase in AE RMS signal during polishing due to the increase in energy required to initial material removal, whereas less-hard materials such as copper exhibit lower AE RMS signal during polishing.[194] However, the frictional force does not exhibit the same sharp transition as the AE RMS signal, demonstrating the superior S/N ratio of AE for the material removal process in CMP when compared to frictional force.

8.10.2.4 AE monitoring of surface chemical reactions for copper CMP

With the advent of low-k dielectric materials in semiconductor processing, the copper damascene process is of increasing use as an interconnect technology. To avoid stress-induced defects during the low-k/copper damascene process, the role of electrochemical reactions at the wafer/pad interface is becoming more important than mechanical abrasion during copper CMP. Understanding the nature of surface chemical reactions during copper CMP is of increasing importance for developing abrasive-free or electrochemistry-based polishing systems, and a series of focused AE experiments in an actual CMP environment were conducted for this purpose[195]. It is generally believed that the basic chemical reactions during copper CMP are 1) oxidization (i.e. passivation) of copper, and 2) removal of this oxide layer by a combination of dissolution and mechanical abrasion.

The types and thickness of the copper oxide are dependent on the oxidant concentration in the slurry, and greatly affects the material removal rate (MRR)[196]. The frictional behavior at the wafer/pad interface (including wear mechanisms) in copper CMP depend on the nature of the oxide films, and the energy associated with interface frictional affects oxide layer formation as well[197]. Understanding the mechanisms of the formation and removal of the oxidized copper layer during CMP is a key component for a comprehensive model of the copper CMP process, and an in-situ monitoring technique for characterization of these reactions during the actual CMP operation is a key issue for model development.

A series of unpatterned blanket copper wafers were used for the AE monitoring experiments. 700Å of Ti was deposited onto 4" Si wafers via e-beam evaporation, followed by 1000Å of Cu as a seed layer. 3μm of copper was then deposited by electroplating. For the formation of surface oxide, the copper-plated wafers were first dipped into 0.1 wt% oxalic acid ($HOOCCOOH.2H_2O$) solution for 1 minute to remove any native oxide on the surface, then treated in 30% hydrogen peroxide for 5 minutes. For the CMP operation, a Toyoda ultra precision float polishing machine (model SP46) was used with a specially designed wafer head for sensor integration. All experiments were conducted with an IC1400 pad with ex-situ pad conditioning. Down pressure was maintained at 2psi, and a constant pad rotation speed of 20 RPM was used for all experiments. Slurry of varying chemical composition and pH was supplied at a constant rate of 25 ml/min. All chemicals were pre-mixed and supplied as needed into a slurry delivery system for baseline signal generation, dissolution of pre-existing oxide layer, and re-oxidation.

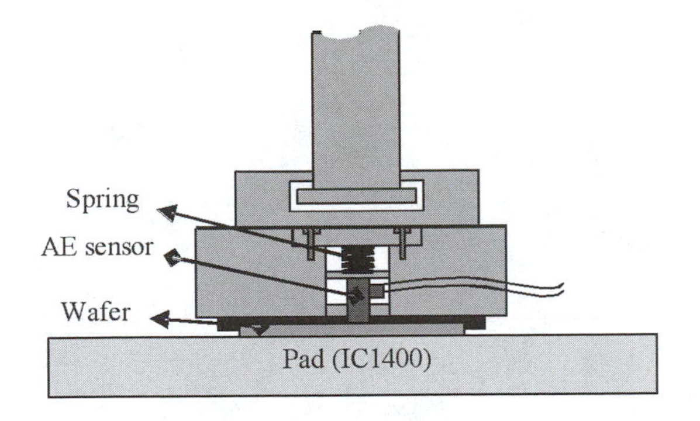

Figure 8.32. Wafer head with integrated AE sensor.

A DECI SE25 AE sensor was used due to its relatively flat frequency response (relative to other commercial sensors) over the 50-500 kHz range. The AE sensor was integrated into the wafer holder, directly coupled to the backside of the wafer with a spring-loading mechanism (see Figure 8.32) to minimize signal attenuation. The raw AE signal was preamplified by 50 dB, filtered through a high pass filter at a cutoff frequency of 50 kHz to reduce ambient system noise, and subsequently amplified by another 100 dB. The filtered signal was then processed by a root-mean-square (RMS) filter with time constant of 1 millisecond. A National Instruments DAQScope PCMCIA data acquisition card was used to acquire the signal at a sampling rate of 100 Hz within a Labview software environment, and data postprocessing was conducted with MATLAB.

The AE RMS signal during copper CMP operation with varying slurry composition is shown in Figure 8.33. Three different levels of AE were observed. First, a relatively high AE RMS signal (~1.3V) during the polishing of the hydrogen peroxide-treated surface with DI water was established. After switching to a 0.1 wt% oxalic acid solution, the signal dropped to ~0.44V. After one minute of polishing with the oxalic acid solution, a 30% hydrogen peroxide solution was supplied, and the AE RMS signal increase again to ~0.84V. The surface conditions during these variations are shown schematically in Figure 8.34.

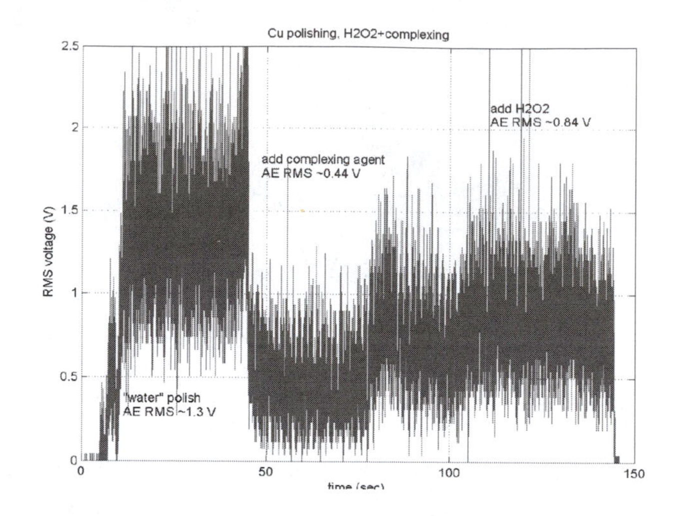

Figure 8.33. In-situ AE signal during copper CMP operation.

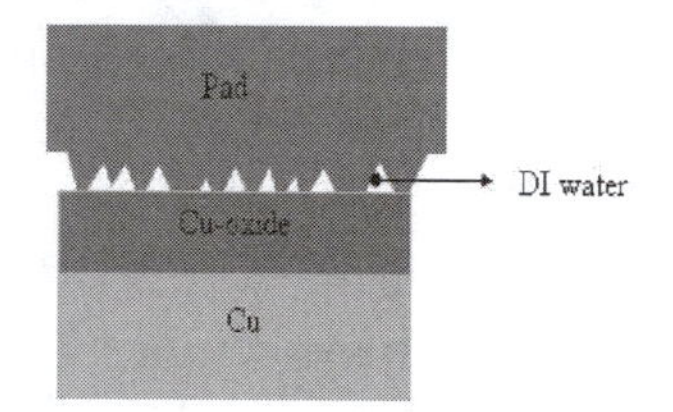

(a) Cu oxide with DI water

(b) Dissolution of cu-oxide with oxalic acid solution

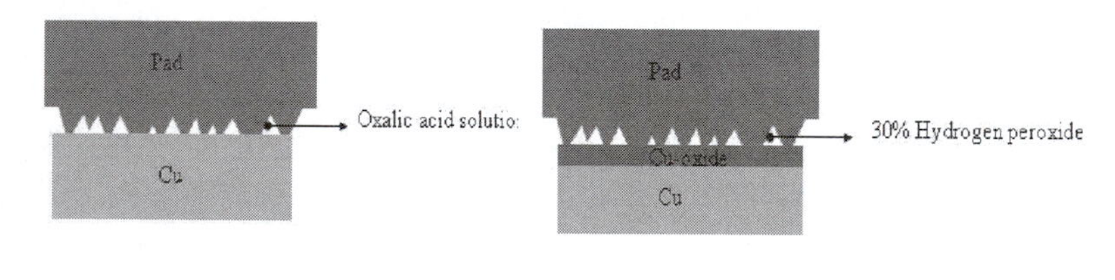

(c) Cu with oxalic acid solution

(d) Re-oxidized layer with hydrogen peroxide

Figure 8.34. States of AE with varying slurry chemical composition.

First, the pre-existing oxidized copper layer (through hydrogen peroxide treatment) and pad asperities interact together with DI water under mechanical motion (see Figure 8.34a). Then, when the oxalic acid solution is introduced, the oxidized copper layer begins to be abraded away through a combination of mechanical motion and chemical reaction with the oxalic acid solution (see Figure 8.34b), and corresponds to the sharp drop in AE RMS signal in Figure 8.37. The AE RMS signal remains around 0.44V during the surface interaction between the newly-generated copper surface, pad asperities, and oxalic acid solution (see Figure 8.34c). Finally, a 30% hydrogen peroxide is supplied to re-oxidize the surface (see Figure 8.34d). The AE RMS signal increased to approximately ~0.84V (i.e. lower than the first AE RMS signal associated with the oxidized surface state in Figure 8.34a).

It is initially believed that the variation in AE RMS signal between the states in Figures 8.34a and 8.34d is due to the varying surface properties of the copper wafer (mainly, frictional characteristics) as a function of the slurry chemistry. A significant dependency in the hydrophobic or hydrophilic characteristics of a surface on its coefficient in friction has been previously established, with hydrophobic surfaces exhibiting an extraordinarily low coefficient of friction with respect to water[198, 199]. As AE RMS signal has been demonstrated as being representative of the energy consumed during the material removal process[192], assuming all other process parameters are identical, an increased AE RMS signal is expected to occur when there is an increased coefficient of friction between wafer and pad. Likewise, decreasing frictional characteristics at the wafer/pad interface will be reflected by a decrease in the AE RMS signal, indicating the decrease in energy consumption as the two surfaces slide over one another more easily. As the slurry composition changes from DI water to oxalic acid, the resulting transition from an oxidized hydrophilic surface (case of increased friction) to a native copper hydrophobic surface (and decreased friction) manifests itself as a decrease in resultant AE signal. Hence, the observed shift in the AE signal is believed to directly correlate to the shift in the frictional characteristics between the pad and wafer. In addition, the AE RMS signal difference observed during polishing of different types of oxide formed under different conditions (i.e. hydrogen peroxide treatment for oxidized copper layer formation with and without mechanical motion) is initially believed to be due to a difference in the surface energy (and possibly frictional characteristics as well) between the two oxidized layers.

The effect of pH (by varying oxalic acid concentration) was also investigated. AE RMS variation during oxidized copper dissolution under CMP conditions with three different oxalic acid solutions of varying concentration and pH (0.1 wt%:pH 2.31, 0.05 wt%:pH 2.49, 0.02 wt%:pH 2.65) is plotted together in Figure 8.35 (time averaged values of AE RMS with a time constant of 3 sec. are used for clarification). Figure 8.35 clearly shows the effect of oxalic acid concentration (and pH) on the speed of dissolution of copper oxide, with 15 seconds for the transition to take place from high to low values with the 0.1 wt% oxalic acid solution. With the 0.05 wt% solution, the AE RMS signal dropped to a lower value over a much longer period of time; approximately 75 seconds. However, with the 0.02 wt% solution, the signal didn't level off within the experiment time (120 seconds); it only demonstrated a slight drop in RMS signal level to about 0.25V. These transitions from high to low values of AE RMS correspond to the state shown in Figure 8.38b where the oxidized copper layer is polished off. During the transition from (a) to (c) in Figure 8.34, the total oxidized copper surface area continues to decrease, with a corresponding increase in newly-generated copper surface area. The oxidized copper area generates higher AE RMS signal and copper area generates lower AE RMS signal due to the friction hypothesis explained in the previous section. Hence, during the surface state transition from oxidized copper to plain copper, the total signal drops. Mathematically, the rate of change in surface area of oxidized copper and plain copper can be expressed as

$$\frac{dA_{ox}}{dt} = -A_{ox}k[Acid]^n \tag{8.23}$$

$$\frac{dA_{cu}}{dt} = (A - A_{cu})k[Acid]^n \tag{8.24}$$

where the total oxidized copper surface area is A_{ox}, and the total copper surface area is A_{cu}. The area of oxidized and plain copper can then be expressed as

$$A_{ox} = Ae^{-k[Acid]^n t} \tag{8.25}$$

$$A_{cu} = A(1 - e^{-k[Acid]^n t}) \tag{8.26}$$

where A is total area, [Acid] is the concentration of oxalic acid, k is a parameter for area sensitivity, and n is a fitting parameter for the acid concentration effect on oxide area variation during the CMP operation. If it is assumed that the total AE signal is linearly proposi-tional to the source area, and the total AE signal can be modeled as a linear sum of AE from each area

$$AE_{RMS} = \alpha A_{ox} + \beta A_{cu} \tag{8.27}$$

where α is a fitting parameter for the higher AE RMS signal associ-ated with oxidized copper and β is a fitting parameter for the lower AE RMS signal associated with plain copper. Combining equations 8.25, 8.26, and 8.27, the AE RMS signal can be modeled as

$$AE_{RMS} = A\left[(\alpha - \beta)e^{-k[Acid]^n t} + \beta\right] \tag{8.28}$$

This model is plotted along with the raw AE RMS data in Figure 8.36 for comparison. With the limited experimental dataset, this model reasonably explains the AE RMS transition, and parameters in this model can be used as a means for characterization of oxidized copper dissolution and removal during the CMP operation.

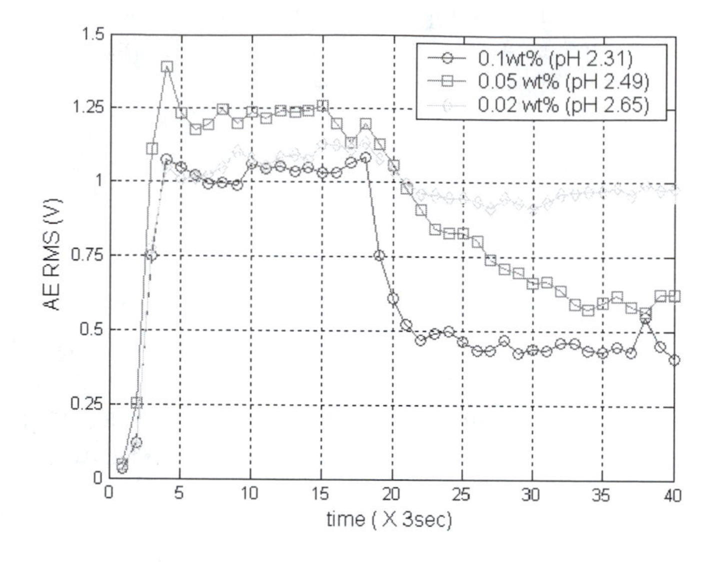

Figure 8.35. AE RMS variations during oxidized copper dissolution with varying oxalic acid composition.

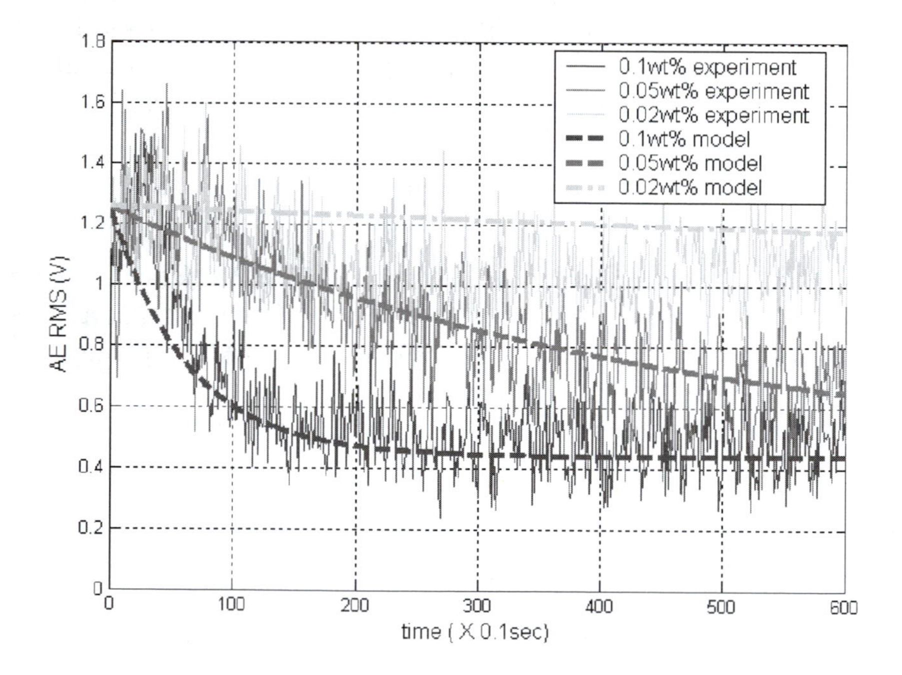

Figure 8.36. Comparison between modeled and experimental data in the AE RMS signal transition regime (50-110 sec).

8.10.2.5 AE characteristics of oxidation and dissolution in copper CMP

To further investigate the sensitivity of AE to chemical transitions and slurry effects in CMP, a tests were done with complementary friction force measurements as a means of cross-calibration between the two signals. Test wafers with 6 microns of copper (deposited via electro-plating) were treated in a 30% H_2O_2 solution for 3 minutes to fully oxidize the surface. A GNP Poli400 CMP machine with a similar AE sensor setup to that used in Section 9 was used, yet with an integrated custom-built Kistler 9317A load cell for in-situ AE and friction force measurements (see Figure 8.37).

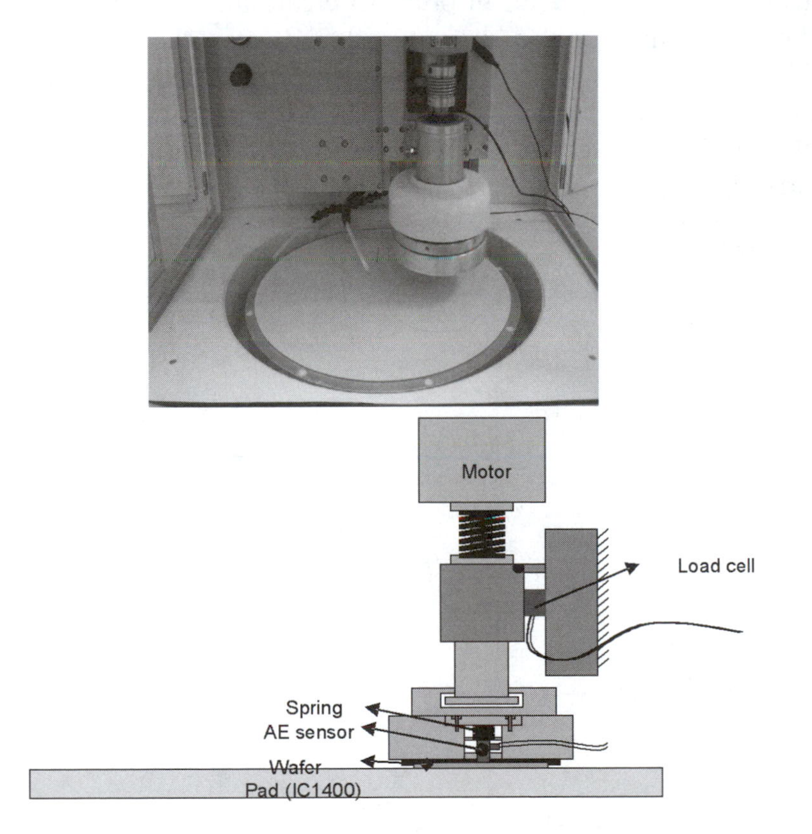

Figure 8.37. Experimental setup on GNP Poli400 CMP machine.

In the first series of experiments, the CMP operation was carried out by altering the slurry composition (without abrasives) in a series of discrete steps to isolate AE and friction force characteristics to individual components of the slurry chemistry. First, a polishing operation was conducted with plain DI water to establish a baseline for the AE signal (see Figures 8.38 and 8.39). A glycine solution was then introduced into the slurry-delivery system, and a slight drop in the AE RMS signal was observed, whereas no significant variation in the friction force was observed. After about 2 minutes of polishing with the glycine solution, a 30% H_2O_2 solution was introduced to reoxidize the surface, with slight increases in the AE and friction force signal being observed. This variation in the slurry chemistry was conducted in order to create three distinct polishing states; the fully oxidized surface with DI water, the copper surface with glycine, and re-oxidized copper surface with H_2O_2, with a different AE signal associated with each. However, the friction force only has limited sensitivity in detecting the various transitions in slurry chemistry.

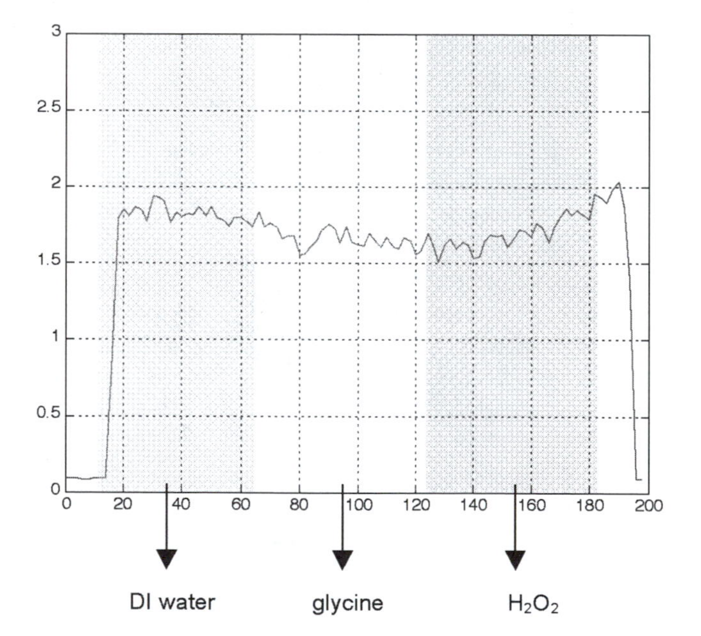

Figure 8.38. AE signal for non-abrasive slurry chemistry transitions.

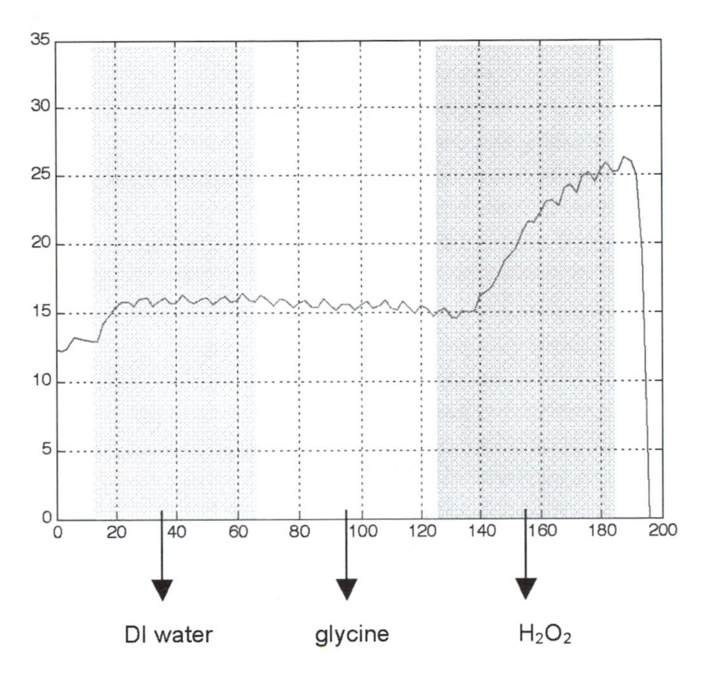

Figure 8.39. Friction Force Signal for Non-abrasive Slurry Chemistry Transitions.

In the second experiment, the exact same slurry chemistries were used, but now with 50 nm alumina abrasives added to the slurry. While the AE signal demonstrates a greater variation with slurry chemistry transition with the addition of abrasives, the variation in friction force is hardly noticeable (see Figures 8.40 and 8.41), further demonstrating the improved sensitivity of AE in detecting phenomena at the ultraprecision scale over other conventional sensors. It is believed that variations in the surface energy of the oxidized and non-oxidized surface of the polished wafer can explain the distinct difference in the recorded AE signal, with the presence of abrasive particles serving to "amplify" the AE signal in some manner, although the exact mechanism through which this occurs is still not well understood.

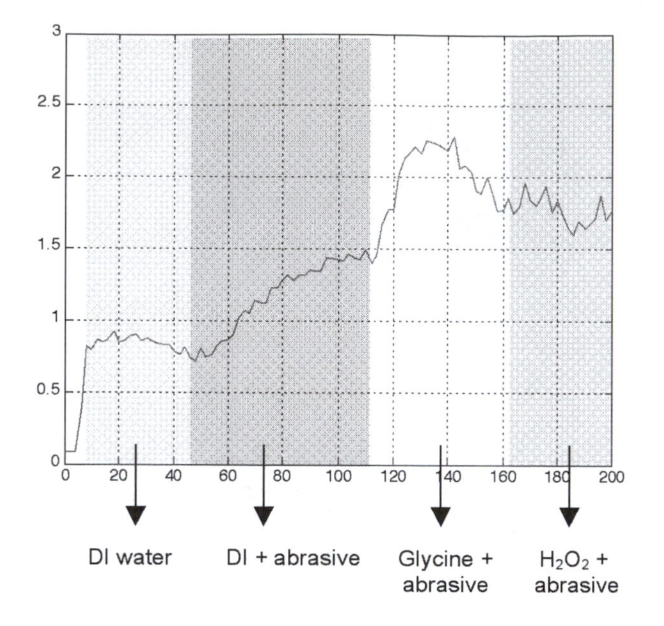

Figure 8.40. AE signal for abrasive slurry chemistry transitions.

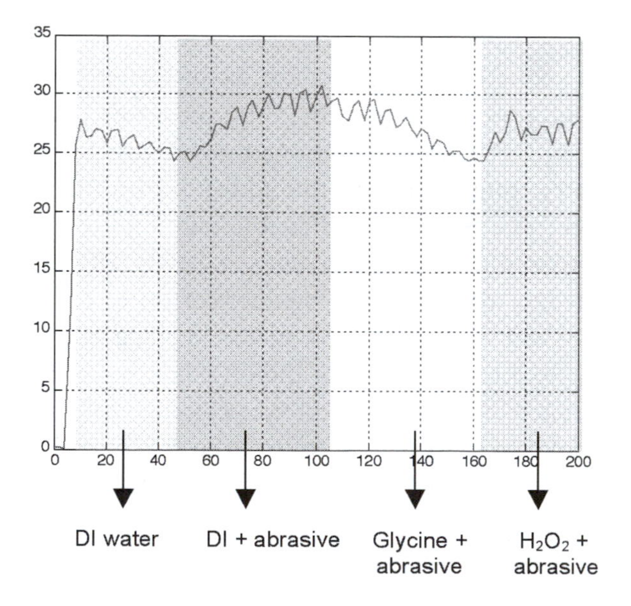

Figure 8.41. Friction force signal for abrasive slurry chemistry transitions.

One potential explanation for the variation in the AE signal is proposed in Figure 8.42. At the initial stage where the pad is in contact with the wafer surface during DI water polishing, a baseline AE signal is generated through solid-solid interaction between the polyurethane pad and wafer surface, which is mostly from elastic stick-slip interactions between pad asperities and the wafer surface without material removal. When abrasives are added to the DI water, a significant increase in the AE signal is observed due to abrasion by the particles, which are an additional AE source (see Figure 8.42, step 2). The total AE RMS at this stage is elastic stick-slip interaction and particle-induced abrasion at the pad asperities. As the slurry chemistry transitions to glycine and abrasives, the AE signal again changes sharply due to the dissolution of the oxidized surface (see Figure 8.42, step 3). This can be explained with the different level of AE generation by different stick-slip and abrasion on copper surface than copper oxide surface. Finally, when peroxide and abrasives are added, the AE signal drops as the copper surface is re-oxidized (Figure 8.42, step 4).

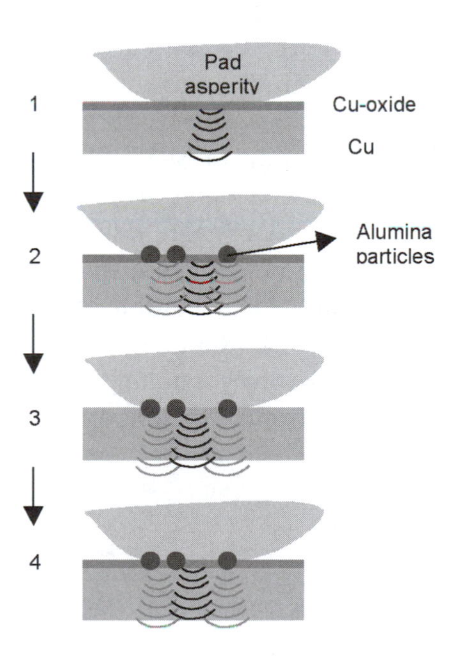

Figure 8.42. Transition in sources of AE at the asperity scale.

8.10.2.6 Monitoring of precision scribing

To examine the effect of crystallographic orientation during ultra-precision machining, a series of nearly-parallel slow speed scribes on the surface of an oxygen-free high-conductivity (OFHC) poly-crystalline copper workpiece (250 micron average grain size) was taken. A 0.274 mm nose radius single crystal diamond tool was chosen as the tool of choice since the tool tip was considerably smaller than the average grain size of the workpiece. The work-pieces were clamped onto the spindle of the lathe, and the spindle was alternately rotated and locked to allow for slow-speed scratches. A scratch speed of 0.7 mm/sec was used, and the infeed (depth of cut/DOC) setting was set at a constant value of 10 microns through-out the experiment. After each scribe, the spindle was unlocked and rotated slightly (~1 degree) to produce a radial pattern of scribes (al-though the scribe pattern can be approximated as a raster scanning pattern over this small angle of rotation). A similar DAQ system to that shown in Figure 8.29 was used, with both cutting force and AE RMS signal collected during each scribing operation.

After collecting data from a series of scribes (typically on the order of 15 quasi-raster pattern scribes), the data was reduced, and a color intensity mapping function in MATLAB was used to plot the cutting force and AE signal as a function of position, with the color map representing the respective magnitude of the signal. Unfortu-nately the color is not shown in the figure so some detail is lost. Fig-ure 8.43 shows a graphical representation of the cutting force and AE RMS signal for a series of 15 scratches, along with a micrograph of the workpiece surface before scratching. Figure 8.44 shows the cutting force and AE RMS for the 5th scribe.

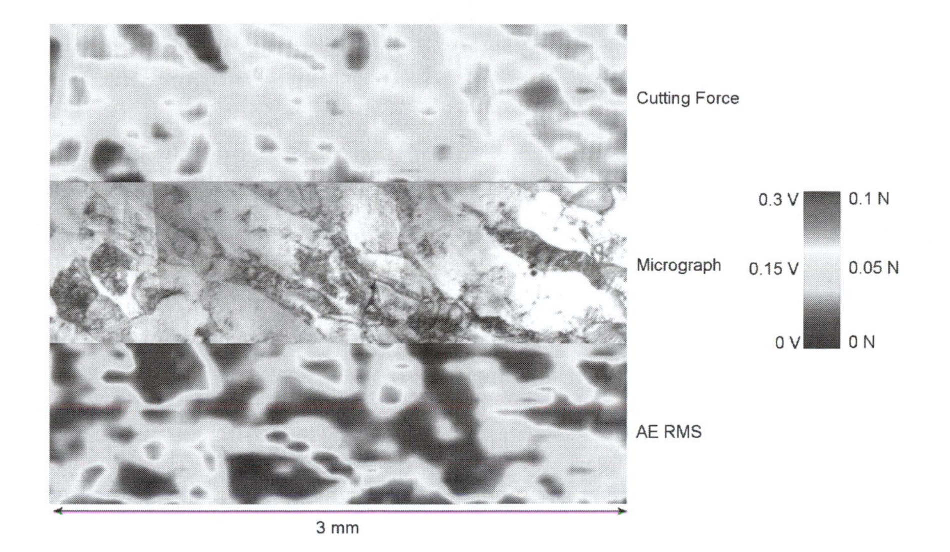

Figure 8.43. Force/AE response for 'quasi-raster' scratch pattern.

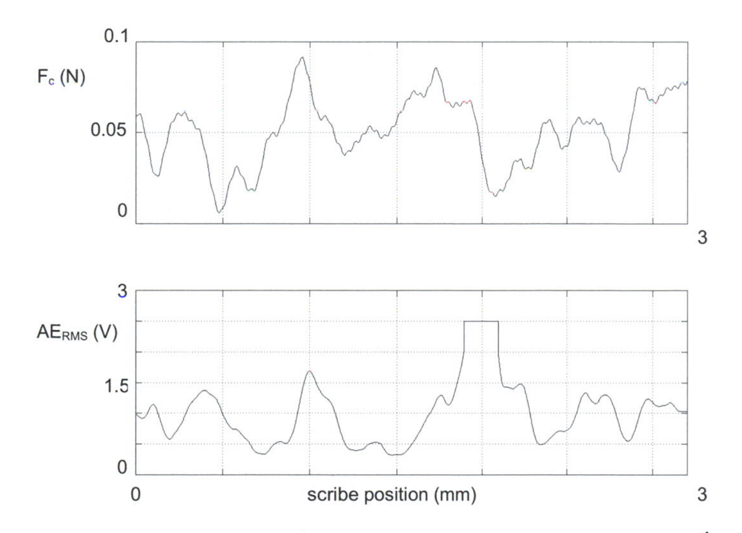

Figure 8.44. Individual Force/AE Response for Single Scribe (5th scribe).

Both the cutting force and AE RMS signal reproduce a crude representation of the grain structure of the material. The variation in force and AE RMS signal is largely due to the fact that each grain has a particular crystallographic orientation, so as the tool passes from one grain to another, a new slip system in the grain is being activated, which changes the amount of applied stress (and cutting force) required to initiate deformation. If the cutting speed and tool cross section are constant, then the AE RMS is simply proportional to the energy (and cutting force) required to initiate deformation. The activation of different slip systems as a function of grain orientation causes the energy of the resulting AE RMS signal to fluctuate accordingly. Of particular note is the good match found between the mappings for the force and AE RMS signal, indicating that advances in load cell technology can allow for improved sensitivity to process mechanisms at the precision scale.

8.10.2.7 Monitoring of ultraprecision turning of single crystal copper

The sensitivity of AE to crystallographic orientation was also tested for single crystal materials. Figure 8.45a shows a sample trace of AE RMS data for a single revolution of cutting on a <100> workpiece. An approximate sinusoidal variation in AE RMS can be seen in this data set. Each trace of AE RMS data is then collected in a data array in LABVIEW, and is represented graphically as an intensity plot in Figure 8.45b. This intensity plot shows AE RMS signal as a function of spindle revolution, with subsequent traces of data for each spindle revolution progressing from left to right for a total of about 80 revolutions for a single face turning pass of the single crystal workpieces. Signal intensity or voltage levels are plotted according to a color intensity map, with a black color corresponding to zero signal, signal saturation at 3.5 V corresponding to a white color, and intermediate signal values appearing as shades of grey.

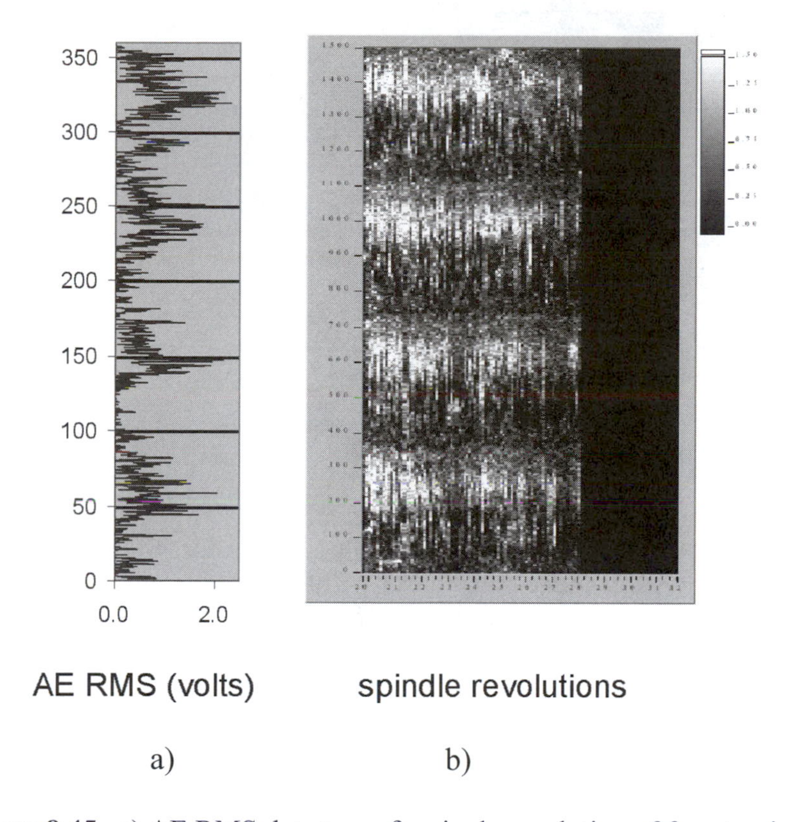

Figure 8.45. a) AE RMS data trace for single revolution of face turning of <100> workpiece, b) Cartesian intensity plot of AE RMS data for face turning.

A Cartesian-polar transform allows for the data in Figure 8.45 to be replotted on a polar intensity map as shown in Figure 8.46a. In the polar intensity map, the AE RMS voltage is plotted vs. physical position on the workpiece, giving an indication of the variation of cutting energy as a function of crystallographic orientation. Figures 8.46b and 8.46c show the theoretical variation in Taylor factor for a <100> crystal as a function of orientation, and an image of the chemically etched surface of the <100> workpiece after machining, respectively. A good correlation can be seen between the theoretical and experimental polar mappings, and variation in the surface finish of the chemically etched workpiece. Due to the black and white images some detail is lost.

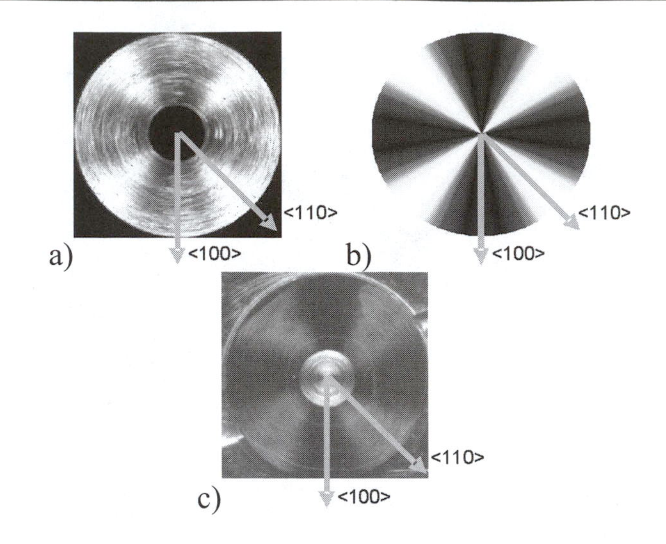

Figure 8.46. a) Experimental AE RMS polar map for <100> workpiece, b) Taylor factor-based theoretical AE RMS polar map for <100> workpiece, c) Chemically etched surface of <100> workpiece after machining.

Figure 8.47 shows SEM images of the machined surface before chemical etching. Figure 8.47a shows the surface corresponding to areas that demonstrated low AE RMS signal levels of approximately 0.5 V, and Figure 8.47b shows the surface of areas with AE RMS values of approximately 2 V. Dornfeld et al. postulated that this variation in AE signal corresponded to the variation in the orientation-dependent Taylor factor (representative of the yield stress, and consequently, specific energy required to initiate plastic deformation in a single crystal material) of the workpiece[200]. These regions of high and low AE RMS signal correspond to regions that demonstrate high and low Taylor factors (ranging from 2.4 to 3.7), respectively. Crystallographic orientations that have a relatively high Taylor factor yield machined surfaces that are significantly rougher than areas with a relatively low Taylor factor. Figure 8.47c shows a relatively smooth surface finish of 42.4 nm R_a (measured with a Wyko surface interferometer) while Figure 8.47d shows the rougher surface with a surface finish of 75.6 nm R_a.

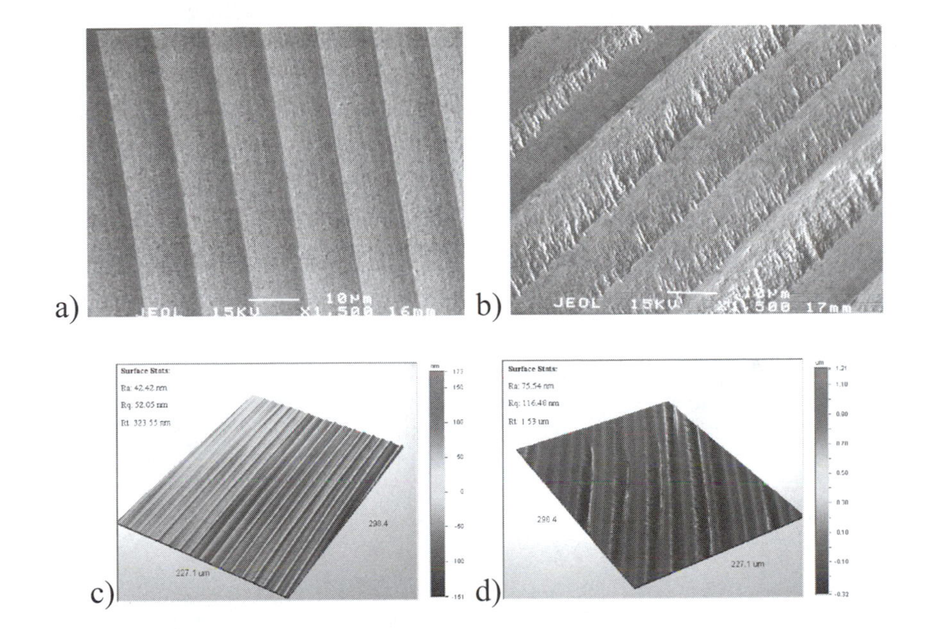

Figure 8.47. a) SEM image for <100> workpiece, <100> cutting direction, "smooth" region, b) Image for <100> workpiece, <110> cutting direction, "rough" region, c) Wyko image for <100> workpiece, <100> cutting direction, "smooth" region, d) Wyko image for <100> workpiece, <110> cutting direction, "rough" region.

8.10.2.8 Monitoring of ultraprecision turning of polycrystalline copper

Figure 8.48a shows an AE polar map for a polycrystalline OFHC copper workpiece. Although this piece was supposed to have an average grain diameter of 250 μm, several large grains were still observed on the workpiece. The AE polar map in Figure 8.48a shows good correlation with the chemically-etched workpiece surface, shown in Figure 8.48b. While small grains below 100 μm cannot be resolved in the AE polar map due to DAQ limitations, the larger grains appear very clearly in both the chemically-etched surface and AE polar map. Because these large grains can serve as defects that affect the homogeneity of the surface finish, the AE polar mapping technique, in addition to serving as a tool contact sensor, provides a very convenient means of detecting potential trouble spots or defective areas on the workpiece[200]. A perfectly homogenous and isotropic workpiece would mostly likely result in little variation in the AE signal and polar map, so variations in the AE signal can serve as useful feedback in a fully-automated manufacturing environment.

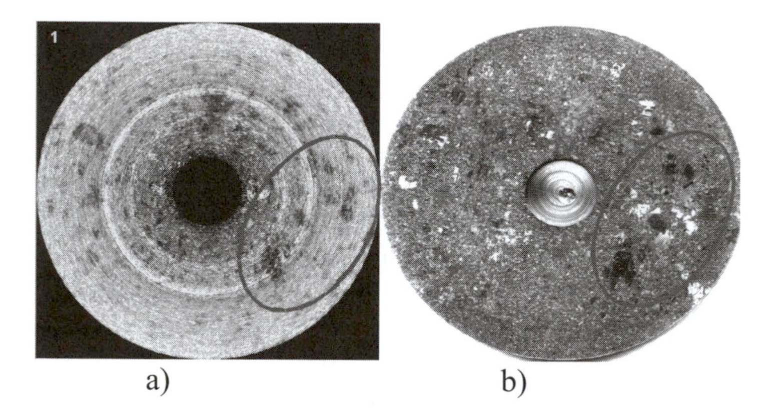

Figure 8.48. a) AE RMS polar map for polycrystalline OFHC copper workpiece, b) Micrograph of chemically-etched workpiece.

8.11 Summary

We have presented here a comprehensive introduction to the application of sensing systems in manufacturing and, at the end, a broader review of sensor technologies with some special attention to semiconductor sensing systems. These are the "sensors of the future." One important factor in the implementation of sensors in manufacturing is clearly the rapid growth of silicon micro-sensors based on MEMS technology. This technology already allows the integration of traditional and novel new sensing methodologies onto miniaturized platforms, providing in hardware the reality of multi-sensor systems.

Further, since these sensors are easily integrated with the electronics for signal processing and data handling, on the same chip, sophisticated signal analysis including feature extraction and intelligent processing will be straightforward (and inexpensive). This bodes well for the vision of the intelligent factory with rapid feedback of vital information to all levels of the operation from machine control to process planning.

While it would be desirable, it is impossible to cover this topic without focusing on major sensor groups or major manufacturing processes. Consequently, we have restricted the tutorial portion of the paper to examples of tool condition monitoring in metal cutting using acoustic emission and force sensors. The most detailed source characterization and sensor design discussions have focused on acoustic emission sensing technology which has been a major interest of our research group for the past 15 years. However, with very few exceptions, our discussion here can be easily altered to include other sensing technologies or processes. Several tables have shown the application of sensors to differing tasks. All have distinct advantages and limits and our discussion has reviewed criteria for assessing those. Similarly, we have seen the broad range of processes that the sensing technology can be applied to. In the case if acoustic emission, we identified the source of the signal in each process. Whereas the sources or mechanisms may vary from sensing technology to technology, similar analyses can be done. The reader is encouraged to be open-minded about the crosslinking of the material presented here to their favorite sensor or process.

We have consciously ignored a major application of sensing in manufacturing because of the need to focus mentioned above. This application is machine and machine tool characterization. Basically, in addition to knowledge of the process being implemented on a machine (milling, for example) it is critical that the basic behavior (in view of static and dynamic loads, thermal distortion, kinematic errors in axis translation or rotation, etc.) be well understood. There is a rich literature associated with this field of study and it reflects the analytical tools and sensors used for machine characterization. NIST researchers have played a strong role in developing tools used in this area and the reader is referred to, for example, Donmez[40, 201-203], for details.

IX PROCESS PLANNING FOR PRECISION MANUFACTURING

9.1 Manufacturing system characteristics

At a very basic level, there are key elements of a process that determine what its capabilities are, for example number of degrees of freedom of the work created (where a milling process has more degrees of freedom than a drilling process.) This can be represented as in Figure 9.1 which shows achievable dimensions (for both surface finish and tolerance) for a range of manufacturing processes from normal to ultraprecision manufacturing. By contrast, Figure 9.2, one of the famous "Ashby Charts" from Ashby[204] is a chart used for selecting processes to meet the combined specifications of tolerance and surface roughness over a range of materials. The tolerance indicated is the permitted allowance in the dimension of the part. The surface roughness R is the root mean square (RMS) amplitude of the surface irregularities. The basic requirement is that the tolerance must be greater than twice the roughness. For a specific surface with a measured RMS roughness, it is known that the maximum peaks of the irregularities are about 5 x RMS roughness. The chart also shows relative processing costs increasing as tolerance and roughness requirements become more stringent. Although this is typically true, one can find ample examples to the contrary in high production manufacture of products. In fact, you are probably aware of many examples where, due to efficient manufacturing, very high precision products are available at relatively low cost — think about the disk drive in your laptop for starters!

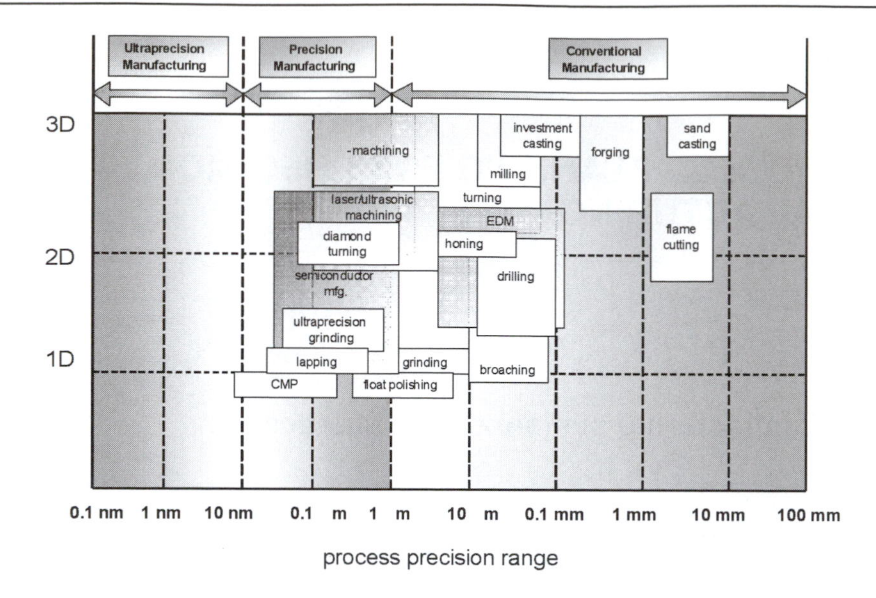

Figure 9.1. Achievable dimensions (for both surface finish and tolerance) for a range of manufacturing processes from normal to ultraprecision manufacturing.

In a paper written in 1971, the author, Jim Bryan, discusses the "myth" that tighter tolerances necessarily mean higher costs[205]. First defining close tolerances as …"one that is smaller than that currently in use in the field of question and is generally regarded as being difficult or expensive to achieve" Bryan lists a number of objections to close tolerances and then illustrates the "life cycle" economies or performance/cost ratios of a number of relatively inexpensive but well made products. One of special interest is his example of a model airplane engine. These engines (selling for $5.98 in 1971; a recent check of the internet shows prices today are in the $20-$60 price range for similar performance) are single cylinder, two cycle engines developing on the order of 0.06 horsepower at 12,500 rpm. The piston has no rings, and is made of free machining leaded steel that is case hardened. The cylinder head is threaded onto the cylinder and no gasket is needed due to the flatness and squareness of the shoulders to the threads. Because there are no piston rings the clearance on the diameter of the piston with the cylinder must be held between 150 μ in (3.8 μm) and 250 μin (6.2 μm). Absolute size on the piston and cylinder is held to ± 75 μin (1.9 μm) for a ± 3 σ statistical limit. Pistons are rough machined on screw machines, centerless ground once, heat treated, then centerless ground in three stages tobetter control distortion due to grinding heat build up to finally

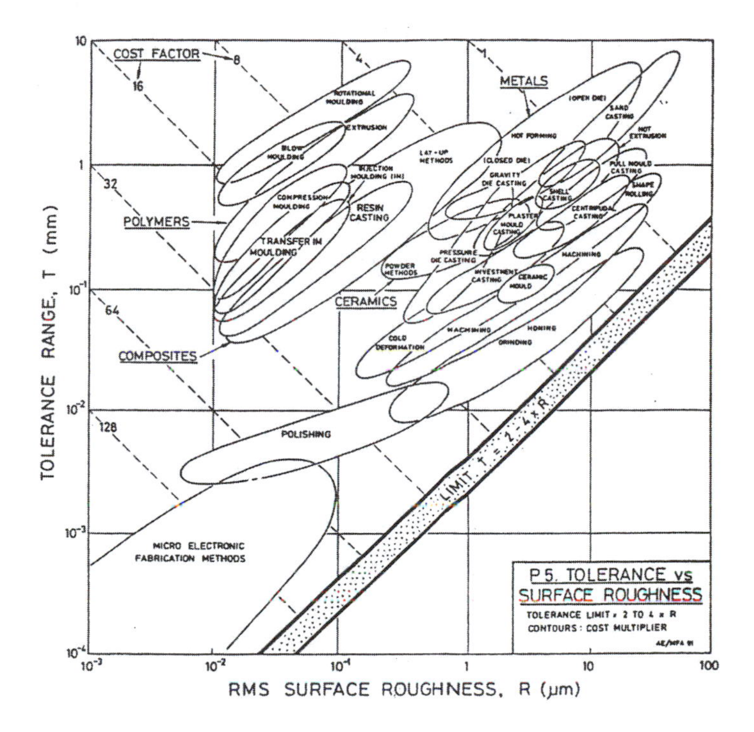

Figure 9.2. Tolerance range vs. surface roughness for a range of processes, from Ashby[204].

achieve their final size tolerance of ± 75 μin (1.9 μm). Some selection of mating parts is done to insure clearance tolerances are met. Bryan gives the following analysis of the performance of the engine. The elimination of the piston rings gives a 25% lower initial cost and means better performance because of lower piston drag and longer life due to reduced cylinder wall wear. The interesting question raised in the analysis is "what happens to the 50 μin (1.2 μm) piston fit tolerance at the normal operating temperature of 300 ° F?" Bryan[205] gives the following:

> "The answer is that the designers have the situation well under control by using the same material for

> cylinder and piston and by taking advantage of the fact that the coefficient of thermal expansion of the piston is slightly reduced as a consequence of the case hardening heat treatment. The operating clearance will therefore tend to increase slightly with temperature, but this has a negligible effect on compression at high speed, and there is a slight advantage of reduced piston drag."

The author finishes the discussion by noting that the price of these engines continues to drop and their quality improves!

For precision manufacturing processes, the design parameters extend from the highest level, the basic dimension, to the lowest level, subsurface damage, for example, Figure 9.3. Reduced amounts of material removed in precision manufacturing results in non-uniform material flow around the edge of the cutting tool, especially when the uncut chip thickness is on the same order as the edge radius of the cutting tool. This will be described in more detail in Chapter 10 during a discussion of processes, and is clearly seen in Figure 9.16 later in this chapter showing the hardness variations in the chip taken from a "quick-stop device." Although our focus is on processes such as diamond turning, this very small chip thickness occurs often in more conventional manufacturing processes as well. Figure 9.4 (from Nakayama[206]) shows the diminishing thickness of cut in turning and milling due to tool geometry and engagement kinematics.

Other affects also influence the "state" of the cutting process in precision machining and, thus, can influence the form, surface or sub-surface characteristics of the work. Workpiece hardness, which can vary through the work due to prior treatment in manufacture (heat treatment or cold work, for example) or a prior manufacturing operation (e.g. machining) will influence the forces generated in machining. Figure 9.5, also from Nakayama[206], illustrates this and shows, for two different rake angle tools, the influence of tool geometry. A difference in cutting force of a factor of two can result due to a rake angle change from 0° to -20° regardless of work piece hardness. Additionally, it is apparent from the figure how the work hardness affects chip formation (shear angle, for example). Finally, cutting tool characteristics (here the composition and surface roughness of the tool) will strongly influence the cutting process (forces and chip formation). Since these affect the degree of residual damage to the work surface or tool deflection (and hence tolerance or form errors) they must also be considered. In Figure 9.6 (also from Nakayama[306]), the data from four tools ranging from as received carbide (surface roughness 1.2 μm) to mirror lapped polycrystalline diamond (surface roughness 0.03 μm) machining copper at a variety of conditions is presented. The influence to tool surface and composition (diamond is an excellent heat conductor so chip-tool temperatures would be lower for the PCD tool and, as seen in the lower figure, the effect on friction forces on the rake face, represented by β here) are dramatic.

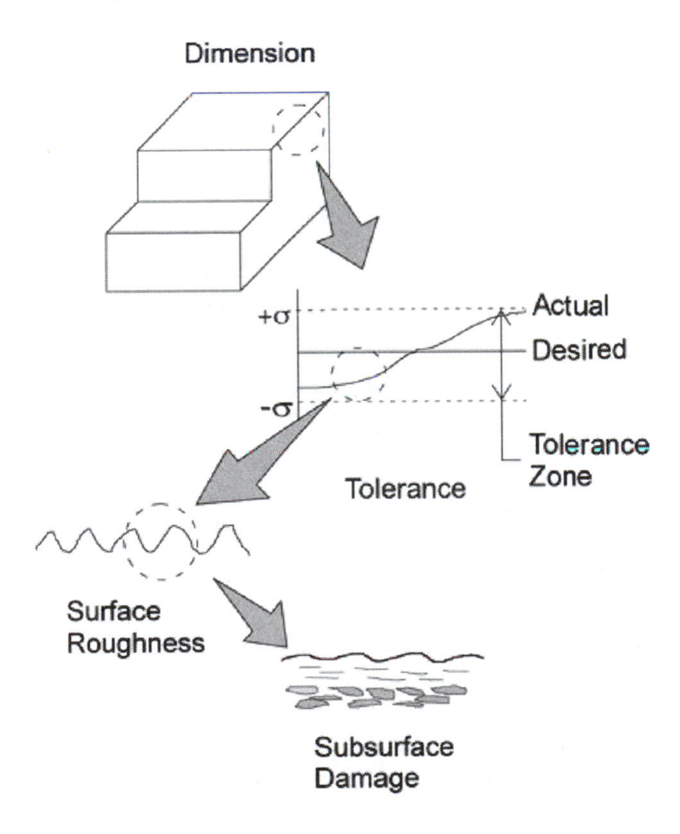

Figure 9.3. Extension of design parameters for precision machining.

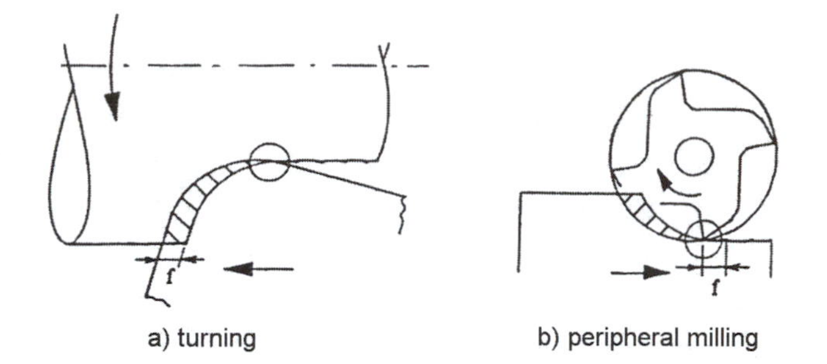

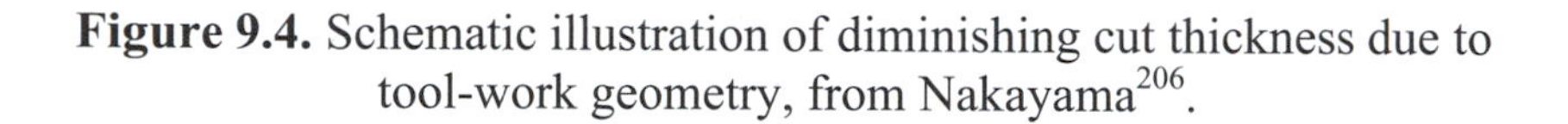

Figure 9.4. Schematic illustration of diminishing cut thickness due to tool-work geometry, from Nakayama[206].

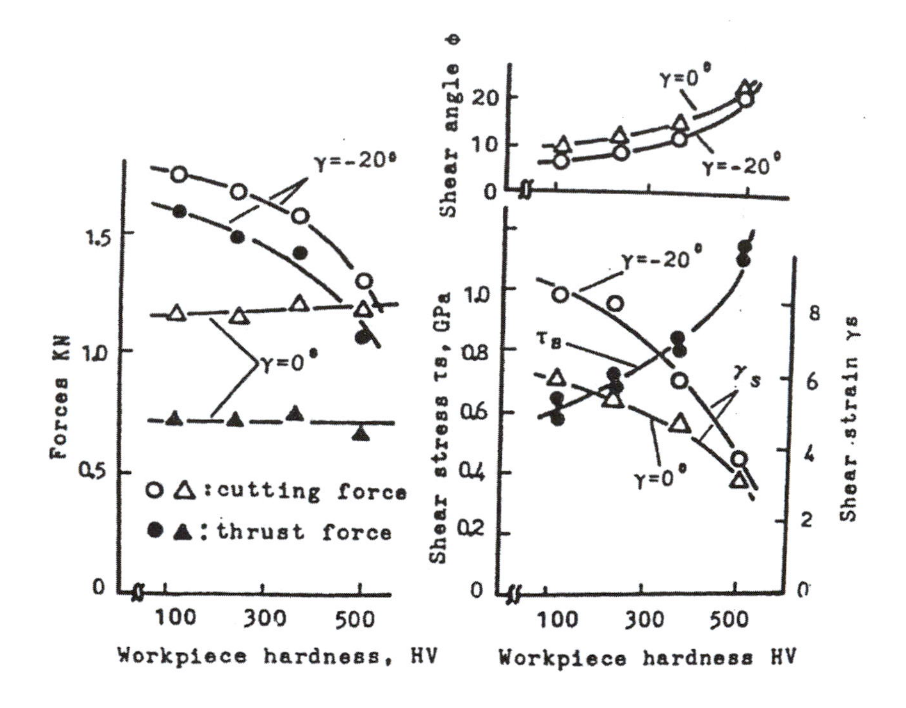

Figure 9.5. Effect of workpiece hardness on the cutting forces and cutting mechanism. Work material: 0.25% C steels of various heat treatment. Cutting tool: HSS, rake angle = 0° and -20°. Operation: orthogonal cutting, f = 0.2 mm, b = 2.0 mm, V = 20 m/min, dry cutting, from Nakayama[206].

There are also other factors to be considered in process planning that will not be dwelled upon here. Other than tolerance, form, roughness and subsurface damage, burrs (residual deformed material adhering to the edges of the workpiece where the tool has exited) are often major problems. Most of the conditions discussed here are features of the "continuous process" — meaning observed while the tool and workpiece are continuously engaged. Things like

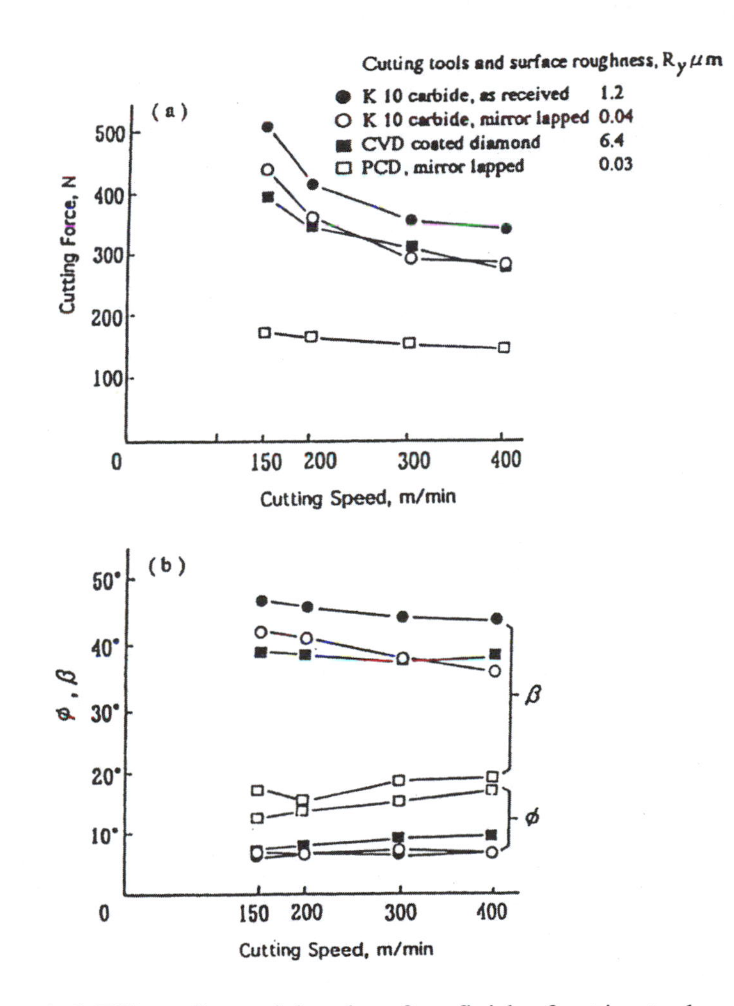

Figure 9.6. Effect of material and surface finish of cutting tool on the cutting force, shear angle φ and friction angle β. Work material: copper, tool geometry: 0, 5, 11, 6, 15, 15, 0.8, a = 2.0 mm, f = 0.05 mm, Nakayama[206].

burrs occur in transition from one stage to the next, accompanying the exit or entrance of the tool in to the workpiece.[207] A simple illustration of this is seen in Figure 9.7 where the sensitivity of burr size to uncut chip thickness is seen for machining test in 65-35 brass over a range of machining conditions. As expected, conditions which minimize the distortion energy in chip formation (high rake angles, for example) show the lowest ratio of burr height to chip thickness. And, this ratio is maintained low at high machining speeds.

It is not necessary here to go into the details of assigning a sequence of manufacturing processes based on the requirements of a design and the capabilities of the process (for example, center drill, drill and then ream to get a hole of desired diameter and cylindricity at a desired location) as there are many excellent undergraduate texts that discuss this. Our interest here is the integration of the requirements of each process with the limitations of precision for the successful completion of a precision product.

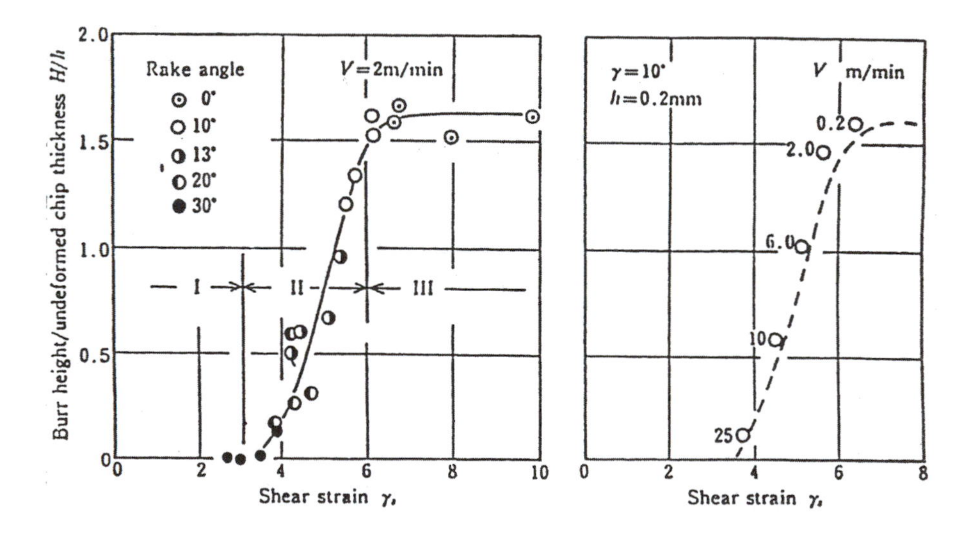

Figure 9.7. Relationship between burr height H, undeformed chip thickness h, and shear strain γ, of the chip. Work material: 65-35 brass, operation: orthogonal cutting. (Numbers in the second figure indicate cutting speed V in m/min), from Nakayama[206].

Investments in Computer Integrated Manufacturing Systems (CIMs) are driven by the desire for high quality components, small batch sizes, agility, and short lead times. In the early stages of part specification, the creation of an integrated CAD/CAM environment is the key to rapid off-line simulation and verification of new part designs. Design, process planning and manufacturing integration occurs at several levels depending upon the objective of the exercise and the degree of "flexibility" (that is to say the degree to which the design or process are fixed). The level of integration attainable between design, manufacturing, and finishing is dependent upon several factors[208]:

1. nature of the design, manufacturing or finishing task;
2. environment in which the task is performed;
3. tools available to assist the designer with the task.

Four distinct levels of integration between the tasks of design, manufacturing and finishing have been identified in present day and future production environment[209]. Levels of integration can be described in terms of the ability at each level to predict, influence and optimize part production objectives at various stages of the total part production process. The objectives encompass a variety of process metrics which influence the goals of maintaining the tolerance of shape, tolerance of form, desired surface characteristics and subsurface damage/residual stress. These four levels of integration from a process planning point of view are illustrated in Table 1 below, using feature quality as a process metric.

At Level I, the highest level of integration, the designer is contemplating the design of the component. At this level, any information that correlates the design, the process plan and manufacturing, can be utilized to improve the overall "Manufacturability" of the component. At a slightly lower level of integration, Level II, the design may be fixed but it is still possible to develop a process plan and a manufacturing configuration to insure that the part specifications are met. At a still lower level, Level III, the design and process plan, as well as the machinery for manufacture, may be fixed. Even at this level, however, it is still possible and useful to consider optimization and fine-tuning of the manufacturing process to accommodate unexpected problems through changes in tool geometry or localized tool paths, for instance. Finally, at Level IV, the lowest level of integration, it may be of interest only to assist in insuring that subsequent manufacturing processes, such as finishing, for example, are efficiently and accurately carried out. As the level of integration between design, manufacturing and finishing increases from Level IV to Level I, the ability of the software tools (and the designer) to influence and optimize the process metric of feature size, shape or quality increases. The result of allowing the designer to work within the highest level of integration is to increase the sensitivity of design decisions for one process metric, such as burr formation, and to enable the designer to optimize and balance opposing process metrics through analysis, simulation and decision evaluation.

Table 9.1. Four levels of integration in the design to fabrication cycle, after Stein and Dornfeld[209].

Integration Level	Process Planning Software Expert and Agent Tasks	Degree of freedom for adjustment
Level I	Feature prediction, control, and optimization in an iterative design and process planning environment	Design: High Manufacturing: High Finishing: High
Level II	Feature prediction, control, and optimization through the selection of a manufacturing plan in an "over-the-wall" design-to-manufacturing environment	Design: Low Manufacturing: High Finishing: High -> low
Level III	Feature prediction and control through limited adjustments to a pre-established manufacturing process	Design: Low Manufacturing: Limited Finishing: High -> low
Level IV	Feature prediction for finishing process planning, finishing tool trajectories and sensor-feedback strategies	Design: Low Manufacturing: Low Finishing: High

9.2 Process planning basics

When discussing the performance of a process we must make a distinction between whether or not the process can stay, statistically, in control (for example, within the bounds of a control chart) or whether or not the process is able to produce parts that meet the engineering specifications. Having control does not necessarily imply good performance. DeVor points out that one never places tolerance/specification limits on an $\overline{X}$ control chart because the control chart is based upon the variation in the sample mean, $\overline{X}$, whereas it is the individual measurements in the sample that should be compared to the specifications[210]. Thus it is misleading. It is possible for a process to produce a number of pieces that do not meet the specifications but are, in fact, under control, statistically (that is all of the points on the $\overline{X}$ chart are within the 3 sigma limits and vary randomly.) DeVor summarizes the possible reasons that a process may be in statistical control but creating parts out of tolerance as follows:

- process is off-center from the nominal (i. e. a bias exists)
- the process variability is too large with respect to the tolerances
- the process is off center and has a large variation

We have previously covered methods for determining the impact, on numbers of defective parts produced, of processes out of statistical control at some level. Process planning is based upon an understanding of the basic form capability of a process (i. e. drills make holes) and the basic process capability representing the ability of the process to make products that consistently stay within the specification limits. We will refer here to the term C_p (and C_{pk} for distributions that are not centered) in our discussion about the latter constraint.

Precision mechanical components produced by material removal, yield a product with uniquely measurable features. As we have discussed earlier, these consist of dimension (as a length, for example), form (as cylindricity or flatness, for example), surface roughness and, often, subsurface damage or residual stresses. The root sources of these measurable features are traceable back to the basic process mechanics, which of course is chip formation in the case of machining, Figure 9.8. This detailed view shows that the successful creation of a workpiece with features within specification (i.e. with a certain C_p) is influenced at all levels of the process, from design to chip removal. In Figure 9.8, after Stein[209] the process plan links the design to the process. Note that the performance of the machine tool (its kinematic and thermal behavior, for example) is lumped in with the process. Unfortunately, it is often impossible or

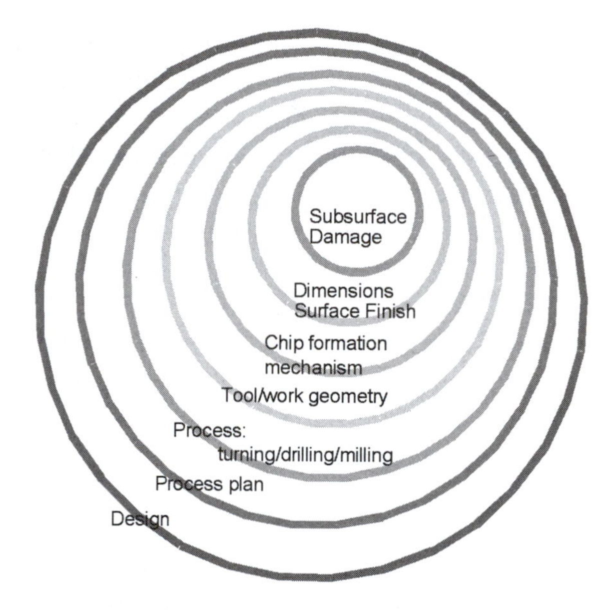

Figure 9.8. Factors which influence precision.

difficult to predict absolutely the output of a specific process *a priori*. Hence, we can often observe a certain error in the creation of a feature on a workpiece relative to some reference. Thus, the location of the centerline of a drilled hole in a component may differ by a small amount from its desired location. This error is termed *systematic* if the cause of the error is linked to understandable, but perhaps not controllable at our level, machine, tool, fixture or work behavior. Errors that are not systematic are often called *random* although this term is hotly contested as most so-called random errors are, in fact, traceable to physical causes if sufficient effort and intelligence is applied to the task. In some cases the errors observed are due to limitations in, or incorrect application of, the measurement method itself.

Process planning for general three-dimensional mechanical components is comprised of several challenging problems[209]. The planning task is complex and is generally broken up into higher level tasks and lower level tasks. At the highest level, it is necessary to identify the machining features that are used to reduce the stock to the desired workpiece. Manufacturing features are broken down into operations — i.e. a machining feature that uses a single tool, a single parameterized tool path, and a single set of process parameters. The complication comes in due to the fact that there are often several alternate sets of machining operations (that is multiple tools and operations are often necessary) that can create the same features. For example, a high-precision hole may involve a center drill, twist drill, and a reamer if the concern is hole cylindricity and diameter. If hole location is of primary concern, a boring tool may be used as a final operation instead of a reamer as it can 'adjust' the final location of the hole center.

In keeping with the strategy of breaking the process planning task into higher and lower level sub tasks, the typical approach is to plan at two independent levels. Macroplanning is the term given to higher level decision making (features, operations sequences, setup grouping, fixturing, for example). Microplanning refers to the selection of the machining operations, tool paths and process parameters (feeds, speeds and cut depths) for each feature. This is often called the "script".[211, 212]

9.3 Process capability

9.3.1 Background

Precision manufacturing has special requirements for process planning. We are looking here to indicate subtle changes in the part features as a result of process parameters as well as time in process. Some processes, polishing, for example, are sensitive to time. Others, like turning or grinding, are less sensitive as the feature creation is associated with a certain tool path. This is fine at a higher level. At a more detailed level, we must consider the elements contributing to loss of accuracy or precision in a process and its influence on our ability to create the feature. A number of excellent references exist elaborating on this, and we have of course also discussed these problems in great detail in earlier chapters of the book and will see examples later. Suffice it to say that major sources of error include:

- thermal influences (external and internal)
- static loads
- process induced structural distortion (cutting loads)
- variable stiffness of machine elements
- dynamic loads (process and machine related as well as environmental)

These error sources, along with the usual uncertainty in the manufacturing process itself, will result in deviations of the part feature from its ultimate desired value. Recall our earlier discussion about process capability and the ability to estimate tolerances for differing machining processes. This is the more traditional approach to process planning and there are many excellent reference texts elaborating these methods. Basically, once the process specifics are categorized (e.g. creates surfaces of revolution as with a lathe, or can achieve a certain dimensional accuracy as in a hole diameter) they are mapped on to the part geometrical features required and a subset of the processes are identified as candidate processes for manufacturing. Then, a string of these processes is assembled using a variety of techniques and the "script" written.

In general this is a reasonable approach. For precision manufacturing, however, there are additional constraints that need to be considered as we have covered. And we need a way of indicating, at a particular stage of the process, if we have successfully created the artifacts. Hence, precision process planning must select a series of processes, the inherited culmination of which will yield a part with the tolerances of dimension, form, surface and subsurface desired. This must be done on both the macro level and micro level. Speaking colloquially, a finely tailored suit is of high quality because each manufacturing step inherits high quality craftsmanship from the previous one: the selection of high grade cloth, perfect cutting on the lay, high grade construction and final hand finishing. And put in another way, hand finishing will not rescue the suit if the inherited culmination of all the factors is not perfect. So too with precision manufacturing.

To be effective in the production of precision components, we must identify the exact mechanism for extracting the key design primitives and mapping them on the manufacturing processes to form the basis for process planning. This requires two things: a means for describing the process capabilities and the requirements from the design point of view for precision. These are, not surprisingly, the components for calculating the C_p of a process. Traditionally, C_p or similar terms are calculated over a wide range of processes comprising the production of a part. The use of C_p is traditionally thought of in a quality control setting, and this is certainly the case. We propose the implementation of this measure of process capability here as it offers some unique advantages with respect to precision manufacturing.

More recently, concepts such as "stream of variation" (SoV) modeling and analysis have been discussed for variation management and reduction in complex manufacturing systems. An excellent reference on this is Shi[213]. This is specially well suited for precision manufacturing applications.

9.3.2 Process capability defined

Process capability is a measure of process performance that is designed to be used by managers, engineers and customers alike as an indication of the ability of the process to reliably manufacture a product or component. Figure 9.9, from DeVor[210], illustrates the relationship between process variability and product specifications. Three scenarios are shown in the figure for different process variation with respect to different specification limits[16]. Process capability is derived by a standard formula which allows comparisons between competing processes. There are two terms of interest here — capability ratio, C_r, and process capability, C_p. In fact, C_p is the inverse of C_r.

The capability ratio is a measure of how much of the tolerance range specified by the designer is consumed by variation in the manufacturing process. That is, how the tolerance on what you can make compares on the tolerance on what you will accept in your product. A C_r ratio is determined by taking the process variation (six standard deviations) and dividing by the tolerance spread (upper specification limits — lower specification limits):

$$C_r = \frac{6\sigma}{USL - LSL} \tag{9.1}$$

where σ = standard deviation
USL = upper specification limit
LSL = lower specification limit

In fact, we would usually estimate σ from the sampled data so that if the estimate of standard deviation is s, we would use that value in the calculation. Recall that 3 standard deviations above and below the mean includes approximately 99.73% of the production.

An example of the application of this ratio is as follows. Assume that a measurement is made of the output of a manufacturing process, say the dimension of a shaft that is being turned in a turning process. The design calls for a shaft of 175 mm with a tolerance of ± 2 mm. A listing of the part diameter measurements in mm is:

Shaft Number	Diameter, mm
1	174.70
2	174.07
3	175.27
4	175.14
5	175.06
6	175.10
7	176.08
8	175.83
9	173.84
10	175.02

In order to determine the capability ratio we need to estimate the mean and standard deviation of this set of data.

Here,

$$\text{mean} = 175.01$$

and

$$\text{standard deviation, } s = 0.69$$

These calculations are made on the assumption that the distribution of the data can be adequately represented by a normal distribution. To test this we could calculate the skew of the distribution. For this data the skew is approximately -0.30. If the skewness coefficient is between -1.0 and +1.0 the distribution is considered to be normal.

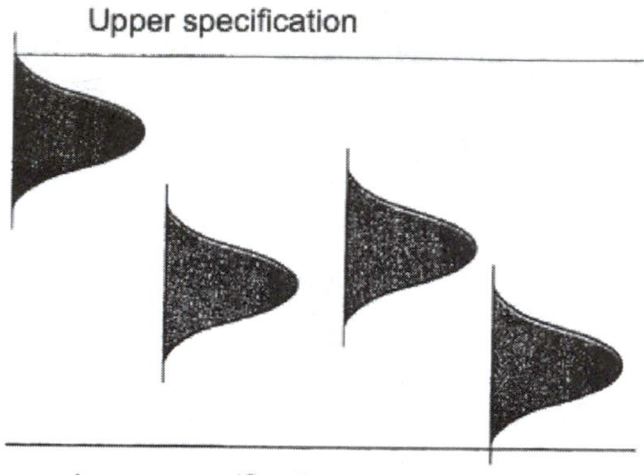

a) Process variation is small relative to the specifications so that the process mean can shift about without causing the Process capability to be jeopardized.

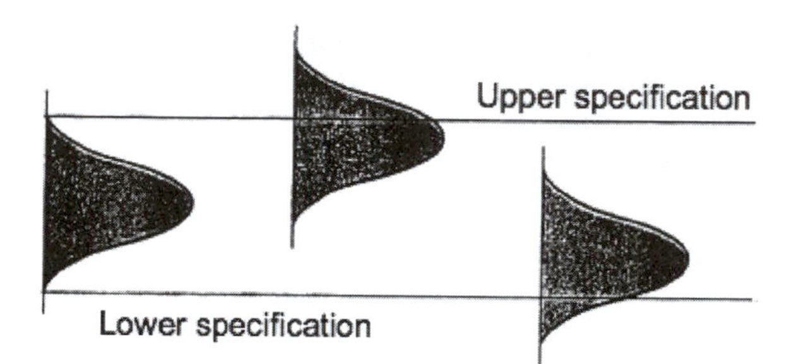

b) Process variation is large relative to the specifications so that the process must remain well centered for the process capability to be maintained.

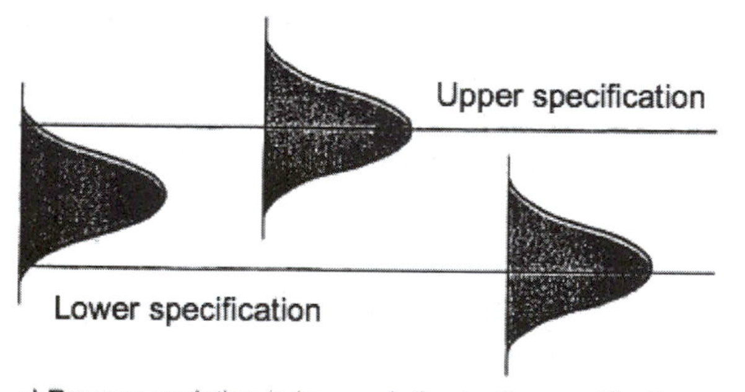

c) Process variation is large relative to the specifications so that the process cannot be considered capable regardless of the process centering.

Figure 9.9. Relationship between process variability and product specifications, from DeVor[210].

Thus the capability ratio is calculated as the ratio of six times the standard deviation, 4.14, to the tolerance spread, 4 mm (177-173 mm). And C_r = 1.035 or 103.5%. If this is considered to be the percentage of the tolerance "used up" by the manufacturing process, it is clear that we will certainly be making products outside of the tolerance. In fact, the smaller the number, the better (usually 75% or less).

The process capability, C_p, is the reciprocal of the capability ratio and is the value more commonly reported as an indication of the quality of the process:

$$C_p = \frac{USL - LSL}{6s} \tag{9.2}$$

C_p for our shaft manufacturing process is 0.966. Here anything less than one is undesirable. And, if you recall the percentage of product falling with in the limits, we really need to insure that the C_p value that reflects a process capability closer to $\pm 6\ \sigma$. Usually, a C_p value of 1.33 or higher is considered to be adequate.

There is one important aspect of the process measurement that is not included in the process capability or capability ratio calculations above — whether or not the distribution is centered about the mean. That is, simply calculating a ratio can give is a good value of the indices even thought the production is biased so that it is entirely outside of the range of the specification! A more appropriate measure to use is C_{pk} which tells us whether the process is capable of meeting the tolerances and whether the process is centered around the desired dimension (target value). We calculate C_{pk} as follows:

- first, the relationship between the process mean, $\bar{x}$, and the specification limits in units of standard deviations is calculated,

$$Z_{USL} = \frac{USL - \bar{x}}{s} \qquad\qquad Z_{LSL} = \frac{USL - \bar{x}}{s}$$

- second, select the minimum of the two values,

$$Z_{\min} = \min\left[Z_{USL}, -Z_{LSL}\right]$$

- third, the C_{pk} index is calculated by dividing this minimum by 3,

$$C_{pk} = \frac{Z_{\min}}{3}$$

and the value should be ≥ 1.00 for the process to be acceptable. If the process is outside of the tolerance range, C_{pk} will be negative indicating that over half of the production will be outside of the specifications. Care must be taken in applying this measure as it is possible to maintain a suitable C_{pk} value by reducing the standard deviation of a process by some improvement but allowing a wider deviation from the mean (i. e. more bias!).

9.4 C_p as a planning metric

On a macro level, using C_p as a metric of success of the process we might represent the total process sequence as shown in Figure 9.10. No transition should occur from one stage of the process to another (Process 1 to Process 2, for example) unless the criteria that $C_{p1} \leq C_{p2}$ *minimum input* to process 2. In other words, a pre-condition to completion of process 1 and beginning of process 2 is that the C_p for the product at the exit of process 1 must be less than the minimum requirement for the start of process 2. A roughing process preceding a finishing process must be, at least, able to create a surface that can be treated by the subsequent finishing process. Else the process may require too much time or be ineffective (the ill-fitting Saville suit!). If this condition is not met, process 1 must continue. Typical examples of this are in grinding operations where additional passes may be needed to accommodate sparkout or in a milling process where a tool may pass over a thin web without in-feed to remove residual elastic deformation.

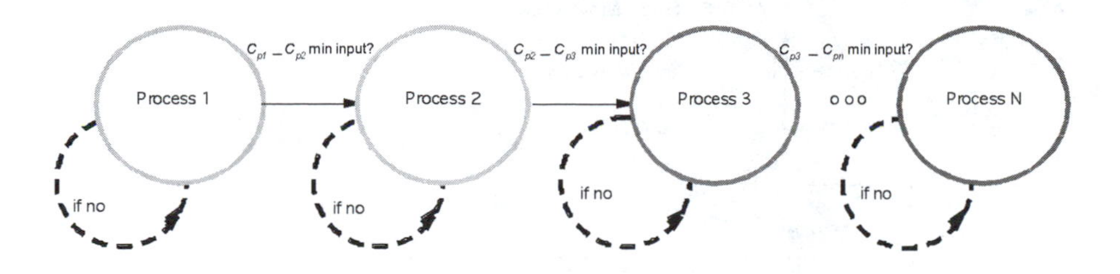

Figure 9.10. Process sequence with transition conditions.

On a micro level, we are concerned with the "real time" determination of the appropriateness of transitions from one process stage to another. This real time control is a key feature of agent-based open architecture manufacturing. We can represent the progress of feature creation in a process as in Figure 9.11. Here, for process 1, we see three parameters representing the status of features on the part. As these parameters change with the process cycle, they ideally tend towards some level which, when reached, indicate successful creation of the feature. So, for the first process we might conclude that when parameter A reaches an acceptance level, a, parameter B reaches an acceptance level, b, and parameter C reaches an acceptance level, c, process 1 is finished with respect to improvement of the features being addressed. The calculation of C_{p1} is based on the values of parameters A, B and C representing features A, B and C, respectively. Intermediate in the process, if a test is done, we would see that the appropriate levels had not been reached and, thus, process 1 must continue. Of course, there is no requirement that all parameters must reach the acceptance level at the same time or, for a process out of control, ever reach the acceptance level.

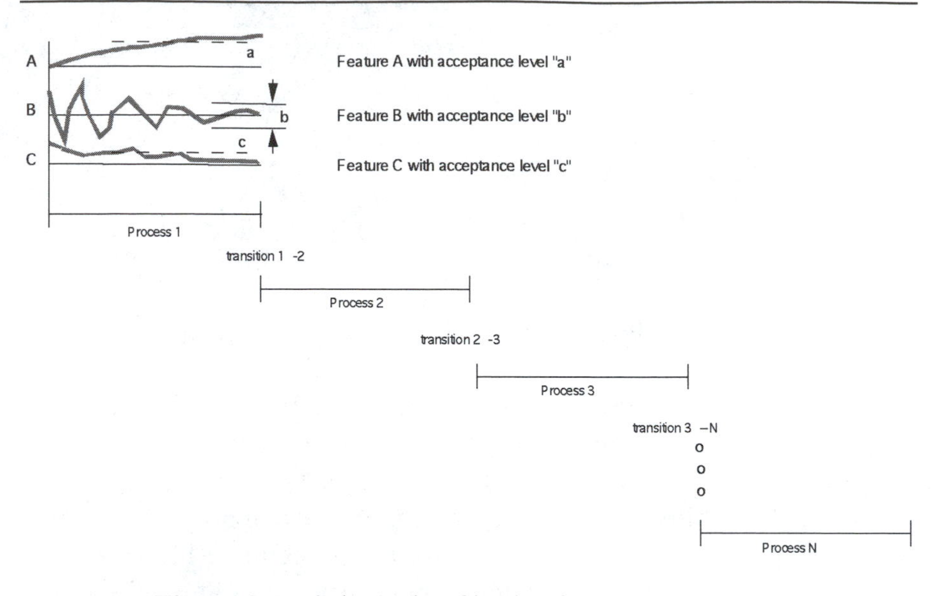

Figure 9.11. Schematic of basis of process transition.

Under normal circumstances we assume processes are under control and that an improvement will occur with process time. In other words, process *error* (defined here as the difference between the desired feature characteristic — a dimension and a tolerance, for example — and its present value) reduces with process time. This can be illustrated as in Figure 9.12. With subsequent processes, we see that the process error reduces in magnitude (as a dimension approaches its final value through subsequent processes — center drill, drill, ream, for example) as well as in variation. Both magnitude and variation of error must be within bounds to conclude that a specific process stage is successful and that the part is appropriate for passing on to the next process, the $C_{p1} \leq C_{p2}$ *minimum input* to process 2 criterion mentioned above.

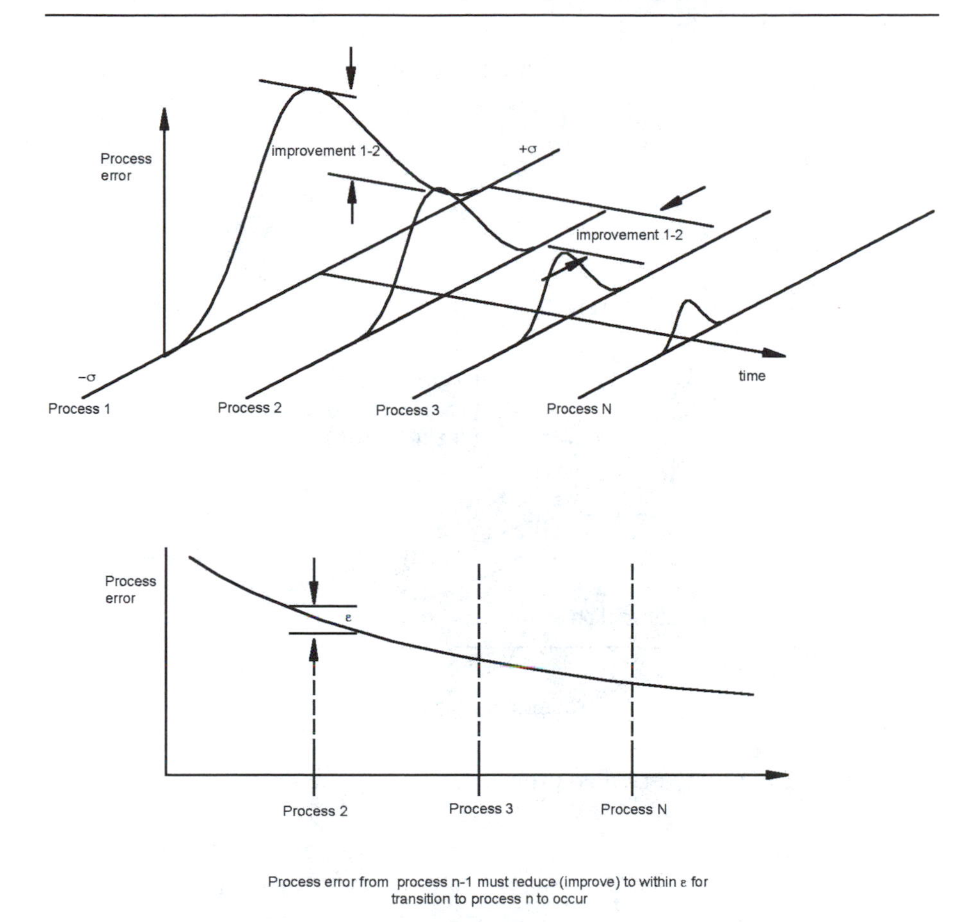

Figure 9.12. Process improvement through a series of stages.

Westkamper[214] illustrates this as the reduction in "non-conformities" as manufacturing proceeds through its process stages... each stage picking up where the last left off, Figure 9.13. Inconsistencies in the next stage's ability to accommodate the non-conformities of the previous stage will cause an error in the process. The "process chain" is under the influence of many disturbances (recall the sensitivity of chip formation to tool materials under identical conditions) that can affect the non-conformities. This can be represented as in Figure 9.14 which details the influences and disturbances in the process sequence based upon setting parameters, E, quality measures, Q, and disturbances, S. The individual stage elements can be further elaborated on and the inter-relationships modeled to represent the influence on time, product cost and quality, Figure 9.15.[214]

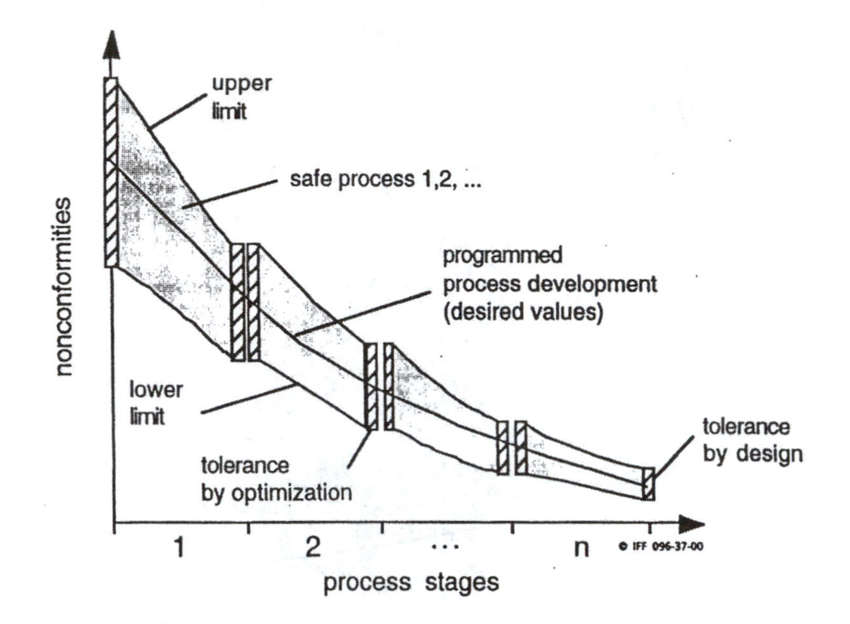

Figure 9.13. Quality oriented process control in the process chain, from Westkamper.[214]

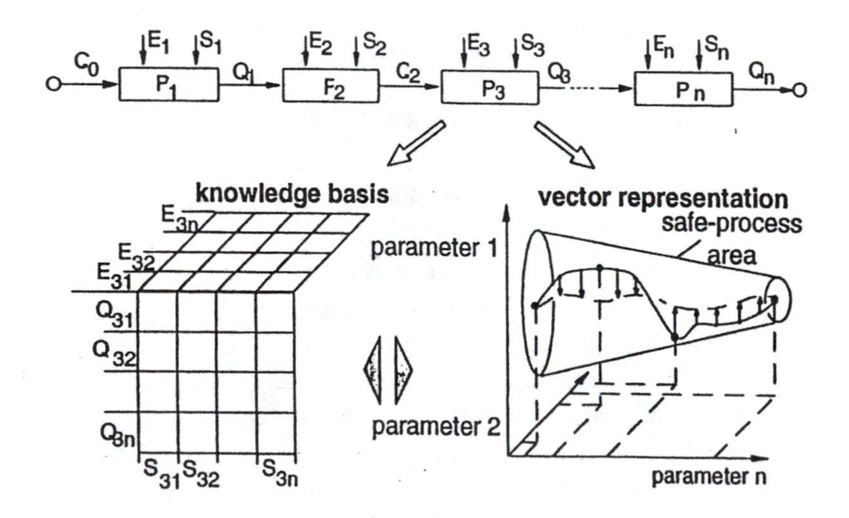

Figure 9.14. Mathematical representation of the influencing and disturbance qualities in the process chain, from Westkamper.[214]

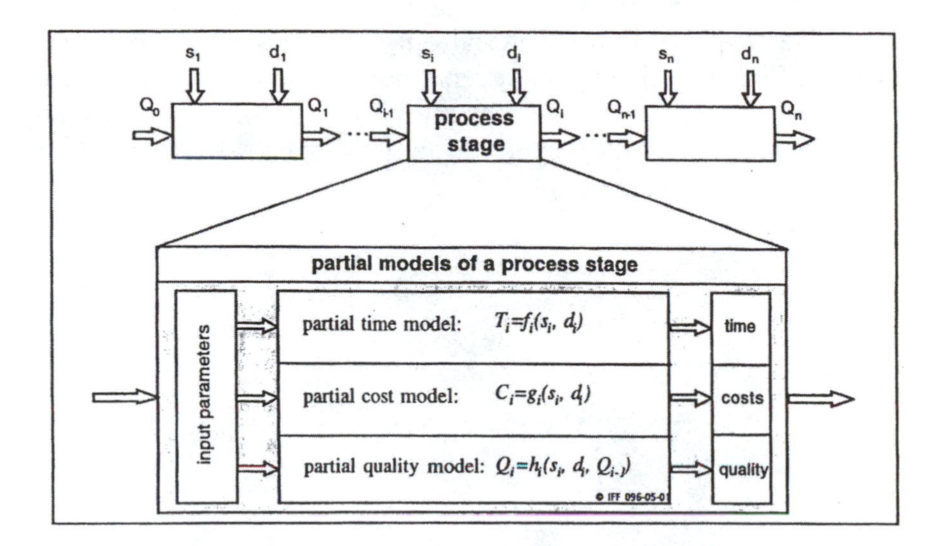

Figure 9.15. Simplified model of a process chain, from Westkamper.[214]

Process planning under these conditions is much more challenging and requires adjustment at the micro-plan level in response to observed changes in both the magnitude and rate of change in process error. The outcome of process n can impact the effectiveness of process n+1 or n+2. A rough machining operation can induce residual stresses in the surface of a part that frustrates finishing operations several stages later. In addition, interactions of processes will affect features on the part and must be considered. A hole made in a surface which will be subsequently machined can create a potential for burrs in the hole requiring expensive finishing operations to remove.

Figure 9.16, from Wright[215], a quick-stop section of a stainless steel chip formation, shows a specific example of the way in which an inherited state is created and how it will transition to the n+1 state, probably needing rectification when dealing with precision manufacturing. The section shows that the "rough turning" operation creates a work hardened layer on the machined surface. For a depth of approximately 0.1mm (approximately half the feed rate in this case), the material is "damaged" by work hardening to a level of

222VPN as opposed to the material bulk hardness of approximately 170VPN. Finish machining and even possibly lapping will be needed if this is a critical component that must have uniform hardness and no work hardened surface layer that might reduce the fatigue life of the precision part in service.

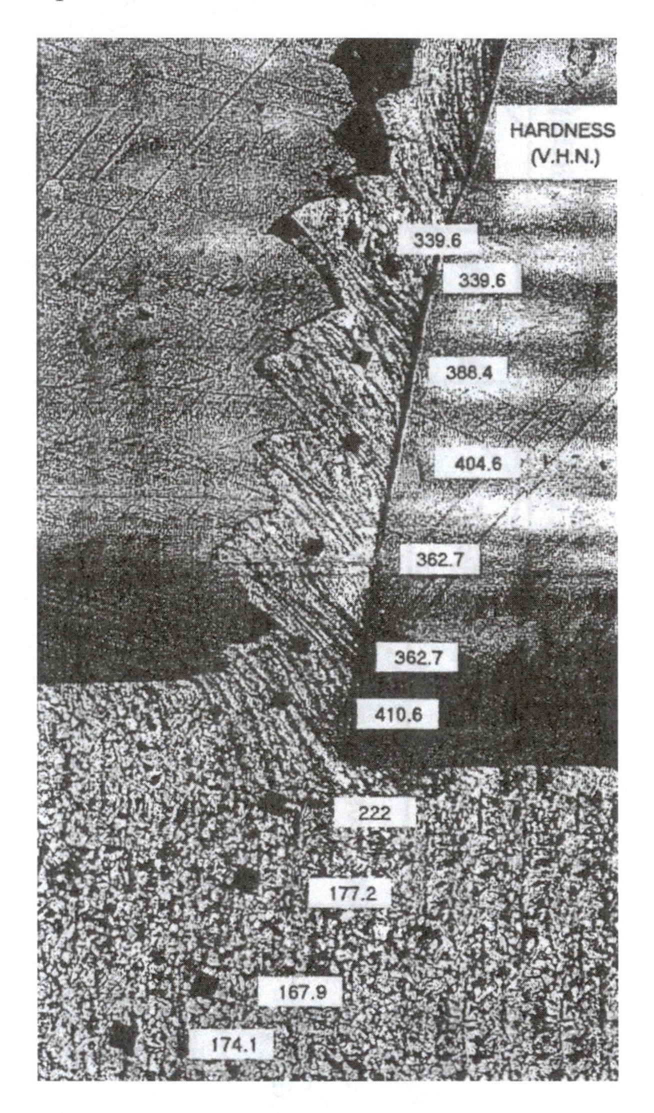

Figure 9.16. Microhardness tests on a stainless steel (AISI 321), machined at 100 m/min, with a feed rate of 0.2 mm, a depth-of-cut in turning of 1.5 mm, and +6° rake angle tool, from Wright[215].

9.5. Legacy-system integration for precision manufacturing

In the past, Java[216], CORBA-based object wrappers[217], the OLE/ ActiveX environment[218], and ACL-based "agents"[219], have served as the software tools for integrated process planning in precision manufacturing. Such software components allow interoperation among different CAD/CAM systems and pre-existing databases that, at first glance, appear difficult to integrate because they run on computing machinery produced by different manufacturers, at different sites in a company, using different languages and operating systems. Specifically the CAD designer might be working with a well-known professional system such as IDEAS or Pro-Engineer and then want to gain access to an old database on the relationship between feed rate, depth-of-cut, rake angle and the depth-of-work hardened layer as shown in Figure 9.16.

Immediately one can see several difficulties that arise in terms of communication between the professional CAD system and this pre-existing old data-structure (often called a legacy or stovepipe system.) Potentially, this old database has some very rich information stored in it for process planning for precision manufacturing, but can the CAD programmer "get to it" in a quick and efficient way? Today, the unfortunate answer is "Probably not." The CAD programmer will be pressed for time and will not want to sift through old-style tables and charts that may not anyway be prepared with parameters or variables that match up to those being used during design.

There is work to be done by the research scientist gathering the information on precision process planning so that CAD programmers "out-there-on-the-Internet" can use it. Researchers at Berkeley, working in conjunction with colleagues at Stanford[220], have used ACL as the communication medium for this type of knowledge exchange but before going into the jargon of ACL, a very brief review of the key issues involved is worthwhile. One main key to communication is to decouple the specific interface or

GUI being used from the actual implementation itself. PostScript was well recognized as such a standard because everyone wanted to use their own favorite word processing format and printer, but still want to share publicly certain text documents. The main problem comes at the next level of sophistication. What happens when it is necessary for two programs to exchange data or information when they do not share the same vocabulary, syntax or semantics of their corresponding languages? Different programs may inconsistently use a word or variable for different things. In the field of machining, there are many inconsistencies around variables such as "rake angle", "feed rate", and "depth-of-cut" that may mean one thing in one program and something else in another. Between CAD systems and machining databases there can be dangerous incompatibilities. We pointed out[221, 222] that the variable "drilled hole" might ignore the sub-variable "point-angle" in one program and not in another causing an error in the part, or a tool breakage during a tapping operation. The precise definition (often called the ontology) of a drilled hole should therefore contain more than just its position on a CAD drawing, and its diameter. And the more we focus on "precision" drilling, as opposed to "normal" drilling, other variables such as the "surface finish" and the "post-drill reaming allowance" obviously come into play. A detailed description of this work is given in .[207]

9.6. Future integration for precision manufacturing process planning

The development of information technology has always pushed manufacturing to new capabilities. Early researcher such as Merchant proposed computer integrated systems with capabilities beyond what the technology could deliver but with the goal of achieving more seamless interoperability[223].

Whereas past interests were focused on legacy integration as discussed above, it is now more common to think of languages such as XML as a common communication format. The creation of a seamless "manufacturing pipeline" from design to production has long been a goal of many industries and we have discussed the value of design decisions linked to manufacturing processes.

The development of digital factory concepts connecting the product and process designer to shop floor/equipment/operation level data and feedback for simulation, optimization and control is moving closer to reality. The challenge is to connect "islands" of technology to make this a seamless link. This is the basic goal of the pipeline concepts for manufacturing systems interoperability. The objective is to provide an extensible standard to enable equipment from multiple manufactures to communicate with each other; allowing them to discover their capabilities, configuration, and geometry as well as subscribing to near-time updates on status and health. An XML-based representation for instrumentation data enables incremental progress towards interoperation without requiring full buy-in in advance. Support for XML in enterprise programming frameworks is ubiquitous and programmers skilled in the use of this technology and its rich tool repertoire are readily found. This will likely be one of the major development in the future of precision process planning.

X PRECISION MACHINING PROCESSES

"The smaller the particles of the substance (abrasives), the smaller will be the scratches by which they continually fret and wear away the glass until it is polished..."

Sir Isaac Newton in *Opticks* 1695[224]

10.1 Introduction

Precision manufacturing processes are defined here as those for which there is a very small amount of material removed (per cutting edge, for example) and for which the surface or feature created can be characterized by stringent tolerances on form, dimension or surface characteristics. We will emphasize here processes with very small chip thickness, a_{cav}, as defined below as well as processes for which no discernible "chip" is produced but features are created never-the-less. These processes include abrasive machining (including lapping, polishing and honing) and can be characterized by either two body or three body abrasive interactions. The nontraditional processes include those Taniguchi referred to as "ultrahigh precision machining" such as atomic bit processing with electron beams or electrolytic polishing, photon sputtering or electric field "removing" or evaporation such as AFM and STM processing.

There are various ways to classify precision material removal processes. We have presented one above, based on the "uncut chip thickness", and will discuss this in more detail. A convenient classification is based upon "energy source" driving the removal process.

Figure 10.1 from Nogowa[225] and represents material removal under the classification of mechanical, chemical, photo-chemical, electro-chemical, electric and optical processes. As we shall see, there can often be two or more processes active in a particular precision material removal operation. For example, Nogowa's item 10, Chemical Mechanical Polishing, in Figure 10.1 is, in fact, a combination of mechanical abrasive action and chemical polishing. Our interest in precision material removal derives from its importance in the manufacture of products and devices with the most stringent tolerances of dimension, form and surface — most often found in semiconductor products and associated computer peripherals.

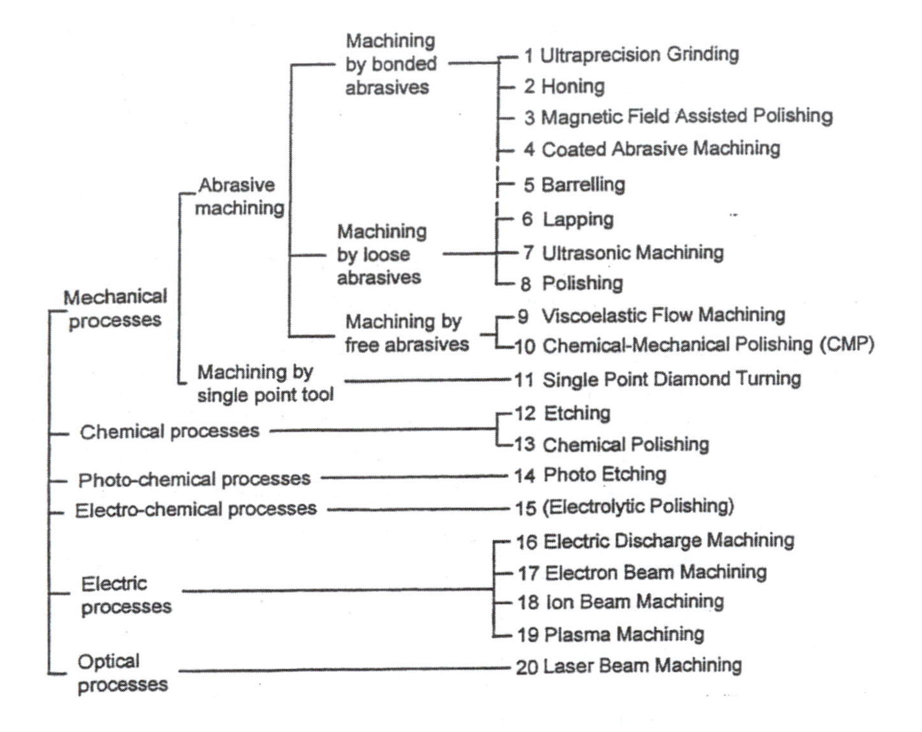

Figure 10.1. Classification of material removal processes in terms of the energy source used, from Nogawa[225].

Although "electronic devices" usually imply manufacturing techniques more associated with lithography and vapor deposition, we shall see that most devices so called require a number of precision material removal processes in their fabrication. In fact, it is these precision processes that often are bottleneck processes because of control or yield problems. Taking ultra-precision finishing for a moment, the ultimate performance of the product reflects the results of a series of stages including the

- machining system
- machining control and inspection/metrology
- the performance of the machine with respect to specifications of dimension, form, surface and sub-surface damage
- product performance specifications

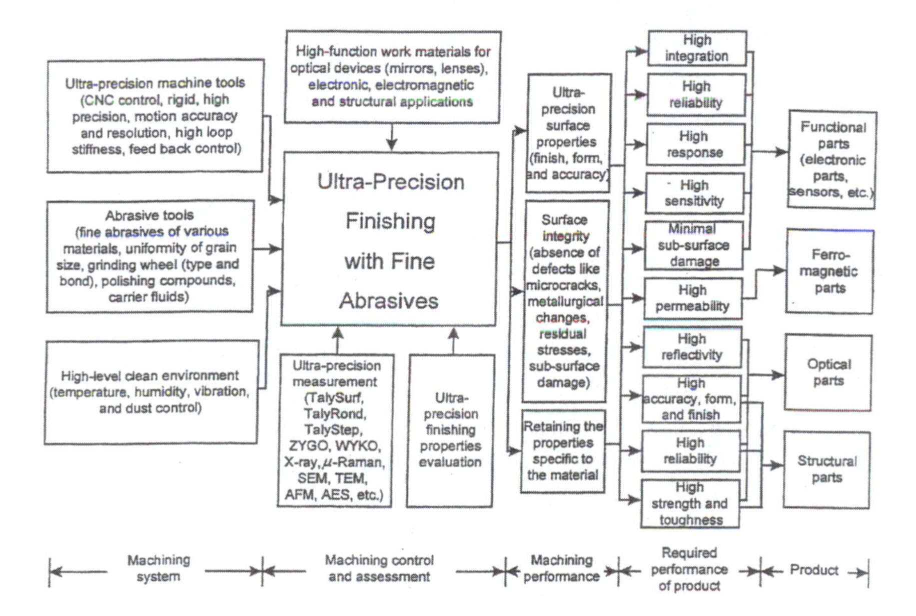

Figure 10.2. Conceptual diagram of ultraprecision finishing with fine abrasives, from Nogawa[225].

Details of this are shown in Figure 10.2, also from Nogowa[225]. Ultimately, the features of the product listed in Figure 10.2, degree of subsurface damage, reflectivity, accuracy, form and finish, reliability and strength and toughness are dependent on the successful application or control of the precision manufacturing process. And, as we have discussed previously, this is dependent on our understanding of the process fundamentals as well as the behavior or the machine tool system. Our discussion and the examples of manufactured products will focus on semiconductor devices and optics and structural elements used in the semiconductor manufacturing equipment in addition to microscale devices including microelectrical mechanical systems (MEMS). We also discuss nanofabrication technologies and applications. These products are fabricated out of a variety of materials from silicon to ceramics and use a number of processes.

A convenient way of viewing the range of processes normally associated with precision manufacturing is to look along the a_{cav} (or grain depth of cut or chip size) axis where appropriate. Even if no traditional chip is being removed, the surface or some other feature of the part is being modified by a finite amount and the size of this "finite amount" is an indication of the achievable size or tolerance on the feature. Recall that the abscissa on the Taniguchi chart is achievable tolerances and is driven lower by, basically, ever smaller sizes of a_{cav}. Figure 10.3, from Boothroyd[226], shows the variation in a_{cav} from 1 micron (.001 mm) to 1 mm and the ranges of machining processes associated with the comparable chip size. In fact, diamond turning, a very common precision manufacturing technique, is actually lower than the lower end of the range for turning and is well into the upper end of the range for grinding. Chip thicknesses in diamond machining are commonly found in the range of 0.1 micron as seen previously in the Taniguchi diagram.

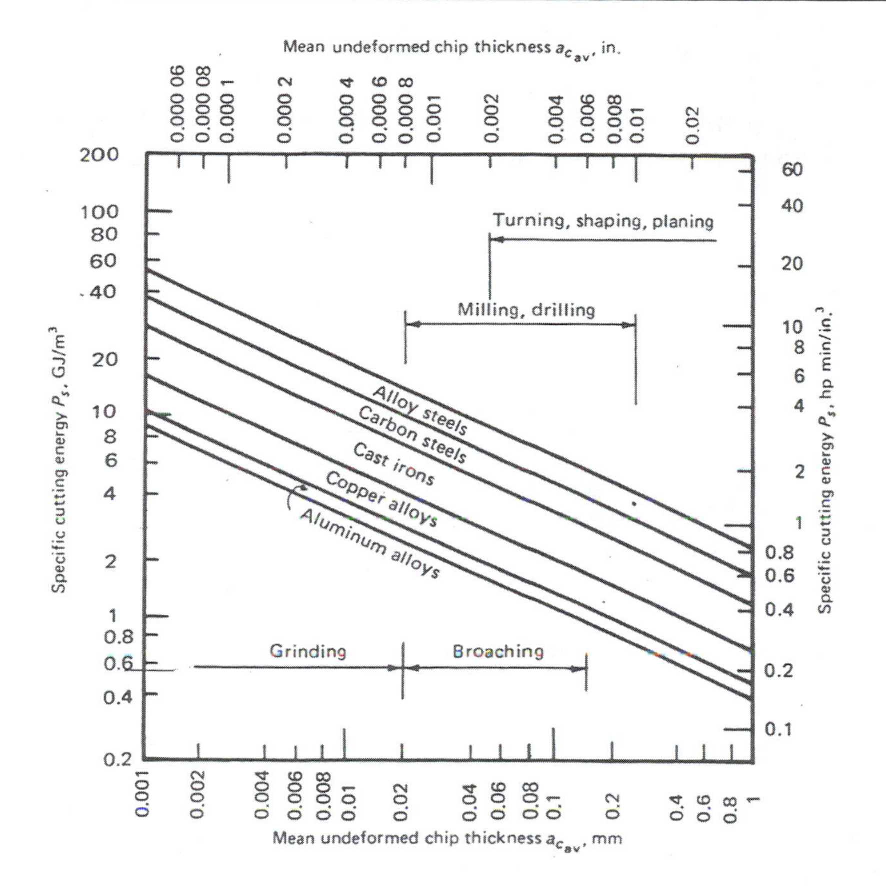

Figure 10.3. Approximate values of specific cutting energy for various materials and processes for differing uncut chip thicknesses, from Boothroyd.[226]

We can clearly see a distinction between material removal at different uncut chip thicknesses by considering as well the "distribution" of the features of the cutting tool(s) used. Abrasive processes tend to have more variance in the height, for example, where as ductile machining in conventional processes (like turning) tend to have narrowly defined height distribution, Figure 10.4 from Whitehouse.[227] An additional element to consider is shown in Figure 10.4, the distinction between ductile and brittle material removal. There is some controversy as to whether any machining is ductile as some evidence exists to indicate that fracture of material plays a role in material removal at all depths of tool engagement. Typically, one can define a critical depth of cut (uncut chip thickness) above which, for brittle materials, the removal has a discrete particle or broken chip-like appearance (small, segmented randomly shaped pieces as opposed to continuous strips of material). This is discussed in more detail in Section 10.6 below.

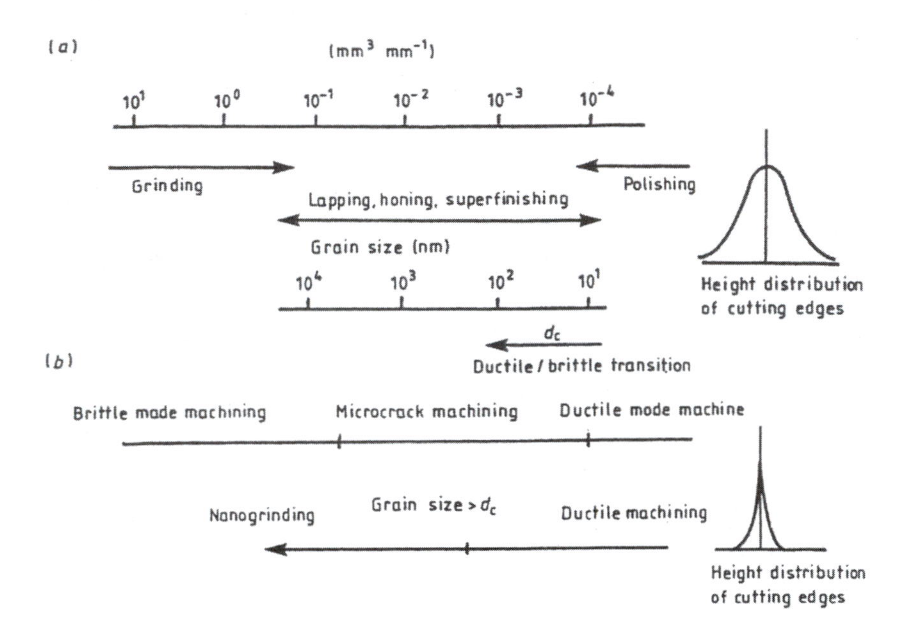

Figure 10.4. Ranges of uncut chip thickness and cutting edge characteristics for conventional machining processes a), versus nano grinding processes b), from Whitehouse.[227]

The grinding region of Figure 10.3 is the transition to uncut chip thicknesses due to non-kinematically defined cutting edges. Figure 10.5 shows the range for a_{cav} below 1 mm extending to less than a nanometer for some electrolytic polishing processes. Our major interest is in the processes like diamond turning relying on more conventional machine tool configurations to yield precision surfaces, loose abrasive processes yielding the next lower range of surfaces and dimensions and, finally, the nontraditional processes. The inherent differences in machining with a well-defined tool geometry and grinding are illustrated in Figure 10.6, from Tönshoff[228].

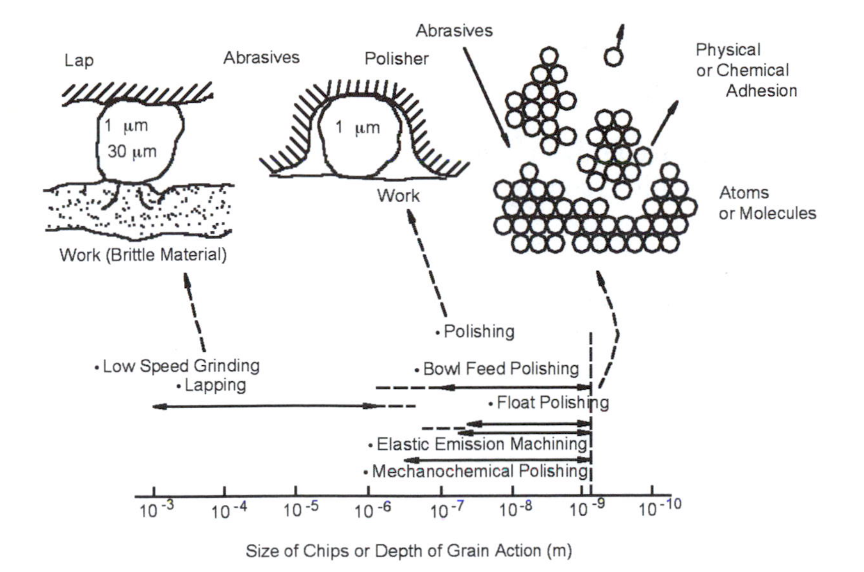

Figure 10.5. Comparison of lapping, polishing, and other processes.

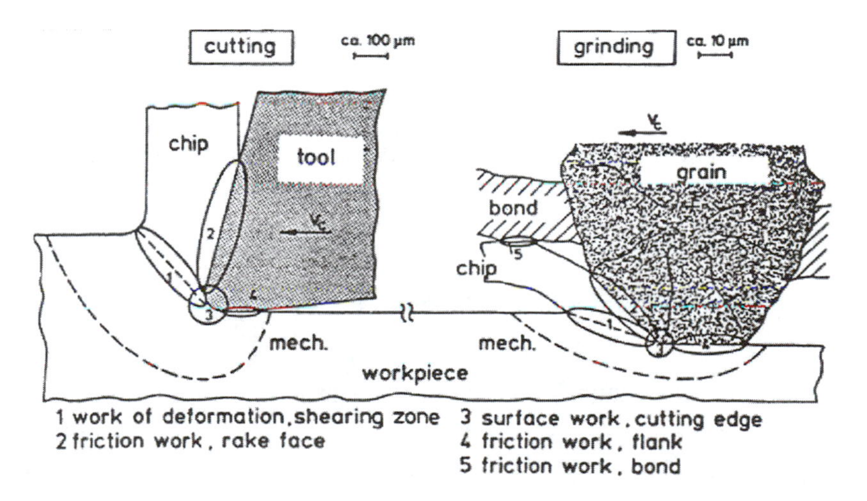

Figure 10.6. Comparison of machining with defined edge and grinding, from Tönshoff[228].

We then can summarize our analysis of precision processes by considering the following categorization from the point of view of the tool and the work piece, and its "reaction" to the tool, in addition to the machine characteristics previously discussed (stiffness, for example). Specifically,

Tool	- fixed (two body interaction) regime of interaction	defined edge
		"random" edge
	- free (three body interaction) regime of interaction	"random" edge
		"oriented" edge

Work Reaction

- ductile

work material orientation effects

- brittle

These will be described in more detail with respect to specific processes and materials.

10.2 Influence of machining parameters, work material, and tool geometry

10.2.1 Influence of uncut chip thickness

The reducing a_{cav} causes changes in the manufacturing process due to two elements: effective tool geometry and the relationship between tool geometry and uncut chip thickness. The "tool geometry" for abrasive particles is hard to define quantitatively. One can achieve a range of sizes and a range of shapes of the abrasive particles but the actual interaction between the abrasive and the work surface will include a number of other factors such as hardness of the abrasive relative to the work or wheel matrix (in grinding) or plate (in lapping, for example), abrasive friability, fluid used, etc. Second, for tool geometries for which the tool edge radius is larger than the

a_{cav} value, the machining efficiency is much lower, Figure 10.7, from Shaw[229]. The figure shows the increasing measurable shear stress (i.e. based on the machining conditions and tool forces measured) during precision machining down to chip sizes of 0.5 micron or lower. This is equivalent to the specific cutting energy we discussed relative to compliance in machining. In fact, we can clearly define three regimes of machining efficiency as follows for tool radius, r, and a_{cav}:

Regime I $a_{cav} >> r$
Regime II $a_{cav} \approx r$, and
Regime III $a_{cav} < r$

Precision processes exist in all three regimes but, for the most part, regime III is most commonly found for abrasive processes. If we more far enough along the ordinate in Figure 10.7 we reach Regime I and it is expected that the efficiency as measured by shear stress will be constant. The edge radius in a carefully prepared diamond cutting tool is in the range of 30-40 nm, thus, the transition from Regime I to II must occur with chip sizes on that order! At this size, however, many other considerations can come into play, such as crystallographic orientation of the work material at the cutting surface, surface chemistry, etc.

The chip-tool interaction in small chip size machining is illustrated in Figure 10.8 showing the diamond mirror turning of soft metals (like aluminum or copper). In the top left image is a situation where the chip size/tool radius is in region II or III on the left (and generating a compressive residual stress) and region I on the right (generating tensile residual stress in the work surface). The figure also shows the ideal surface roughness due to varying nose radii on the tools.

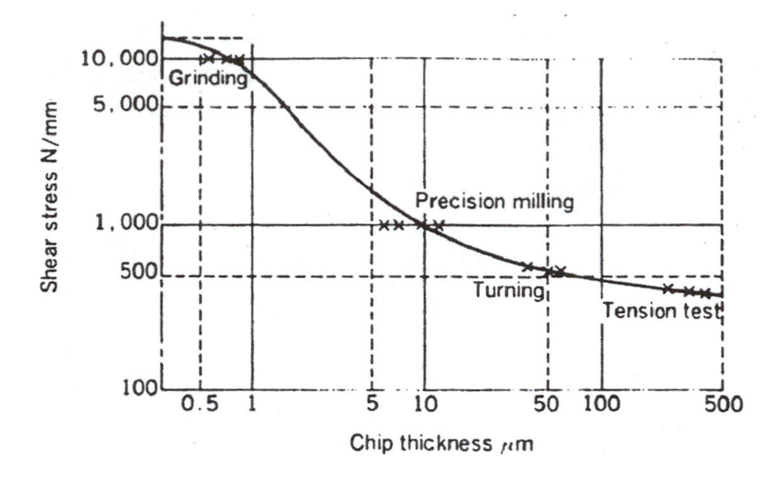

Figure 10.7. Shear stress vs. chip thickness for precision Machining carbon steel 1112, after Shaw[313].

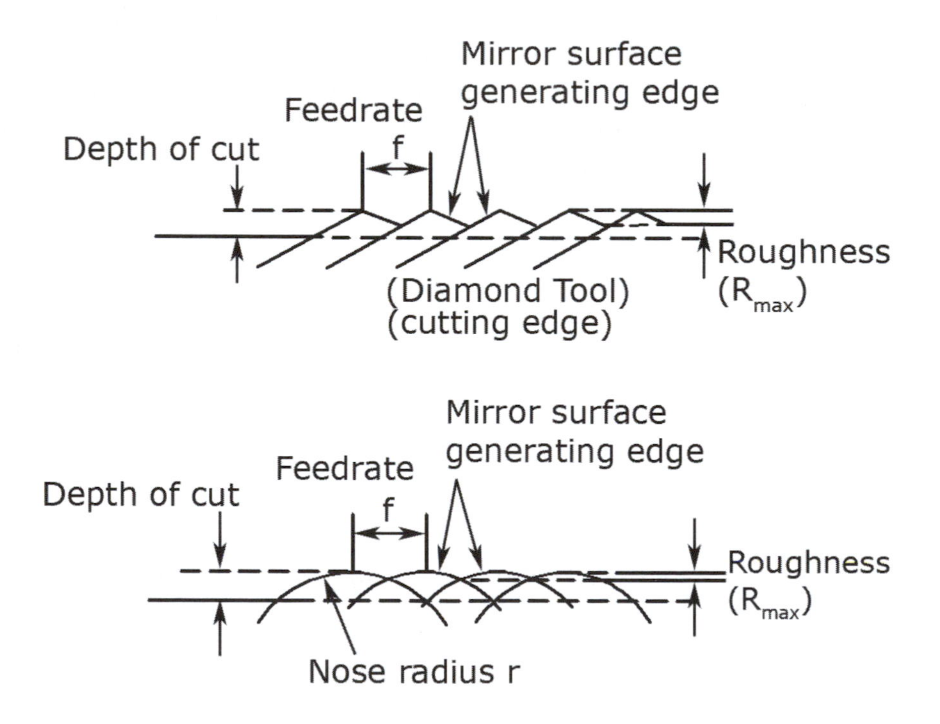

Figure 10.8. Illustration of tool effects in mirror cutting of soft metals.

10.2.2 Machining brittle materials

Machining at very small depths of cut (or uncut chip thicknesses) often causes "brittle" materials (i.e. likely to deform slightly and then fracture due to very low fracture strain characteristics- most ceramics, for example) to machine in a ductile manner. This means that rather than creating surfaces exhibiting large amount of damage due to fracture and crack propagation during machining, the surfaces appear to have been created as in a normal ductile mode machining process. We will discuss this again later in this chapter under abrasive processes. A simple schematic comparison of the behavior of ductile and brittle materials under indentation at varying pressures is seen in Figure 10.9. A measure of "brittleness" often used is the ratio of the modulus of longitudinal elasticity, E (in N/mm^2) to the Vicker's hardness for a material, H_v (N/mm^2). A measure of a material's ability to with stand conditions leading to fracture is the ratio c/a. Here *c* is the radius of the elastic-plastic boundary (as seen in Fig. 10.9, from Lawn and Wilshaw[230]) and *a* is the radius of the indentor used in the indentation test. Figures 10.10 and 10.11 define the geometry of the indentation test (10.10) and the variation in c/a for a variety of materials characterized by their E/H_v ratios (10.11). Hard materials, such as diamond with a E/H_v value of almost zero, have a very low c/a ratio and are primarily brittle in behavior but can exhibit ductile behavior under high hydrostatic stress conditions. Soft metals, aluminum, copper and lead, have high c/a ratios. For these materials, brittle behavior is almost impossible to induce except under very high strain rates.

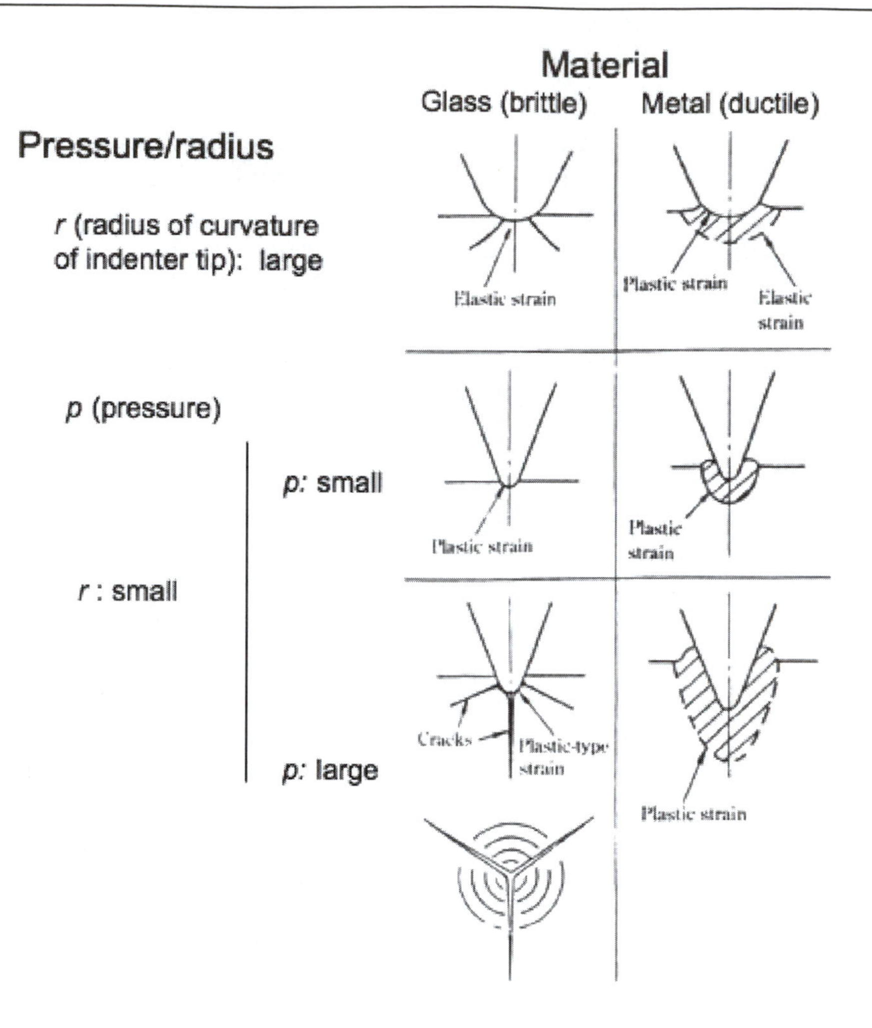

Figure 10.9. Microscopic deformation and fracture of materials, after Lawn and Wilshaw[230].

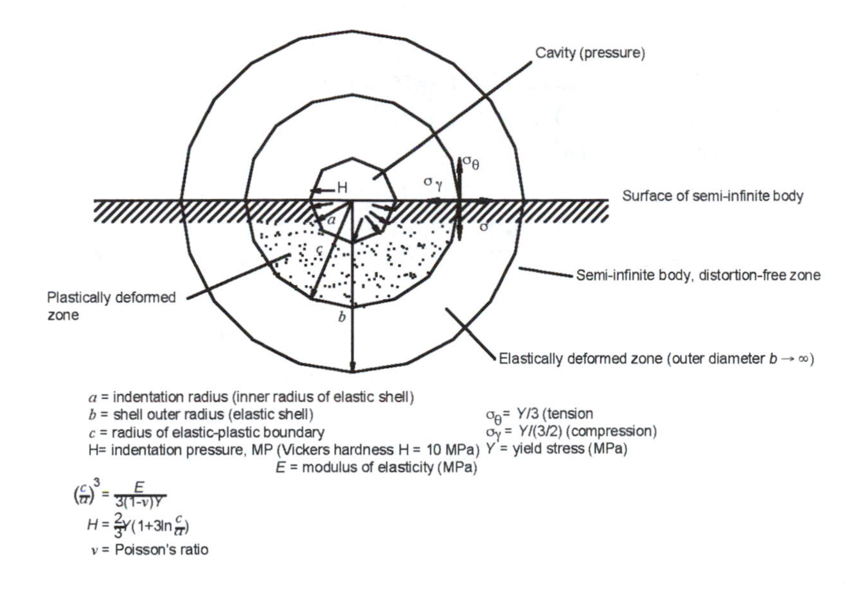

Figure 10.10. Geometry and nomenclature for indentation and resulting deformation, from Nakazawa[19].

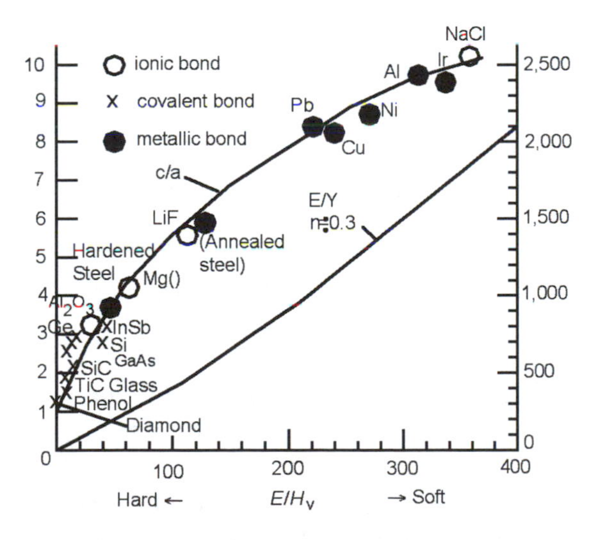

Figure 10.11. Plastically deformed zone (c/a) for various materials, from Nakazawa[19].

Materials do behave differently under different material removal conditions. Many materials exhibit a so-called ductile to brittle transition driven by the depth of cut (or uncut chip thickness depending on how the tool-work geometry is defined). Figure 10.12, from Brinksmeier[231] illustrates an experiment gradually increasing the depth of cut of a diamond tool scribing silicon (Si {111}) cutting in the <221> direction. The regions of tool-work interaction from initial contact and elastic deformation through cutting and the transition from ductile to brittle are illustrated. The graph of thrust force with time (displacement) clearly indicates the transition from ductile to brittle removal. Figure 10.13 (also from Brinksmeier) show AFM images of both the ductile removal area as well as the transition region.

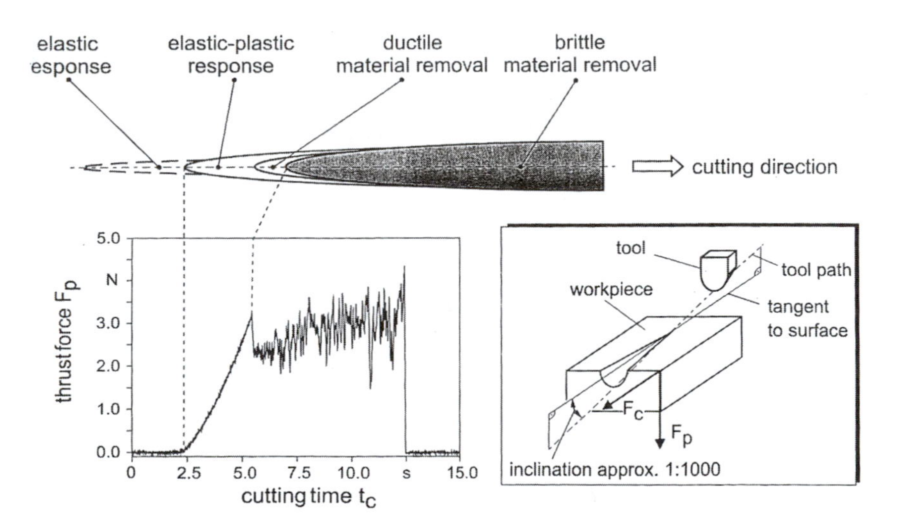

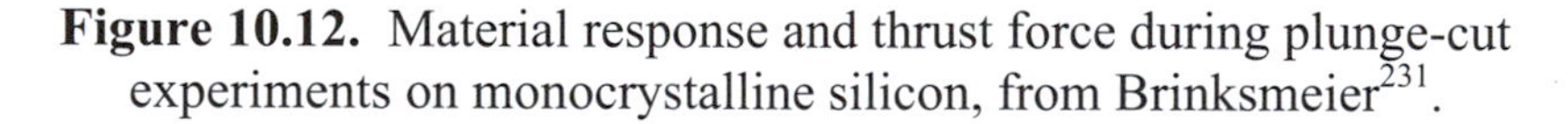
Figure 10.12. Material response and thrust force during plunge-cut experiments on monocrystalline silicon, from Brinksmeier[231].

Bifano[232], based on the earlier work of Lawn and co-authors[230, 233], derived a relationship predicting the critical depth of cut at which this transition would occur for a variety of materials. The transition is considered from the point of view of the "energy" consumed in plastic *versus* fracture processes. For lower depths of cut, plastic flow is favorable to fracture. The energy, E_p, required to plastically deform a volume V_p can be derived using the material yield stress σ_y as

$$E_p = \sigma_y V_p \tag{10.1}$$

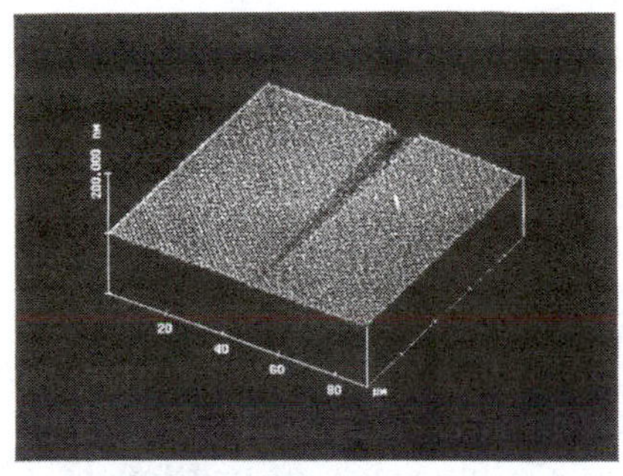

Ductile material removal

Material: Si {111}
Cutting direction: <22$\bar{1}$>
Tool nose radius: r_ε = 0,76 mm
Rake angle: γ = 0°
Cutting speed: v_c = 20 mm/min

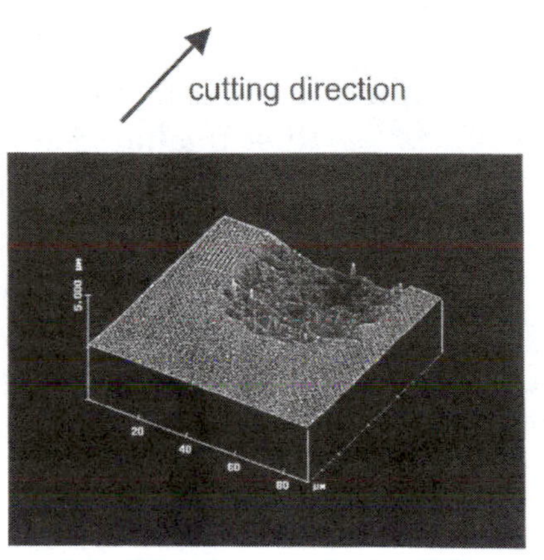

Transition from ductile to brittle material removal

Figure 10.13. AFM images of groove topography generated by plunge-cut experiments, from Brinksmeier[231].

The material property most characterizing its response to fracture is G, Griffith's crack propagation parameter. This describes the fracture energy, E_f, as

$$E_f = G\,A_f \tag{10.2}$$

where A_f is the fracture surface area.

For a machining depth, d, it is reasonable to assume that the volume of material undergoing plastic flow is approximately

$$V_p = d^3 \tag{10.3}$$

and

$$A_f \approx d^2 \tag{10.4}$$

then the ratio of plastic flow energy to brittle fracture energy, E_p/E_f, is proportional to the depth of cut d. The critical cut depth, d_c, at which the transition from ductile to brittle mode material removal (actually indentation depth) occurs at

$$d_c = (E\ E_f)/H_v^{\ 2} \qquad (10.5)$$

where E = elastic modulus
E_f = fracture energy
H_v = Vicker's hardness

Depending on the material, we will use Griffith's analysis for materials with a plastic zone near the crack tip (that is, to determine fracture energy). Otherwise, one can derive a "measure of brittleness"[234] for which

$$E_f \propto K_c^{\ 2} / H_v \qquad (10.6)$$

Hence, the critical penetration depth for indentation, or depth of cut for micro-machining, can be defined as

$$d_c \propto (E/H_v)\ (K_c/H_v)^2 \qquad (10.7)$$

This agrees well with data from indentation tests for a variety of materials. Bifano[232] tested this relationship for ten materials ranging from fused silica to silicon carbide, Figure 10.14, finding overall good agreement between predicted and observed transition depths for grinding process. Two materials, alumina and toughened zirconia did not agree with the predictions. Bifano's predictive relationship is

$$d_c = 0.15\ (E/H_v)\ (K_c/H_v)^2 \qquad (10.8)$$

The major problem noted in applying this relationship for various materials is the difficulty in extrapolating the value for K_c. Data is usually from an indentation scale process (usually around 10 μm) whereas our interest is in the micro-machining (here micro-grinding) scale where depths of cut are more usually < 1 μm).

Other machining parameters affect the critical depth, d_c, as well, Figure 10.15. This data, from Hellmold[235] for machining of silicon, shows the dependency of d_c (called h_{cu} in the figure, in nanometers) on feedrate, cutting velocity, and abrasive grain size (d_g in the diagram).

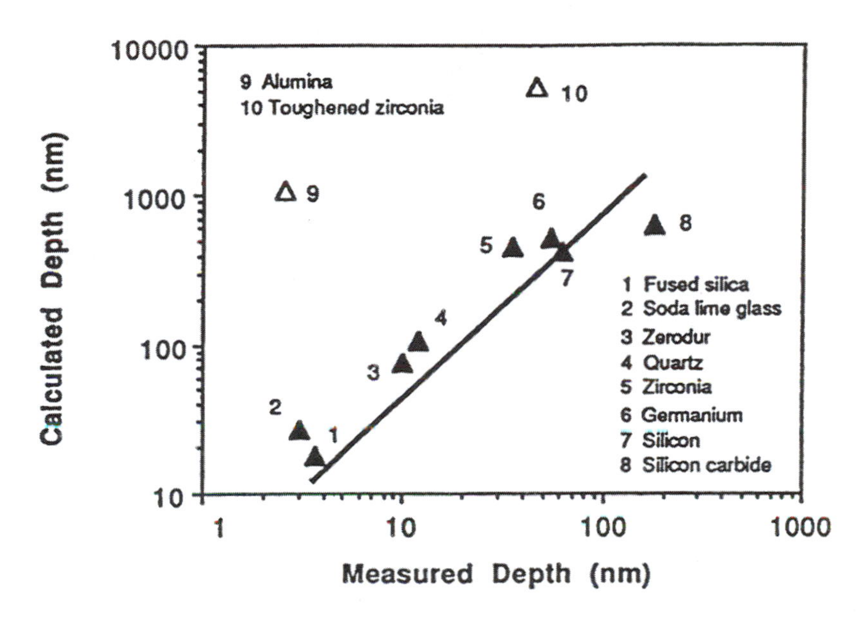

Figure 10.14. Measured grinding in feed rate corresponding to the ductile-brittle transition vs. estimated critical depth of cut for a variety of materials, from Bifano[232].

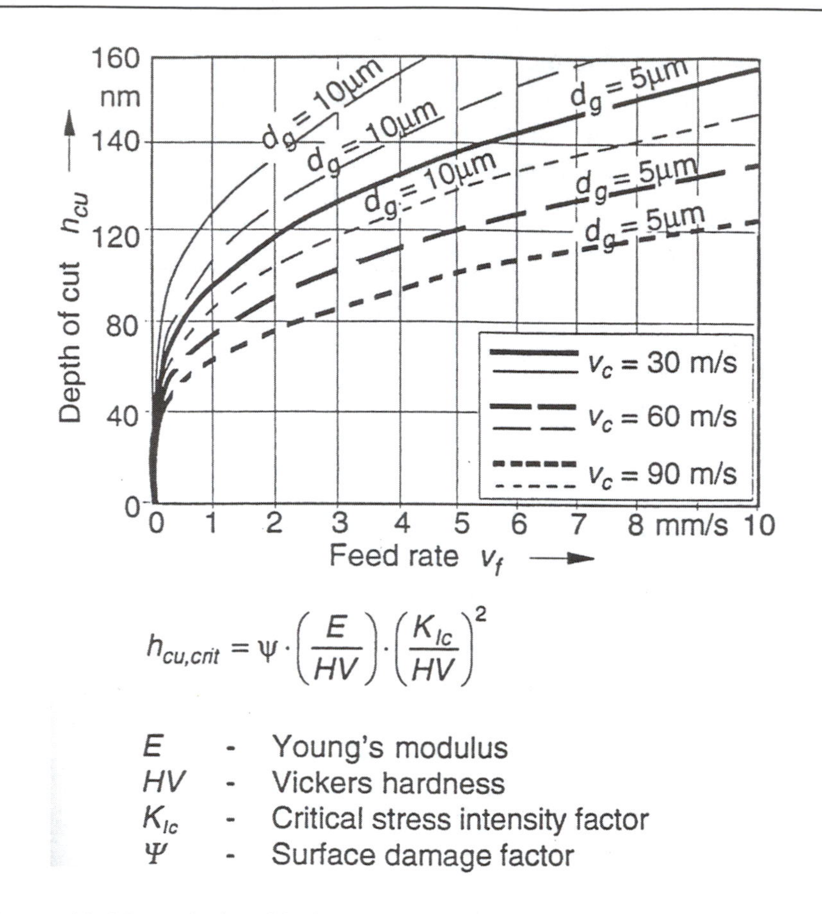

Figure 10.15. Relationship between cutting process parameters and depth of cut using Bifano's relationship, from Hellmold[235].

10.2.3 Effects of work material crystallography/directionality

The response of the material in micro-machining depends on the size and magnitude of the stress field induced in the field by the "tool" — whether abrasive grain or defined edge diamond. Early work on the behavior of crystallographic materials in polishing processes was done by Yoshikawa[236]. Figure 10.16, from Yoshikawa, divides the size of the stress fields into four distinct domains. The sizes, called "working units", range from 10^{-7} to 10^{0} mm and stretch from the atomic domain to grain boundary sized features. The domains are as follows:

> Domain I — material removal is on the order of a few atoms or molecules; since at this level removal could not be by pure mechanical action, chemical action enhanced by mechanically induced stress state and temperature most likely play an important role in the removal process.
>
> Domain II — at this level where no dislocations or cracks are present a crystal behaves mechanically ideally. Dislocations should arise prior to brittle fracture under processing at this level.
>
> Domain III — assuming dislocations present at this level, plastic deformation will occur first and, likely, encourage some crack nucleation in the zone of deformation.
>
> Domain IV — material removal mechanism is dominated by defects present, cracks in this case.

Yoshikawa's analysis is based on the assumption of an ideal single crystal with no pre-existing defects: point defects, dislocations, grain boundaries, second phase, microcracks or voids.

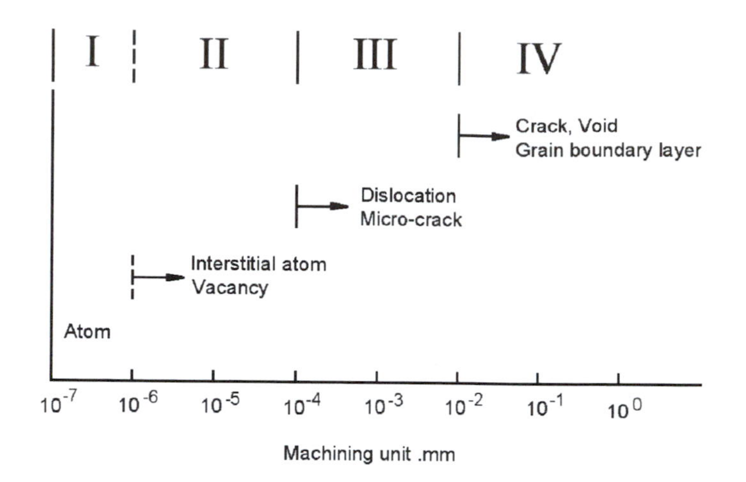

Figure 10.16. Factors affecting deformation and fracture of materials, from Yoshikawa[236].

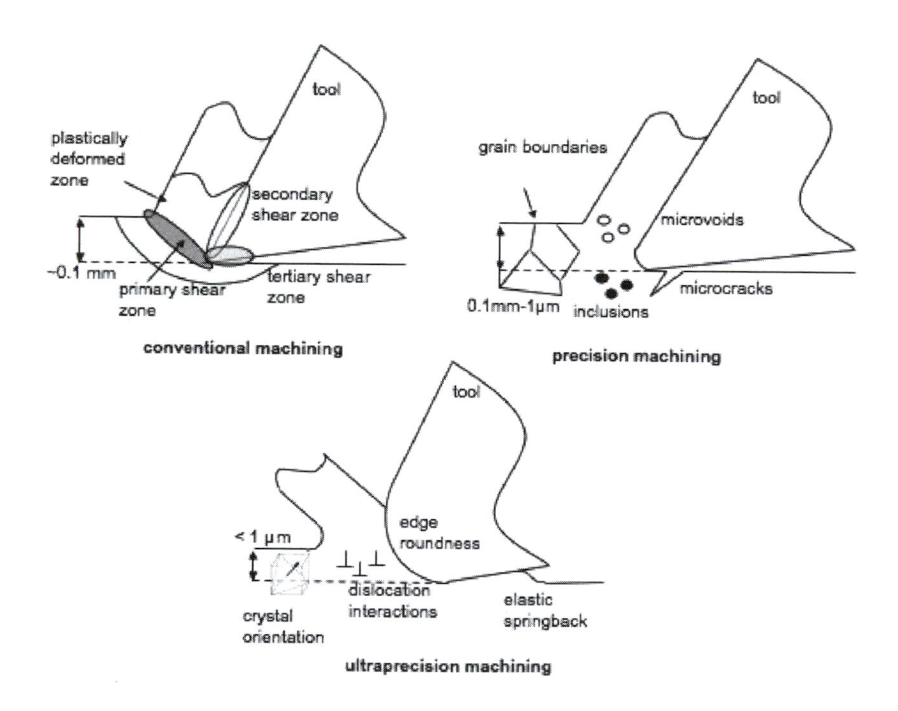

Figure 10.17. Size effect in machining showing different scales of process variables at different uncut chip thicknesses.

A rough sense of the scale of these domains is seen in Figure 10.17 showing a cartoon of a tool-workpiece engagement over several orders of magnitude difference in uncut chip thickness. An excellent example of Domain IV behavior, the influence of grain boundaries on cutting forces during plunge cutting of coarse grained OFHC copper, Figure 10.18, from Brinksmeier.[231] Here, the effect of the grain boundaries on the cutting force during tool motion under the conditions listed in the figure. A cutting force level fluctuation of some 0.5-2 N was observed. It is expected that other sensor information, acoustic emission, for example, would indicate the same grain boundary influence.

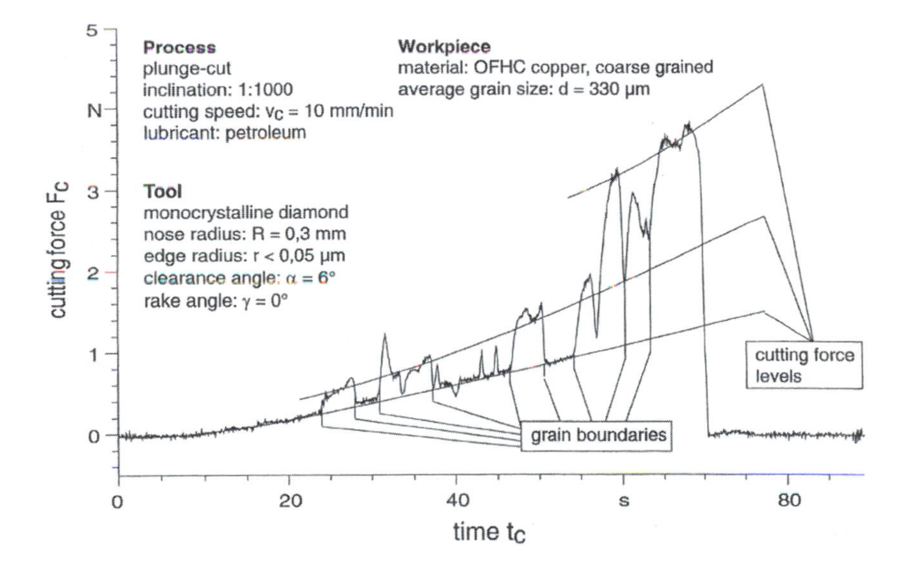

Figure 10.18. Cutting forces in plunge cutting coarse grained OFHC copper, from Brinksmeier[231].

In addition to the more "global" (if the term can be used to describe nanometer scale phenomena — perhaps general is better!) aspects defined by Yoshikawa, there is a predominate effect of crystallographic orientation when machining crystalline materials at these levels. Some crystallographic orientations of turned surfaces are illustrated in Figure 10.19 and, according to this, the potential surface damage for different cutting directions imposed on different orientations of material during diamond machining are shown in Figure 10.20.

In practice, this crystallographic orientation can be seen in chip formation (for ductile machining of single copper) in terms of the measured shear angle, Figure 10.21 (from Moriwaki[237]).

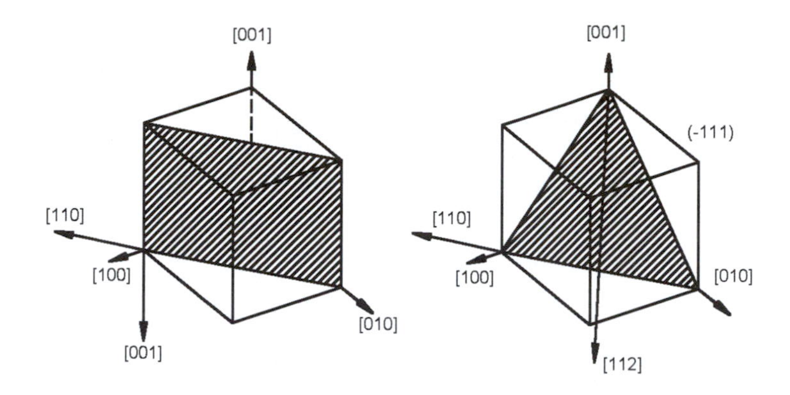

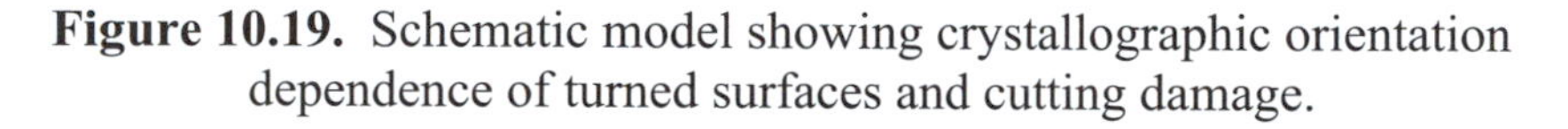
Figure 10.19. Schematic model showing crystallographic orientation dependence of turned surfaces and cutting damage.

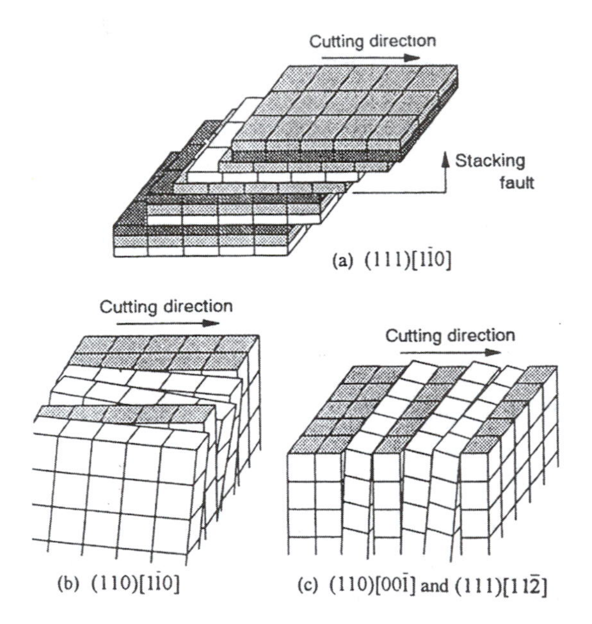

Figure 10.20. Schematic model showing crystallographic orientation dependence of surface damage.

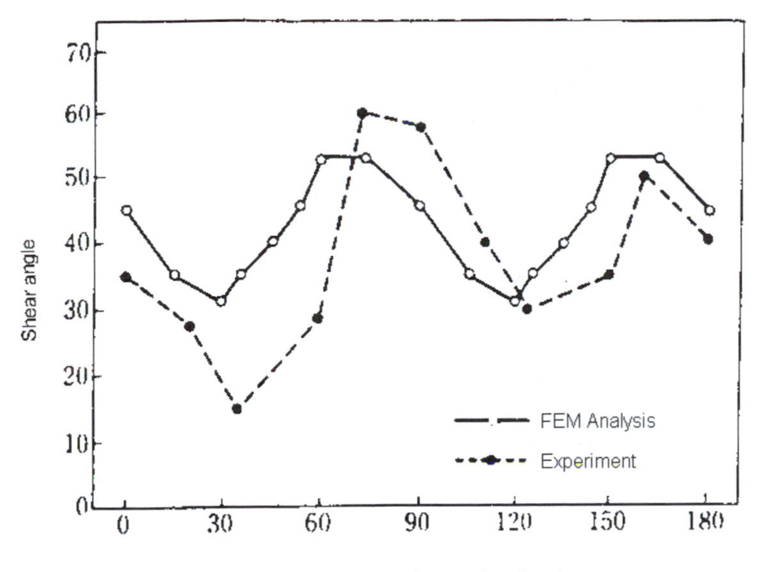

Figure 10.21. Comparison of shear angles obtained by FEM calculation and experiment as affected by crystal orientation angle vs. cutting direction, from Moriwaki[237].

This behavior is also evident in materials like copper, for example. Recall the plunge cutting experiments of Brinksmeier illustrated in Figures 10.12 and 10.13. Observing the surface of the cut groove as the groove intersects and passes through individual grains of OFHC copper (and, naturally, with different grain crystallographic orientation, the ease and difficulty of machining in preferred and non-preferred directions, respectively can be clearly seen, Figure 10.22. The surface condition changes accordingly.

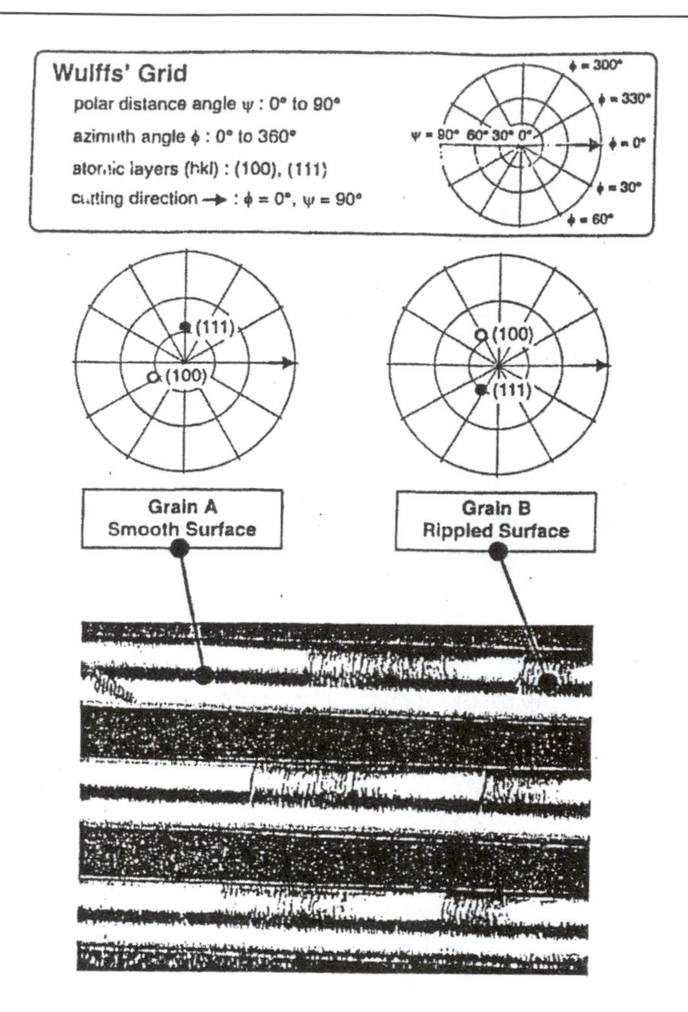

Figure 10.22. Grain orientation and surface texture of coarse-grained OFHC copper, from Brinksmeier[238].

10.3 Process operating conditions

Process operating conditions can be described in many ways depending on what is driving the decision making process. In Chapter 9, we presented two illustrations that showed the variation of process capabilities in terms of achievable tolerances and surface roughness.

In those cases, both surface roughness and tolerance were dependent most strongly on the uncut chip thickness discussed above as it is this feature that drives tool forces (and hence tolerances under conditions of perfect tool positioning, etc.) and surface roughness. We have also seen the ideal dependency of surface roughness on feedrate. Some dependency on material is noticed. Some processes, like resin casting or blow molding, don't have a comparable parameter to uncut chip thickness driving surface finish. Although, in many polymer processing operations, we see the influence of die surface on surface finish of the polymer part and the die is usually manufactured by a sequence of material removal processes.

The table below summarizes the material removal rate (MRR), machining time, and uncut chip thickness for a range of manufacturing processes. For the region II and III processes, calculation of uncut chip thickness is given in the following sections. The nomenclature for the equations in the table is listed below. The basic parameters, MRR, machining time, and mean undeformed chip thickness are derived from the geometry and kinematics of the process. Material removal rate is usually the swept path of the tool at some rate (feedrate, for example). Machining time is related to the path length of the tool over the workpiece (or summation of all path lengths) divided by the feed rate or rate of tool motion. The uncut chip thickness reflects the minute engagement of the tool with the workpiece and is dependent upon the tool geometry, work geometry and the subtleties of the tool motion (recall the discussion in Chapter 10 with illustrations of the varying chip thickness in turning and milling.) Often a maximum value is given as the chip thickness can vary. This is a fundamental parameter for setting up the process.

Table 10.1. Nomenclature

a_e = working engagement	a_p = back engagement
a_t = total depth of material to be removed	b_w = width of surface to be machined
d_m = diameter of machined surface	d_t = diameter of tool
d_w = diameter of the work surface	f = feed
l_t = length of tool	l_w = length of surface to be machined
n_r = frequency of reciprocation	n_t = rotational frequency of the tool
n_w = rotational frequency of the workpiece	N = number of teeth in the cutter
t_s = spark out time	v = cutting speed
v_f = feed speed	v_{trav} = traverse speed
κ = major cutting edge angle	

Table 10.2. Summary of equations for manufacturing processes.

Process	Machining time t_m	Material removal rate Z_w or Z_{wmax}	Mean undeformed chip thickness, a_{cav}
turning	$\frac{l_w}{fn_w}$	$\pi f a_p n_w (d_m + a_p)$	$f \sin\kappa_r$
boring	$\frac{l_w}{fn_w}$	$\pi f a_p n_w (d_m - a_p)$	$f \sin\kappa_r$
facing	$\frac{d_m}{2fn_w}$	$\pi f a_p n_w d_m$	$f \sin\kappa_r$
parting/cut-off	$\frac{d_m}{2fn_w}$	$\pi f a_p n_w d_m$	$f \sin\kappa_r$
shaping and planing	$\frac{b_w}{fn_r}$	$f a_p v$	$f \sin\kappa_r$
drilling	$\frac{l_w}{fn_t}$	$\frac{\pi f d_m^2 n_t}{4}$	$\frac{f}{2}\sin\kappa_r$
slab/horizontal milling	$\frac{l_w + \sqrt{a_e(d_t - a_e)}}{v_f}$	$a_e a_p v_f$	$\frac{v_f}{Nn_t}\sqrt{\frac{a_e}{d_t}}$
face milling	$\frac{l_w + d_t}{v_f}$	$a_e a_p v_f$	$\frac{v_f}{Nn_t}$
end milling	$\frac{l_w + 2\sqrt{a_e(d_t - a_e)}}{v_f}$	$a_e a_p v_f$	$\frac{v_f}{Nn_t}\sqrt{\frac{a_e}{d_t}(1+\frac{a_e}{d_t}}$
broaching	$\frac{l_t}{v}$	$f a_p v_f$ per engaged cutting edge	f
surface grinding-traverse	$\frac{b_w}{2fn_r}$	$f_{ap} v_{trav}$	see notes
surface grinding-plunge	$\frac{a_t}{2fn_r} + t_s$	$f_{ap} v_{trav}$	see notes
internal grinding-traverse	$\frac{a_t}{2fn_r} + t_s$	$\pi f d_m v_{trav}$	see notes
internal grinding-plunge	$\frac{a_t}{v_f} + t_s$	$\pi a_p d_m v_f$	see notes
cylindrical grind-ing- traverse	$\frac{a_t}{2fn_r} + t_s$	$\pi f d_w v_{trav}$	see notes
cylindrical grind-ing- plunge	$\frac{a_t}{v_f} + t_s$	$\pi a_p d_w v_f$	see notes

The sections that follow will discuss the particulars of diamond turning and milling, fixed abrasive processes, loose abrasive processes and nontraditional manufacturing processes. The diamond turning/milling and fixed abrasive (that is, grinding and lapping) will be most closely related to the table above. Other processes have little or no relationship to the parameters set out above except that in each case an equivalent "uncut chip thickness" (or minimum affected material) can be identified.

10.4 Precision mfg. processes-diamond turning/milling

10.4.1 Introduction

Diamond turning and milling has been applied for the fabrication of a wide variety of components and die surfaces. The mirror finish achieved with the single final pass of the tool under very low cutting forces has made this technology popular for finishing large optical elements (as the mirror segments of primary reflectors; Note that many reflectors are now made in segments individually position controlled within an array of segments making up the whole mirror due to difficulties of distortion of one-piece reflectors). A wide range of materials are suitable for machining with diamond with the usual exception of ferrous (i.e. carbon-based) alloys although some advances have been made in machining these as well. The list below gives some of the commonly machined materials; alloys of these are also commonly machined:

Metals	Polymers	Crystals
aluminum	crylic	germanium
copper	nylon	zinc selenide
brass	polycarbonate	zinc sulfide
bronze	polystrene	lithium niobate
tin	polysulfate	cesium iodide
lead	acetyl	potassium dihydrogen
electroless nickel	fluoroplastic	phosphate
zinc		silicon
platinum		potassium bromide
silver		
gold		
magnesium		

Applications include fly-cutting of Fresnel lenses as well plastic injection mold components for production of precision optics. Many innovative products require key components with complex shapes and precise surfaces (e.g. diffractive optics, intra-ocular lenses, new auto light systems, HDTV projection screens, sine wave mirrors or facetted mirrors. The surfaces of optical components fall into two basic classes: surfaces with continuously varying slope and structured surfaces with discontinuous features, e.g. steps and profiles, superimposed onto planar, aspheric or asymmetric envelope surfaces. The shape of the tool can be adjusted to accommodate the specific geometry of the lens or surface. In addition, mounting of the workpiece on the machine is often challenging, as kinematic principles must be maintained even with complex geometries. Some examples of these challenges are given in Chapter 11.

The fundamental elements of single point diamond (SPD) machining (SPDT for turning) are machine design, tool design and alignment, and control and software. Machine design is for a 3 to 5 axis machine structure allowing flexibility in motion without loss of stiffness or accuracy. Tool design centers on novel diamond single point tools for milling and turning and alignment, including fixturing, of the tool and workpiece within several degrees of freedom. Control and software capabilities require multi-axes operation with high resolution and rapid processing of large amounts of data, often representing complex surface geometries. As expected, much of the basic understanding of the structural, accuracy, fixturing and metrology constraints have been discussed earlier in the text. We will review here some of the particular issues related to diamond milling and turning.

10.4.2 Machine tool design

Machine tools for diamond machining are usually characterized by several distinct features:

- air bearing spindles and slides with massive, generally granite, machine bases
- vibration isolation mounts
- control systems with laser alignment systems, in process interferometers for position feedback, piezoelectric transducer for fine positioning and servo positioning motors
- motors and drive belts selected to minimize vibration generation
- balanced mounting geometries; workpieces balanced to minimize cyclic geometry errors; work fixtures designed to hold the workpiece in a "stress-free" attitude to avoid residual stresses after removal from the machine
- operation in a stable, temperature and humidity controlled environment

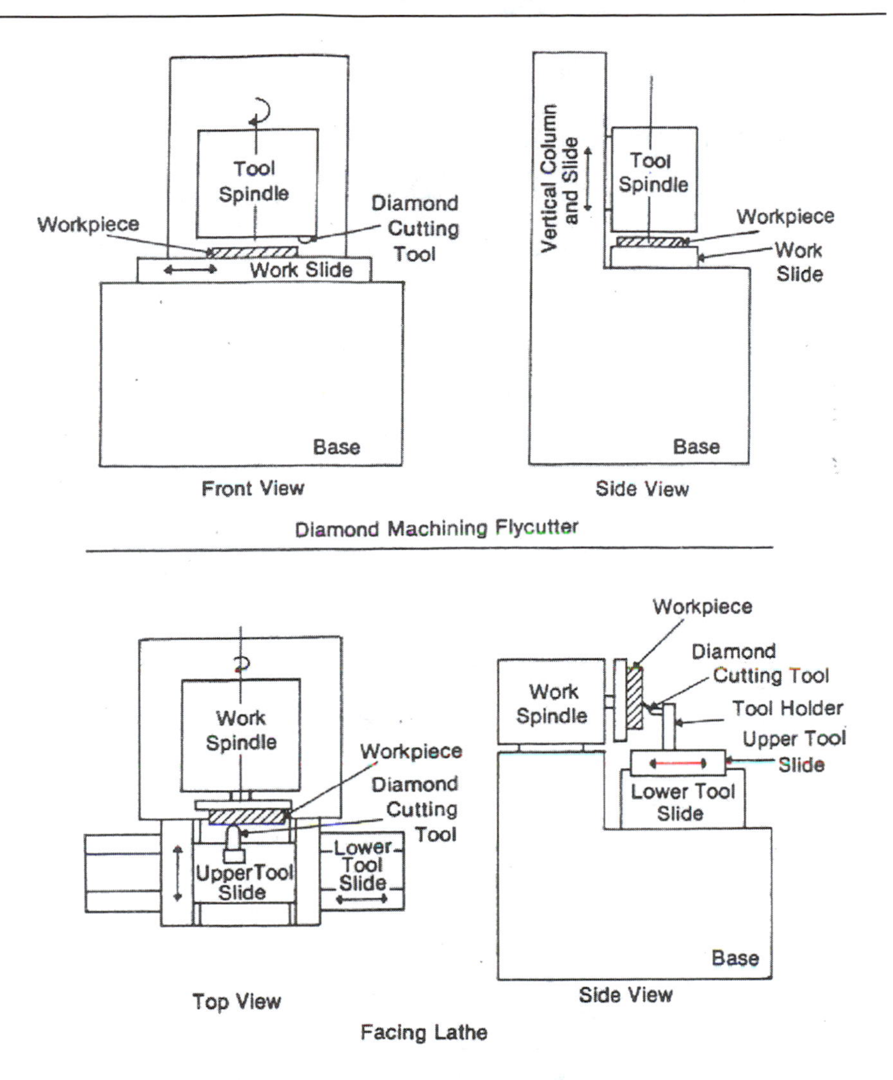

Figure 10.23. Typical fly-cutting diamond machine configurations.

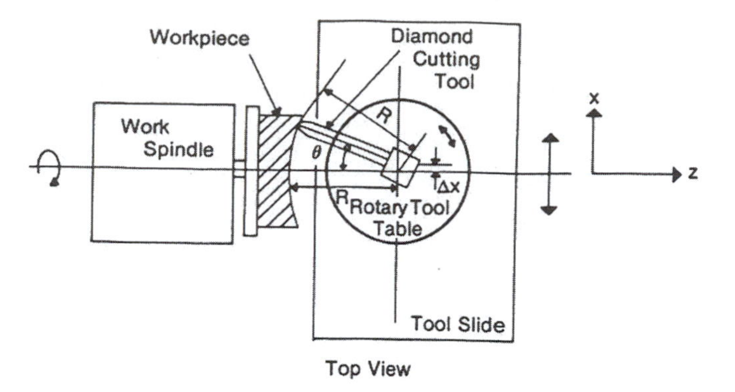

x-θ Contouring Machine

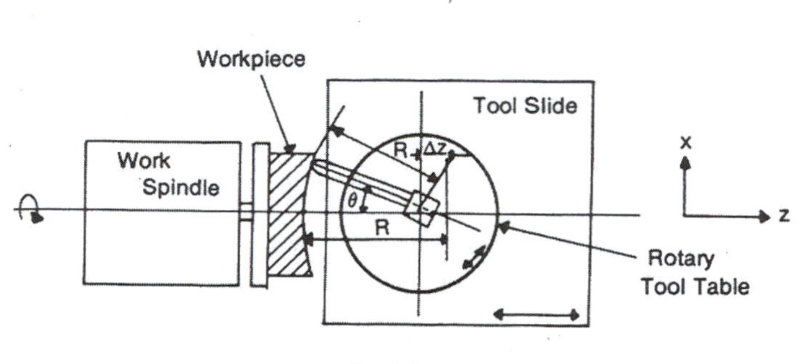

z-θ Contouring Machine

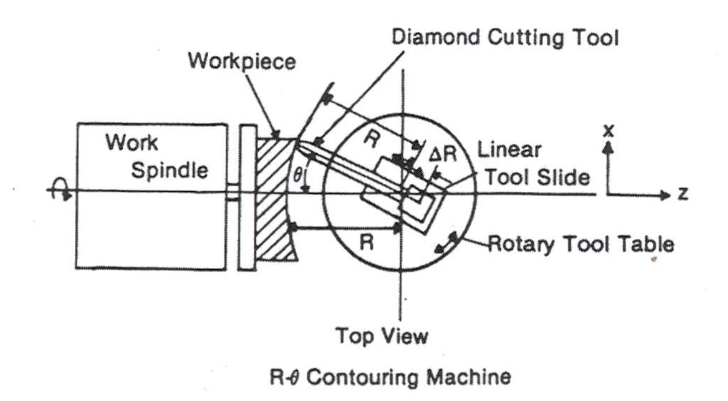

Figure 10.24. Rotating work, stationary tool diamond machine configurations.

Typical diamond machine configurations are shown in Figures 10.23 and 10.24. These configurations are conventional in that either the tool, Figure 10.23 upper image, or workpiece, lower image of Figure 10.23 and Figure 10.24, rotate. The upper image in Figure 10.26 is one configuration of milling (fly cutting). Recently, the configuration shown in Figure 10.25, fly cutting with the tool rotating in a spindle on the machine table, has been applied to the creation of more complex shapes than achievable with the conventional designs. Figure 10.26 illustrates tool motions with this fly cutting configuration in 4 axes and 5 axes milling of complex lens geometries. The tool design will be discussed in the next section.

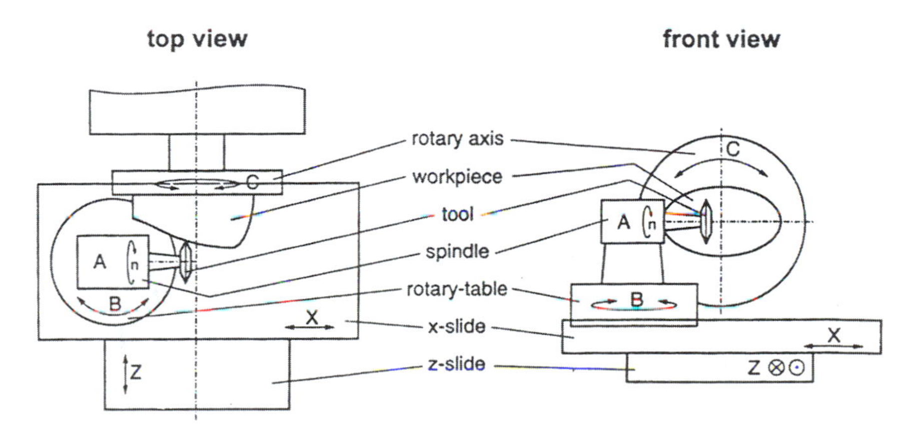

Figure 10.25. Flycutting machine configuration with rotating tool and indexable workpiece, from Brinksmeier[231].

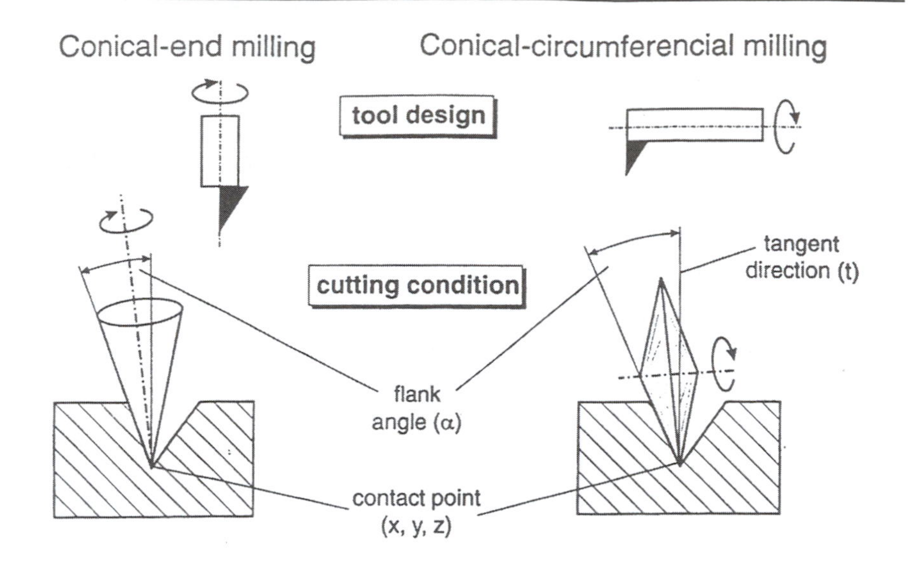

Figure 10.26. Tool-workpiece configurations for conical-circumferential milling, from Brinksmeier[231].

One of the more exciting realizations of these machine design concepts is the machine introduced by Fanuc in 1998 (as the ROBOnano U*i*, now referred to as α-OiB), Figure 10.27. The machine axis configuration is illustrated in Figure 10.28 and the machine's performance specifications, for reference, are shown in Figure 10.29. The machine was designed to machine small super precision parts and/or molds for injection molding. A typical example of an application is the production of encoder disks for rotational position control of servo motors. The machine has grooved disks radially with 14 million individual scribes to create an encoder that, with quadrature software, can indicate some 64 million pulses per revolution.

A machine tool introduced recently by Sodick Company of Japan (http://www.sodick.jp/) offers another example of ultra-precision machine tool design. The UltraNANO 100 machine has a unique design with ceramic aero-static guideways on each linear axes (with dual linear motors on each linear axis) and aero-static rotational axis bearings, Figure 10.30. High accuracy linear scales and

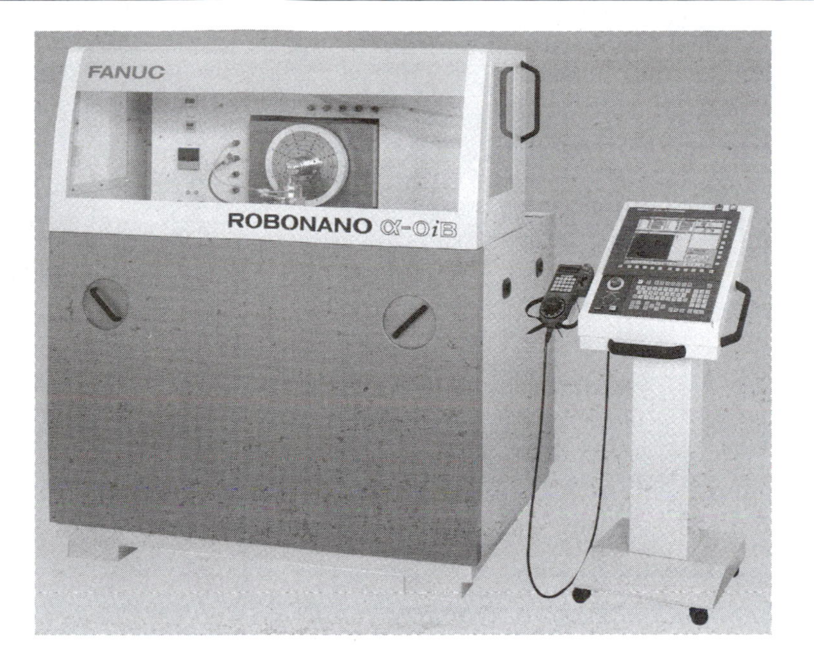

Figure 10.27. FANUC RoboNANO α-OiB super precision micro machining center, from FANUC[239].

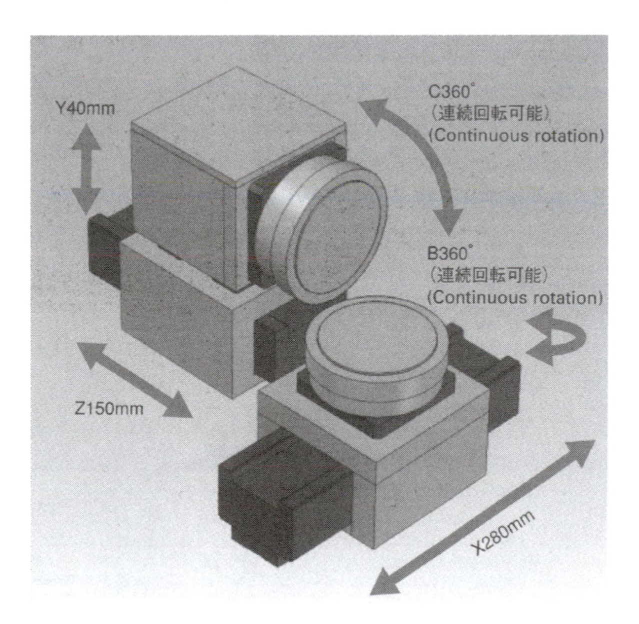

Figure 10.28. Schematic of machine axes configuration, from FANUC[239].

Standard Mechanical Specifications

	Item	Specification
Stroke	X axis (Horizontal linear)	280mm
	Z axis (Horizontal linear)	150mm
	Y axis (Vertical linear)	40mm
	B axis (Horizontal rotation)	360° (Continuous rotation)
	C axis (Vertical rotation)	360° (Continuous rotation)
Command resolution	X、Z、Y axes	1nm
	B、C axes	0.00001°
Work-table area	B、C axes	ϕ210mm
Maximum feed speed	X、Z axes	500mm/min
	Y axis	50mm/min
	B、C axes	250min^{-1}
Straightness	X axis	0.2μm/280mm
	Z axis	0.2μm/150mm
	Y axis	0.2μm/40mm
Run out	B、C axes	0.05μm

Figure 10.29. Standard mechanical specifications for the Robonano α-OiB, from FANUC[239].

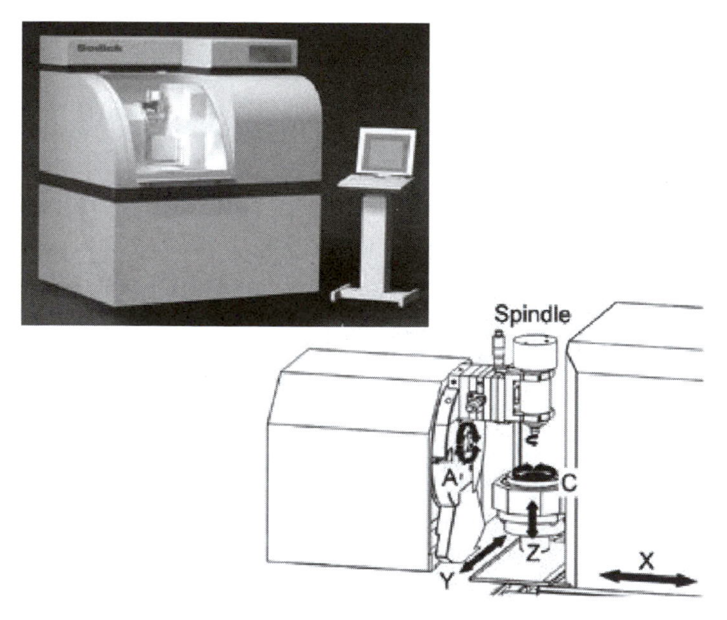

Figure 10.30a. Sodick UltraNANO 100 machine tool, from Sodick[240].

Guide system	X, Y, Z axis	Ceramic-made aero-static guide
	A, C&R axis	Aero-static bearing
Travel	X, Y, Z axis	100 X 100 X 55mm
	A axis	180°
	C & R axis	360° (Endless)
Position detection	X, Y, Z axis	Linear scale 0.07 nm
	A axis	Laser encorder 45,360,000 pulse/rotation
	C axis	Indexing 15,120,000 pulse/rotation
	R axis	Rotation 1,210,000 pulse/rotation
Speed	X, Y, Z axis	Rapid, cutting feed : 3,000 mm/min
	A axis	10 min^{-1}
	C & R axis	Indexing axis (C-axis) 25 min^{-1}
		Rotation axis (R-axis) 1,000 min^{-1}
Spindle		Air turbine, Aero-static bearing
		50,000 min^{-1}

Figure 10.30b. Sodick UltraNANO 100 machine mechanical specifications, from Sodick[240].

encoders allow positioning accuracy detection to 0.07 nm for a 1 nm step resolution. The machine is designed with very close control of temperature gradients to insure thermal stability (temperature variation in critical components is held to +/- 0.1 °C at the work location), Sriyotha[241].

10.4.3 Tool design and alignment

Diamond tools are generally composed of the tool tip element (diamond, polycrystalline diamond, CVD or other synthetic diamond) and the tool holder/shank which secures the tool tip and provides a basis for orienting it to the workpiece relative to the machine. A typical single point diamond tool is shown in Figure 10.31. This geometry is commonly used in facing operations. The tool can either have a radius, R in Figure 10.31 (which can be as large as a kilometer – i.e. almost straight), or the tool can be straight, Figure 10.32.

The figure also shows the profile of the tool tip for the two types measured from a groove made in a soft metal with the tools. With the increased use of "fly-cutting" tool configurations, especially for four and five axis machine tools as discussed above, new tool configurations have been developed. Figure 10.33, from Brinksmeier[231], shows both conical end milling and conical circumferential tool configurations and the accompanying tool-work intersections.

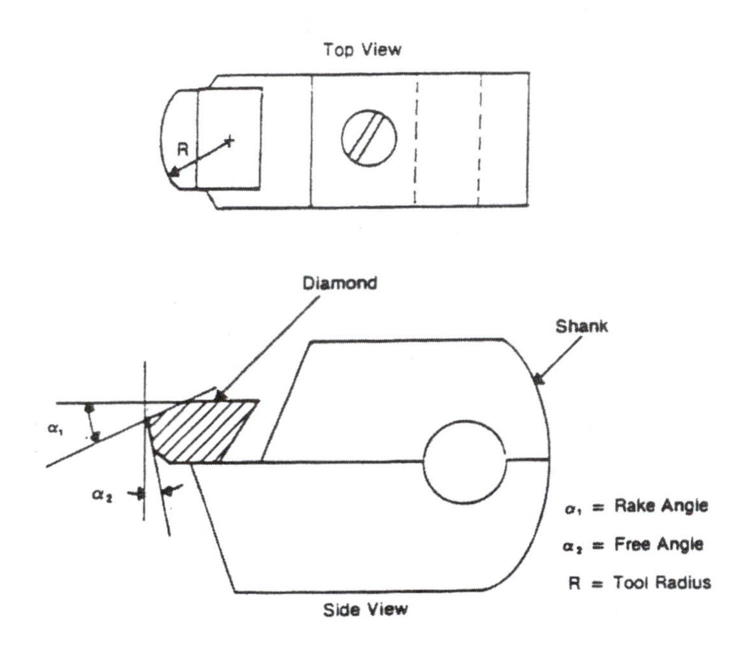

Figure 10.31. Top and side views of a typical single-point diamond tool geometry.

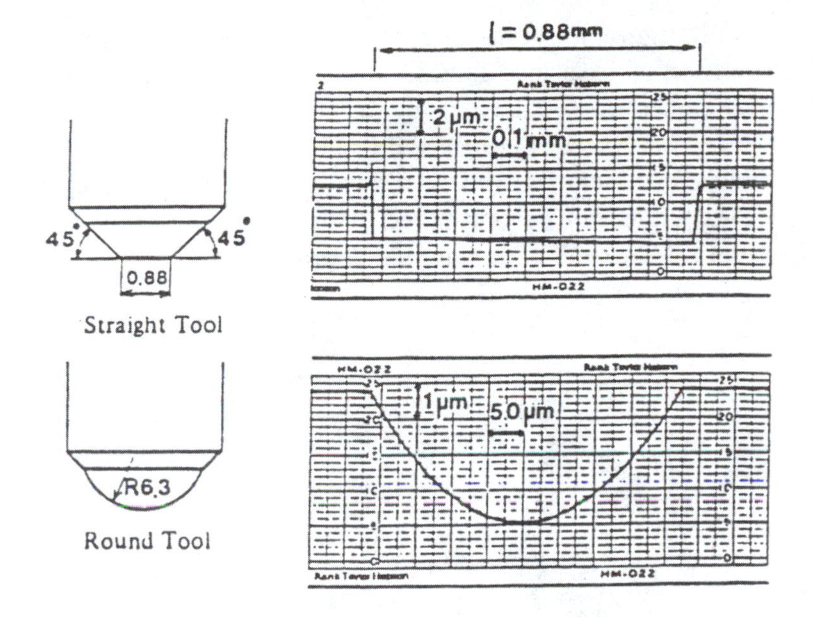

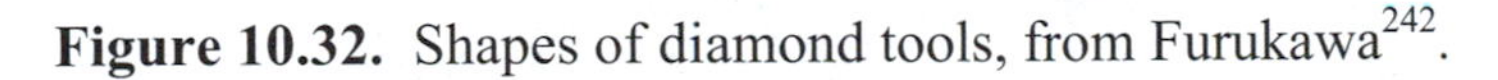

Figure 10.32. Shapes of diamond tools, from Furukawa[242].

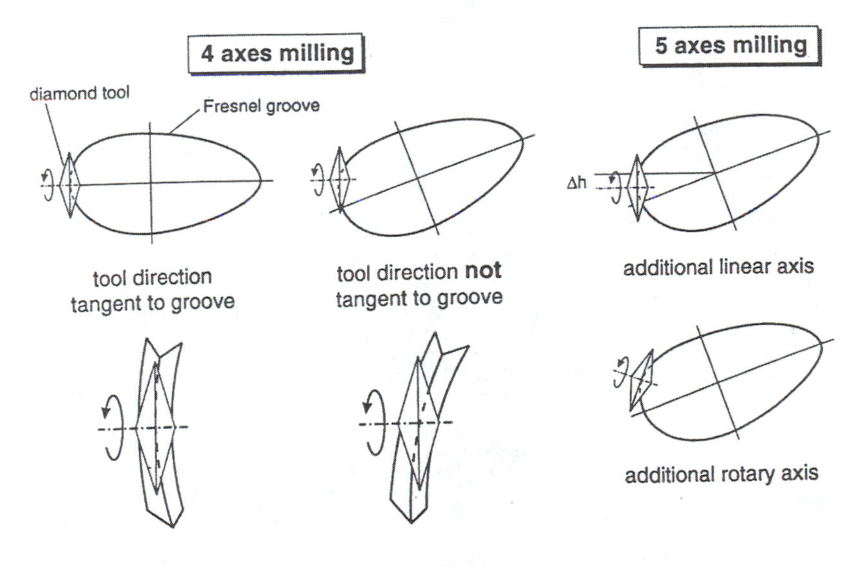

Figure 10.33. Conical milling tool designs and applications, from Brinksmeier[231].

Tool alignment with respect to the workpiece is important to guarantee acceptable surface conditions. The "tool setting angle," which can be set as neutral, positive, or negative, can affect chip formation and surface roughness, Figure 10.34. This plays an important role for surface finish and surface defects, burrs for example, Figure 10.35. This angle can be optimized with respect to surface roughness, Figure 10.36. Notice that the angular difference is minimal (approximately -1° to +1°).

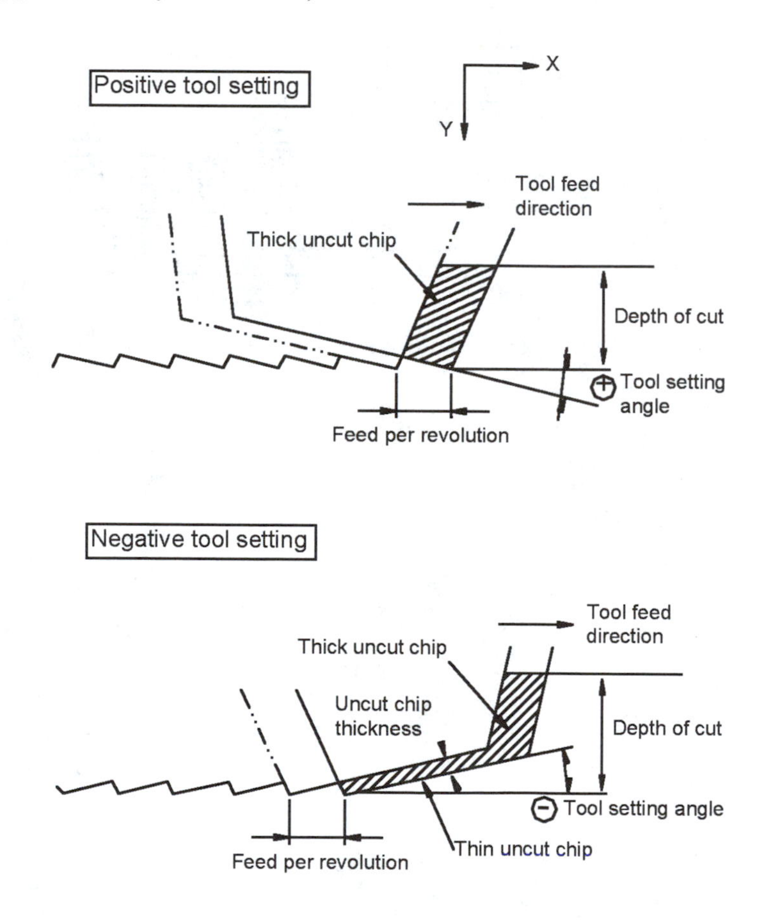

Figure 10.34. Definitions of tool setting angle of a straight tool, from Furukawa[242].

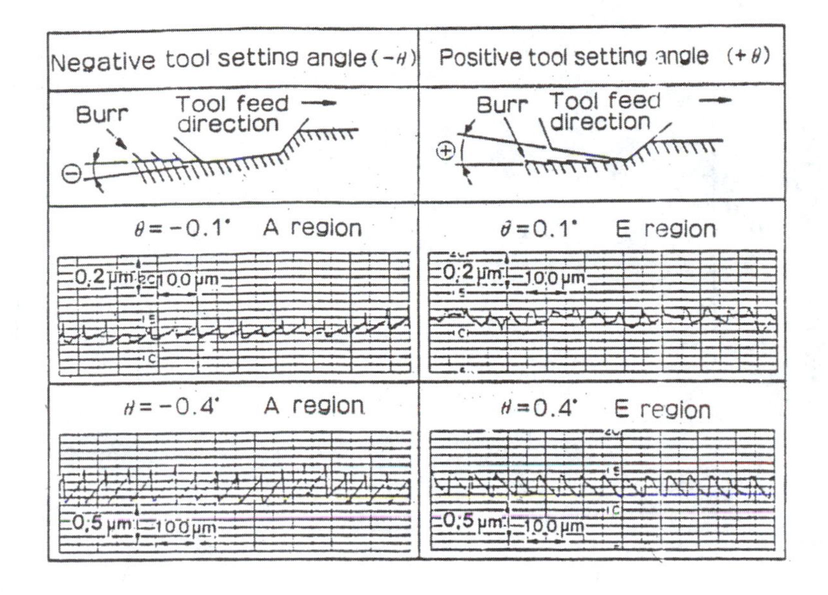

Figure 10.35. Effect of tool setting angles on surface texture and burrs, from Furukawa[242].

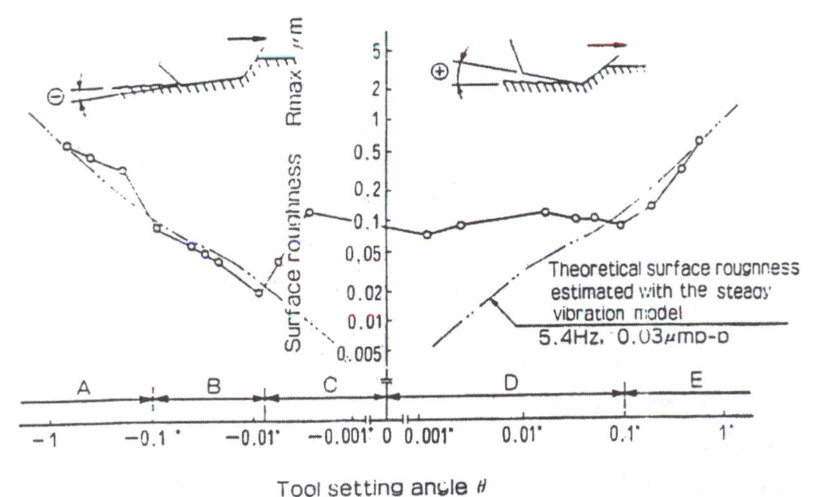

Figure 10.36. Effect of tool setting angle on machined surface roughness, from Furukawa[242].

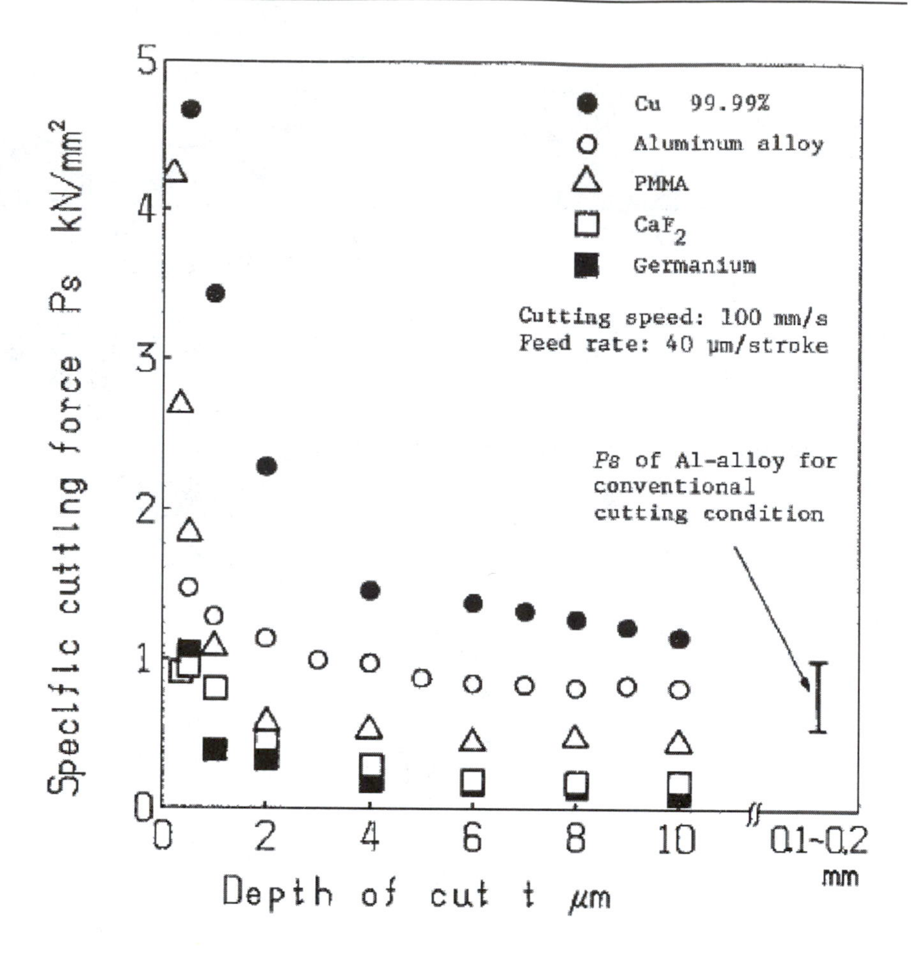

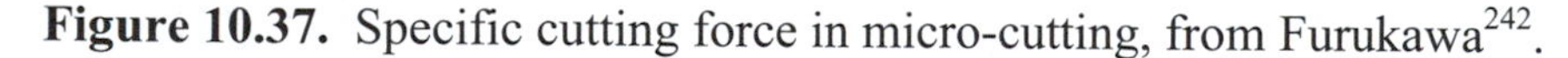

Figure 10.37. Specific cutting force in micro-cutting, from Furukawa[242].

10.4.4 Chip formation and process mechanics

Diamond machining process conditions generally fall into the Regime II or III previously defined with respect to the uncut chip thickness, a_c. This means that, from the point of view of specific cutting energy, we can expect an increase at lower chip thicknesses, Figure 10.37, for a normalized depth of cut. In fact, the nature of the tool-work engagement will also affect the specific energy. Figure 10.38 shows the specific energy as a function of the "aspect ratio" of engagement of a diamond tool with an Al-alloy. Here the aspect ratio is defined as the ratio of the length of contact of the tool edge with the workpiece to the maximum uncut chip thickness. This tool-work contact geometry is quite common in diamond machining operations.

Because of tool-work deflections and tool edge geometry effects the actual and theoretical uncut chip thicknesses differ. Figure 10.36 illustrates the regions of stable chip production and unstable removal and the relationship between the nominal and effective

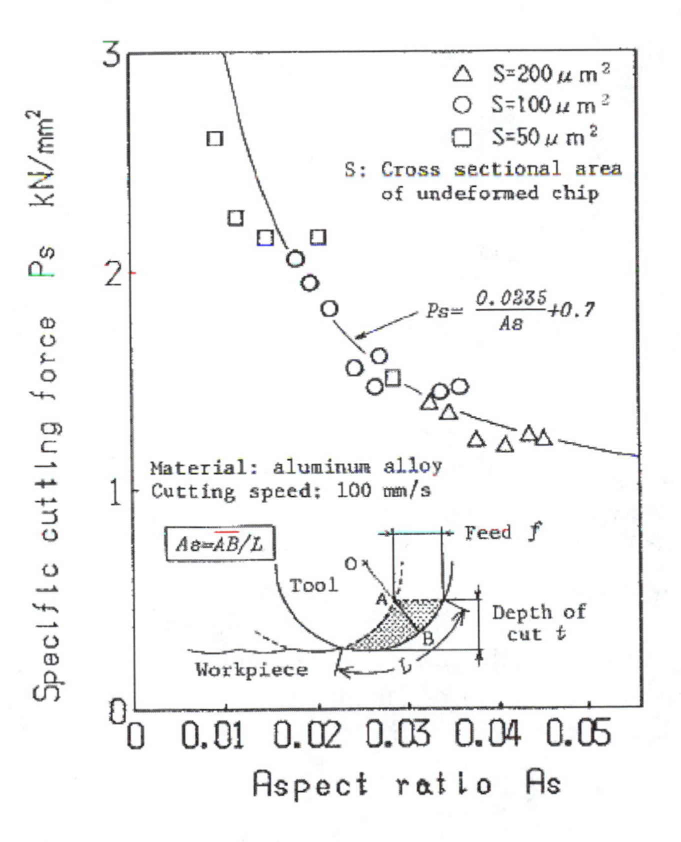

Figure 10.38. Relationship between aspect ratio and specific cutting force, from Furukawa[242].

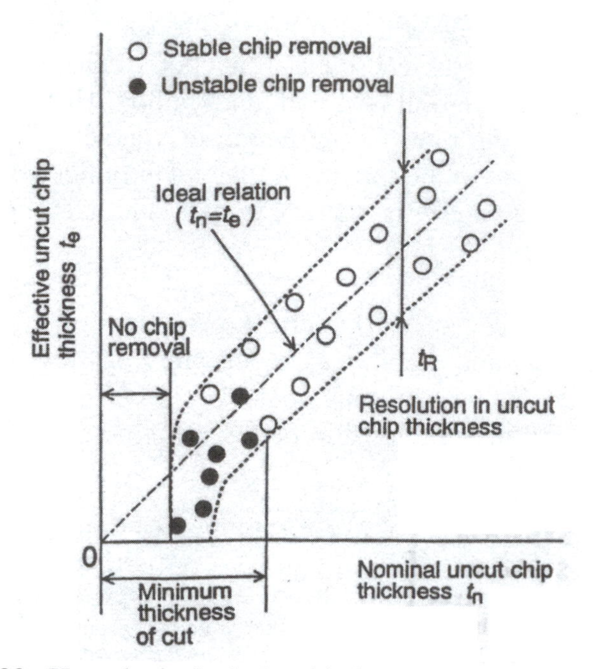

Figure 10.39. Hypothetical relationship between nominal and effective uncut chip thickness, from Ikawa[243].

uncut chip thickness[243]. In practice, chip formation can be extended to the creation of chips with nanometer-level uncut thicknesses, Figure 10.40, from Ikawa[243]. The cutting ratio, chip thickness/nominal chip thickness, varies as the uncut chip thickness reduces as predicted in Figure 10.39. The range of chip thicknesses and cutting ratios for the chips in Figure 10.40 is given in Figure 10.41. The chip images in Figure 10.40 show a peculiar "buckled" appearance common of chips at those thicknesses. Ikawa and co-authors proposed a mechanism explaining this chip buckling based on conventional buckling theory and the role of friction along the chip-tool interface, Figure 10.42. The ratio of length of unbuckled portion of chip to thickness, h/l, is related, Figure 10.43.

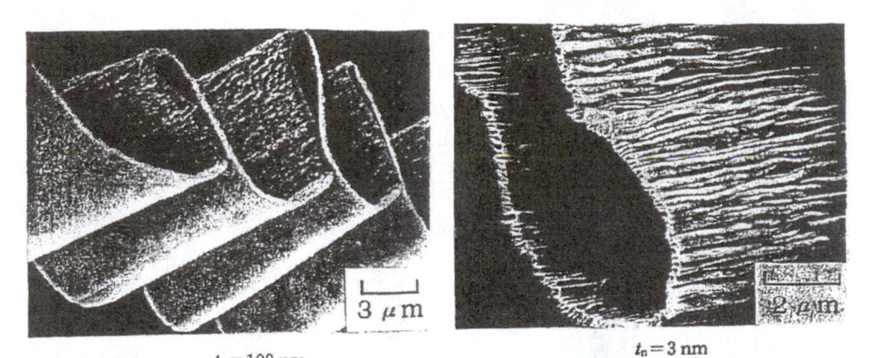

$t_n = 100$ nm

$t_n = 3$ nm

$t_n = 30$ nm

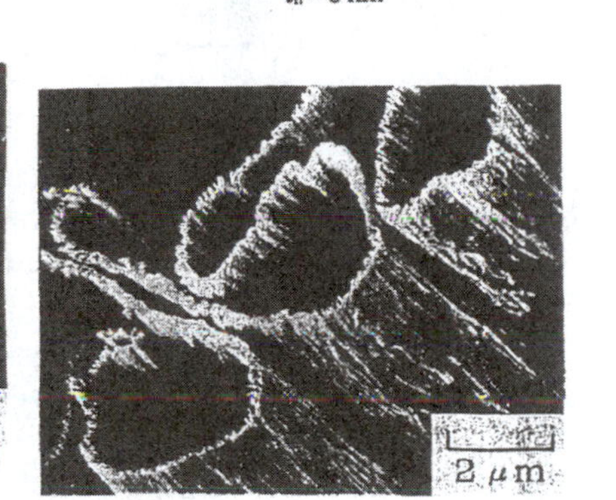

$t_n = 1$ nm

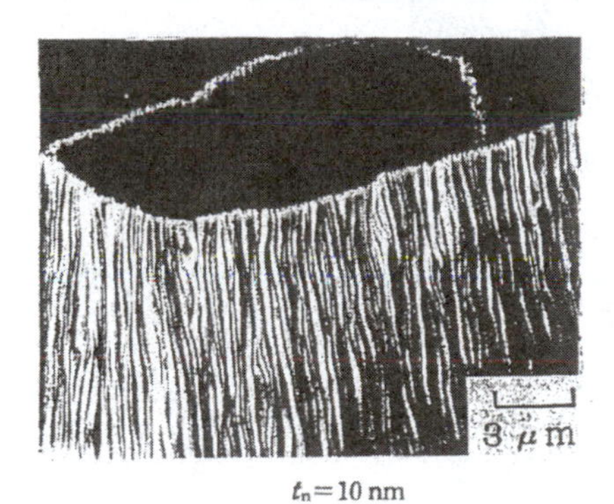

$t_n = 10$ nm

Figure 10.40. SEM micrographs of chips removed at nanometric uncut chip thicknesses, from Ikawa[243].

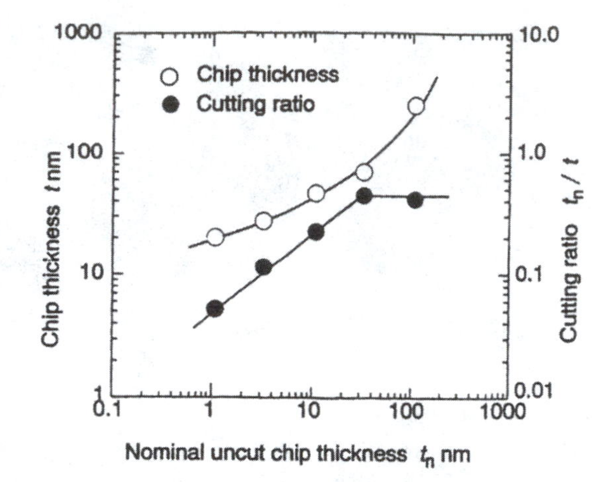

Figure 10.41. Change in chip thickness and cutting ratio with nominal uncut chip thickness, from Ikawa et al[243].

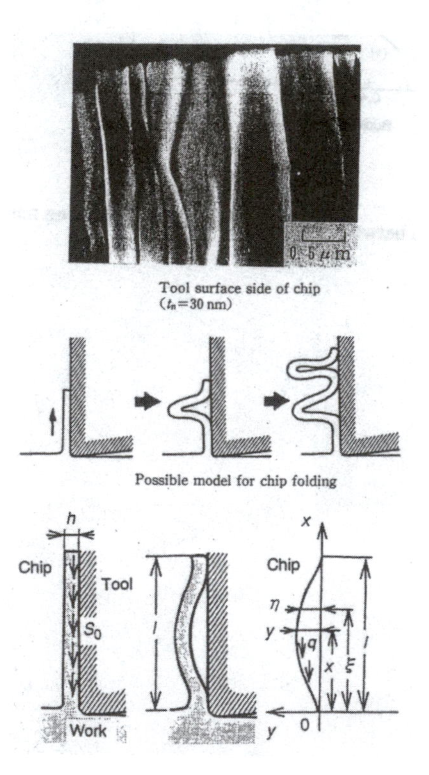

Figure 10.42. Analytical model for chip buckling, from Ikawa et al[243].

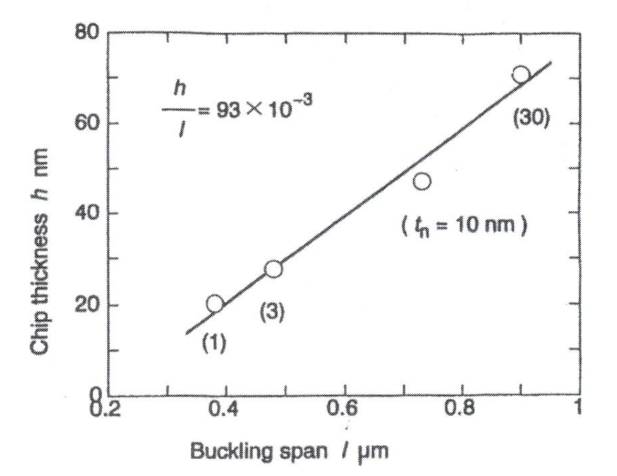

Figure 10.43. Relationship between buckling span and thickness on nanometric chips, from Ikawa et al[243].

Materials are machined with single point diamond turning with conventional responses in terms of surface finish, for example. Surface roughness is proportional to feedrate in machining silicon under the conditions shown in Figure 10.44. Machining a variety of materials often requires other accommodations be made. For example, our previous discussions of ductile-brittle transition behavior in materials, that is ductile at low cutting depths and brittle at higher, has encouraged schemes for insuring maximum removal rate while maintaining reasonable surface roughness and minimal subsurface damage, Figure 10.45, from Blake[244]. Here we see how the chip-tool contact in SPDT allows the final tool contact with the surface to occur at a depth of cut (uncut chip thickness) below the ductile-brittle transition insuring no surface damage in machining of germanium and silicon. At the same time, the heavier removal, with accompanying brittle damage, occurs in the bulk of the material being removed so not affecting the work surface. However, as observed earlier, the crystallographic orientation of the workpiece will affect the "machinability" of the work. In the application of this technique by Yan[245], the variation of the critical depth of cut as a function of orientation of the work was documented, Figure 10.46 and accompanying variation in surface roughness, Figure 10.47. In other cases, for which the workpiece is fragile or easily deformed under cutting loads, the use of a low frequency (20 kHz) oscillation of the cutting tool has shown promise in reducing the cutting forces in diamond turning of OFHC copper, Figure 10.48[231].

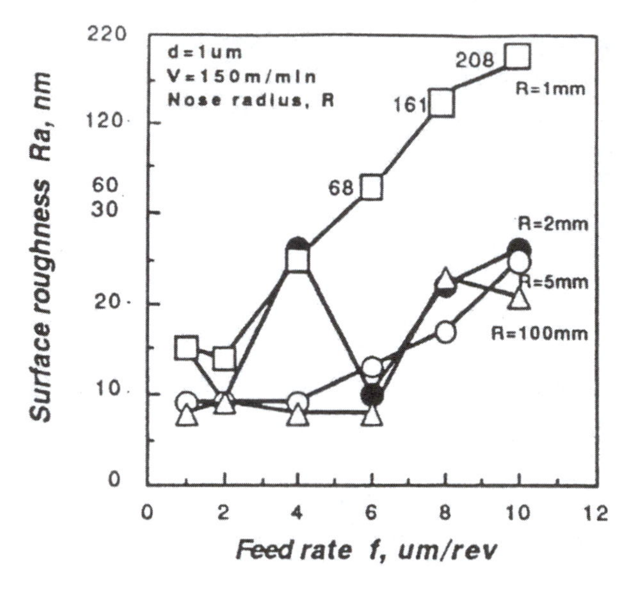

Cutting Conditions and Tool Dimensions in Single Point Diamond Cutting of Silicon Material

Feed Rate, f	1,2,4,6,8,10 mm/rev
Tool Nose Radius, R	1,2,5,100 mm
Shank Dimensions	8 X 7.5 X 32 mm
Cutting Speed, V	150 m/min
Depth Of Cut, d	1mm
Rake Angle	0°
Relief Angle	6°
Coolant	Dry

Figure 10.44. Surface roughness relationship with federate and tool nose radius of silicon material with cutting conditions shown, from Blake[244].

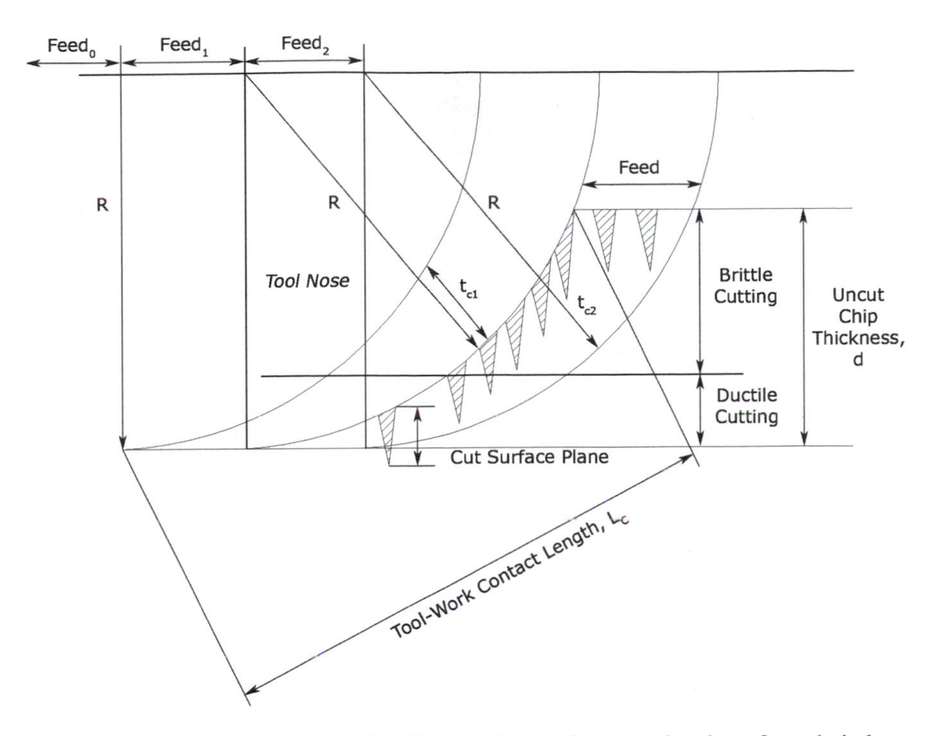

Figure 10.45. Proposed ductile-brittle cutting mechanism for a brittle material applied in the SPDT process, after Nakajima[246].

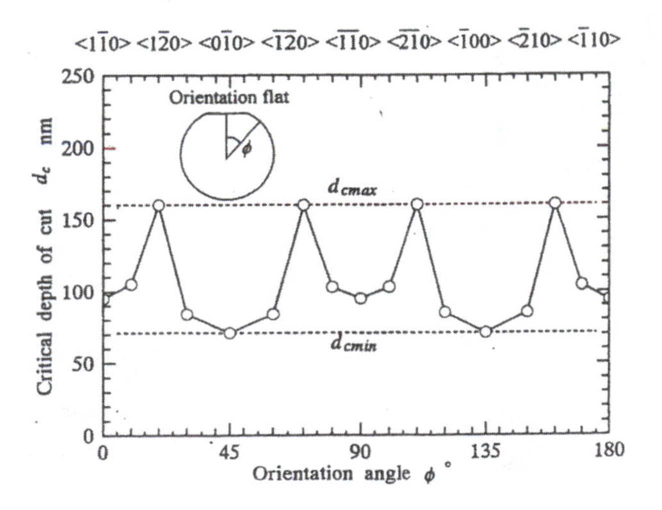

Figure 10.46. Variation of d_c with crystalline orientation, from Yan[245].

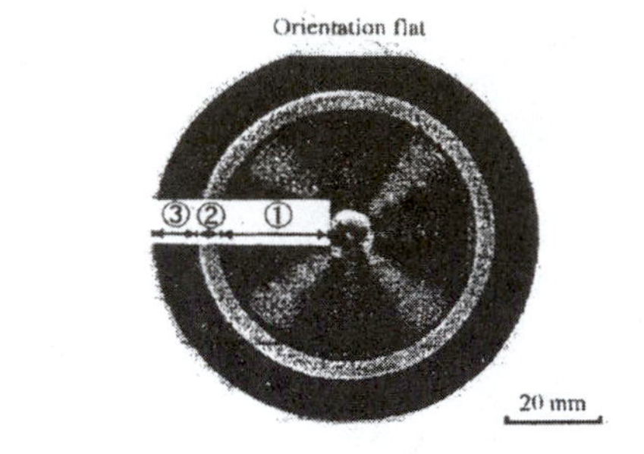

(a) Variation of surface feature with h

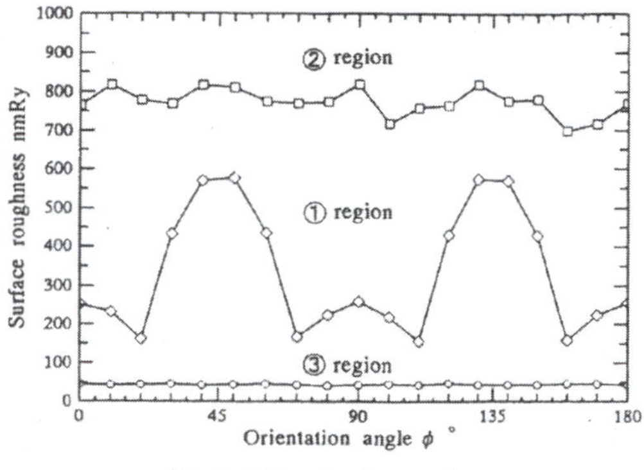

(b) Variation of surface roughness

Figure 10.47. Variation of surface feature and surface roughness with crystalline orientation, from Yan[245].

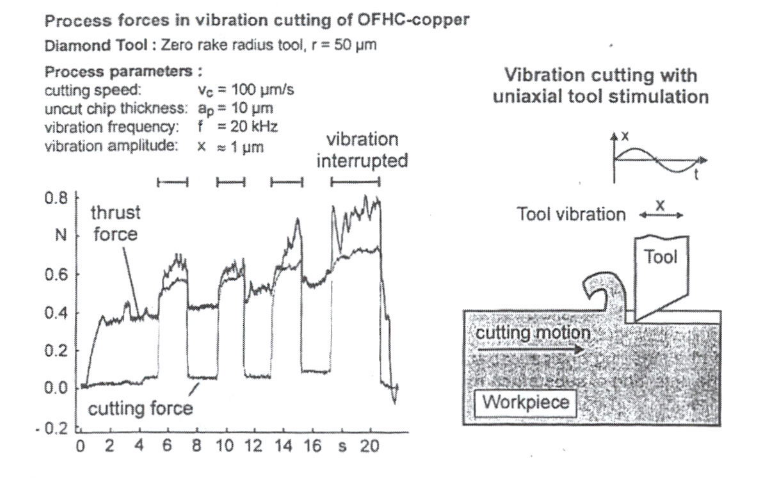

Figure 10.48. Cutting principle and process forces in linear ultrasonic vibration cutting, from Brinksmeier[231].

10.5 Abrasive processes – fixed and loose

10.5.1 Fixed abrasive processes

We can borrow terminology from tribology and generally characterize the interactions between surfaces as either "two-body" or "three-body". Fixed abrasive processes are an example of two body interaction as the grit is fixed in the grinding wheel and contacts the work surface directly. Grinding with fixed abrasives (usually abrasives bonded into a wheel or other carrying mechanism, paper or cone, for example) is commonly applied in a wide variety of manufacturing applications. Several excellent references are available for details of fixed abrasive processes in conventional manufacturing including Malkin[247] and Tönshoff et al[228]. Tonshoff's contribution reviews models proposed for various aspects of grinding (power consumption, chip size, surface roughness, etc.) Our objective here is to review grinding as it applies to precision surface or feature generation. The variables influencing abrasive machining are reviewed in Figure 10.49[248]. The grinding wheel holds the abrasive in place and is the controlling structure for imparting form to the workpiece. The wheel is usually "trued" or "dressed", called conditioning, on the grinding machine before production. The conditioning process has a decisive influence on the final work shape and efficiency of material removal.

10.5.1.1 Material Removal Mechanisms

The size of chips removed in grinding is much smaller than in cutting providing better surface finish and machining accuracy. The order of chip thickness in grinding is much less than 0.1 mm compared to larger than 0.1 mm in conventional machining. The mechanism of material removal with abrasive cutting edges is basically the same as with kinematically defined cutting edges discussed earlier except that other mechanisms are often involved due to the unique interaction between the abrasive grain and the work surface, Figure 10.50.

In metal machining material is removed primarily through shear deformation and this is basically the same for grinding. Grinding other materials, ceramics, for example, employ additional material removal mechanisms such as brittle mode processes. Interactions in the grinding zone are illustrated in Figure 10.50, from Subramanian[248] and show the three possible types of interaction of abrasive with work — sliding (elastic interaction), plowing (material displacement without removal) and cutting (material removal). At higher magnification, Figure 10.51a, from Subramanian[248], the grits appear as indenters and the classical chip formation appearance seems reasonable. Figure 10.51b shows the variation between ductile and brittle contributions to material removal as a function of chip thickness (recall Bifano's critical depth of cut?) and increasing force per abrasive grain and abrasive grain radius.

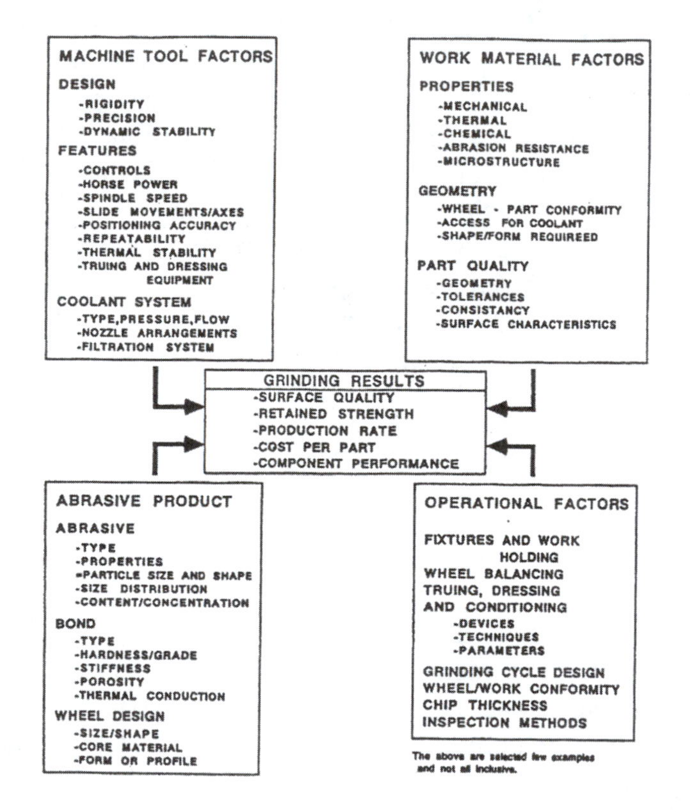

Figure 10.49. Selected variables influencing the abrasive grinding system, from Subramanian[248].

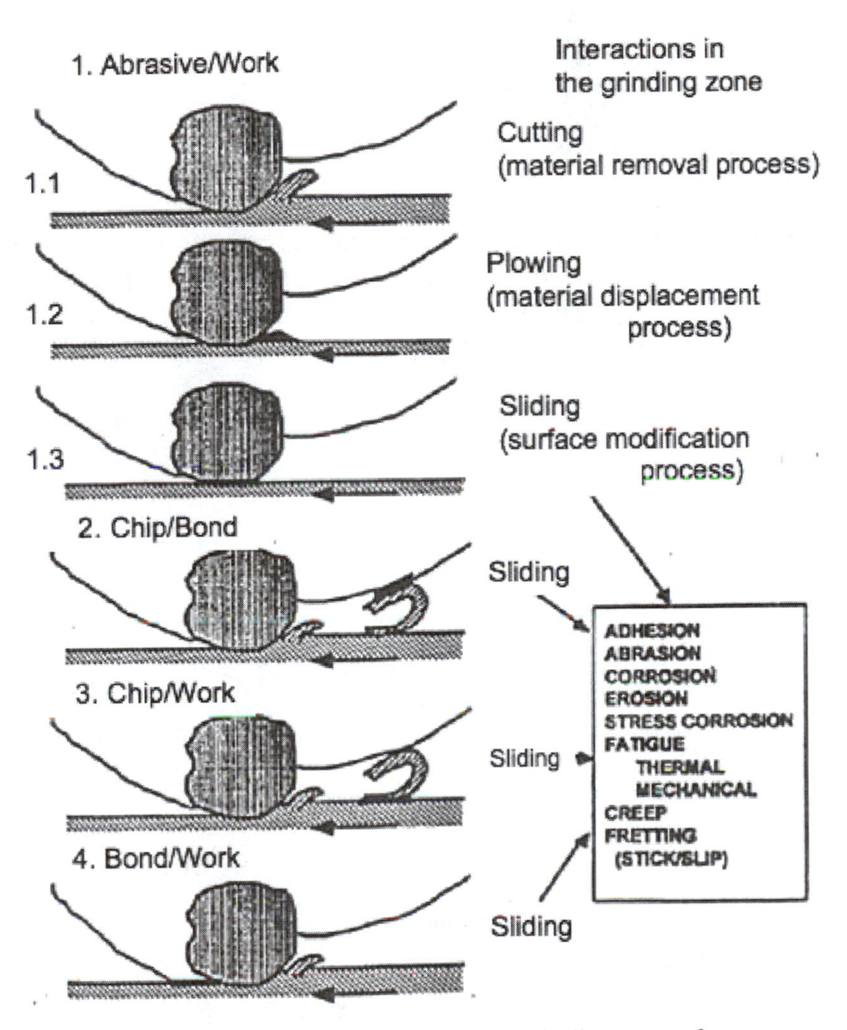

Figure 10.50. Interactions in the grinding zone, from Subramanian[248].

The average cross sectional area, A, of a chip in grinding can be estimated as follows taking the example of surface grinding, Figure 10.55, from Inasaki[249]. The material removal rate q is

$$q = bv_w a \tag{10.1}$$

where b is the grinding width v_w is the workpiece speed, and a is the depth of cut. The number of chips produced per unit time N is

$$N = Cbv_s \tag{10.2}$$

assuming that each cutting edge on the wheel surface produces a chip. Here v_s is the grinding wheel surface speed and C is the number of cutting edges per unit area on the wheel surface. One can add an additional coefficient to reduce the number of active cutting edges if it is known that not all edges are active. The average chip volume V is, therefore,

$$V = \frac{q}{N} \tag{10.3}$$

$$= \frac{v_w a}{Cv_s} \tag{10.4}$$

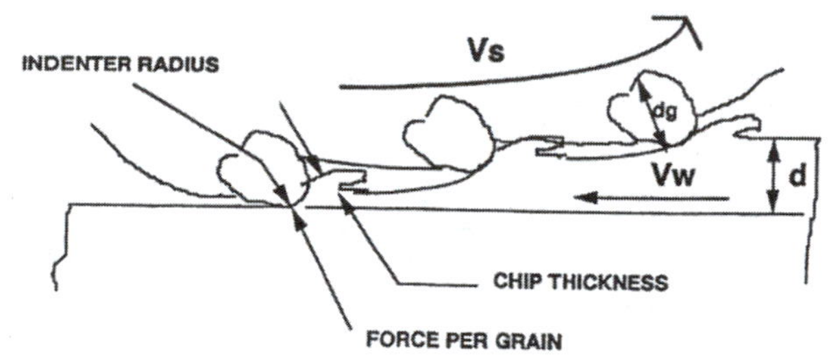

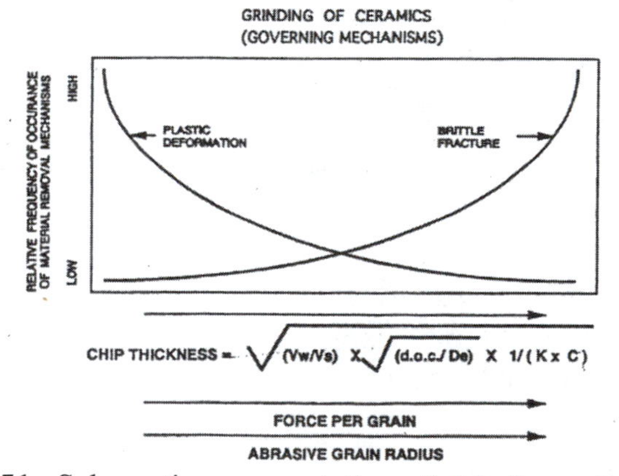

Figure 10.51. Schematic representation of chip formation with abrasive grit and influence of grit diameter/chip thickness on material removal mechanism, from Subramanian[248].

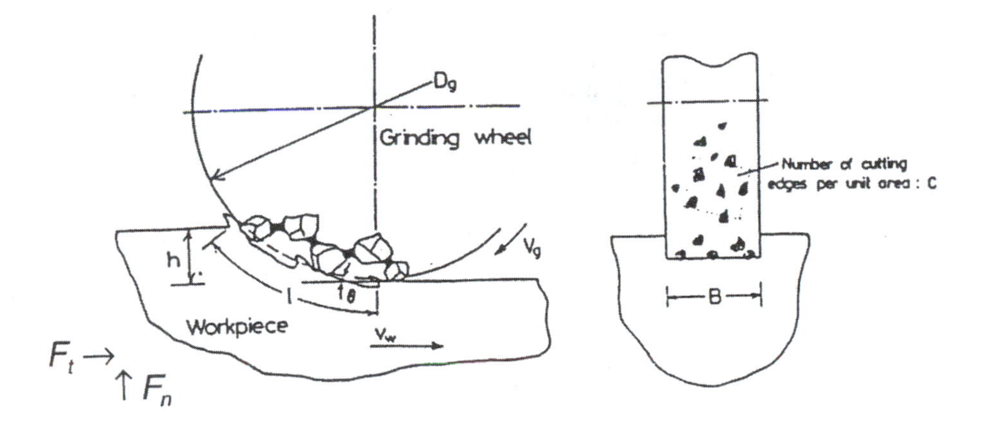

Figure 10.52. Parameters in surface grinding, from Inasaki[249].

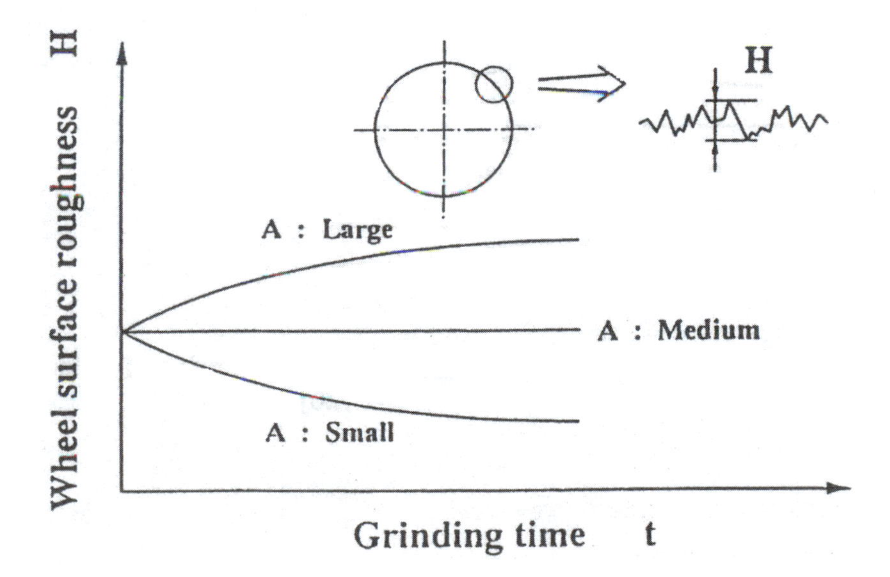

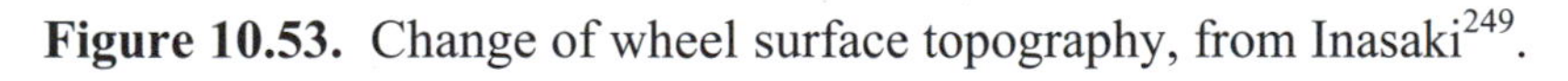

Figure 10.53. Change of wheel surface topography, from Inasaki[249].

If we assume that the chip length is equal to the contact length between the grinding wheel and the workpiece, the chip length can be expressed as

$$l = \sqrt{d_s a} \tag{10.5}$$

where d_s is the wheel diameter. Other expressions can be derived for the chip length as well but this will suffice for this discussion. From eqs. (10.3, 4, 5) the average cross section area of chips is obtained as

$$\begin{aligned} A &= \frac{V}{l} \\ &= \frac{v_w a}{C v_s \sqrt{d_s a}} \end{aligned} \tag{10.6}$$

Equation 10.6 can be generalized for cylindrical as well as internal using the equivalent wheel diameter d_e given by

$$\frac{1}{d_e} = \frac{1}{d_s} \pm \frac{1}{d_w} \tag{10.7}$$

where

d_w = workpiece diameter
$+$ = cylindrical grinding
$-$ = internal grinding
$d_w \rightarrow \infty$ = surface grinding

Thus,

$$A = \frac{v_w}{C v_s} \sqrt{\frac{a}{d_e}} \tag{10.8}$$

The average chip cross sectional area given by eq. 10.6 is very useful for understanding qualitatively the influence of setup parameters on the grinding process. When A is increased through the increase of the workpiece speed or depth of cut, the cutting force acting upon the each cutting edge will increase resulting in the breakage or dislodging of the abrasive particles from the wheel (binder). This effect is referred to as "self-sharpening". As grits dull, force increases as well, resulting in further dislodging or fracturing of the abrasive. Some abrasives are more prone to fracture than others (called friable) and some binders are weaker and thus release the abrasive more easily. Thus, some wheels will "self-sharpen" more readily than others. This effect prevents the grinding forces from increasing with time as is usual in the case of no or limited self-sharpening. Unfortunately, the surface roughness of the workpiece deteriorates with self sharpening. By contrast, if the chip thickness A is decreased by, for example, reducing the wheel surface speed, then the attritious wear of the abrasives becomes significant. Heat generation increases whereas the surface roughness is reduced. In general, from eq. 10.6, we can say the chip cross section A increases with the material removal rate, q. Chip cross section is one of the most important variables to set for successful grinding.

The average cross sectional area of the chips is also affected by varying the number of cutting edges per unit area C. This can be altered by using grinding wheels of different composition (grain size, bond type or "openness") or by changing the dressing conditions. If the wheel is dresses d roughly with a large lead (similar to feed rate in turning), C will decrease causing an increase in cutting force on each abrasive grain. Wheel surface roughness will naturally change over time (faster for less self-sharpening wheels and slower for self-sharpening wheels). The change of wheel surface topography with time is illustrated in Figure 10.53 for two different chip thickness values.

In addition to the number of cutting edges an important factor in the efficiency of grinding is the openness of the wheel, the presence of chip pockets on the wheel surface to accommodate the chips generated. Otherwise, the chips tend to load the wheel and can lead to thermal buildup and burning of the surface. The volume of chip pockets, a function of the porosity of the wheel, is approximately inversely proportional to the number of cutting edges on the wheel surface. In order to increase the material removal rate, it is important to insure that grinding wheels are sufficiently open to accommodate the chips generated.

10.5.1.2 Grinding forces, power and specific energy

Forces are developed between the wheel and the workpiece as a result of the interaction between the abrasive grits and work generating the chips discussed above. For plunge-type grinding operations as illustrated in Figure 10.52 the total force vector exerted by the workpiece against the wheel can be separated into a tangential (to the wheel velocity at the contact point) component, F_t, and a normal component F_n. For grinding with a traverse feed there would be an additional force component in the direction parallel to the wheel axis. The grinding power associated with the force components is

$$P = F_t(v_s \pm v_w) \tag{10.9}$$

The difference indicated by the plus and minus signs is for up-grinding (wheel and work velocities in opposite directions) and down grinding (wheel and work velocities in same directions), respectively. Since v_w is usually much smaller than v_s, the total power can be simplified as

$$P = F_t v_s \tag{10.10}$$

A fundamental parameter derived from the power and machining conditions is the specific energy, defined as the energy per unit volume of material removal. This parameter is the same as the specific power which is the power per unit volumetric removal rate.) The specific grinding energy, E, is calculated as follows

$$E = P/q \tag{10.11}$$

The material removal rate, q, is determined in terms of the grinding parameters

$$q = v_w h\, B = \pi D_g v_f B$$

where v_f is the in-feed rate of the wheel into the workpiece in a plunge grinding operation.

We can also represent the specific energy empirically as in Inasaki[249]

$$E = E_o\, a_m^{-\varepsilon} \tag{10.12}$$

where E_o and ε ($\approx$ 0.2-0.5) are constants.

The presence of a so-called "size-effect" is also assumedin eq. 10.12 meaning that for smaller chip cross sections there is a larger specific energy. This can also be represented as

$$E = E_o\, C^{\varepsilon} v_s^{\varepsilon} v_w^{1-\varepsilon} a^{1-\varepsilon/2} d_e^{\varepsilon/2} \tag{10.13}$$

The tangential grinding force is then obtained by P/v_s. If we assume that the exponent ε equals 0, meaning there is no size effect, then the tangential force is

$$F_t = E_o b\, (v_w/v_s)\, a \tag{10.14}$$

The normal grinding force, which is a significant fraction of the tangential force, can be expressed as

$$F_n = F_t/\lambda \tag{10.15}$$

where λ is a constant.

The size effect is due to the following combination of sources of energy dissipation due to the interaction between the abrasive grain and the work surface. The specific energy consists of components due to chip formation (cutting), chip plowing (displacement of material without cutting) and sliding. At very small uncut chip thicknesses, the component due to chip formation is small and the bulk of the energy is consumed in plowing and sliding. This occurs to some extent in normal machining (for example single point diamond turning) but the edge geometry of the diamond tool (at least in regions I and II) is small enough to insure that mostly chip formation occurs. In grinding, the process can be operating in regions II and III exclusively due to the small and "random" shapes and orientations of the grits (tools). The increase in "non-chip forming" contributions to the specific energy is seen clearly in Figure 10.54, from Malkin.[247] We could read the horizontal axis as uncut chip thickness as well, decreasing toward the origin.

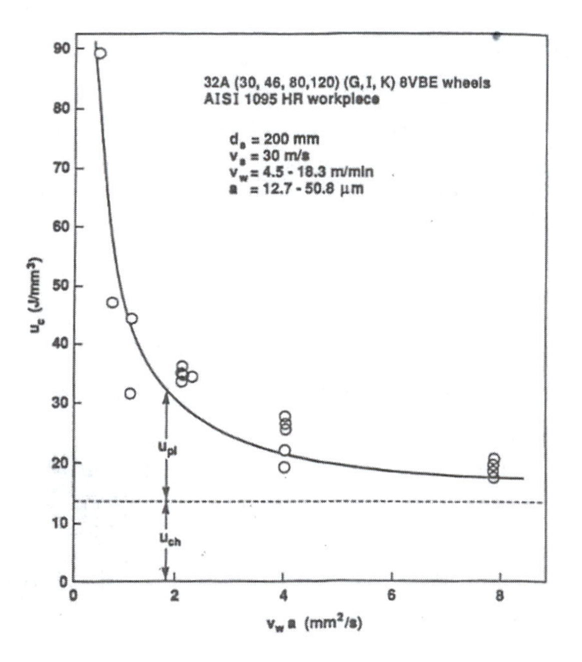

Figure 10.54. Specific cutting energy versus volumetric removal rate per unit width in straight plunge cutting, from Malkin[247].

The understanding of the nature of the approach of the abrasive grit to the surface also sheds light on whether plowing, cutting or sliding occurs. A fascinating area of study derived from tribology investigates the interaction of asperities (such as the abrasive grits in grinding) with surfaces. Tribologists are concerned with wear of the surfaces. We are more interested in controlled removal of material from a surface. Differing attack angles of abrasive grits (or asperities, or other particles) will cause differing effects[250], Figure 10.55, from Williams[251]. The figure defines the geometries for both pyramidal and conical asperities and the relationship between the type of tool-work interaction as a function of the dihedral and attack angles of the indentors. Although this idealizes the shape of the abrasive grit, it lends a lot of insight into the nature and regimes of material removal in abrasive processes. The mechanical aspects of surface damage (and less to thermal effects since velocity is not included as an independent variable) are shown in Figure 10.56, from Childs[252].

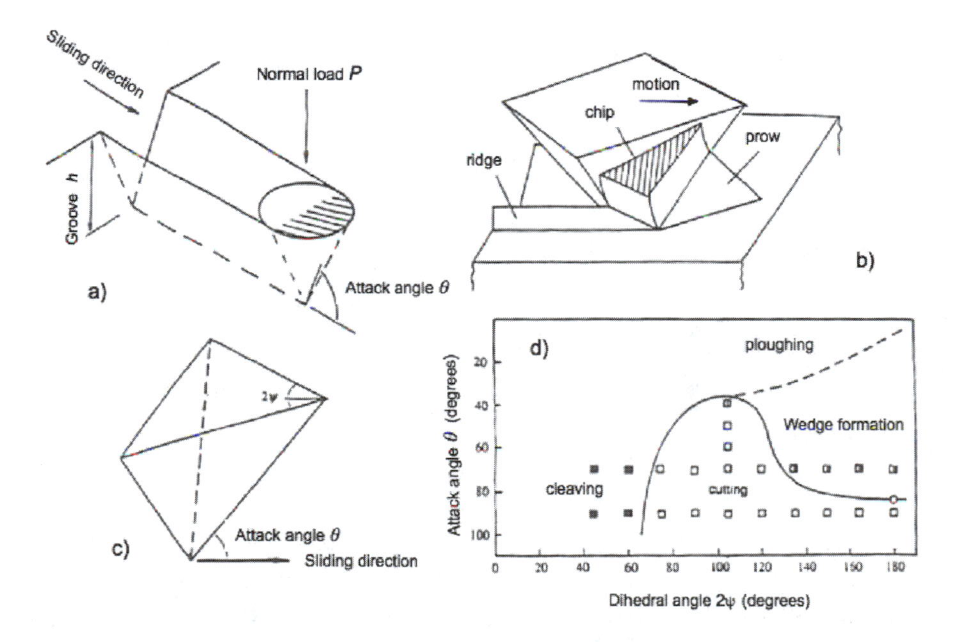

Figure 10.55. a) A simple conical wear indenter presenting an attack angle θ and ploughing a groove in a soft material of depth h, b) a pyramidal indenter moving through previously undeformed material, c) The geometry of the indenter is described by the two angles ψ and θ, d) The wear mode diagram, from Williams[251].

This shows the regimes of wear on axes representing the shear strength of the interface τ between the two materials plotted linearly (and normalized by k the shear strength of the weaker material) versus the logarithm of some roughness parameter such as the angle representing the average slope of asperities on the harder abrading surface, θ. Processes of abrasive machining are represented here analogously to abrasive wear. Clearly, abrasives in grinding processes that can be oriented to fall generally with in certain angles of attack are more likely to remove material. Alternatively, abrasives whose attack angles (or shapes) fall into other regimes are likely to cause damage to the surface or subsurface.

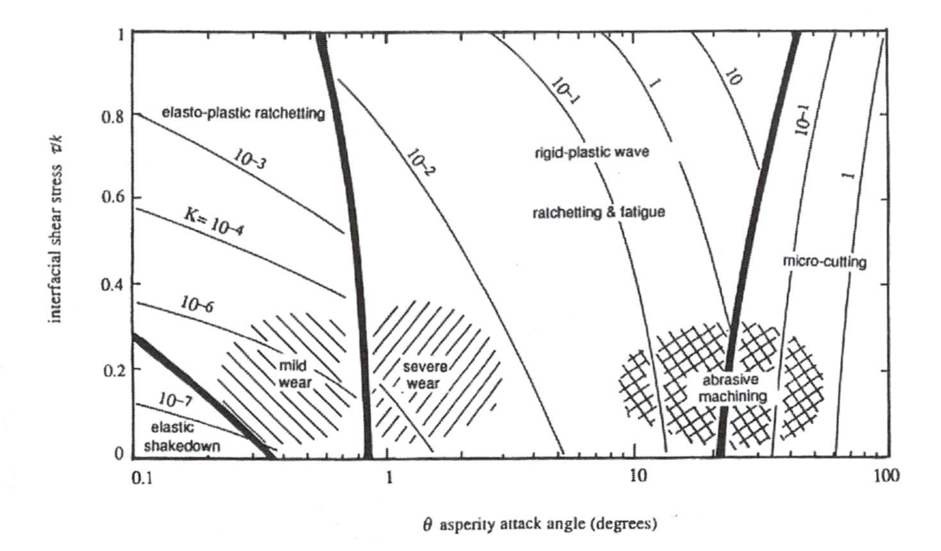

Figure 10.56. Wear mechanism map for a soft surface abraded by a harder, rough counterface. The ratio τ/k represents the relative strength of the interface and θ the mean slope of the rough surface (or the attach angle of a single asperity). Thicker lines delineate different wear mechanism and thin lines represent contours of equal wear rate, from Childs[252].

10.5.1.3 Grinding stiffness, contact siffness and process time Constant

From Eqs. 10.14 and 10.15 we can derive an important parameter which has a decisive influence on the accuracy of the grinding process. Recall the discussions we have had in Chapter 7 on compliance. Here we apply similar concept s to the grinding process. Differentiating the normal grinding force with respect to the depth of cut, the increase in the grinding force due to a unit increase in the depth of cut can be determined. This is referred to as the grinding stiffness (as the cutting stiffness in Chapter 7) and is represented by k_g. It is really a stiffness as it represents the ratio of force to penetration of the wheel into the workpiece. The actual "spring constant" is nonlinear in the case of grinding as we have seen in Eq. 10.13 although for simplicity here we will assume it is linear of the form

$$\mathrm{k} = \frac{1}{\lambda} E_0 b \frac{v_w}{v_s} \tag{10.16}$$

For the surface grinding process, an intermittent infeed grinding process, or cylindrical grinding (and following the development in Chapter 7) the actual depth of cut a is given by

$$a = a_n - \delta \tag{10.17}$$

where a_n is the nominal depth of cut and δ is the elastic deformation of the wheel and machine tool. The elastic deformation is obtained by dividing the normal grinding force by the static stiffness of the machine, k_m. Thus

$$\delta = F_n/k_m \tag{10.18}$$

Combining eqs. 10.16, 17, and 18, we can derive the relationship between the actual depth of cut and the nominal depth of cut as

$$\frac{a}{a_n} = \frac{1}{1 + \frac{k_g}{k_m}} \tag{10.19}$$

In addition to the grinding stiffness we must also consider the contact stiffness k_c between the grinding wheel and the workpiece. The contact stiffness (machine – wheel stiffness) determines the elastic deformation of the grinding wheel surface due to the normal force. Needless to say, it affects the actual depth of cut along with the stiffness of the mechanical system. Its influence can be considered in series with the mechanical system stiffness as

$$\frac{1}{k} = \frac{1}{k_c} + \frac{1}{k_m} \tag{10.20}$$

The contact stiffness is a nonlinear value as it is affected by the bonding material and wheel characteristics (grade, for example) and has a strong effect on the stability of the grinding process (chatter) as well as the process "time constant." For continuous grinding processes as cylindrical or surface grinding where the process is basically a plunge grinding operation the system stiffness will determine the time constant (or amount time is takes for the desired surface dimension to be reached.) If we consider a simple cylindrical grinding cycle consisting of rough grinding and spark-out grinding as shown in Figure 10.57, the workpiece radius is different from the nominal infeed motion due to the elastic deformation of the mechanical system.

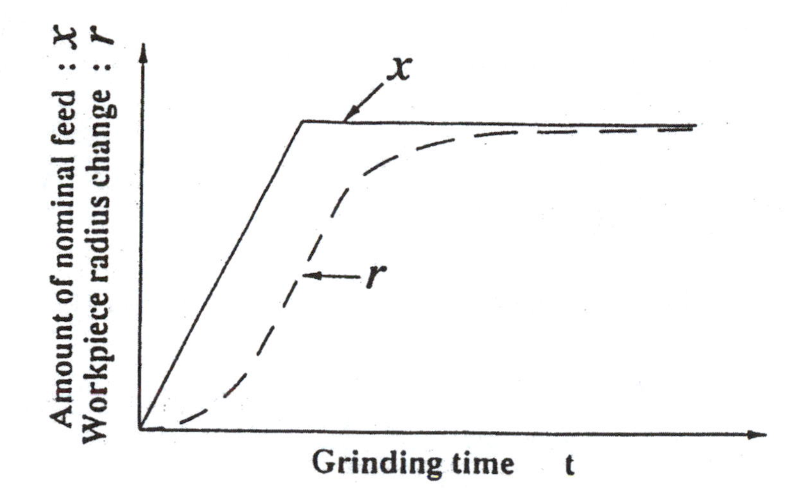

Figure 10.57. A simple grinding cycle for cylindrical grinding, from Inasaki[249].

This will affect not only the cycle time for grinding but the surface roughness and accuracy as well because the infeed rate changes with time[249]. The actual change in the radius of the work-piece r is

$$r = x - \delta \tag{10.21}$$

where x is the amount of the nominal infeed, and v_f the infeed speed as before. Then, with the definition of elastic deformation from eq. 10.18 we have

$$r + \frac{F_n}{k_m} = v_f t \tag{10.22}$$

Assuming a plunge grinding process in which the material removal rate is q and given by

$$q = \pi d_w \frac{dr}{dt} \tag{10.23}$$

where *dr/dt* is the reduction rate of the workpiece radius. For the normal grinding force proportional to the material removal rate, eq. 10.23 becomes

$$r + \alpha\pi \frac{d_w}{k_m}\frac{dr}{dt} = v_f t \tag{10.24}$$

where α is a constant. The solution of this first order linear differential equation is

$$r(t) = v_f t - v_f T\left\{1 - e^{\frac{1}{t}}\right\} \tag{10.25}$$

where *T* is the time constant of the system and is equivalent to

$$T = \alpha\pi \frac{d_w}{k_m} \tag{10.26}$$

In general, *T* should be as small as possible for achieving a reliable grinding process. The time constant increases as the mechanical stiffness decreases and the constant α increases. Although it is hard to define α quantitatively, difficult to machine materials have a large α and, consequently, a larger time constant. If conditions in grinding change during the process, like wheel loading for example, the constant α can change as well due to change in difficulty in grinding, Inasaki[249].

10.5.1.4 Nanogrinding

Nanogrinding is an ultraprecision machining process useful in challenging surface machining applications such as advanced ceramics. The process kinematics are based on the lapping process and the workpieces are held against a rotating plate[253]. Lapping uses a loose abrasive whereas for nanogrinding, the abrasive grain (diamond of a size 1.5 to 3 μm) is completely embedded in a soft metallic plate, thus becoming the grinding tool. It is first necessary to create the grinding plate and then the workpiece can be machined, Figure 10.58, from Gatzen[253]. The grinding plate is first conditioned (roughened) by use of a pumice in a fluid. During this process the pumice is embedded in the plate and remains there throughout the process. Between pumice particles, the basic soft metallic plate material forms into plateaus into which the diamond grains are embedded by a conditioning ring. The summits of the plateaus are aligned co-planar with the surface of the plate. The effect is ductile mode machining of brittle materials (due to low depth of cut) with little subsurface damage and excellent surface finish.

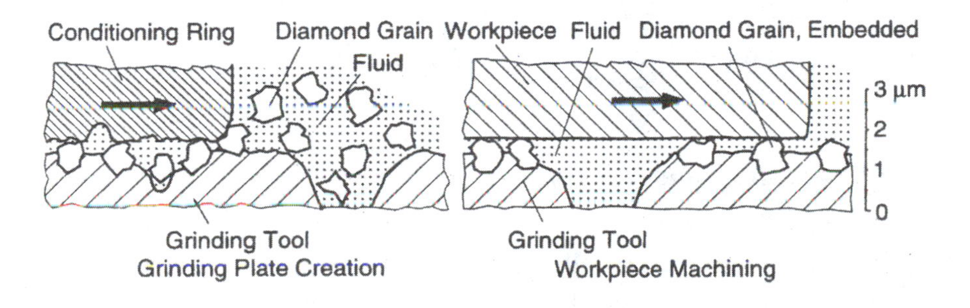

Figure 10.58. Schematic of the nanogrinding process, from Gatzen[253].

10.5.2 Loose abrasive processes

Loose abrasive processes (lapping, polishing, CMP, etc.) are examples of three-body processes where in the tool, abrasive, and work make up three individual elements of the process and the abrasive is not part of the tool as in grinding. An excellent paper by Komanduri, et al[254] entitled "Technological Advances in Fine Abrasive Processes" reviews a host of fine abrasive processes including most loose abrasive ones specially those operating at small a_{cav} values. The reference also includes an extensive literature review and is a valuable source of state of the art information on a range of processes capable of achieving work surface roughness from 1 nm R_a to 60 nm R_a. Loose abrasive processes are distinguished from the abrasive (fixed) processes reviewed above in the last section which create surfaces which are on the order of 10x rougher.

Our interest here is the class of processes defined in Figure 10.1 by Nogowa as machining by loose abrasives and machining by free abrasives. This includes polishing and lapping as well as chemical mechanical polishing (or planarization), abbreviated CMP, which is discussed in a later section below. We have also previously seen Figure 10.2, also from Nogowa, which details some of the system elements of fine abrasive machining. In general, we are talking about mechanical processes of material removal. However, in the case of CMP, the chemical influence of the liquid carrying the abrasive particles plays a critical role in the effectiveness of the process.

10.5.2.1 Polishing and lapping

These two processes actually straddle the fence between loose and fixed abrasive processes. The processes themselves go back well over two thousand years as evidence of loose abrasive polishing of the death masks of the Egyptian Pharaohs exists, specifically that of King Tutankhamun. Polishing can be done with either free abrasive in a slurry carried over a rotating plate against the work surface or with an abrasive imbedded (although not necessarily permanently) in the plate. In the case of Tutankhamun's death mask the polishing was done with fine particles suspended in fluid — the longer the suspension time the finer the particle size in the solution as the larger (hence heavier) particles tended to settle out first. More contemporary parameters influencing the polishing process are seen in Figure 10.59, from Klocke[255]. The form of the geometry to be machined is determined by the machine type since only the shape of the polishing wheel can be reproduced on the work surface. Figure 10.60, from Klocke[255], illustrates the basic polishing process for a double-sided polishing machine. The same principle applies if only one sides polishing is undertaken. The workpieces themselves are guided in revolving cages between two polishing wheels which are rotating in opposite directions. A suspension of abrasive grit in a fluid (as simple as water as with Tutankhamun's mask polishing, or, often de-ionized water) is fed through the upper wheel. This results in material removal characterized by various mechanisms in the contact zone between the polishing wheel and the workpiece.

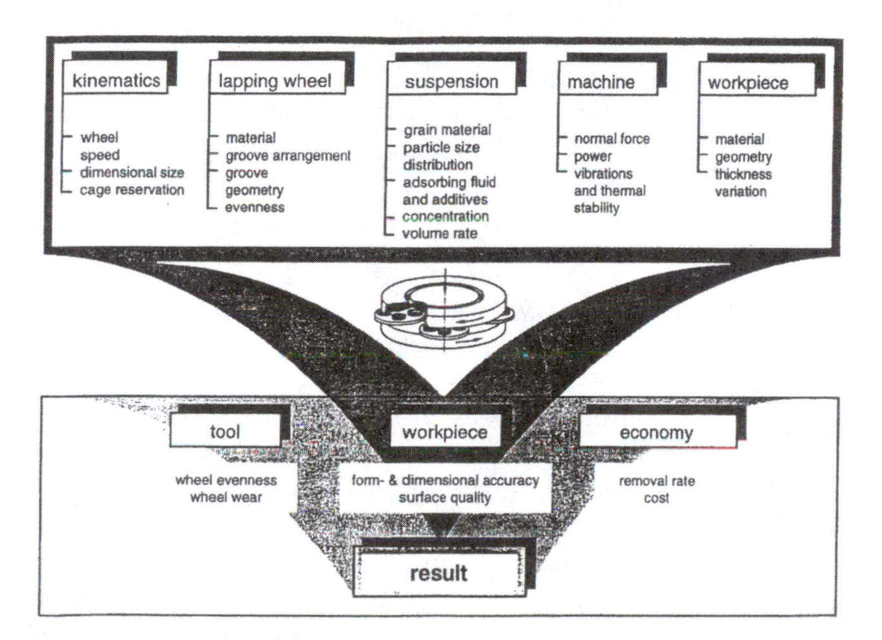

Figure 10.59. Parameters influencing the polishing process, from Klocke[255].

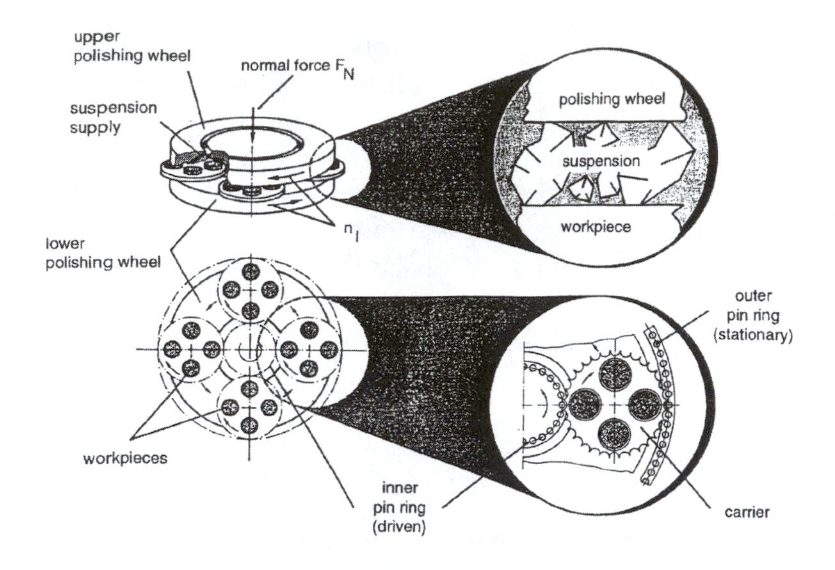

Figure 10.60. Principle of two-wheel polishing with guided workpiece, from Klocke[255].

The removal mechanisms in mechanical polishing are characterized by abrasive grits that are temporarily anchored in the polishing wheel. The loose abrasive grains fed in the slurry between polishing wheel and work surface attach themselves to the relatively soft polishing wheel or move freely in the interface between the two, Figure 10.61, from Klocke[255]. Depending on the grain penetration depth into the workpiece surface, elastic followed by plastic deformation (microgrooves and micro-fatigue) occurs. With increasing penetration depth, micro-chipping and finally micro-cracking can occur in brittle materials. There are two sets of phenomena at work here — one related to the behavior of the material as either a ductile or brittle material and responding to the grain depth of cut as previously discussed and the other relating to the specifics of the interaction between the abrasive grain (secured in the plate or not) and its shape and attack angle. The various modes of material removal can be summarized as in Figure 10.62, after Chang[256]. The matrix is comprised of, basically, ductile and brittle modes and two or three body interactions. Both can occur in a typical polishing process, Figure 10.63, from Xie[257]. Wear rates are inversely proportional to workpiece hardness in two body abrasion when the abrasives are harder than the workpiece. In three body abrasion, for example with a soft pad, this is not necessarily true. Never-the-less, the combination results in material removal with cutting and plowing the dominant material removal modes. In general, experimental data indicates that when the ratio of abrasive particle hardness, H_a, to workpiece hardness, H_w, is greater than 1, the wear rate is approximately inversely proportional to workpiece hardness (or the wear coefficient is approximately constant). If the ratio is less than 1, the wear rate (or wear coefficient) decreases rapidly with the decrease in hardness ratio. In the region $H_a/H_w > 1$ the wear coefficient can slightly increase with an increase in H_a/H_w. Also, the material removal rates in some three-body abrasion situations can be an order of magnitude lower than those for two-body abrasion due to the presence of loose abrasives that can be rolling in the interface rather than sliding and abrading the work surface. With time, an abrasive slurry comprised of abrasive grains of an average size will degrade into grains of much smaller size and, without refreshing, result in a reduction in material removal rate. Typical abrasive materials include diamond (synthetic monocrystalline, density 3515 kg/m^3, hardness 78-102 GPa, elastic modulus 900-1050 GPa, diameters 0.12-3 μm typical) and Al_2O_3 (α-Al_2O_3 structure, density 3980 kg/m^3, hardness 20 GPa, elastic modulus 390 GPa, diameters 0.1-9 μm typical).

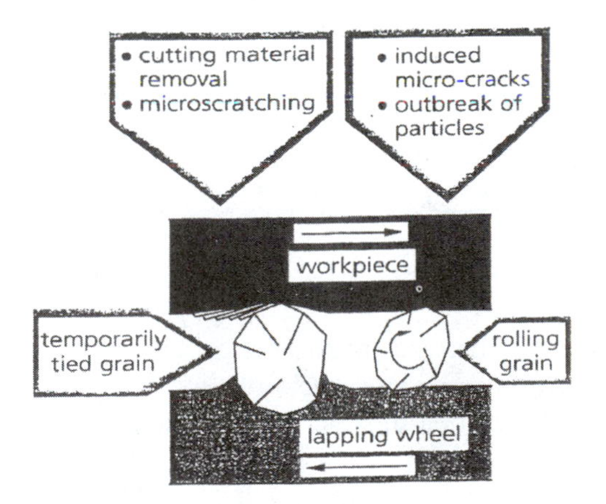

Figure 10.61. Principle of polishing a plane surface, from Klocke[255].

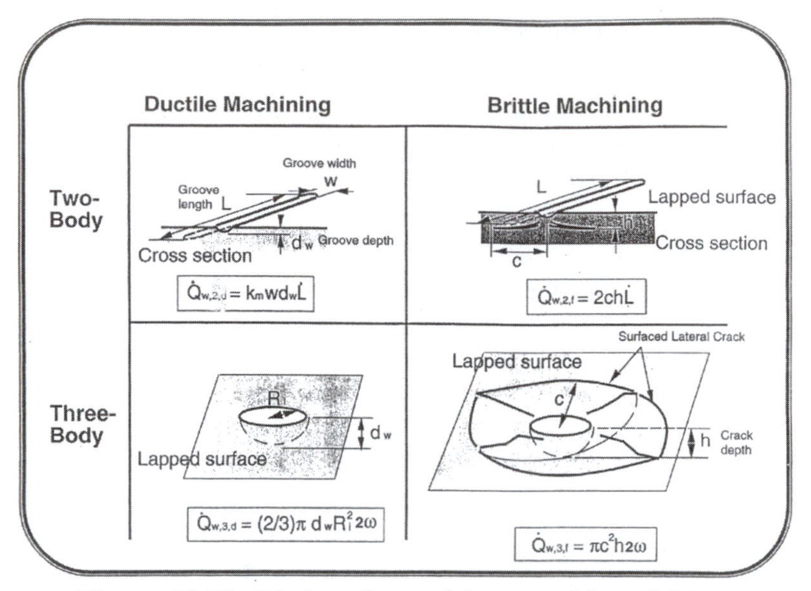

Figure 10.62. Modes of material removal in polishing.

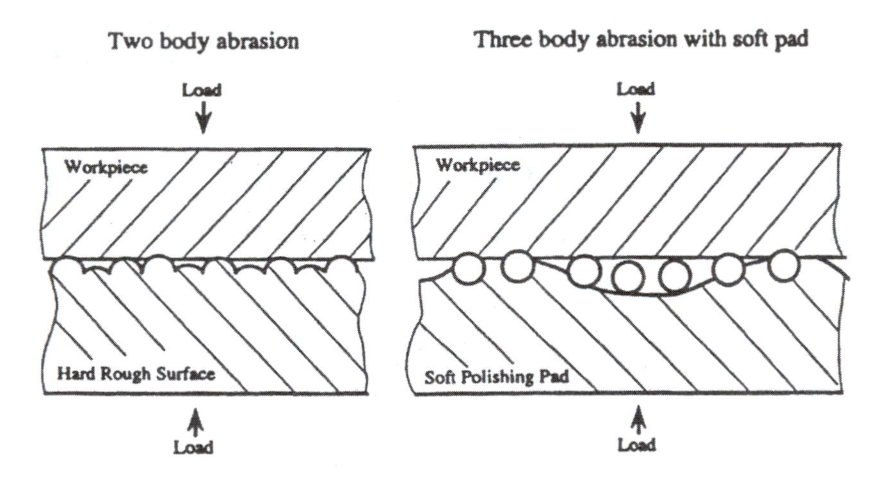

Figure 10.63. Contact between a hard rough surface and a soft flat surface in two-body abrasion, and contact between a soft polishing pad and a flat surface with spherical particles in three-body abrasion, from Xie[257].

Depending upon the combination of tool characteristics and abrasive characteristics, differing effect on the work surface can be accomplished. The variables are size of abrasives (fine or coarse) and tool hardness (hard or soft). According to these variables, different finishes are possible on different workpieces as follows:[258]

Abrasive Size	Tool	Application and Finish
Fine	Hard	Lapping for block gages; mechano-chemical polishing for sapphire Mirror finish
Fine	Soft	Optical polishing for glass lens Mirror finish
Coarse	Hard	Lapping of various materials Matte finish
Coarse	Soft	Finishing of die materials Mirror finish Finishing of magnetic ferrite Semi-mirror finish

The chemo-mechanical polishing (or planarization) process will be discussed in a separate section below.

In optics manufacturing, pitch polishing is a commonly used process for preparing fine surface finish on transmitting or reflecting optical materials, mirrors for example[254]. A sequence of processes is used including rough grinding with a cylindrical-shaped fixed abrasive SiC or diamond wheel. Contour accuracy of 100 μm can be obtained at this stage with surface roughness of 100 μm approximately. This is followed by grinding with loose abrasives, such as SiC or Al_2O_3 or diamond using a spherically shaped segmented metal lap (cast iron or brass), ceramic or other glass-like material. A typical procedure would involve the use of progressively finer abrasives (60 μm, 40 μm, 30 μm, 12 μm and 3 μm) stages. Contour accuracy at the end of this sequence would be on the order of 25 μm or better and surface roughness in the range of 1 – 3 μm. The final stage utilizes a polish on a soft pitch lap using fine abrasives such as iron oxide (Fe_2O_3), aluminum oxide, rare earth oxide (CeO_2), or diamond. Again, a series of steps with progressively smaller abrasive sizes ranging from 2 μm down to 0.8 μm is used. As a result of this sequence of processes, final contour accuracy for large mirrors (2-5 m diameter) is about one wavelength of visible light (653 nm) with a surface roughness of 1-2 nm RMS.

Pitch polishing referred to above uses a soft pad material (pitch, for example, or synthetic materials such as polyurethane, foam, felt, and Teflon) into which some of the abrasive particles become imbedded or fixed. The smoothing or polishing occurs at low pressures (1-3 kPa for glass) and at low velocity (0.05-0.25 m/s) to minimize frictional heating. Most modern polishing machines utilize a combination of reciprocating and circular motions if the polishing pad or lap and the workpiece. Although some loose particles are present, pitch polishing processes are often classified as fixed abrasive processes.

Generally, a "lap" refers to a plate in which a distinct groove pattern has been machined. The slurry of abrasive and liquid separates the lap plate from the work surface. In some cases the lap plate has specially designed grooves to insure that a hydrodynamic film is maintained between the plate surface and the work surface to prevent contact between plate and work, Figure 10.64, from Watanabe[259]. The polisher is designed such that the lap surface has alternating flat and sloping surfaces similar to a hydrodynamic thrust bearing, Figure 10.65, from Watanabe[259]. It was reported that Si wafers were polished with this machine to a flatness of 0.3 μm over 76.2 mm diameter except for the outer 3 mm and a surface roughness of 1 nm.

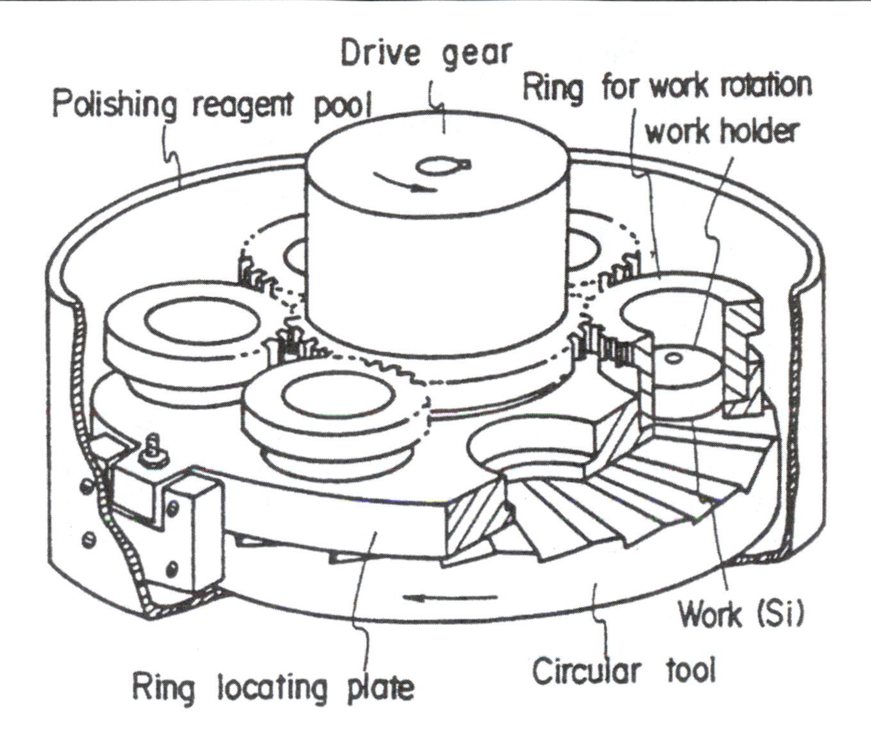

Figure 10.64. High precision polisher using a hydrodynamic fluid film, from Watanabe[259].

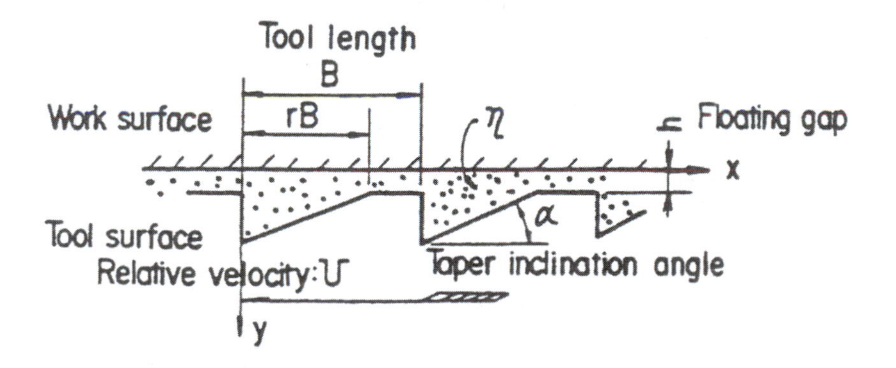

Figure 10.65. Tool-work interface for the high precision polishing machine, from Watanabe[259].

An extension of the more conventional lapping process is a recent development called float polishing[260]. This was first developed for the ultra-fine finishing of sapphire crystals using a free abrasive (a combination of colloidal SiO_2, 4-7 nm grain size, CeO_2 and Al_2O_3) in de-ionized water on an ultraprecision diamond turned tin lap supported on a rigid support base, Figure 10.66, from Namba[261]. The concentration of abrasive particles in water ranged from 2-8% by weight. The results are quite exceptional — removal rates of approximately 1 x 10^{-5} mm^3/sec over a 100 mm^2 area and surface roughness of ≤ 1nm RMS were routinely achieved. Unique characteristics of this float polishing technique include:

1. relatively soft material (tin in this case) used for the lap instead of pitch or synthetics used in conventional polishing; in addition, the lap plate is diamond machined on the polishing machine for desired surface finish and to maintain form relative to work carrier,
2. hydrodynamic film fluid gap many times greater than the particle size is maintained between lap and work surface, and
3. very high polishing speeds (1.5-2.5 m/sec) between lap and workpiece are used (recall the Stribeck curve and the role of relative speed in this type of "bearing").

This method has also been applied to the polishing of other materials including optical materials (borosilicate glasses and Zerodur). Apparent pressures between the lap and the workpiece are some order of magnitude higher than conventionally used in pitch polishing processes. Surface roughness is in the range of 1-2 Angstrom RMS and flatness to 0.03 μm (λ/20).

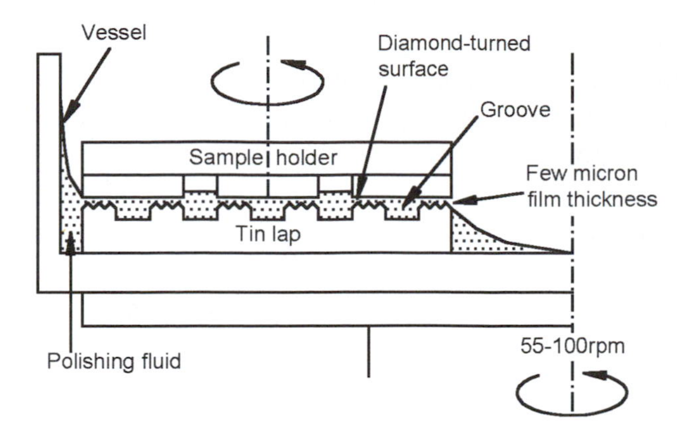

Figure 10.66. Schematic of float polishing machine using a diamond turned tin lap, from Namba[261].

Finally, computer numerical control (CNC) systems for polishing were introduced in the 1980's, usually for aspheric, higher cost optics than for flats and spheres. Figure 10.67 illustrates a CNC grinding/polishing machine, from Stowers.[262] There are a number of means of employing these machines. More complicated machines have up to 9 independent axes of which 6 can operate simultaneously under computer control. The polishing pad is pneumatically applied to the surface and the contour error data is used to determine the polishing dwell time required to remove the contour error. With these types of machines it is possible to produce strongly aspheric components with less than 1 arc second of tangential error.

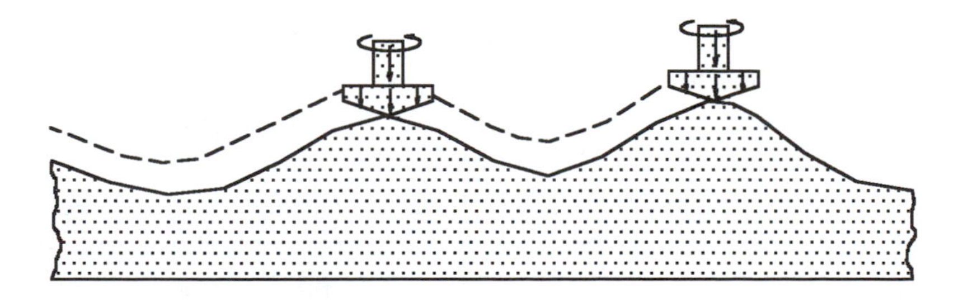

Figure 10.67. Schematic view of CNC grinding/polishing Process, from Stowers[262].

Modeling of polishing and lapping processes are usually based on a simple formulation known as Preston's equation, after an early paper on optics polishing written by F. W. Preston in 1927[363]. Preston's equation is

$$\text{Removal Rate} = C_p\, P\, V$$

where P is pressure and V is velocity of work surface relative to the pad, plate or surface of the tool.

The coefficient in Preston's equation, C_p, is usually experimentally obtained and its sensitivity to the peculiarities of each polishing process and setup emphasize the limitations of Preston's equation. In addition, the Preston's equation can vary with time and processing conditions over the cycle of a polishing process. It is in many ways similar to the wear coefficient discussed earlier with respect to the mechanics of material removal in abrasive processes. There are a number of "enhancements" to Preston's equation reflecting specific polishing regimes but, in general, this is used as a key descriptor of the process.

10.5.2.2 Chemical mechanical planarization (CMP)

CMP is a special, but increasingly important category of loose abrasive finishing processes that is heavily used in semiconductor manufacturing. CMP has become one of the key bottleneck or roadblock issues in semiconductor manufacturing today[188]. It is used to insure that interconnects between multilayer chips are achieved reliably and that the thickness of dielectric material is uniform and sufficient. In that sense it is a "planarization" process as well as polishing and the "P" in CMP refers to both terms. These must be accomplished over, now, a 300 mm diameter wafer achieving a surface roughness on the order of 1-2 nm R_a and global planarity in the order of sub 0.5 micron to meet the requirements of lithography tools (recall that for line widths on the order of 0.25 micron the depth of field of the lithography machinery is very limited and the wafer flatness or planarity must be held to stringent tolerances). And all this must be done at higher production rates to maintain a low cost of process ownership. The decreasing line widths of semiconductor devices require new materials, such as copper and the so-called low-k dielectrics, which further challenge the process. Preferential polishing rates of adjacent materials or surface features resulting from previous manufacturing steps often lead to defects such as dishing frustrating efforts to obtain planarity. The abrasive slurry can cause contamination on the surface as well as cause scratches on the surface, residual slurry, etc. Figure 10.68 below from *MICRO*[188] illustrates the range of defects that occur in tungsten and oxide CMP.

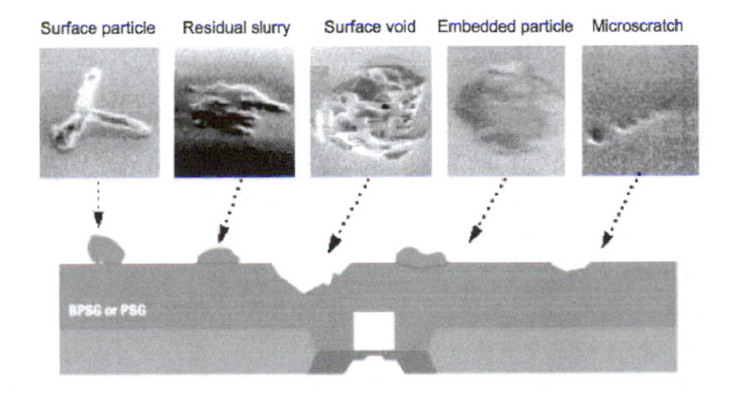

a. Five types of oxide CMP defects

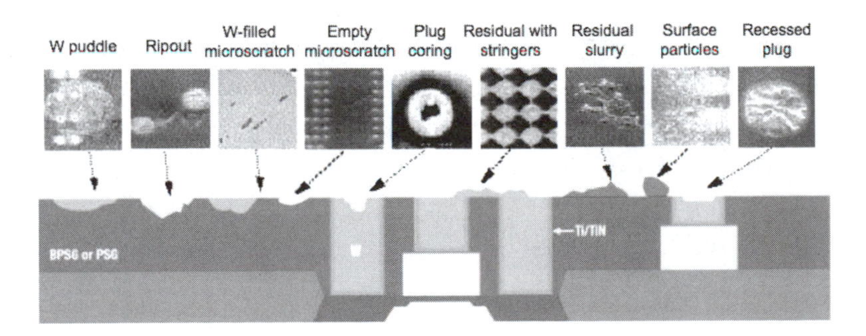

b. Nine types of tungsten CMP defects

Figure 10.68. Illustration of critical CMP defects that occur in oxide and tungsten polishing, from *MICRO*[188].

Chemical mechanical polishing is a planarization technology suitable for logic and DRAM devices with feature sizes in the sub half-micron range. As a result of the interaction of an abrasive slurry with specific chemical properties, a polishing pad with specific density and texture with the surface of a semiconductor device in wafer form. CMP planarization of interlayer dielectric (ILD), basically, is a combination of chemical reaction and free abrasive machining in which the abrasives are allowed to rotate between the ILD surface and polishing pad and remove material by micro indentation or three body abrasion. The pad "holds" and enhances the motion of the abrasive particles in slurry, composed of, for example 5-7 nm fused silica in an aqueous solution with pH between 8.5-11 (this will vary dramatically depending upon the material being polished and abrasive used) and transmits the abrasive/fluid load to the wafer surface. The polishing operation is comprised by the creation of a silica layer chemically which is then removed by the mechanical abrasive action. The mechanism differs with other materials, polymers for example. The chemically-enhanced removal of layers of surface material including oxides, metals and polymers to produce a "planarized" surface is a unique output of CMP. Figure 10.69, below, is a schematic of the CMP process. The process is similar to polishing processes for glass and other metals dating back thousands of years and is roughly governed by Preston's equation which predicts the material removal rate as a function of the polishing pressure, relative pad-wafer speed and a constant. It is not possible to review in great detail here the CMP process. An excellent recent reference text on CMP is by Steigerwald, et al[264]. In addition, Komanduri, Lucca and Tani[254] mentioned earlier gives an excellent review of the process and places it in perspective with other conventional abrasive processes along with Evans[265].

There are a significant number of input variables to the CMP process. These include:[264]

- Slurry chemicals (pH, concentration, buffering agents, oxidizers, dielectric constant, isoelectric point, zeta potential, stability of the suspension, etc.)

- slurry flow rate
- abrasive (including hardness, composition, size, shape, concentration)
- temperature
- pressure
- velocity and kinematic influences on velocity of pad and wafer

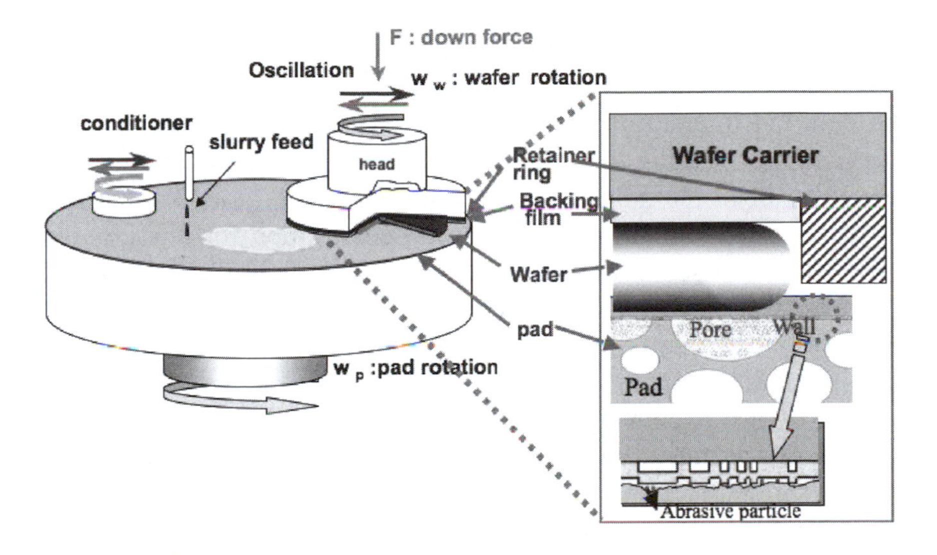

Figure 10.69. Schematic of the CMP process.

- frictional forces/lubrication
- pattern geometries including feature size and pattern density
- pad including fiber structure, height, pore size, compressibility, elastic and shear modulus, hardness, thickness, surface features/embossing/perforations, conditioning, etc.
- wafer geometry including curvature and mounting
- film stack, stress, and hardness
- wafer cleaning sequence, and
- wafer size.

The process outputs of interest to the manufacturer include, also from Steigerwald, et al:[264]

- polish rate
- polish rate of underlying film
- planarization rate and "efficiency"
- polish rate uniformity
- feature size dependency including polish rate, planarization rate and damage
- surface quality including roughness, particles and corrosion resistance, see also Figure 10.68.
- surface damage, see also Figure 10.68

The detailed interaction between the wafer, pad and abrasive (i.e. with the goal of tying the inputs to the outputs in some quantitative sense) has been the subject of research for some time. The earliest reference to an attempt to model the process, albeit for glass, is due to Preston discussed above. It is commonly applied to Si wafer CMP even though, physically, the actual process mechanics are much more complex. The coefficient in Preston's equation, C_p, is usually experimentally obtained and its sensitivity to the peculiarities of each CMP setup emphasize the limitations of Preston's equation. Runnels and Eyman[266] and Runnels[267] looked in more detail at the mechanical interactions and fluid mechanics, respectively. In the first study the normal and shear stresses involved in polishing were considered and Preston's equation modified as:

$$\text{Material Removal Rate (MRR)} = C_p\, \sigma_n \tau\, V$$

where σ_n is the normal stress and τ is the shear stress due to slurry flow.

In the second study by Runnels, the flow of the polishing slurry (in two dimensions) at the velocity of the fluid between the pad and the wafer and the stresses induced by the flowing slurry on the features on the surface of the wafer were calculated and used in an erosion model. The chemical aspects of CMP were investigated by Cook[268] and he proposed several mechanisms governing the rate of surface removal. He also considered the mechanical aspects of the removal process as a travelling indentor moving along the wafer surface.

In fact, the problem with most models is that, depending upon the velocity and slurry characteristics, there are three differing types of interaction between the pad, wafer and slurry. At high relative velocities the wafer will move over a fluid pad as with a hydrostatic bearing so that no contact exists between the pad and the wafer. The influence and action of the abrasive includes erosion and impact as well. At lower velocities there may be some solid-solid contact in addition to support on a fluid layer. In this case the action of the abrasive can appear as either two-body or three-body depending on the action of the pad. Finally, at the lowest speed (or highest pressure) there can be direct wafer-pad contact where the entire load is supported on the solid structure. The abrasive action in this case, for the mechanical elements of CMP, is most likely primarily two-body abrasion due to asperity contact.

There are any number of issues related to CMP that have been identified as "roadblocks" or "showstoppers" driven by the decreasing size of device features, approaching 0.18 micron now, and the problems with contamination control that accompany such feature sizes. Any perusal of recent *Semiconductor International* magazine, the industry chronicler, refers to the new materials, productivity requirements, yields, wafer sizes, etc. required to maintain the competitiveness of the industry. This occurs in the presence of an expected 10-15% a year growth in the personal computer semiconductor for the future and at some point, usually in reference to interconnect technology, CMP is mentioned as a major concern. The concerns relate to process uniformity and planarization efficiency, defects and contamination.

The Semiconductor Industry Association (SIA) *2005 National Technology Roadmap*[269] (with update for 2006) outlines the significant challenges to the industry out to the year 2013 when devices with features as small as 0.032 microns (32 nm) are scheduled to go into production. Previous roadmaps had identified six "Grand Challenges" demanding special attention including affordable scaling, affordable lithography (100nm or less), new materials and structures, Ghz frequency operation on and off chip, metrology and test, and what they referred to as the R&D challenge. Two of these relate specifically to CMP, new materials and structures and affordable scaling. The first is most obvious as it affects the interconnect technology CMP is critical to. Industry needs to develop new gate dielectric and gate electrode materials as well as new interconnect materials and processes. This will pose challenges for CMP due to the unequal removal rates of differing materials, the effect on localized removal rates of increasingly dense patterns on the wafer surfaces. As circuit dimensions drop below the 1.0 micron level and additional layers of metal are added the requirements for surface smoothing increase. The effect of multiple layers of metal on planarity (or lack of it) is cumulative. The non-planar topography along with finer dimensions leads to problems in processing such as step coverage. So, in general, planarization requirements increase with reduced dimensions and planarity over the surface of increasing large (next 300 mm diameters from 200 mm in practice now) wafers. Figure 10.70, from Steigerwald[372] shows the degrees of surface planarity starting with the "as received" wafer for CMP and the progressive stages of planarization. Planarization efficiency refers to the rate and degree to which the process can yield a planar surface given the surface pattern of the wafer. Figures 11.71 from Bibby[373], show the changes in removal rate and thickness, respectively, as a function of time during the planarization process. The differences are a result of differing removal rates and pad/wafer interaction due to the materials and patterns present. The final surface, at t_2 in both figures, must meet the ultimate specifications for the process including the depth of field requirements of lithography tools.

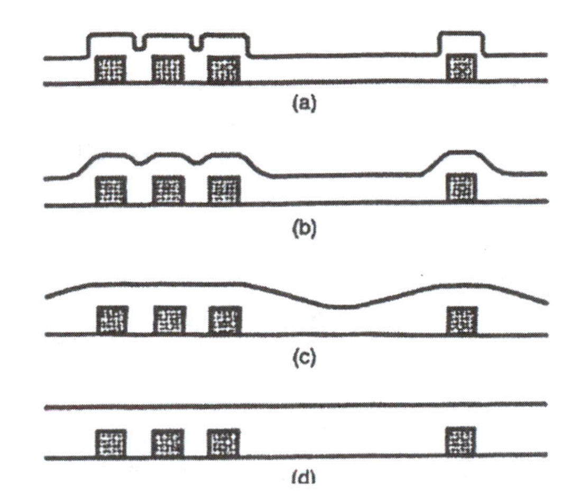

Figure 10.70. Degrees of surface planarity, a) unplanarized, b) surface smoothing, c) local planarization, d) global planarization, from Steigerwald.[264]

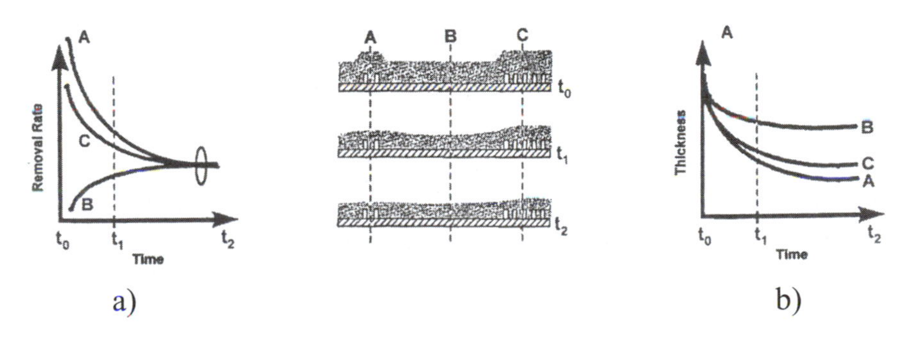

Figure 10.71. CMP pattern effects on a) removal rate, and b) remaining thickness, from Bibby.[270]

The second grand challenge related to CMP is in the production of optics for the lithography machines used to achieve the increasingly reduced feature sizes on the wafers. Polishing technologies, especially with respect to pad or lap plate groove design can have a major influence of polishing efficiency and achievable surface quality.

10.5.2.3 Process modeling in CMP

The CMP process is known as a very effective polishing process providing excellent global planarization (usually represented as within wafer non-uniformity, WIWNU) and local planarization which is critical in decreased dimensions of integrated circuits. It has been widely used for the planarization of interlevel dielectric (ILD), shallow trench isolation (STI), tungsten plug, and damascene metallization process[271-274].

Compared to the lapping process in which material removal is mainly caused by scratch, fracture, and crack through direct indentations[275-279], the general mechanism of material removal in CMP is chemical etching and mechanical abrasion[264, 280]. The material removal mechanism of silicon dioxide is known as a sequence of diffusion, dissolution, and adsorption of dissolution made by the interaction between silicon dioxide and abrasive slurry[264]. In metal CMP, material removal is made by the action of a metal etchant and a metal passivating agent with abrasive particles[264].

As presented above, the basis for the CMP process model is Preston's wear equation. However, Preston's equation does not take into effect any of the kinematic characteristics of the process (velocity variation on the wafer surface, for example), temperature effects, or the fluid mechanics at the pad/wafer interface. Further, depending on abrasive size, shape and hardness (relative to the wafer and pad), differing mechanical removal behavior occurs[271]. Since CMP has become a popular technique many mathematical models for CMP process have been developed. A model dealing with the degree of nonplanarity on material removal has been developed[280] and the actual pattern has also been considered for the polishing rate mainly based on phenomenological study[281].

The material removal mechanism of CMP process can be approached from two aspects: chemical and mechanical, even though the actual material removal in CMP is caused by the combined action. Compared to conventional abrasive machining, lapping, polishing, and grinding, the role of the chemical elements in the abrasive slurry in CMP is more significant. Chemical action causes a defect-free wafer surface and low surface roughness with the help of the mechanical action of the abrasives in the slurry. Even though the role of abrasives in lapping and grinding has been well known as the major cause for material removal, the mechanism of material removal by abrasives is not clear in CMP. This is due to the mixed chemical and mechanical action in material removal, the large gap thickness between wafer and pad comparing with the size of the abrasive particle, and the surface roughness of the polishing pad comparable to the size of the particle. In this proposal, the effect of slurry film thickness on CMP is investigated.

The slurry film thickness in CMP is important. It has been previously shown that the major parameter determining the slurry film thickness is the Hersey number defined as viscosity times velocity divided by pressure. Thus, the slurry film thickness increases as the relative velocity of the wafer and the viscosity of the slurry increase and normal pressure on the wafer decreases. Due to the fact that slurry film thickness determines the characteristics of lubrication, boundary lubrication, elasto-hydrodynamic lubrication, and hydrodynamic lubrication, the Hersey number is also known to be a key parameter which determines the wafer-pad contact mode (direct contact, semi-direct contact, and hydroplane sliding as described earlier). The slurry film thickness has been measured in-situ by using a capacitance probe apparatus. By using hydrodynamic lubrication theory, the gap thickness was calculated. It is known that the gap thickness has an effect on the planarization and material removal rate in CMP.

By measuring the friction force variation between the silicon wafer and the polishing pad the slurry film thickness can be measured and correlated with the Hersey number. Based on the friction force variation at the interface and the Stribeck curve, Figure 10.72, the characteristics of lubrication in CMP are determined and the theoretical slurry film thickness is calculated based upon the characteristic of the lubrication. The effect of slurry film thickness variation on CMP performance such as material removal, planarization, surface defects (Figure 10.71), surface roughness, material removal mechanism, and Preston's equation can be determined. To compare the absolute amount of material removed with respect to the slurry film thickness, the material removal per sliding distance is determined instead of material removal rate, which is the standard method of measuring the material removal for CMP. It has been shown that the material removal is heavily dependent upon the slurry film thickness variation (wafer-pad contact mode), as shown in Figure 10.73.

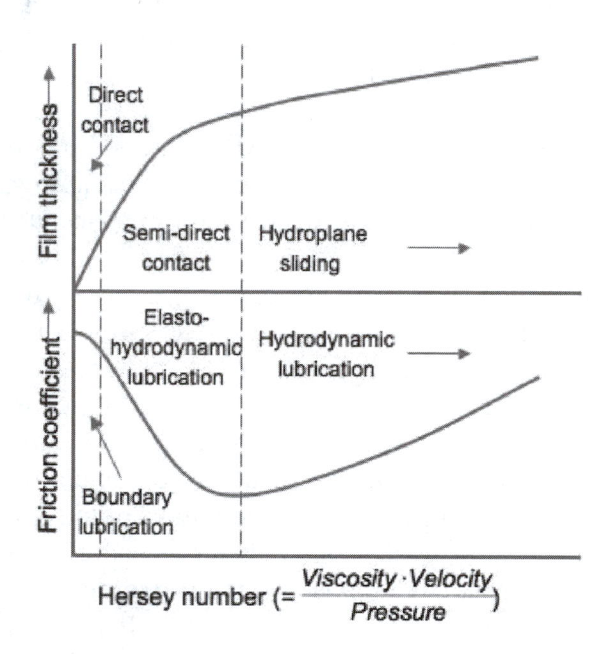

Figure 10.72. Stribeck curve; friction coefficient and film thickness vs. Hersey number.

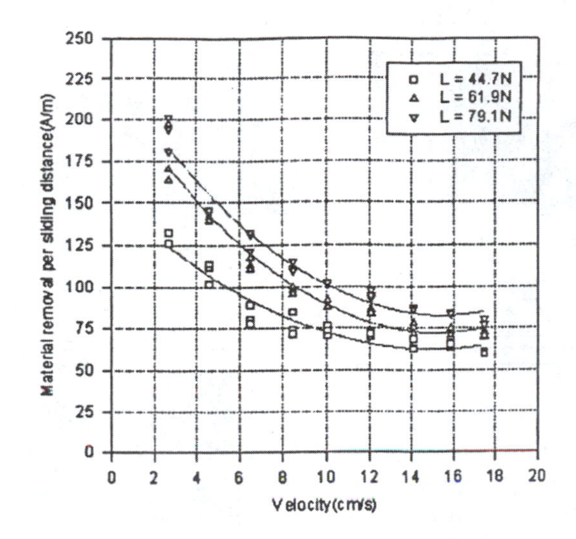

Figure 10.73. Material removal per sliding distance on oxide wafer for different polishing loads.

In Figure 10.73, the x-axis indicates the relative velocity of the wafer and the y-axis indicates the material removal per sliding distance. Three normal loads were used. As the relative velocity increases, the slurry film thickness also increases, resulting in variation in wafer-pad contact mode. Contrary to the prediction from Preston's equation that the material removal per sliding distance is constant with respect to the velocity, in this case it decreased as velocity increased, which also increases the slurry film thickness. Surface defects such as micro-scratching, Figure 10.74, were more frequently observed when the slurry film thickness was relatively small and the pad-wafer gap approaches the mean particle size. It is believed that the slurry film thickness is a strongly correlated with the material removal mechanism in CMP, Figure 10.75.

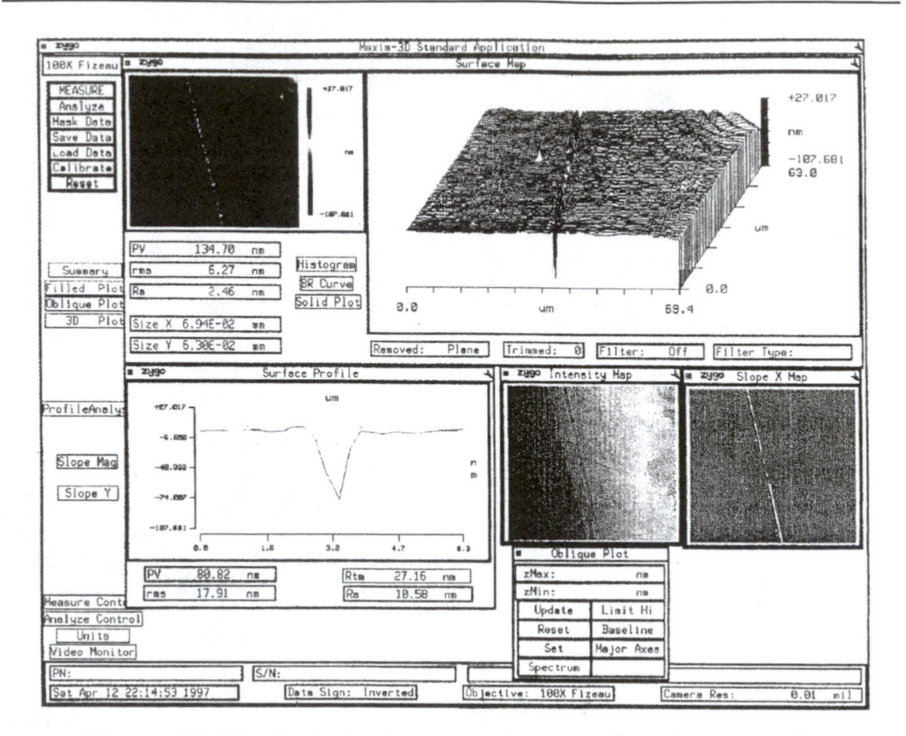

Figure 10.74. Optical microscope images of a micro-scratch.

Preston's wear equation was modified based upon the relationship between Preston's coefficient and friction coefficient in CMP. The relationship was obtained analytically and verified experimentally. It was verified that Preston's coefficient is linearly proportional to the friction coefficient, Figure 10.76. It was also found that the material removal is linearly proportional to the friction force between the workpiece and the polishing pad and to the work done by friction in CMP.

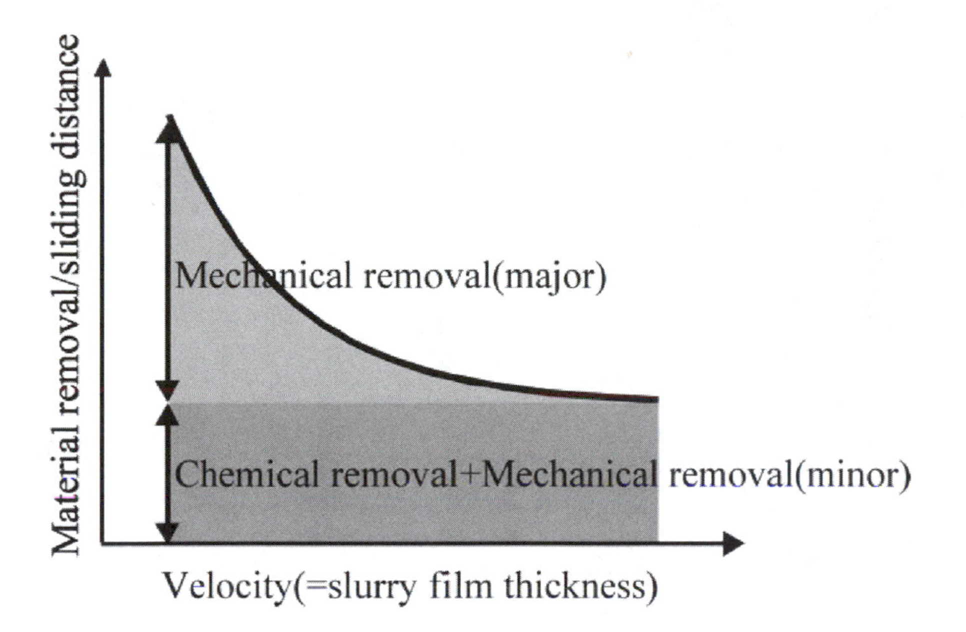

Figure 10.75. Mechanical and chemical removal for slurry film thickness.

The actual contact area of the workpiece must be considered in modeling the CMP process. Experimental results showed that the material removal rate was strongly dependent upon the surface profile, Figure 10.77, and on the wafer, Figure 10.78. The variation of the material removal rate based on the actual contact area, determined analytically, was not explained by Preston's equation. A modified Preston's equation was proposed based upon the dependence of the material removal rate on the actual contact area.

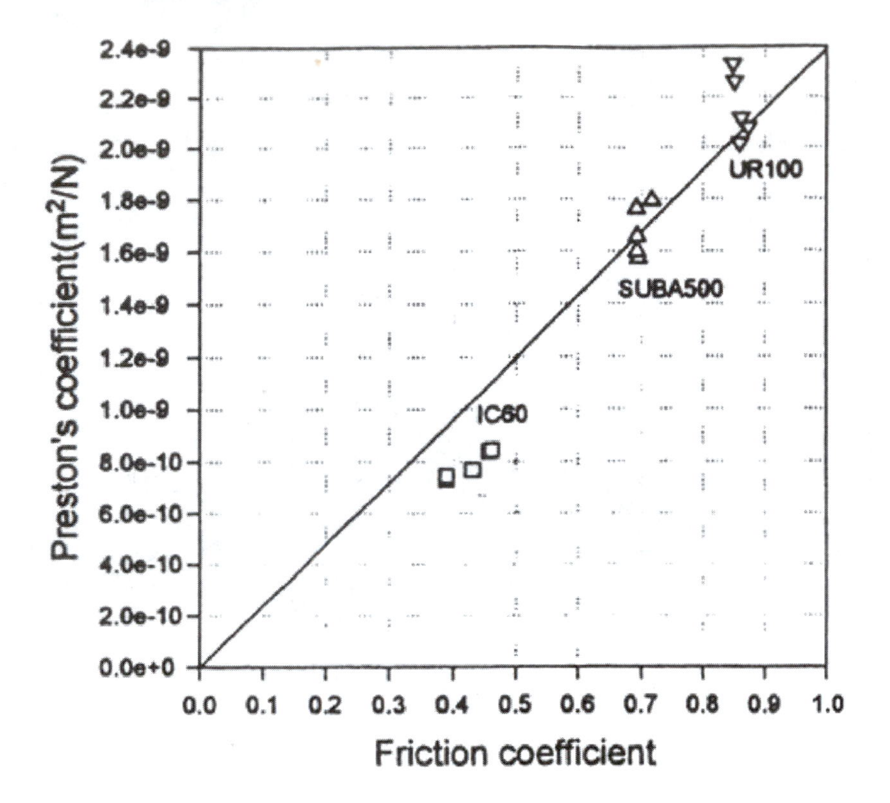

Figure 10.76. Preston's coefficient variation with friction coefficient.

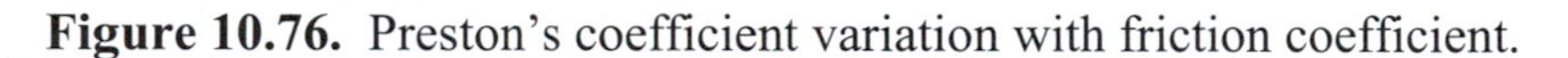

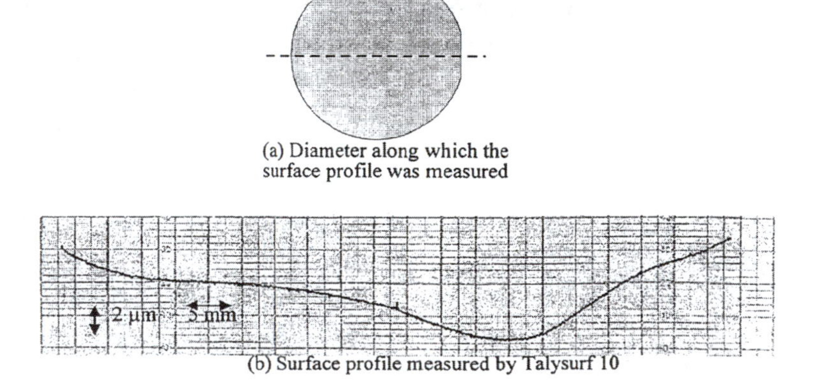

Figure 10.77. Surface profile of a silicon wafer.

If the contact mode between the wafer and pad is either direct or semi-direct, material removal in CMP is aggressive and mechanical removal (nano-plowing and nano-scratching) is dominant compared to the chemical removal (corrosion). Due to the predominant mechanical removal, a greater possibility of micro-scratching on the wafer surface exists and a rough surface is generated compared to a surface finished by dominant chemical action. However, due to the semi-solid contact with the pad, the global planarization of wafer is superior. As the contact mode between wafer and pad varies from direct or semi-direct contact mode to hydroplaned sliding mode, the material removal per sliding distance decreases. This indicates that chemical removal (corrosion) by slurry plus minor mechanical removal (erosion) becomes the major material removal mechanism. Due to dominant chemical removal and reduced mechanical removal, the surface roughness is improved and there is less possibility for surface defects such as micro-scratches. The global planarization is less effective in this mode since the chemical removal is the primary material removal which is more isotropic in directionality.

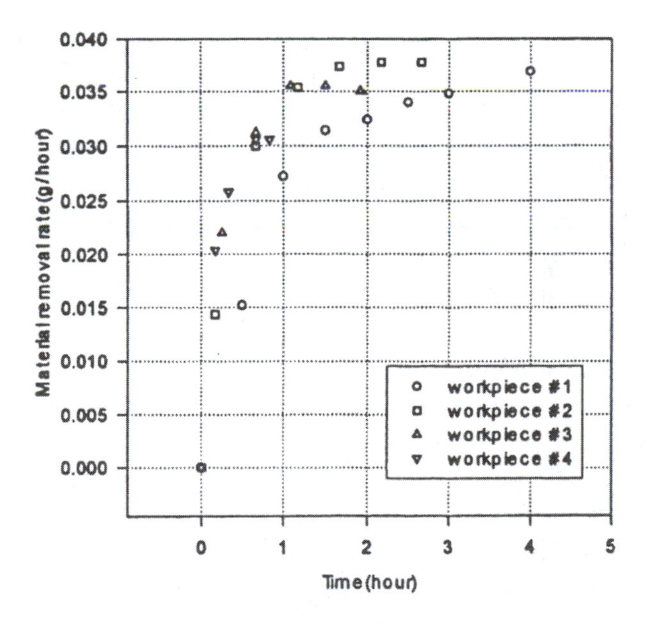

Figure 10.78. Variation of material removal rate of a silicon wafer.

The role of abrasive particles in CMP has received less attention compared to lapping and grinding processes. This is due to the smaller size of abrasives relative to the other dimensions such as slurry film thickness and surface roughness of the polishing pad. As the dimension of line width and depth in IC devices decreases and the requirement for planarization becomes more stringent, the importance of understanding the behavior of the abrasive particles in the slurry becomes critical. As discussed earlier in the chapter, in lapping, the mechanical removal process is known to be the dominant material removal process. Abrasion, scratching, and fracture through indentation by abrasives are examples of this action. The material properties and geometry of abrasives are considered to be major factors determining material removal of the workpiece. One of the main interests in CMP is developing a method to prevent surface defects such as micro-scratch. This is partly due to the irregular size distribution of abrasives.

Along with the abrasive slurry, the polishing pad, which is principally composed of polyurethane with either open or closed cells, is one of the most important elements and has a direct impact on the material removal mechanism and process performance as measured by material removal rate and global planarization. Although the polishing pad is known to play a critical role in CMP, the interaction of pad with abrasive and wafer is not fully understood. The influence of polishing pad characteristics on CMP has been investigated by a number of researchers. Material removal rate is inversely proportional to the density and shear modulus of the polishing pad, and the shear modulus of polishing pad is affected by the variation of temperature. The polishing pad usually has an open cell structure and the mechanical properties of the cellular solids or foam are related to the properties of the cell wall material as well as the cell geometry. The polishing pad also has the ability to remove contaminants as a post-polishing process. Typically, the polishing pad is deformed after being used for a long period of time, causing a decrease in the material removal rate. As shown in 3-D FEM simulations, the non-uniformity on the wafer is related to the variation of Von Mises stress and shear stress distribution.

The dependence of material removal rate on the pad density and compressibility has been tested, Figure 10.79. Three polishing pads with differing characteristics, SUBA500, IC60, and UR100, were utilized. Figure 10.80 shows typical pad surface features and construction details. The results of the investigations suggest that the material removal rate is strongly dependent upon the pad density and compressibility. It is believed that the variation in the material removal is caused by the actual contact area of the polishing pad and the wafer depending on pad deformation patterns. In general, it is known that the lower the density, the larger the deformation. Also distinct types of material removal may be caused by different pad surface profiles.

There is much that is not known about the CMP process even though it is a key process in semiconductor processing. As the number of transistors and metal layers increases and the die size increases the effect is to push a steadily decreasing feature size (line width) and depth of focus (hence flatness requirements). The ILD or damascene within die (WID) non-uniformity budget is about 25% of the depth of focus (DOF). The CMP WWID budget will drop to less than 500 Angstroms by the time line widths reach 0.1 μm in the early 2000's. In addition, new processes utilizing Cu for interconnects will require some additional process development because of

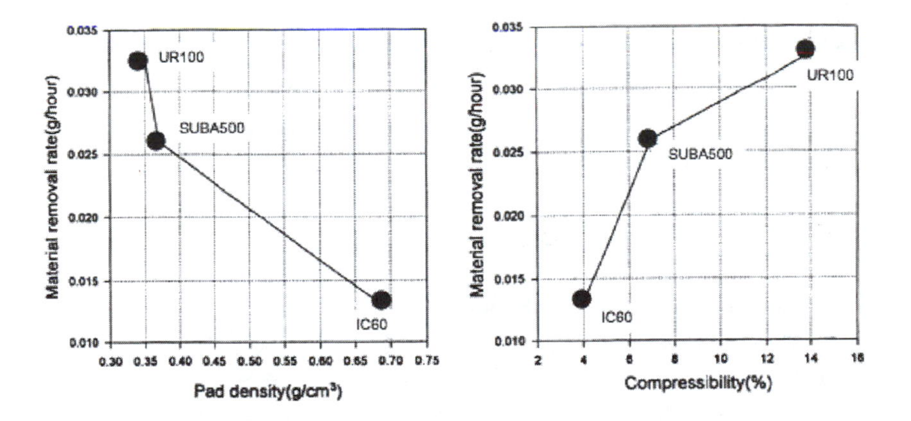

Figure 10.79. Material removal rate variation with a) pad density, and b) pad compressibility for three pads.

the peculiarity of Cu, difficulty of determining endpoint, development and disposal of slurries and improved polishing pads. In addition, a new host of low-k dielectrics (polymers for example) will offer additional polishing challenges.

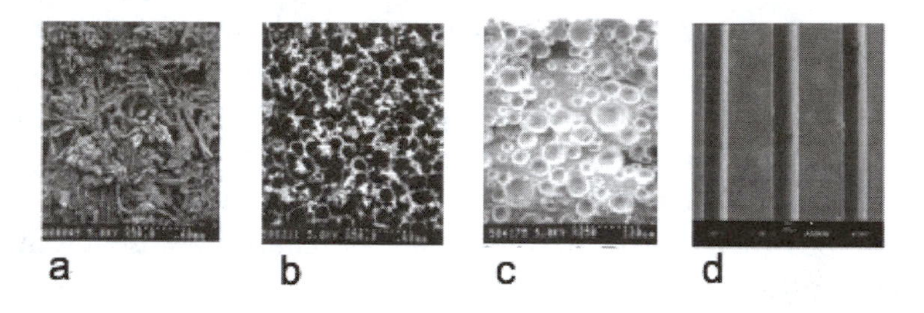

a. felted fibers impregnated with polymeric binder
b. porous layer on a supported substrate
c. microporous polymer sheet
d. nonporous polymer sheet with macro texture (grooves)

Pad	a	b	c	d
Microstructure	Continuous channels between fibers	Vertically oriented, open pores	Closed cell foam	None
Inherent Microtexture	High	High	Medium	Low
Slurry Holding	Medium	High	Low	Minimal
Examples of Commercial Pads	Suba™, STT 711™	Politex™, Surfin™	IC1000™, FX9™	OXP4000™, NCP-1™
Compressibility	Medium	High	Low	Very Low
Stiffness/Hardness	Medium	Low	High	Very High
Typical Applications	Tungsten CMP	Tungsten CMP, post-CMP buff	ILD CMP, STI, metal damascene CMP	ILD CMP, STI

Figure 10.80. Comparison of polishing pads and structural features.

Finally, a number of researchers have been focusing on comprehensive models of material removal in CMP. Some of these efforts are reviewed in Steigerwald[264] and Evans[265]. A very comprehensive discussion of CMP process modeling is included in Luo and Dornfeld[282] and details on pad effects modeling in Wang[283].

10.6 Non-traditional processes

There are a number of processes that are considered "non-traditional" in the realm of precision manufacturing. We recall the Taniguchi diagram seen in the first chapter and it's inclusion of processes capable of producing features of dimensions on the order of nanometers, parts of nanometers and Angstroms. It is not possible to review all of the processes capable of such fine dimensions. The reader is referred to excellent reviews such as Komanduri et al[254], and Taniguchi.[284] Many of these processes don't fit well into the analogy of uncut chip thickness although the typical size of material removed by the process can be quantified. Often these processes are referred to as "atomic bit processes" and are characterized by removing, deforming and consolidating at the atomic level. In atomic bit processing, removing or processing atoms at a processing point must be held in the state of extremely high potential energy density ($10^4 - 10^6$ J/cm^3), which corresponds microscopically to the atomic lattice bonding energy of each stock material or elastic energy density at the elastic limit of defect-free materials[358]. This density is called the extreme threshold specific machining or processing energy. Table 10.2, from Taniguchi[284], lists the specific volumetric lattice bonding energies U_b (J/cm^3) of various materials as well as the atomic bonding energies E_b (J/cm^3). The threshold values for specific processing energy must, in practice, include surplus energy corresponding to the surface barrier potential energy E_s which is greater than the specific volumetric lattice bonding energy U_b or atomic bonding energy E_b, in order for the atom to escape from the work surface.

A number of different kinds of energy beam processes using high energy elementary particles, such as protons, electrons, ions, chemically reactive atoms or clusters as well as neutral beam atoms have been developed to perform atomic bit processing. A comprehensive list of the basic processes is given in Table 10.2. Sketches of the implementation of many of these processes are shown in Figure 10.81, also from Taniguchi[284].

Table 10.2. Specific volumetric lattice bonding energy U_b and atomic bonding energy E_b, from Taniguchi[284].

Materials	U_b (MJ/m^3)	E_b (J/atom)	Remarks	Bonding
Fe	2.6×10^5	1.6×10^{-20}	for tensile breaking	metal bond
	(1.03×10^5)	(8×10^{-21})	for shear slip	
Al_2O_3	6.2×10^5	5.26×10^{-18}		ionic bond
Si	7.5×10^5	1.59×10^{-17}	for tensile breaking	covalent bond
SiC	1.38×10^6	1.1×10^{-17}	- and so on -	covalent bond
cBN	2.09×10^6	1.7×10^{-17}	(cubic boron nitride)	covalent bond
B_4C	2.26×10^6	1.8×10^{-17}		covalent bond
Diamond I	5.64×10^6	4.5×10^{-17}	abundant N_2 (natural)	covalent bond
Diamond II	1.02×10^7	8.2×10^{-17}	N_2 free	covalent bond
Ref. P_s (surface tension to He): Fe 1.74 J/m^2, SiC 20.27 J/m^2		U_b: Plendle, J. N. Phys. Rev., 123, 4, (1961) 125, 3, (1962) E_b: the author's estimation		1 eV = 1.60×10^{-19} J

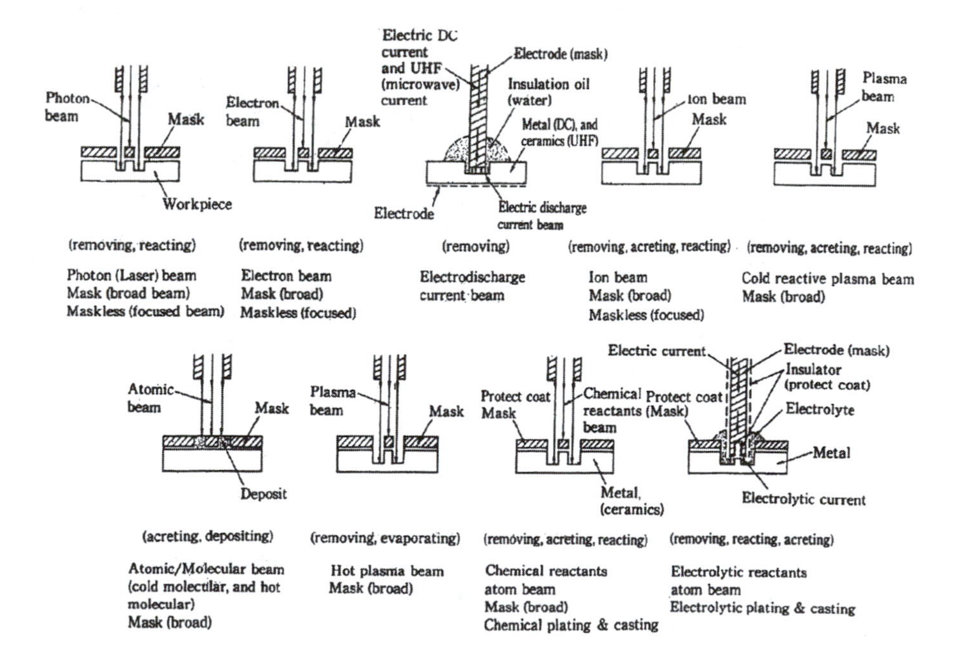

Figure 10.81. Sketches of energy beam processing methods for 2D stock removal, from Taniguchi[284].

One technique of interest is designed to remove atomic size amounts of material (the ultimate "uncut chip" in machining) using mechanical methods to obtain a finished surface. The surface is not only a complete mirror but also remains crystallographically and physically undisturbed. The process is termed elastic emission machining (EEM) and was introduced by Mori and co-workers[285]. Briefly, the process uses vibration to induce collision between small particles of abrasive and the surface being machined. Using single crystals of Si and Al as the work material, Mori and his associates removed material using alumina particles of 2, 8, and 20 µm size. The process is based on atomic scale fracture induced elastically so that the finished surface remains undisturbed at the crystallographic as well as physical level. The process is based on a surface energy phenomenon in which each abrasive particle removes a number of atoms after coming in to contact with the surface. The primary difference between this technique and, say float polishing, is that a rotating sphere, made from a relatively soft polyurethane material, is used for the lap as compared to a hard tin disk used in float polishing. Figures 10.82 and 10.83 show a schematic of the process mechanics as well as a view of a process industrially implemented, respectively. Here, a polyurethane ball, 56 mm in diameter, is mounted

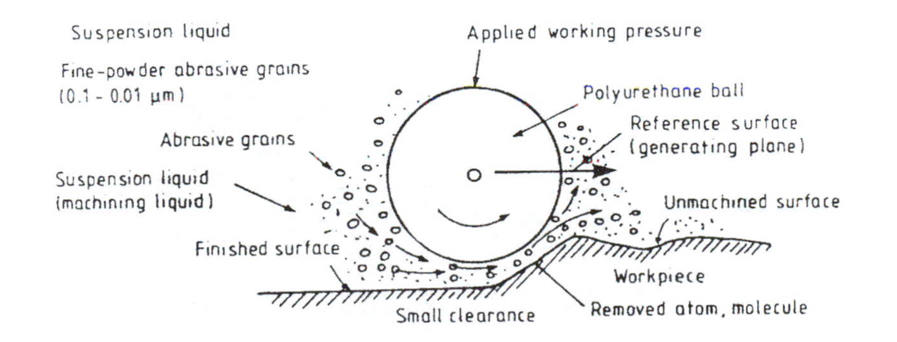

Figure 10.82. Elastic emission machining with rotating elastic sphere loaded against a work surface in the presence of abrasive Particles, from Whitehouse[227].

on a shaft driven by a variable speed motor with axis of rotation approximately 45° relative to the work surface. The work is submerged in a slurry of ZrO_2 or Al_2O_3 abrasive particles ranging in size from 20 nm to 20 μm. In most cases a fluid gap on the order of 1μm is maintained between the sphere and the work surface being polished and the abrasive particles range in size from 0.1 – 0.6 μm. Removal rate is approximately linear with dwell time.

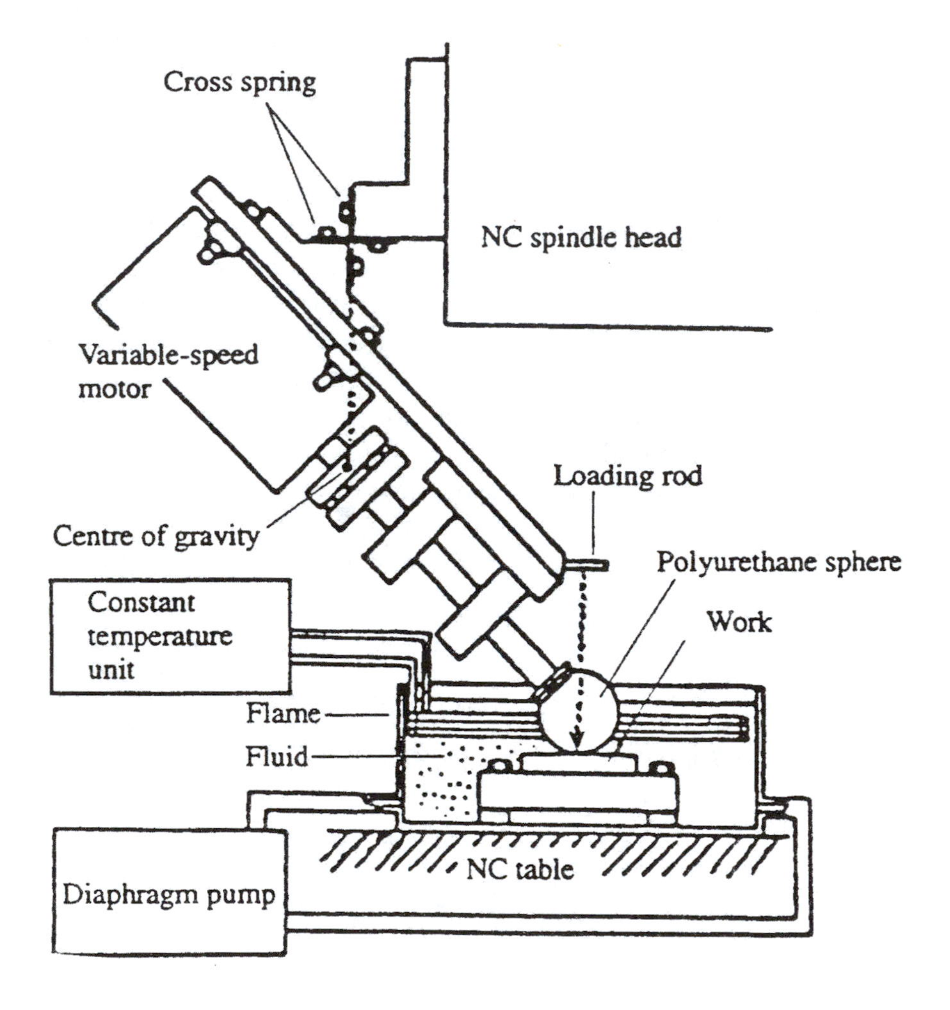

Figure 10.83. Schematic of EEM apparatus used on a NC machine, from Tsuwa[286].

XI PRECISION MANUFACTURING APPLICATIONS AND CHALLENGES

11.1 Introduction

We will focus here on precision manufacturing processes and equipment related to unique applications for which the manufacturing capability is the enabler for the product, starting with the semiconductor industry. This is logical as this industry segment accounts for tremendous growth and sales that drive, literally, the rest of the economy. It continues to influence new product development having been the basis of the creation of micro electro-mechanical systems (MEMS) and, more recently, nanotechnology. The Semiconductor Industry Association (SIA) reports that the "chip market" worldwide approached the $230 billion mark in 2005 (and the US market share is about 60%). In 2005, the semiconductor industry made over 90 million transistors for every man, woman and child on Earth, and by 2010, this number should be 1 billion transistors.[287] Other industries certainly contribute to the US and global economy. But semiconductors drive the development of key process technology that, eventually, ends up impacting other industries, autos, for example. And this is not counting the significant usage of computers and related control devices in autos and other products already. Recently, the growth in alternate energy sources (photovoltaics, for example) is linked to capability in basic semiconductor processes. Finally, the growth in nanotechnology and applications will drive precision manufacturing as well. Much of that process technology is based in semiconductor-like processes. So, we will consider here the semiconductor industry the source or headwaters of the stream of precision technology flow and, often slowly but surely, the flow influences the more traditional industries.

What drives this? If you recall the discussions in the early part of the text regarding the role of important pieces of technology in the development and control of certain markets, then we can extend that analysis directly to this industry. So the industry (or company) that connects best the mastery of the technology with the insight into the drivers will be the most successful. Two notable and somewhat controversial examples are Intel and Microsoft. In fact, Microsoft has been so successful and commands so much of the market (and, hence, the market for follow-on products) that the US government considered anti-trust restraints against it[288]. The argument is, interestingly, that the traditional role of the marketplace to "restrain" successful companies from becoming monopolies may not apply to the high-tech industry when certain companies (and their technology) become so ubiquitous.

If you read any of the semiconductor industry trade magazines (e.g., *Semiconductor International*) it is clear that the pressure for change in product design is continuous and strong, first in microprocessors and then for other devices. This translates into steady reductions in size but increases in capacity. In 2005, the industry continued to maintain the pace predicted by Moore's Law – the doubling of transistors every two years even in the face of continued technical challenges. Transistor speed continued to improve at the historical improvement rate of 17 percent per year[289]. This implies smaller features to be created in manufacturing as well as tighter tolerances on size (line width for example), form (line straightness or feature parallelism for example), surface roughness and, increasingly important, subsurface damage. Figure 11.1 below illustrates the steady reduction in area/function required for DRAM devices[290]. Thus, the products as well as the machines that fabricate the products require higher and higher levels of precision and repeatability in performance to guarantee manufacturing tolerances are met and product specifications can be obtained. And, on top of it all the industry requires that this be done at faster and faster speeds. SEMATECH, the semiconductor industry research and development organization predicts that the next ten years will see tremendous changes in industry productivity but will face a few "show stoppers" as physical dimensions approach 0.15 μm, the limit of optical lithography. There are others as well related to materials and ease of processing – for example, the so-called "low-k" dielectrics.

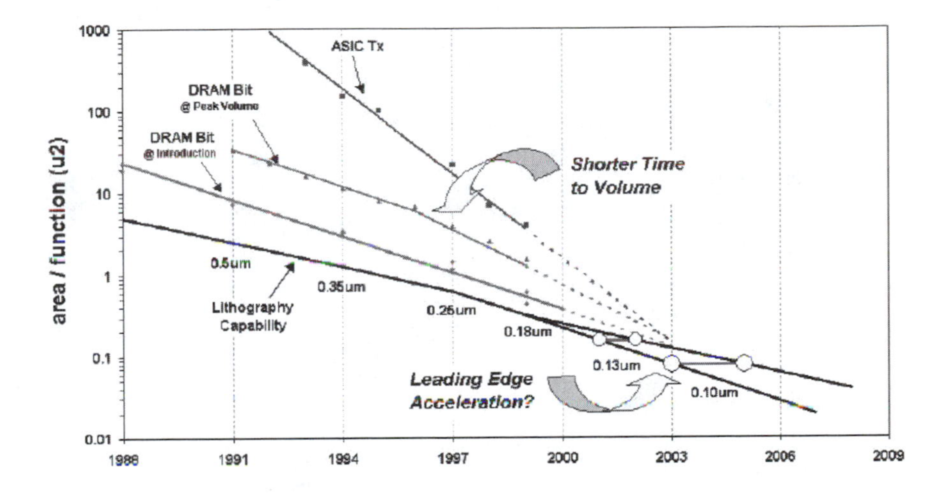

Figure 11.1. Required reduction in size pushed by "technology acceleration", from Landler[290].

The impressive advances are well illustrated in Figure 11.2 from *Semiconductor International*, 1997[287]. Here the decrease in cost per function over time (where cost per function is defined as cost per bit in DRAM's, cost per million instructions per second, MIPS, for microprocessor devices) for semiconductor manufacturing is a result of the combined benefits of reducing feature size (recall Taniguchi?!), increasing the size of wafers, improving process yields and improving equipment productivity. According to the report, these have traditionally accounted for a 25-30% reduction in the cost of functions.

Design rules (read as the size of the minimum feature created by lithography) drive this in the first place. The steady decrease in feature sizes have put the present capability near the limit of optical lithography. Other technology is being developed based on e-beam, X-ray and extreme-UV. Following line widths, the other key driver is wafer size.

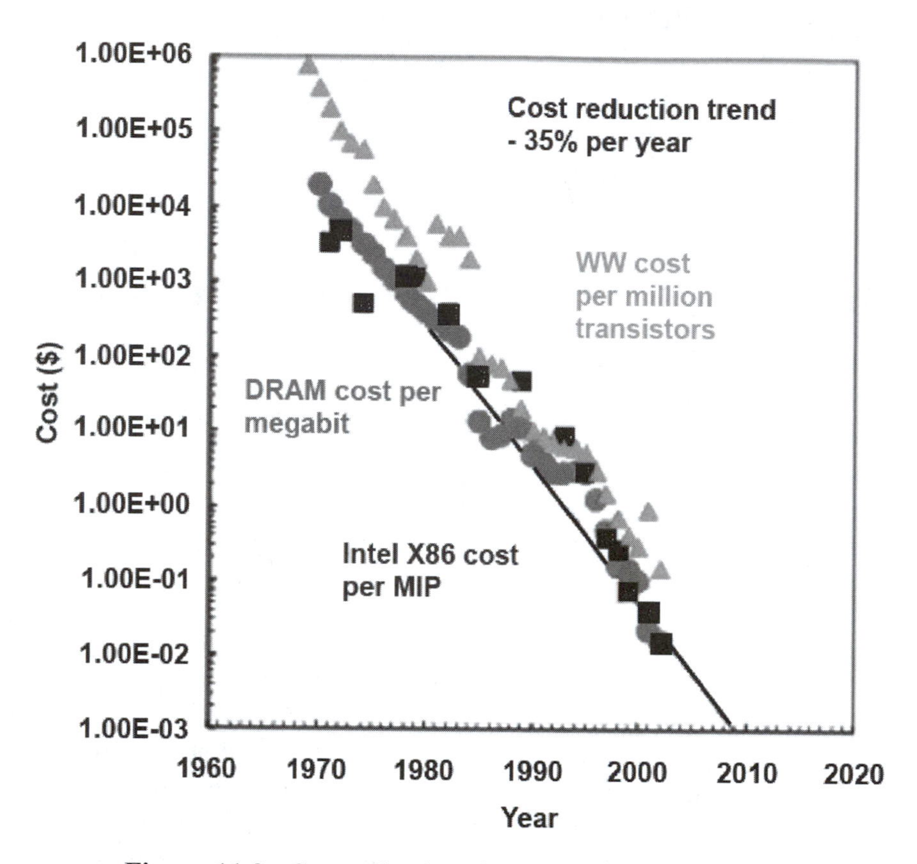

Figure 11.2. Cost reductions in basic building blocks of electronic devices, from ITRS[287].

Appreciate the fact that over the past 25 years we have seen wafer sizes increase from 75 mm (roughly 3 inch diameter) to the present 300 mm wafer with the 450 mm (16 inch) wafer to follow around 2008. This poses interesting challenges in processing – specifically in maintaining focus during lithography over such a large surface. Ditto for deposition and planarization over such a large area.

The changes in lithography technology and wafer size are of interest to us here as they impact such precision manufacturing processes as optics manufacturing and planarization processes, such as chemical-mechanical polishing. We will discuss some of these later in this chapter. In addition, computer peripheral device manufacture, such as read-write heads, are following the same trends of miniaturization, need to increase yields and process throughput. This chapter will look first at some of the fundamentals of semiconductor device manufacturing (definition of terms) and then some basics and processes related to two other related technologies: MEMS and nanoscale device technology.

11.2 Basic semiconductor device manufacturing

11.2.1 Introduction

December 23, 1947 marks the "birth" of the first point contact transistor (so it is over 50 years old) and the transistor in the form we know today (bi-polar) was developed a month later, both events occurring at Bell Labs thanks to John Bardeen, Walter Brattain and William Shockley, respectively. The integrated circuit (IC) appeared some ten years later in 1958 at Texas Instruments due to the work of Jack Kilby. Schockley went on to start his own firm on the West coast in what was to become Silicon Valley and this firm spawned a number of small start ups, some 13 in 1968 alone started by former employees. Perhaps the best known of this group of spin offs is one small start up at the time named Intel. And the rest, as they say, is history. The history of the development of this technology and then field is fascinating. Many good references exist for this.[390]

Growth of the industry was fast and continuous as the techniques for fabrication, applications and design technology improved. Transistors cost on the order of $1 a piece in 1960 and are now around $1 for a million thanks to the sophistication of the IC. The growth has followed quite a consistent pattern, now called "Moore's Law" after Gordon Moore, a co-founder of Intel and one of the original employees of Shockley's little company. Moore's Law says that the number of transistors per chip doubles every 18 months. Moore's "second law" states that fabrication facilities for IC chips double in cost with each new generation. And the generations can be as close as 4 years apart! So, there is great pressure on reducing the production costs of these devices. This is accomplished by increasing the yield of the process, increasing the throughput, and increasing the wafer size. If the same growth in performance pattern observed by Moore was applied to other industry products we'd see aircraft

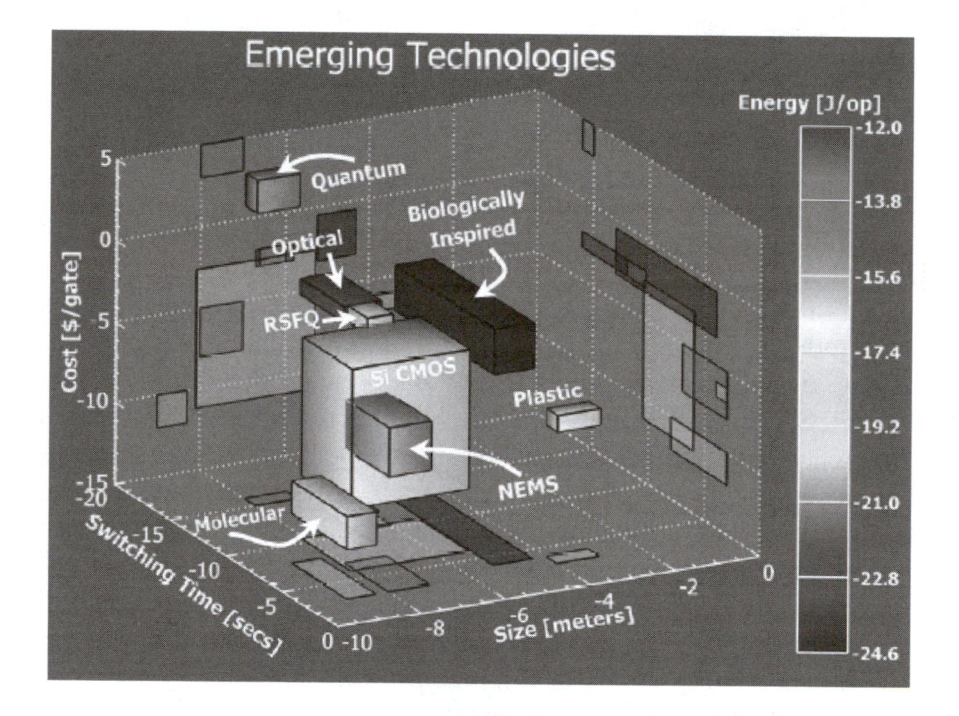

Figure 11.3. Parameterization of Emerging Technologies and CMOS – Speed, Size, Cost, and Switching Energy, from ITRS [291].

traveling at the speed of light and automobiles consuming as little as a thimble full of gas for the same distance traveled previously on a conventional tankful!

One way to visualize where thing are going is to look at the technology from different perspectives, Figure 11.3, from ITRS Roadmap[291]. The figure highlights the relative positions of the emerging technologies in the CMOS application space. It shows the benefit to be derived from heterogeneous integration of the emerging technologies with silicon to expand its overall application space.

11.2.2 So, what are they anyway and how are they made?

11.2.2.1 Microfabrication: background and overview

The term “microfabrication” is used interchangeably to describe technologies that originate from the microelectronics industry as well as small tool machining for mechanical parts production. In this chapter we will cover both but first focus on microfabricated devices including integrated circuits (“microchips”), microsensors (e.g. air bag sensors), inkjet nozzles, flat panel displays, laser diodes, and similar products. Similarly, the term “micromachining” is used for semiconductor processing as well as mechanical machining at the micron scale. Overall, microelectronic fabrication, semiconductor fabrication, MEMS fabrication and integrated circuit technology are terms used instead of microfabrication, but microfabrication is the broad general term. There are a number of references that go into some detail on these topics that the reader is referred to[292-299].

Traditional machining techniques from the conventional scale of processes like milling and turning to the “non-conventional” such as electro-discharge machining, spark erosion machining, and laser drilling have been scaled from the millimeter size range to micrometer and, now, nanometer range. But, they do not share the main idea of microelectronics-originated microfabrication: e.g. replication and parallel fabrication of hundreds or millions of identical predominantly two dimensional structures. This parallelism is present in various imprint, casting and molding techniques which have successfully been applied at the microscale. For example, roller printing of structures for fuel cell membranes and injection molding of compact discs involves fabrication of micrometer-sized spots on the disc. Injection molding of lenses involves creation of micron scale form features on optical devices.

Microfabrication refers to a set of technologies utilized to produce microdevices. Many of the technologies are derived from very different processes and "arts", often not connected to manufacturing in the traditional sense. For example, lithography derives from early printing techniques using etched plates to transfer patterns to paper. Planarization technology, formerly referred to as only polishing, comes from optics manufacturing dating back to the time of early astronomers and physicists. Much of the vacuum techniques also come from 19th century physics research. Electroplating is also a 19th century technique adapted to produce micrometer scale structures, as are various stamping and embossing techniques.

In the fabrication process for microdevics, a number of processes must be performed, in a defined sequence, often repeated many times. In the fabrication of memory chips, some 30 lithography steps, 10 oxidation steps, 20 etching steps, 10 doping steps, etc. are carried out as part of this process, Figure 11.4 below, from IBM[300]. Typical process steps include:

- Photolithography
- Etching (microfabrication), such as RIE (Reactive-ion etching) or DRIE (Deep reactive-ion etching)
- Thin film deposition, see e.g. sputtering, CVD Chemical
- vapor deposition, evaporation
- Epitaxy
- Thermal oxidation
- Doping by thermal diffusion or ion implantation bonding

- Chemical-mechanical planarization (CMP)
- Wafer cleaning also known as "surface preparation"

The complexity of microfabrication processes can be described using a number of measures but "mask count" is typical. Mask count refers to the number of different pattern layers that will make up the final microelectronic device, Figure 11.4. Modern microprocessors are made with 30 masks while only a few masks may be used for a microfluidic device or a laser diode. The fabrication process is not unlike multiple exposure photography in that many individual patterns (each on a mask) must be aligned with each other in the various layers of the process to create the final structure. In between the stages of fabricating these layers a number of other critical process steps occur as seen in Figure 11.5 (for example,

Figure 11.4. Cross-section of 64-bit high-performance microprocessor chip with CMOS technology with Cu-low-k wiring, Courtesy of International Business Machines Corporation. Unauthorized use not permitted[300].

etch/strip and chemical mechanical planarization, CMP). The masks used in photolithography constitute a major portion of the cost of processing the microdevice and, recently, a number of so called "maskless techniques" relying on writing processes without the mask have been discussed[301].

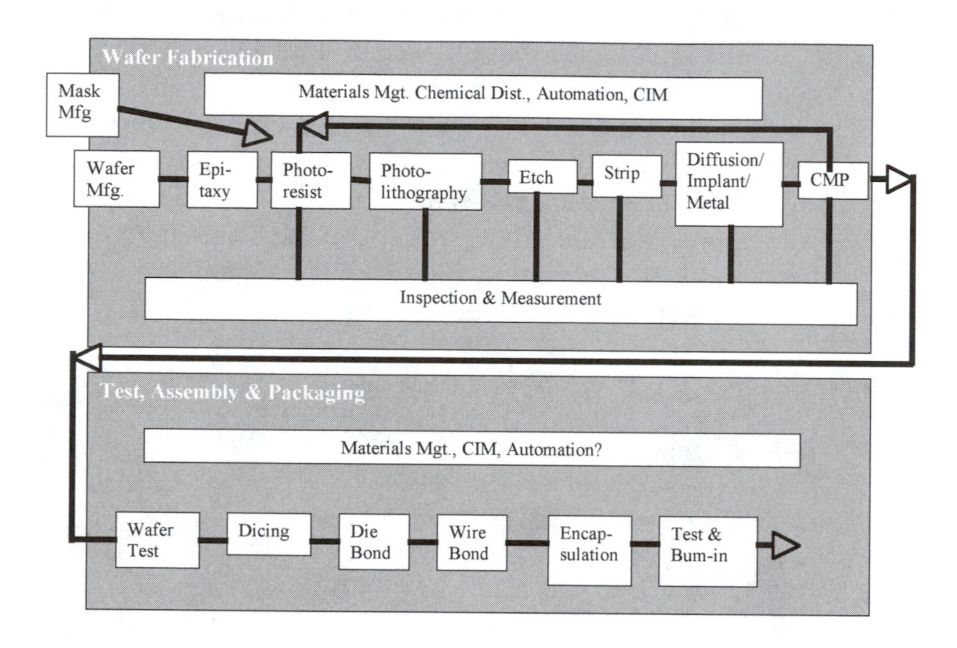

Figure 11.5. Diagram of semiconductor manufacturing process steps.

We cannot cover in great detail all of the steps in the figure above but will concentrate on a few more prominent process steps having a strong precision machine or manufacturing basis. We have already discussed chemical mechanical planarization (CMP) in chapter 10.

11.2.2.2 Lithography

A major component of semiconductor fabrication is photolithography. The lithography process is the means whereby patterns are transferred onto a substrate (e.g. silicon, Ga As, etc.). The pattern is used to isolate areas for subsequent etching to create trenches for interconnects and lines or to protect the substrate from etching. The patterns are written on glass plates called reticles, much like the glass slides used in earlier forms of photo presentations with projectors. These are the masks referred to in the previous section. Lithography is used because it allows exact control over the shape and size of the features created, and because it can create patterns over an entire surface simultaneously. The main disadvantages are that it is primarily used for creating 2-D (i.e. "flat") structures, and, as with other semiconductor processes, requires extremely clean operating conditions. In a complex integrated circuit, (for example, CMOS) a wafer will go through the photolithographic cycle up to 50 times. Lithography machines are designed to enhance throughput but necessarily require sophisticated mechanical structures, control and metrology to maintain pattern quality at high exposure speeds.

Photolithography involves a number of steps in a series of often repeated combinations including:

- substrate preparation
- photoresist application
- soft-baking
- exposure
- developing
- hard-baking
- etching

and various other chemical treatments (thinning agents, edge-bead removal etc.) in repeated steps on an initially flat substrate, see figure 11.6 below.

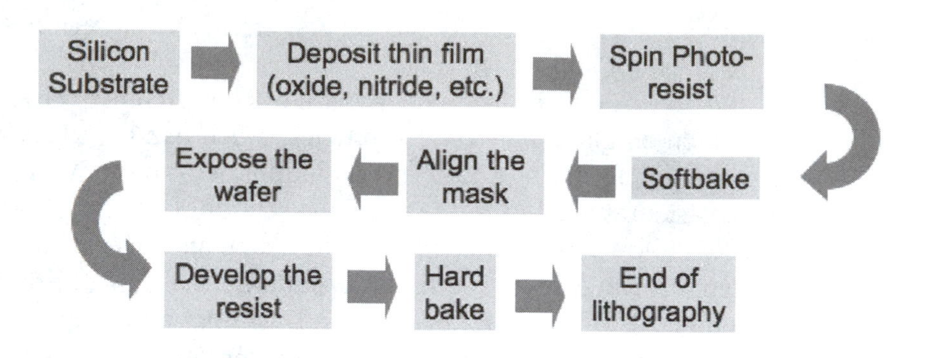

Figure 11.6. Basic process steps in photolithography.

A typical cycle of silicon lithography would begin with the deposition of a layer of conductive metal several nanometers thick on the substrate[302, 303]. A layer of photoresist – a chemical that 'hardens' when exposed to light (often ultraviolet) – is applied on top of the metal layer by spinning the substrate under a stream of photoresist. The mask, basically a transparent plate with opaque areas printed on it, is placed between a source of illumination and the wafer, selectively exposing parts of the substrate to light. The photoresist is then developed during which areas of unhardened photoresist undergo a chemical change. After a hard-bake, a series of subsequent chemical treatments etch away the conductor under the developed photoresist, and then etch away the hardened photoresist, leaving the conductor exposed in the pattern of the original photomask, see figure 11.7 below.

A characteristic of of photolithography cleanroom environments is that the filtered fluorescent lighting contains no ultraviolet or blue light to prevent accidental exposure of the photoresist. Most types of photoresist are available as either "positive" or "negative" resists. With positive resists the area that is opaque (masked) on the photomask corresponds to the area where photoresist will remain upon developing (and hence where conductor will remain at the end of the cycle). Negative resists will create the inverse – any area that is exposed will remain, while any areas that are not exposed will be developed. After developing, the resist is usually hard-baked before subjecting to a chemical etching stage which will remove the metal underneath.

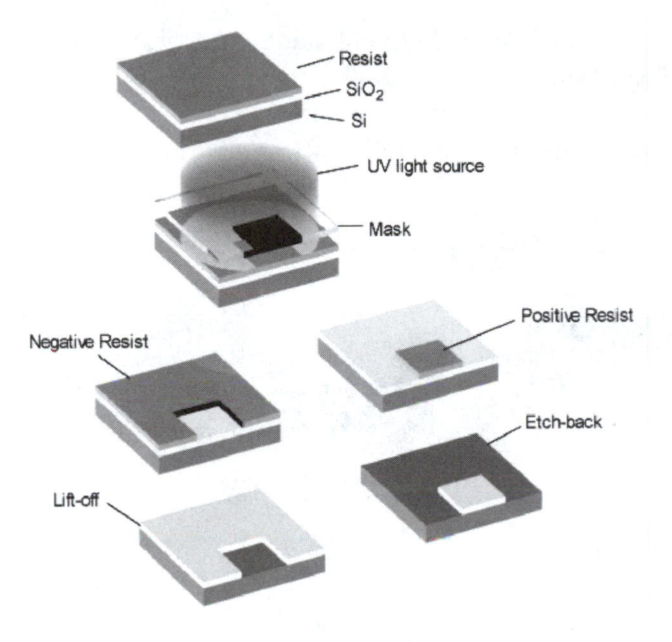

Figure 11.7. Principle process steps in photolithography, from ICKnowledge[304].

It is expected that immersion lithography will be used by 2009 to print 45 nm lines and spaces. As of 2006, immersion lithography has been introduced in the 45 nm node by most chipmaking companies, including IBM, Toshiba, STMicroelectronics and others[407, 409]. The interest in this technology is driven by the influence of a water "lens" element on the effective numerical aperture capability of the process. The numerical aperture is the product of the refractive index of refraction and the maximum angle of propagation of light, relative to the surface. The numerical aperture has a maximum possible value equal to the refractive index. In an air-based optical system, the numerical aperture has a maximum value of 1.0. Hence, the advantage of using immersion is the ability of the numerical aperture to go above 1.0, thus enhancing the resolution without changing the light source. The absolute resolution limit is given by a quarter-wavelength divided by the N.A. The current maximum N.A. is 1.35 for water-based immersion.

Figure 11.8 shows typical stage control metrology setup and Figure 11.9, both from ICKnowledge[305] shows the immersion lithography stage setup. Figure 11.10 shows a commercial photolithography machine introduced in 2007 for immersion lithography.

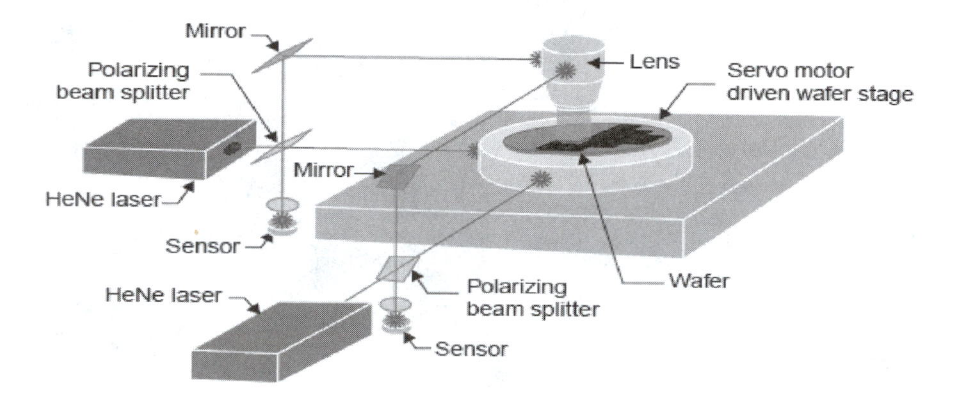

Figure 11.8. Typical stage control metrology, from ICKnowledge[305].

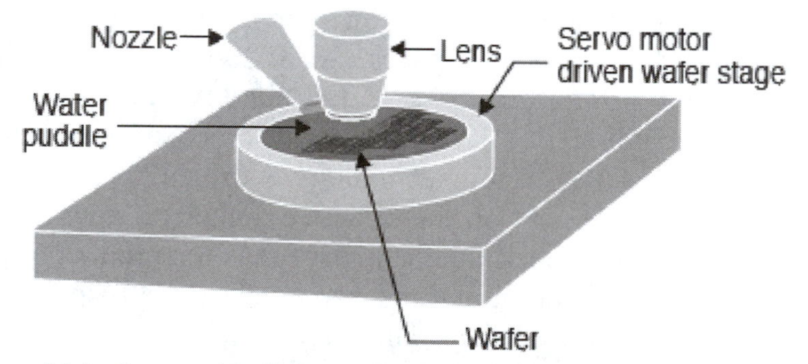

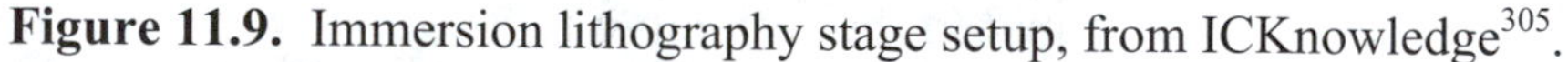

Figure 11.9. Immersion lithography stage setup, from ICKnowledge[305].

After introduction of immersion lithography, additional enhancements will extend the use of the technology to smaller features following Moore's Law. Enhancements such as the use of higher refractive index materials in the final lens, immersion fluid, and photoresist are likely.

An alternative to using a mask, is to directly expose the resist using an excimer laser, electron beam or ion beam.[304] These systems do not require a mask and are known as Direct Write to Wafer methods (DWW). These methods are also able to address the challenges of increasing density of features and the requirement of finer lithographic techniques. A comparison of a number of lithography methods is given below:

Method	Feature Size (μm)
UV Photolithography	1
Laser DWW	1-2
Electron Beam	0.25-0.1
Ion Beam	0.05-0.1
X-Ray Lithography	<0.1

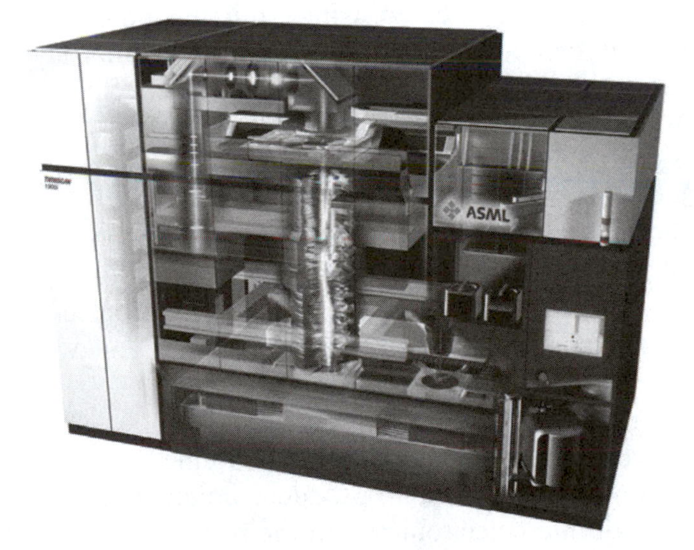

Figure 11.10. ASML Photolithography machine, Twinscan XT:1900i, from ASML[306].

11.3 Applications of semiconductor manufacturing – MEMS

We next review two current areas of application for semiconductor manufacturing technologies (or "semiconductor-like" processes), microelectromechanical systems (MEMS) and nanotechnology. Both topics are extensively covered in reference texts and the technical press/websites on their own and this discussion serves to introduce the basics of these processes. The reader is referred to the sources cited for more detailed information. We start with MEMS.

Micro-Electro-Mechanical Systems (MEMS) is, as the name suggests, the integration of mechanical elements, sensors, actuators, and electronics on a common silicon substrate usually through microfabrication technology[307, 308]. The field of MEMS has grown substantially in the last 2 decades. Early work on MEMS technology and applications was centered at the Berkeley Sensor and Actuator Center (BSAC), www-bsac.eecs.berkeley.edu, which has been active in MEMS since its inception in 1986. The micromechanical components in MEMS are fabricated using "micromachining" processes that selectively etch away parts or add new structural layers of the silicon wafer to form the mechanical and electromechanical devices. The electronics are fabricated using traditional integrated circuit (IC) processes (e.g., CMOS, Bipolar, or BICMOS processes).

The real advantage of MEMS, and now nanotechnology, is the use of IC-based fabrication technology, which usually result in 2-D "flat" structures, to make complex 3-D electromechanical systems and integrating these electromechanical elements with adjacent electronics.

MEMS is often referred to as an enabling technology since it allows the development of smart products, augmenting the computational ability of microelectronics with the perception and control capabilities of microsensors and microactuators and expanding the space of possible designs and applications. The BSAC website lists applications and products resulting from MEMS including[308]:

- Gyro Inertial Sensors and Accelerometers
- Lamb Wave Acoustic Sensors
- Acoustic Wave and Fluidic Micropumps and Mixers
- Comb-Driven MEMS Actuators
- MEMS Micropositioning Components & System for Hard Disk Drive
- Surface-Micromachined Gears, Cranks and Springs
- Anti-Stiction Elements, Dimpled Structures, and Surface Treatments
- MEMS Micro-Vibromotors
- Hinged, Fold-Out Micromachined Out-of-plane Robotic Structures
- X- and Y-Rastered Real-Time Projected Display System
- MEMS Based Free-Space Collimated Beam Communications/ Optics
- Piezioelectric MEMS Silicon-Diaphragm Microphone
- MEMS-Scanned Barcode Reader
- Microfluidic Host-Fueled Glucose Microbial Power Cell
- Room temperature, directed growth of Multiwalled carbon nanotubes and Silicon nanowires
- Localized thermal bonding for micropackaging and fluidic/ biosample encapsulation, and
- Precision controllable arrays of polymer lenses and mirrors for adaptive optics and imaging applications

These electro-mechanical microsystems are often able to sense and control the environment they operate in. Recent examples include sensors that gather information from the environment through measuring mechanical, thermal, biological, chemical, optical, and magnetic phenomena. The integral electronics then process the information derived from the sensors and, utilizing logic or embedded decision making capability, control the response of the actuators' motion, position, or other action (e.g. pumping, or filtering) to affect some desired outcome or purpose. Since the fabrication of MEMS devices relies on the same batch manufacturing techniques used for integrated circuits, the same functionality, quality, reliability, and sophistication can be placed on a silicon chip at a reasonable cost. And, many of the same design tools utilized in IC design can be applied to MEMS.

As can be seen from a number of elements in the above list there has been a natural extension of MEMS technology into nanotechnology. In fact, it is sufficient to regard nanotechnology as an outgrowth (miniaturization?!) of MEMS since the two are so closely linked now. We will go into nanotechnology in substantially more detail in the next section.

11.4 Nanotechnology

11.4.1 Background and definitions

Nanoscale features and capabilities have long been common in precision engineering and precision manufacturing. Concerns about dimensional accuracy of lens and features on machines and instruments at the nanometer level can be found over the history of precision engineering. The Tangiuchi curve, Figure 1.2, shows achievable machining accuracies for ultra-precision processes approaching one nanometer in the 1980's. Spectroscopists, astronomers and other physics researchers were interested in diffraction gratings for measuring, among other things, the lengths of light waves for astronomical observation. Specifications for straightness and location of these gratings were better than one millioneth of an inch (that's 0.000025 mm, or 25 nm) as "late" as 1947, and the best gauge blocks (Johansson "Jo" blocks) are accurate to +/- 5 nanometers- those have been with us since the late 1800's[309].

The McGraw-Hill Dictionary of Scientific and Technical Terms defines *nanotechnology* as,

> "Systems for transforming matter, energy, and information that are based on nanometer-scale components with precisely defined molecular features."[310]

Just after that definition the dictionary adds the following:

> "Techniques that produce or measure features less than 100 nanometers in size."

Sounds like precision manufacturing!

Nanotechnology traces its roots to a lecture given at the American Physical Society annual meeting by Nobel-laureate physicist, Richard Feynmann, in 1959 entitled "There's plenty of room at the bottom."[311] In that lecture, Feynmann proposed the possibility of "manufacturing" objects that move and function at the level of biological system components…that is, at the nano-scale. These objects can be built of nanostructured materials, chemical or biological elements and are fabricated either using "top down" approaches (as with conventional semiconductor manufacturing – lithography, deposition, growth, etc.) or, more recently, "bottom up" approaches involving self-assembly of constituent components into larger structures. Nano-systems include, at a smaller scale, most of the capabilities previously associated with micro-scale systems such as micro electro-mechanical systems (MEMS). The term NEMS is now applied to nano-scale equivalents. Applications range from biological nano-systems for chemical and biochemical analysis (lab-on-a-chip), minimally invasive surgical techniques, micro-instrumentation and micro-manipulators, nano-wires for even smaller versions of sensors and analytical tools developed based on MEMS technology and building transistors and memory. There are several excellent references on this whole field, the most comprehensive being the *Springer Handbook of Nanotechnology* edited by Bharat Bushan with the 2nd edition published in 2007[293].

A common building block of nano-scale systems is the carbon nanotube. Graphenes are the 2-D counterparts of 3-D graphite. Single wall carbon nanotubes, SWNT, are graphene cylinders with nanometer diameters. They are single layers of carbon atoms densely packed into a benzene-ring structure. Some carbon nanotubes have hemispherical graphene caps at each end comprised of 6 pentagons. In addition to carbon nanotubes, many carbon-based materials, such as graphite and large fullerines (e.g., "Bucky balls" named after R. Buckminster Fuller) have graphene properties[312].

We will not discuss details of the formation of SWNTs here and refer the reader to reference texts, such as Bushan, cited above. These small tube structures have diameters typically in the 1.0-1.5 nm range. The achievable diameter appears to be due to a balance between stress and formation energy considerations and synthesis techniques. Length is not constrained but can be affected by synthesis conditions – for example, thermal gradients. Typical lengths are reported from a few microns to a millimeter. Thus, these are essentially single molecules of very high aspect ratio (length/diameter), Figure 11.11 below, from Dresselhaus[313].

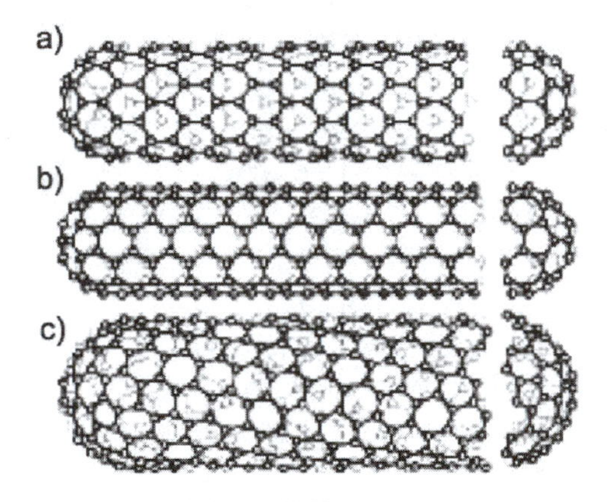

Figure 11.11. Sketch of three different SWNT structures as examples for (a) a zig-zag-type nanotube, (b) an armchairtype nanotube, (c) a helical nanotube, from Dresselhaus[313].

Synthesis challenges for SWNTs include controlling the quality or purity, and the form or configuration (the helix nature is often referred to as "chirality"). This is difficult because many of the details of the mechanism of nanotube growth are not clearly understood.

Three processes are currently used for SWNT production: laser ablation, solar energy and electric arc. All three techniques employ high temperatures (between 1000 and 6000°K) and rely on the erosion of solid graphite as the source of carbon. All three methods utilize the energy transfer resulting from the interaction between either a target material and the radiation source (either laser or solar energy) or an electrode and a plasma (for the electric arc). The resulting energy transfer in all three methods drives the sublimation of graphite and subsequent atomic recombination eventually resulting in the formation of carbon nanotubes. Either continuous or pulsed lasers are used to vaporize graphite to be re-deposited as soot on a suitable target. The electric arc method vaporizes carbon in the presence of catalysts contained in the anode (usually in the form of graphite powder and catalyst in a pocket in the anode) in an inert gas. SWNTs are deposited in various locations of the reactor device containing the plasma. Both single wall and multiwall nanotubes are formed. Solar furnace techniques are similar to the electric arc reactor. In a solar furnace a mixture of graphite powder and catalysts is placed in a graphite crucible at the focal point of the rays of a solar collector in a controlled atmosphere. Variables affecting quality and rate of growth include temperature, gas composition and pressure/vacuum, catalysts used, and location of tube formation.

11.4.2 Nanostructured materials

In addition to their very small size and structures, nanomaterials such as SWNT have impressive properties. The table below is a relative comparison of the mechanical properties of various materials from nanotubes to wood. Specific alloys of some materials are not given. This table gives a simple comparison across some common properties[314].

Table 11.1. Comparison of carbon nanotubes to other typical materials.

Material	Young's modulus (GPa)	Tensile Strength (GPa)	Density (g/cm^3)
Single wall Nanotube	1054	150	-
Multi wall Nanotube	1200	150	2.6
Carbon Steel	208	0.4	7.8
Epoxy	3.5	0.005	1.25
Wood	16	0.008	0.6

With respect to carbon nanotubes, a detailed list of specific carbon nanotube characteristics, including equilibrium structure and thermal, electrical and mechanical properties, is listed below[314].

Equilibrium Structure

Average Diameter of SWNT's	1.2-1.4 nm
Distance from opposite Carbon Atoms (Line 1)	2.83 Å
Analogous Carbon Atom Separation (Line 2)	2.456 Å
Parallel Carbon Bond Separation (Line 3)	2.45 Å
Carbon Bond Length (Line 4)	1.42 Å
C-C Tight Bonding Overlap Energy	~2.5 eV
Group Symmetry (10, 10)	C5V

Lattice: Bundles of Ropes of Nanotubes:
Triangular Lattice(2D)
Lattice Constant 17 Å
Lattice Parameter:
(10, 10) Armchair 16.78 Å
(17, 0) Zigzag 16.52 Å
(12, 6) Chiral 16.52Å
Density:
(10, 10) Armchair 1.33 g/cm3
(17, 0) Zigzag 1.34 g/cm3
(12, 6) Chiral 1.40 g/cm3

Interlayer Spacing:
(n, n) Armchair 3.38 Å
(n, 0) Zigzag 3.41 Å
(2n, n) Chiral 3.39 Å

Optical Properties
Fundamental Gap:
For (n, m); n-m is divisible by 3 [Metallic] 0 eV
For (n, m); n-m not divisible by 3 [semiconducting] ~0.5 eV

Electrical Transport
Conductance Quantization n x (12.9 kW)-1
Resistivity 10-4 W·cm
Maximum Current Density 1013 A/m2

Thermal Transport
Thermal Conductivity(Room Temperature) ~ 2000 W/m•K
Phonon Mean Free Path ~ 100 nm
Relaxation Time ~ 10-11 s

Elastic Behavior
Young's Modulus (SWNT) ~ 1 TPa
Young's Modulus (MWNT) 1.28 TPa
Maximum Tensile Strength ~30 GPa

This information is provided as reference on the basic building block of nanostructured materials.

We will now go into some detail on some common fabrication techniques for nanostructured materials and devices.

11.4.3 Nanofabrication techniques

Nano-scale manufacturing may be characterized by top-down or bottom-up approaches. Top-down refers to machining or etching larger parts down to size, while bottom-up describes forming materials or devices additively from smaller components (individual molecules or atoms). Top-down manufacturing approaches can be further categorized into photolithographic processes, imprint lithography and single point operations. Bottom up approaches include vapor processes, plasma or ion processes, as well as single point operations. The details of each of these classes of nano-scale manufacturing processes are described briefly below and then, in sections, some are more completely covered.

TOP-DOWN

Photolithography

Photolithographic processes refer to those commonly used to manufacture semiconductor products. These processes include etching, furnace steps, ion implantation, chemical vapor deposition (CVD), metallization and cleaning in addition to photolithography itself.

The resolution attainable by photolithographic processes is rapidly increasing, in accordance with the International Technology Roadmap for Semiconductors (ITRS), and gate length after etching is projected to reach 20nm by 2009, from 32nm today.

Imprint Processes

Nano-scale imprint processes are also referred to as nanoembossing lithography or nanoimprint lithography. Imprinting involves pressing a hard (metal) mold into a soft layer of thermoplastic or photoresist that can be hardened with heat or UV light respectively. The hardened layer may be left as is or lifted off once another material is deposited in the pattern gaps.

Molds are generally made using electron beam lithography or photolithography. Molds are often used between 1 and 30 times, though degradation appears over time. Imprint lithography is capable of resolutions on the order of 1 to10nm, with less equipment, and corresponding cost, when compared to photolithography.

Single Point Operations

Electron beam lithography, plasmonic imaging , mechanical cutting, and material removal using scanning probe microscopy can all be classified as single point operations. These are distinguished from other top-down approaches in that material is removed in a serial rather than parallel fashion. This approach can be more time intensive, though it can offer sub-wavelength resolution, from 10nm to less than 1nm.

BOTTOM-UP

Vapor Processes

Chemical vapor deposition (CVD), the deposition of reactive gas on the surface of a substrate, is used in bottom-up as well as top-down manufacturing. Nanowires, composite nanowires, and carbon nanotubes (CNT) all may be produced using CVD (or PECVD), often in the presence of a catalyst.

Vapor-liquid-solid (VLS) processes involve a gaseous feed material being introduced to a liquid catalyst to form a solid crystal. VLS occurs at atmospheric pressure and is used to form nanowires, ultrafine lines composed of a single-material.

Molecular beam epitaxy (MBE) entails the direction of a flow or flows of ultrapure gases at high temperature under ultrahigh vacuum onto a single crystal substrate, and is used to form quantum dots.

Plasma/Ion Processes

Single ion implantation uses a low dose, low energy beam to implant ions into a crystal surface in ultrahigh vacuum and may be used to form quantum dots.

Plasma enhanced CVD (PECVD) employs the same mechanism as CVD but, by using a charged gas, allows films to form at lower temperatures.

Single Point Operations

Scanning probe microscopy, or proximity probing, has also been used to form three dimensional nanostructures, a primary example being the formation of the letters "IBM" with individual atoms of xenon. Such processes occurr in ultrahigh vacuum at extremely low temperature, using an atomic force microscopy (AFM) tip to locate each atom. Probes have also been used to form nanofibers by pulling liquid polymers.

We now go into more substantive detail on these processes. The remainder of the material in this section 11.4 is based on and extracted substantially from Bushan, 1st ed., *Springer Handbook of Nanotechnology*, Springer, 2004, pp. 170–180 n(see ref. 293). It has been edited for content and format.

The microfabrication techniques discussed so far were mostly geared toward fabricating devices in the 1mm to 1μm dimensional range using a variety of more "conventional" processes. Recent years have witnessed a tremendous growth of interest in fabricating sub-micro (1μm–100 nm) and nanostructures (1–100 nm range). This interest arises from both practical and fundamental view points. At the more scientific and fundamental level, nanostructures provide an interesting tool in studying electrical, magnetic, optical, thermal, and mechanical properties of matter at the nanometer scale. These include important quantum mechanical phenomena (e.g., conductance quantization, band-gap modification, coulomb blockade, etc.) arising from the confinement of charged carriers in structures such as quantum wells, wires, and dots, Fig. 11.12. On the practical side, nanostructures can provide significant improvements in the performance of electronic/optical devices and sensors. In the device area, investigators have been mostly interested in fabricating nanometer-sized transistors in anticipation of technical difficulties forecasted in extending Moore's law beyond 100 nm resolution. In addition, optical sources and detectors having nanometer-sized dimensions exhibit improved characteristics unachievable in larger devices (e.g., lower threshold current, improved dynamic behavior, and improved emission line-width in quantum dot lasers). These improvements create novel possibilities for next-generation computation and communication devices. In the sensors area, shrinking dimensions beyond conventional optical lithography can provide major improvements in sensitivity and selectivity. One can broadly divide various nanofabrication techniques into top-down and bottom-up categories.

The first approach starts with a bulk or thin film material and removes selective regions to fabricate nanostructures (similar to micromachining techniques). The second method relies on molecular recognition and self-assembly to fabricate nanostructures out of smaller building blocks (molecules, colloids, and clusters). The top-down approach is an offshoot of standard lithography and micromachining techniques. On the other hand, the bottom-up approach has more of a chemical engineering and material science flavor and relies on fundamentally different principles. In this section, we will discuss four major nano-fabrication techniques. These include:

i) e-beam and nano-imprint fabrication,
ii) epitaxy and strain engineering,
iii) scanned probe techniques, and
iv) self-assembly and template manufacturing.

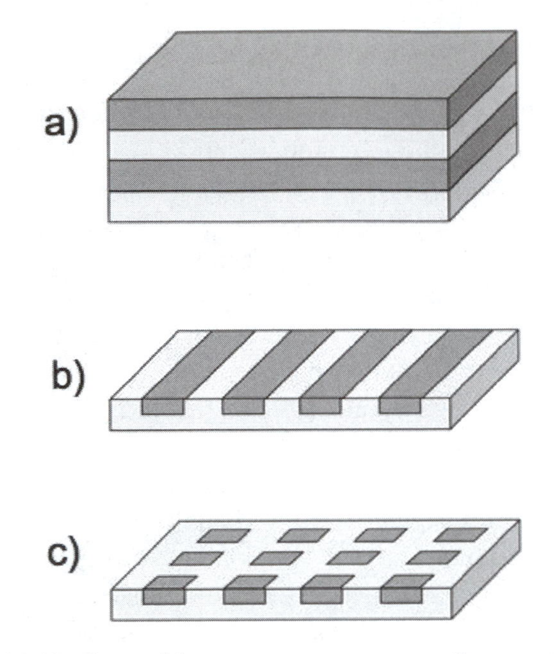

Figure 11.12. Several important quantum confinement structures, (a) quantum well, (b) quantum wire, and (c) quantum dot

11.4.3.1 E-beam and nano-imprint fabrication

In previous sections, we discussed several important lithography techniques used commonly in MEMS and microfabrication. These include various forms of UV (regular, deep, and extreme) and X-ray lithographies. However, due to the lack of resolution (in the case of the UV), or the difficulty in manufacturing mask and radiation sources (X-ray), these techniques are not suitable for nanometer-scale fabrication. E-beam lithography is an alternative technique for fabricating nanostructures that is gaining acceptance. It uses an electron beam to expose an electron-sensitive resist such as polymethylmethacrylate (PMMA) dissolved in trichlorobenzene (positive) or poly chloromethylstyrene (negative).[315] The e-beam gun is usually part of a scanning electron microscope (SEM), although transmission electron microscopes (TEM) can also be used. Although electron wavelengths on the order of 1Å can be easily achieved, electron scattering in the resist limits the attainable resolutions to > 10 nm. The beam control and pattern generation are achieved through a computer interface. E-beam lithography is serial and, hence, it has a low throughput. Although this is not a major concern in fabricating devices used in studying fundamental microphysics, it severely limits large scale nanofabrication. E-beam lithography, in conjunction with such processes as lift-off, etching, and electro-deposition, can be used to fabricate various nanostructures. An interesting new technique that circumvents the serial and low throughput limitations of the e-beam lithography for fabricating nanostructures is the nanoimprint technology[316]. This technique uses an e-beam fabricated hard material master (or mold) to stamp and deform a polymeric resist. This is usually followed by a reactive ion etching step to transfer the stamped pattern to the substrate. This technique is economically superior, since a single stamp can be used repeatedly to fabricate a large number of nanostructures. Figure 11.13a shows a schematic illustration of nano-imprint fabrication. First, a hard material (e.g., silicon or SiO2) stamp is created using e-beam lithography and reactive ion etching. Then, a resist-coated substrate is stamped, and, finally, an anisotropic RIE is performed to remove the resist residue in the stamped area. At this stage, the process is complete and one can either etch the substrate or, if metallic nanostructures are desired, evaporate the metal and perform a lift off. The resist used in nanoimprint technology can be thermal plastic, UV-curable, or thermal-curable polymer.

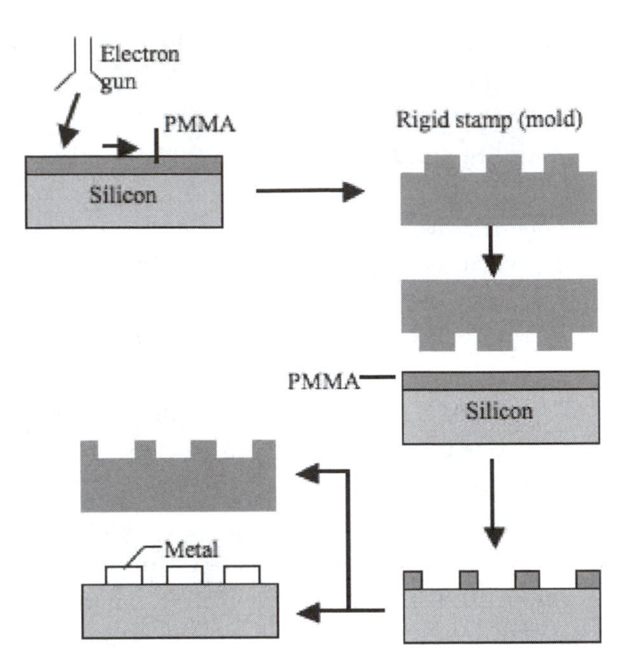

a. Schematic illustration of nano-imprint fabrication.

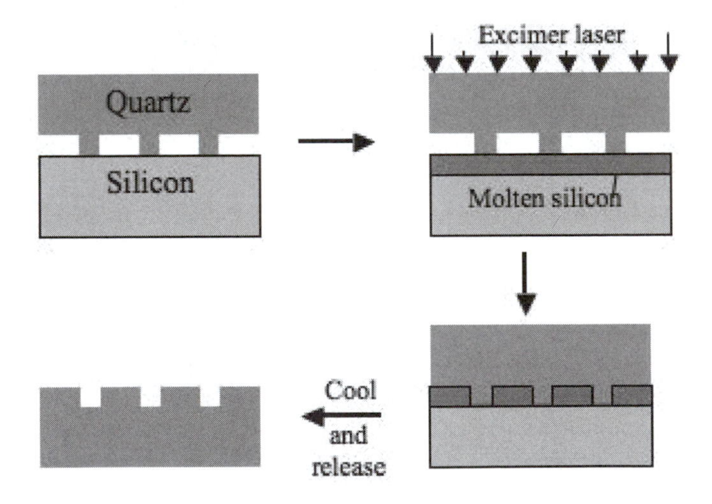

b. Ultrafast silicon nano-imprinting using an excimer laser.

Figure 11.13. Schematics of nano-fabrication techniques.

For a thermal plastic resist (e.g., PMMA), the substrate is heated to above the glass transition temperature (Tg) of the polymer before stamping and is cooled to below Tg before the stamp is removed. Similarly, the UV and thermal-curable resists are fully cured before the stamp is separated. Nano-imprint technology resolution is limited by the mold and polymer strengths and can be as small as 10 nm. More recently, the nano-imprint technique has been used to stamp a silicon substrate in less than 250 ns using a XeCl excimer laser (308 nm) and a quartz mask (laser assisted direct imprint, LADI), Fig. 11.13b[317].

11.4.3.2 Epitaxy and strain engineering

Atomic precision deposition techniques such as molecular beam epitaxy (MBE) and metallo-organic chemical vapor deposition (MOCVD) have proven to be effective tools in fabricating a variety of quantum confinement structures and devices (quantum well lasers, photodetectors, resonant tunneling diodes, etc.).[318-320] Although quantum wells and superlattices are the structures that lend themselves most easily to these techniques (see Fig. 11.14a), quantum wires and dots have also been fabricated by adding subsequent steps such as etching and selective growth. Fabrication of quantum well and superlattice structures using epitaxial growth is a mature and well developed field and, therefore, will not be discussed in this chapter. Instead, we will concentrate on quantum wire and dot nanostructure fabrication using basic epitaxial techniques[321, 322].

11.4.3.2.1 Quantum structure nanofabrication using epitaxy on patterned substrates

There have been several different approaches to the fabrication of quantum wires and dots using epitaxial layers. The most straight forward technique involves e-beam lithography and etching of an epitaxial grown layer (e.g., InGaAs on GaAs substrate)[323]. However, due to the damage and/or contamination during lithography, this method is not very suitable for active device fabrication (e.g., quantum dot lasers). Several other methods involving regrowth of epitaxial layers over nonplanar surfaces such as step-edge, cleaved-edge, and patterned substrate have been used to fabricate quantum wires and dots without the need for lithography and etching of the quantum confined structure[322, 324]. These nonplanar surface templates can be fabricated in a variety of ways such as etching through a mask or cleavage along crystallographic planes. Subsequent epitaxial growth on top of these structures results in a set of planes with different growth rates depending on the geometry or surface diffusion and adsorption effects. These effects can significantly enhance or limit the growth rate on certain planes, resulting in lateral patterning and confinement of deposited epitaxial layers and formation of quantum wires (in V grooves) and dots (in inverted pyramids). Figure 11.14a shows a schematic cross section of an InGaAs quantum wire fabricated in a V-groove InP.

As can be seen, the growth rate on the sidewalls is lower than that of the top and bottom surfaces. Therefore, the thicker InGaAs layer at the bottom of the V-groove forms a quantum wire confined from the sides by a thinner layer with a wider bandgap. Figure 11.14b shows a quantum wire formed using epitaxial growth over a dielectric patterned planar substrate. In both of these techniques, it is relatively easy to create quantum wells. However, in order to create quantum wires and dots one still needs e-beam lithography to pattern the groove and window templates.

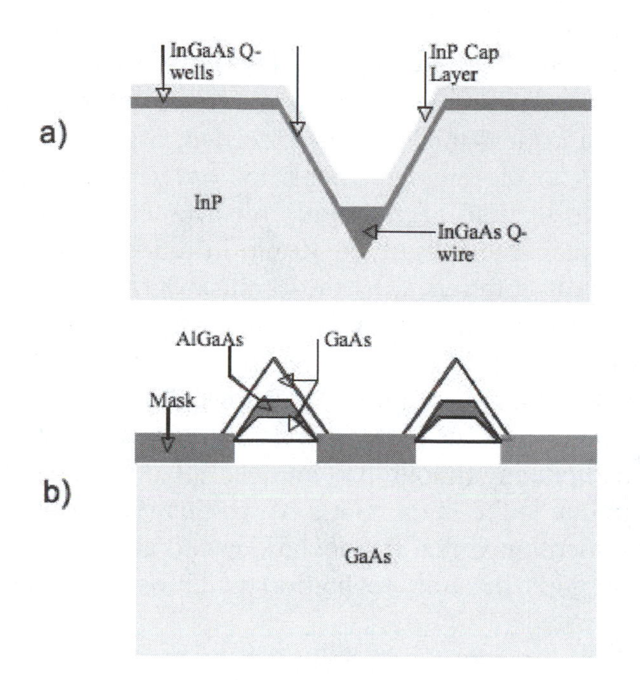

Figure 11.14. (a) InGaAs quantum wire fabricated in a Vgroove InP and (b) AlGaAs quantum wire fabricated through epitaxial growth on a masked GaAs substrate.

11.4.3.2.2 Quantum structure nanofabrication using strain-induced self-assembly

A more recent technique for fabricating quantum wires and dots involves strain-induced self-assembly[322, 325]. The term self-assembly represents a process whereby a strained 2-D system reduces its energy by a transition into a 3-D morphology. The most commonly used material combination for this technique is the InxGa1−xAs/ GaAs system, which offers a large lattice mismatch (7.2%between InAs and GaAs)[326, 327], although recently Ge dots on Si substrate have also attracted considerable attention[328]. This method relies on lattice mismatch between an epitaxial grown layer and its substrate to form an array of quantum dots or wires. Figure 11.15 shows a schematic of the strain-induced self-assembly process. When the lattice constants of the substrate and the epitaxial layer differ considerably, only the first few deposited monolayers crystallize in the form of an epitaxial strained layer in which the lattice constants are equal. When a critical thickness is exceeded, a significant strain in the layer leads to the breakdown of this ordered structure and the spontaneous formation of randomly distributed islets of regular shape and similar size (usually < 30 nm in diameter).This mode of growth is usually referred to as the Stranski-Krastanow mode. The quantum dot size, separation, and height depend on the deposition parameters (i.e., total deposited material, growth rate, and temperature) and material combinations. As can be seen, this is a very convenient method to grow perfect crystalline nanostructures over a large area without any lithography and etching. One major drawback of this technique is the randomness of the quantum dot distribution. It should be mentioned that this technique can also be used to fabricate quantum wires by strain relaxation bunching at the step edges.

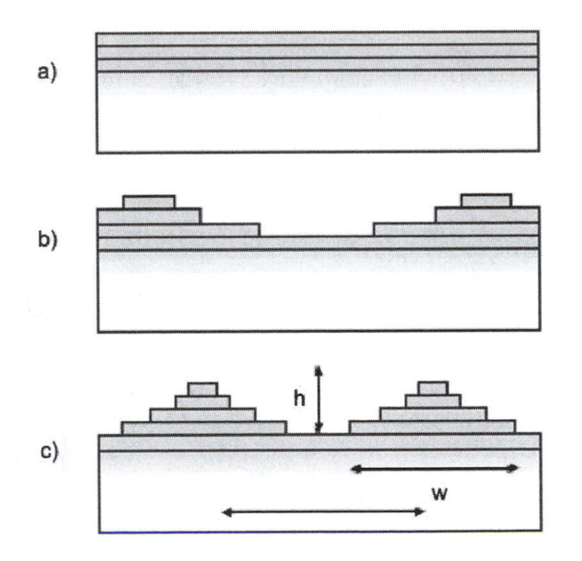

Figure 11.15. Stranski-Krastanow growth mode, (a) 2-D wetting layer, (b) growth front roughening and breakup, and (c) coherent 3-D self-assembly.

11.4.3.3 Scanned probe techniques

The invention of scanned probe microscopes in the 1980s revolutionized atomic-scale imaging and spectroscopy. In particular, scanning tunneling and atomic force microscopes (STM and AFM) have found widespread applications in physics, chemistry, material science, and biology. The possibility of atomic-scale manipulation, lithography, and nanomachining using such probes was considered from the beginning and has matured considerably over the past decade. In this section, after a brief introduction to scanned probe microscopes, we will discuss several important nano-lithography and machining techniques that have been used to create nanometer-sized structures.

The scanning probe microscopy (SPM) systems are capable of controlling the movement of an atomically sharp tip in close proximity to or in contact with a surface with subnanometer accuracy. Piezoelectric positioners are typically used to achieve such accuracy. High resolution images can be acquired by raster scanning the tips over a surface while simultaneously monitoring the interaction of the tip with the surface. In scanning tunneling microscope systems, a bias voltage is applied to the sample and the tip is positioned close enough to the surface, so that a tunneling current develops through the gap (Fig. 11.16a). Because this current is extremely sensitive to the distance between the tip and the surface, scanning the tip in the x-y plane while recording the tunnel current permits the mapping of the surface topography with resolution on the atomic scale. In a more common mode of operation, the amplified current signal is connected to the z-axis piezoelectric positioner through a feedback loop, so that the current and, therefore, the distance are kept constant throughout the scanning. In this configuration the picture of the surface topography is obtained by recording the vertical position of the tip at each x-y position. The STM system only works for conductive surfaces because of the need to establish a tunneling current. The atomic force microscopy was developed as an alternative for imaging either conducting or nonconducting surfaces. In AFM, the tip is attached to a flexible cantilever and is brought in contact with the surface (Figure 11.16b).

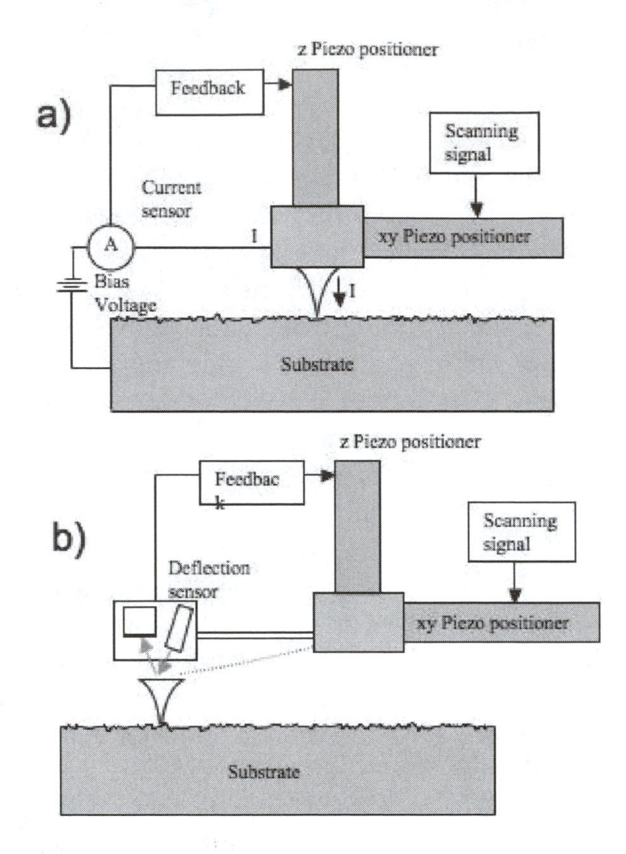

Figure 11.16. Scanning probe systems: (a) STM and (b) AFM.

The force between the tip and the surface is detected by sensing the cantilever deflection. A topographic image of the surface is obtained by plotting the deflection as a function of the x-y position. In a more common mode of operation, a feedback loop is used to maintain a constant deflection, while the topographic information is obtained from the cantilever vertical displacement.

Some scanning probe systems use a combination of the AFM and STM modes, i.e., the tip is mounted in a cantilever with an electrical connection, so both the surface forces and the tunneling currents can be controlled or monitored. STM systems can be operated in ultrahigh vacuum (UHVSTM) or in air, whereas AFM systems are typically operated in air. When a scanning probe system is operated in air, water adsorbed onto the sample surface accumulates underneath the tip, forming a meniscus between the tip and the surface. This water meniscus plays an important role in some of the scanning probe techniques described below.

Scanning Probe Induced Oxidation

Nanometer-scale local oxidation of various materials can be achieved using scanning probes operated in air and biased at a sufficiently high voltage. Tip bias of −2 to −10V is normally used with writing speeds of 0.1–100μm/s in an ambient humidity of 20–40%. It is believed that the water meniscus formed at the contact point serves as an electrolyte such that the biased tip anodically oxidizes a small region of the surface[329]. The most common application of this principle is the oxidation of hydrogen-passivated silicon. A dip in HF solution is typically used to passivate the silicon surfaces with hydrogen atoms. Patterns of oxide "written" on a silicon surface can be used as a mask for wet or dry etching. Ten-nm line width patterns have been successfully transferred to a silicon substrate in this fashion[330]. Various metals have also been locally anodized using this approach such as aluminum or titanium[331]. An interesting variation of this process is the anodization of deposited amorphous silicon[332]. Amorphous silicon can be deposited at low temperature on top of many materials. The deposited silicon layer can be patterned and used as, for example, the gate of a 0.1μm CMOS transistor[333], or it can be used as a mask to pattern an underlying film. The major drawback of this technique is poor reproducibility due to tip wear during the anodization. However, using AFM in noncontact mode has overcome this problem[329].

Scanning Probe Resist Exposure and Lithography

Electrons emitted from a biased SPM tip can be used to expose a resist the same way e-beam lithography does (Fig. 11.17) [333]. Various systems have been used for this lithographic technique. These include constant current STM, noncontact AFM, and AFM with constant tip-resist force and constant current. The systems using AFM cantilevers have the advantage of performing imaging and alignment tasks without exposing the resist. Resists well characterized for e-beam lithography (e.g., PMMA or SAL601) have been used with scanning probe lithography to achieve reliable sub-100-nm lithography. The procedure for this process is as follows. The wafers are cleaned and the native oxide (in the case of silicon or poly) is removed with a HF dip. Subsequently, 35–100 nm-thick resist is spin-coated on top of the surface. The exposure is done by moving the SPM tip over the surface while applying a bias voltage high enough to produce an emission of electrons from the tip (a few tens of volts). Development of the resist is performed in standard solutions following the exposure. Features below 50 nm in width have been achieved with this procedure.

Dip-Pen Nanolithography

In dip-pen nanolithography (DPN), the tip of an AFM operated in air is "inked" with a chemical of interest and brought into contact with a surface. The ink molecules flow from the tip onto the surface as with a fountain pen. The water meniscus that naturally forms between the tip and the surface enables the diffusion and transport of the molecules, as shown in Fig. 11.18. Inking can be done by dipping the tip in a solution containing a small concentration of the molecules followed by a drying step (e.g., blowing dry with compressed difluoroethane). Line widths down to 12 nm with spatial resolution of 5 nm have been demonstrated with this technique[334]. Species patterned with DPN include conducting polymers, gold, dendrimers, DNA, organic dyes, antibodies, and alkanethiols. Alkanethiols have also been used as an organic monolayer mask to etch a gold layer and subsequently etch the exposed silicon substrate.

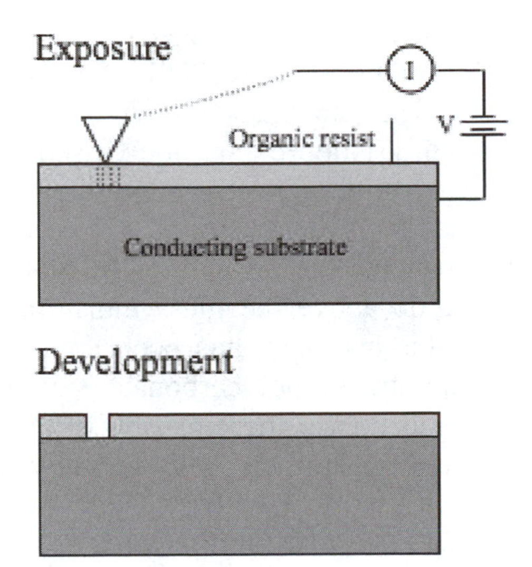

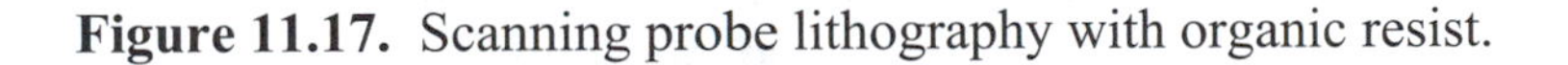

Figure 11.17. Scanning probe lithography with organic resist.

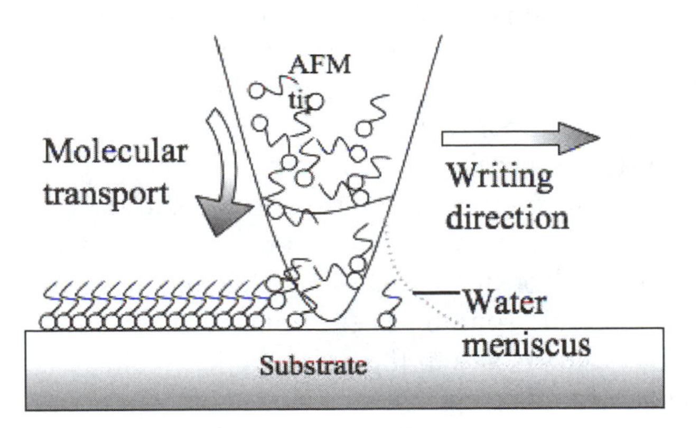

Figure 11.18. Schematic representation of the dip-pen nanolithography working principle.

Other Scanning Probe Nanofabrication Techniques

A great variety of nanofabrication techniques using scanning probe systems have been demonstrated. Some of these are proof of concept demonstrations, and their utility as a viable and repeatable fabrication process has yet to be evaluated. For example, a substrate can be mechanically machined using STM/AFM tips acting as plows or engraving tools[335]. This can be used to directly create structures in the substrate, although it is more commonly used to pattern a resist for a subsequent etch, liftoff, or electro-deposition step.

Mechanical nanomachining with SPM probes can be facilitated by heating the tip above the glass transition of a polymeric substrate material. This approach has been applied to SPM-based high-density data storage in polycarbonate substrates[336]. Electric fields strong enough to induce the emission of atoms from the tip can be easily generated by applying voltage pulses above 3V. This phenomenon has been used to transfer material from the tip to the surface and vice versa. Ten- to 20-nm mounds of metals such as Au, Ag, or Pt have been deposited or removed from a surface in this fashion[337]. The same approach has been used to extract single atoms from a semiconductor surface and re-deposit them elsewhere[338]. Manipulation of nanoparticles, molecules, and single atoms on top of a surface has also been achieved by simply pushing or sliding them with the SPM tip[339]. Metals can also be locally deposited by the STM chemical vapor deposition technique[340]. In this technique, a precursor organometallic gas is introduced in the STM chamber. A voltage pulse applied between the tip and the surface dissociates the precursor gas into a thin layer of metal. Local electro-chemical etch[341] and electrodeposition[342] are also possible using SPM systems. A droplet of suitable solution is first placed on the substrate. Then the STM tip is immersed into the droplet and a voltage is applied. In order to reduce Faradaic currents the tip is coated with wax such that only the very end is exposed to the solution. Sub-100-nm feature size has been achieved using this technique. Using a single tip to serially produce the desired modification in a surface leads to very slow fabrication processes that are impractical for mass production. Many of the scanning probe techniques developed thus far, however, could also be performed by an array of tips, which would increase their throughput and make them more competitive compared with other parallel nanofabrication processes. This approach has been demonstrated for imaging, lithography[343], and data storage[344] using both 1-D and 2-D arrays of scanning probes. With the development of larger arrays with individual control of force, vertical position, and current advances, we might see these techniques become standard fabrication processes in the industry.

11.4.4 Self-assembly

Self-assembly is a nanofabrication technique that involves aggregation of colloidal nano-particles into the final desired structure[345]. This aggregation can be either spontaneous (entropic) and due to the thermodynamic minima (energy minimization) constraints, or chemical and due to the complementary binding of organic molecules and supramolecules (molecular self-assembly)[346]. Molecular self-assembly is one of the most important techniques used in biology for the development of complex functional structures. Since these techniques require that the target structures be thermodynamically stable, they tend to produce structures that are relatively defect-free and self-healing. Self-assembly is by no means limited to molecules or the nano-domain and can be carried out on just about any scale, making it a powerful bottom-up assembly and manufacturing method (multiscale ordering). Another attractive feature of this technique relates to the possibility of combining self-assembly properties of organic molecules with the electronic, magnetic, and photonic properties of inorganic components. In the following sections, we will discuss various important self-assembly techniques currently under investigation.

Physical and Chemical Self-Assembly

The central theme behind the self-assembly process is spontaneous (physical) or chemical aggregation of colloidal nano-particles.[347] Spontaneous self-assembly exploits the tendency of mono-dispersed nano- or submicrocolloidal spheres to organize into a face-centered cubic (FCC) lattice. The force driving this process is the desire of the system to achieve a thermodynamically stable state (minimum free energy). In addition to spontaneous thermal self-assembly, gravitational, convective, and electrohydrodynamic forces can also be used to induce aggregation into complex 3-D structures. Chemical self-assembly requires the attachment of a single molecular organic layer (self-assembled monolayer, or SAM) to the colloidal particles (organic or inorganic) and subsequent self-assembly of these components into a complex structure using molecular recognition and binding.

Physical Self-Assembly. This is an entropic-driven method that relies on the spontaneous organization of colloidal particles into a relatively stable structure through non-covalent interactions. For example, colloidal polystyrene spheres can be assembled into a 3-D structure on a substrate that is held vertically in the colloidal solution, Figure 11.19[348, 349].

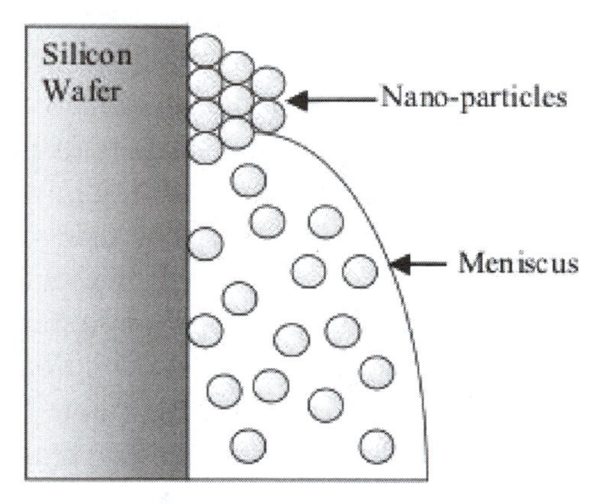

Figure 11.19. Colloidal particle self-assembly onto solid substrates upon drying in vertical position.

Upon the evaporation of the solvent, the spheres aggregate into a hexagonal close packed (HCP) structure. The interstitial pore size and density are determined by the polymer sphere size. The polymer spheres can be etched into smaller sizes after forming the HCP arrays, thereby altering the template pore separations[350]. This technique can fabricate large patterned areas in a quick, simple, and cost-effective way. A classic example is the natural assembly of on-chip silicon photonic bandgap crystals[348], which are capable of reflecting the light arriving from any direction in a certain wavelength range[351]. In this method, a thin layer of silica colloidal spheres is assembled on a silicon substrate. This is achieved by placing a silicon wafer vertically in a vial containing an ethanolic suspension of silica spheres. A temperature gradient across the vial aids the flow of silica spheres. Figure 11.20 shows the cross-sectional SEM image of a thin planar opal template assembled directly on a Si wafer from 855 nm spheres.

Figure 11.20. Cross-sectional SEM image of a thin planar opal silica template (spheres 855 nm in diameter) assembled directly on a Si wafer.[348]

Once such a template is prepared, LPCVD can be used to fill the interstitial spaces with Si, so that the high refractive index of silicon provides the necessary bandgap. One can also deposit colloidal particles into a patterned substrate (template-assisted self-assembly, TASA)[352, 353]. This method is based on the principle that when aqueous dispersion of colloidal particles is allowed to dewet from a solid surface that is already patterned, the colloidal particles are trapped by the recessed regions and assembled into aggregates of shapes and sizes determined by the geometric confinement provided by the template. The patterned arrays of templates can be fabricated using conventional contact-mode photolithography, which gives control over the shape and dimensions of the templates, thereby allowing the assembly of complex structures from colloidal particles. The cross-sectional view of a fluidic cell used in TASA is shown in

Figure 11.21. The fluidic cell has two parallel glass substrates to confine the aqueous dispersion of the colloidal particles. The surface of the bottom substrate is patterned with a 2-D array of templates. When the aqueous dispersion is allowed to slowly dewet across the cell, the capillary force exerted on the liquid pushes the colloidal spheres across the surface of the bottom substrate until they are physically trapped by the templates. If the concentration of the colloidal dispersion is high enough, the template will be filled by the maximum number of colloidal particles determined by the geometrical confinement. This method can be used to fabricate a variety of polygonal and polyhedral aggregates that are difficult to generate[354].

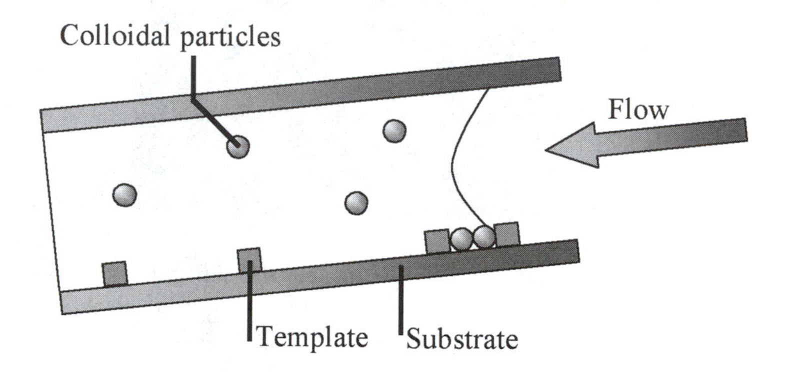

Figure 11.21. A cross-sectional view of the fluidic cell used for template-assisted self-assembly.

Chemical Self-Assembly. Organic and supramolecular SAMs play a critical role in colloidal particle self assembly. SAMs are robust organic molecules that are chemically adsorbed onto solid substrates[355]. Most often, they have a hydrophilic (polar) head that can be bonded to various solid surfaces and a long, hydrophobic (nonpolar) tail that extends outward. SAMs are formed by the immersion of a substrate in a dilute solution of the molecule in an organic solvent. The resulting film is a dense organization of molecules arranged to expose the end group. The durability of a SAM is highly dependent on the effectiveness of the anchoring to the surface of the substrate. SAMs have been widely studied, because the end group can be functionalized to form precisely arranged molecular arrays for various applications ranging from simple, ultrathin insulators and lubricants to complex biological sensors. Chemical self-assembly uses organic or supramolecular SAMs as the binding and recognition sites for fabricating complex 3-D structures from colloidal nano-particles. Most commonly used organic monolayers include: 1) organosilicon compounds on glass and native surface oxide layer of silicon, 2) alkanethiols, dialkyl disulfides, and dialkyl sulfides on gold, 3) fatty acids on alumina and other metal oxides, and 4) DNA. Octadecyltrichlorosilane (OTS) is the most common organosilane used in the formation of SAMs, mainly because of the fact that it is simple, readily available, and forms good, dense layers[356, 357]. Alkyltrichlorosilane monolayers can be prepared on clean silicon wafers whose surface is SiO2 (with almost 5×1014 SiOHgroups/cm2). Figure 11.22 shows the schematic representation of the formation of alkylsiloxane monolayers by adsorption of alkyltrichlorosilanes from solution onto Si/SiO2 substrates. Since the silicon chloride bond is susceptible to hydrolysis, a limited amount of water has to be present in the system in order to obtain good quality monolayers. Monolayers made of methyl- and vinyl-terminated alkylsilanes are autophobic to the hydrocarbon solution and hence emerge uniformly dry from the solution, whereas monolayers made of ester-terminated alkylsilanes emerge wet from the solution used in their formation. The disadvantage of this method is that if the alkyltrichlorosilane in the solvent adhering to the substrate is exposed to water, a cloudy film is deposited on the surface due to formation of a gel of polymeric siloxane. Another important organic SAM system is the alkanethiols (X(CH2)nSH, where X is the endgroup) on gold[355, 358-360]. A major advantage of using gold as the substrate material is that it does not have a stable oxide, and thus it can be handled in ambient conditions. When a fresh, clean, hydrophilic gold substrate is immersed (several minutes to several hours) into a dilute solution (10−3 M) of the organic sulfur compound (alkanethiols) in an inorganic solvent, a close-packed, oriented monolayer can be obtained. Sulfur is used as the head group, because of its strong interaction with gold substrate (44 kcal/mol), resulting in the formation of a close-packed, ordered monolayer. The end group of alkanethiol can be modified to render hydrophobic or hydrophilic properties to the adsorbed layer.

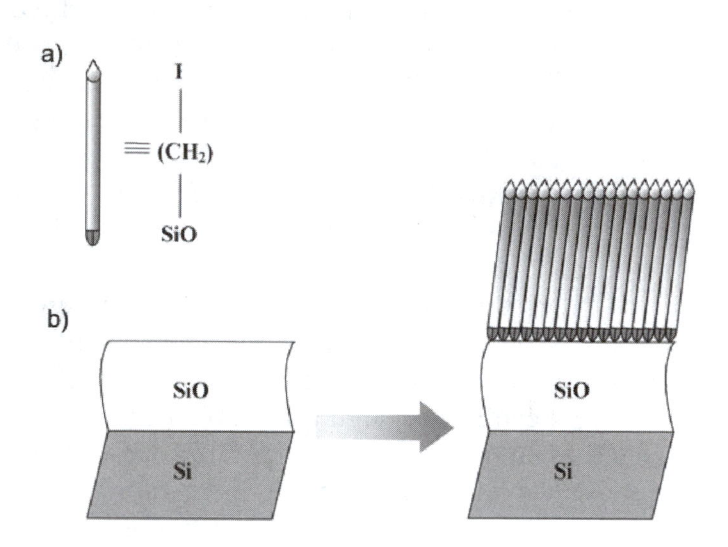

Figure 11.22. a) Alkylsiloxane formed from the adsorption of alkyltrichlorosilane on Si/SiO2 substrates, b) Schematic representation of the process.

11.5 MEMS and nanotechnology applications

The potential applications for MEMS are numerous and many products are now on the market utilizing MEMS technology. Some of the more prominent areas current interest include:

- Biotechnology: for example, biochips for detection of hazardous chemical and biological agents, and microsystems for high-throughput drug screening and selection.
- Communications: for example, fabrication of electrical components such as inductors and tunable capacitors using MEMS and nanotechnology for better integration yielding reduced total circuit area, power consumption and cost. One key product area includes mechanical switches for use in various microwave circuits.

- Accelerometers: MEMS accelerometers are widely used for crash air-bag deployment systems in automobiles. MEMS and nanotechnology has made it possible to integrate the accelerometer and electronics onto a single silicon chip at a cost suitable for automotive application.

11.5.1 Nanotechnology applications

We've seen that the peculiar nature of carbon in combination with the molecular perfection of buckytubes (or single-wall carbon nanotubes) gives them with exceptionally high material properties such as electrical and thermal conductivity, strength, stiffness, and toughness[361]. In most materials, however, the values of material properties that are found experimentally are usually substantially lower due to the occurrence of defects in their structure. For example, high strength steel typically fails at about 1% of its theoretical breaking strength. Buckytubes, however, have property values very close to their theoretical limits because of their perfection of structure. They offer a lot of promise relative to their application in materials, electronics, chemical processing and alternate energy sources.

Specific areas of application include:

- field-emission flat-panel displays. Building on their exceptional electrical and heat conductivity and mechanical properties they are excellent electron field-emitters. Since they are molecularly perfect, they are free from the limitations of defects seen in other materials. The flat panel displays build using Buckytubes utilize a separate Buckytube as an electron gun for each pixel in the display in contrast to the more typical single electron gun as in a traditional cathode ray tube display.
- conductive plastics. Most plastics are excellent insulators. This can be changed by loading plastics up with conductive fillers, such as carbon black and conventional graphite fibers.

This addition of results in added (undesirable) weight and degradation of structural properties. Buckytubes are ideal as fillers, since they have the highest aspect ratio of any carbon fiber and have a natural tendency to form polymer-like ropes resulting in long conductive pathways even at ultra-low loadings. This makes such filled plastics ideal for EMI/RFI shielding composites and coatings; electrostatic dissipation (ESD), and antistatic materials and coatings; and radar-absorbing materials.

- energy storage and alternate energy sources. Having good electrical conductivity and high surface areas, Buckytubes are well suited for use as electrodes in batteries and capacitors. They also have applications in a variety of fuel cell components since their high surface area and thermal conductivity make them useful as electrode catalyst supports in PEM fuel cells. And their high strength and toughness for their weight gives them useful durability in these complex systems.
- conductive adhesives and connectors. The features of Buckytubes making them useful in conductive plastics make them similarly attractive in this application are, specially for adhesives and solder-like materials.
- fibers and fabrics. Due to the tremendous strength to weight ratio and aspect ratio of Buckytubes, a natural application area is in super strong fibres. These are expected to be useful in body and vehicle armor, transmission line cables, woven fabrics and textiles.
- biomedical applications. This is an area showing tremendous promise. Buckytubes appear non-toxic to living cells in many environments and have been used to encourage cell growth. The cells do not adhere to the Buckytubes suggesting applications such as coatings for prosthetics and anti-fouling coatings for structures. The sidewalls of buckytubes can be chemically modified enabling biomedical applications such as vascular stents, and neuron growth and regeneration. Dual wall Buckytubes with "sandwiched" chemical materials also show potential in this area.

- other applications. Researchers, buoyed by the results of application of Buckytubes are investigating a wide range of applications in cosmetics, photovoltaics, etc. The table below, after Bhushan[293] summarizes the principal application areas for nanotubes in mechanical, electrical, thermal and thermo-mechanical fields of application.

Table 11.2. Applications of carbon nanotube materials, after Bhushan[293].

		Mechanical			Electrical			Thermal		Thermo-mechanical	
Fiber Fraction	Applications System	Strength/ Stiffness	Specific Strength	Through-thickness strength	Static Dissipation	Surface Conduction[a]	EMI Shielding	Service[b] Temp.	Conduction/ Dissipation[c]	Dimensional Stability[d]	CTE reduction[e]
Low Volume Fraction (fillers)											
Elastomers	Tires	X			X				X		
Thermo-Plastics	Chip Package								X		
	Electronics/ Housing	X			X		X	X	X		
Thermosets	Epoxy products	X	X	X		X				X	X
	Composites			X						X	
High Volume Fraction											
Structural Composites	Space/aircraft Components		X	X							
High Conduction	Radiators	X							X	X	
	Heat Exchangers	X						X	X		X
Composites	EMI Shield	X					X				

[a]For electrostatic painting, to mitigate lightning strikes on aircraft, etc.

[b]To increase service temperature rating of product

[c]To reduce operating temperature of electronic packages

[d]Reduces warping

[e]Reduces microcracking damages in composites

11.6 Micro-machining and small scale defects

11.6.1 Introduction

Increasingly, small, so-called *meso* scale devices (because they are smaller than conventional machined components but larger than MEMS-scale devices) are utilized in commercial products. These can vary from small actuators, flow control devices for mL size quantities of fluids, and mechanical mechanisms in valves, etc (see Figure 11.23 and Figure 11.24). There is also a lot of application for mold making for injection molding plastic components.

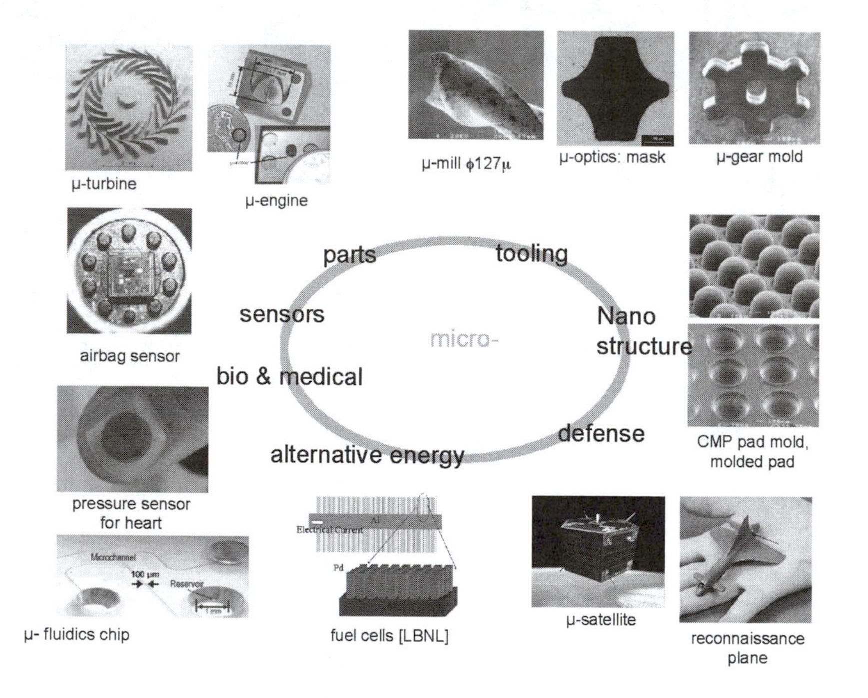

Figure 11.23. Applications for micromachining.

They are created using conventionally shaped tools but with much smaller scale and higher speeds (for example diameters of milling tools down to 50 μm and speeds of 50-60K rpm). These processes are either based on conventional large scale machine tools or on miniature scale machine tools. Since the chip loads are small, forces are small so it is desirable to be able to have low mass (hence high speed and acceleration/deceleration) tables. The size of components manufactured is generally small (a few centimeters on edge or larger if a mold).

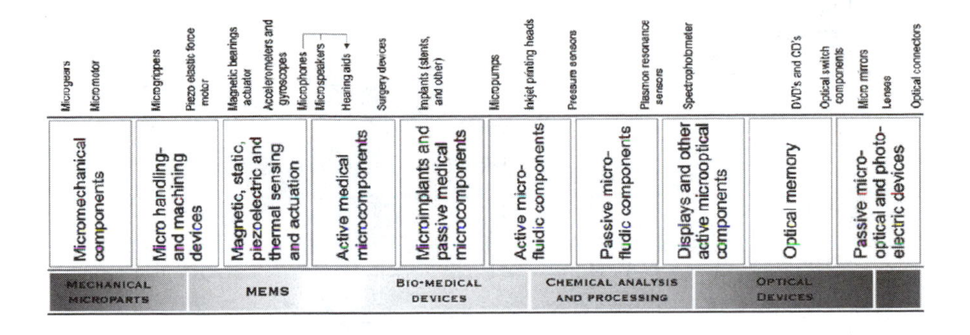

Figure 11.24. Micro product categories, after Hansen, et al[362].

The motivation for the fabrication of smaller and smaller workpieces has been essentially the same since manufacturing was first established as an art/science – new applications, better performance, less expensive and higher quality. Machining processes have always played an important role in manufacturing of workpieces and have seen their capability for precision machining steadily improve. There are a number of good reviews of the development of micromachining technology, for example Masuzawa and Tönshoff[363] and Masuzawa[364] discussed micromachining capabilities and defined micromachining relative to parts that are "too small to be machined easily" and where it fits relative to other microfabrication processes. Alting, et al[365] covers a broad range of fabrication techniques for small scale parts and placed microfabrication processes relative the broader field of "microengineering" and discussed the design of microproducts. Other references covering a broad range of topics on micromachining include Liu[366], and, for a US perspective, the WTEC Panel on Micromanufacturing[367]. Finally, a comprehensive overview is given in work conducted by Dornfeld et al covering machines, modeling and related topics[368].

In addition, micromachining can serve as an extremely important parallel to other fabrication techniques, and can add significantly to the fabrication engineer's suite of capabilities available to fabricate features at the sub-millimeter to micro- scale (see Figure 11.25).

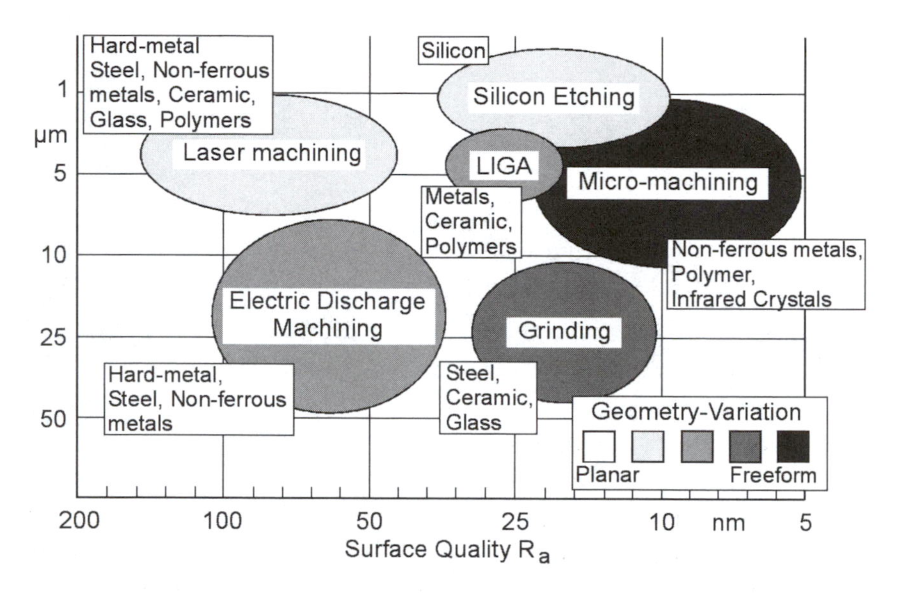

Figure 11.25. Comparison of micromachining with other manufacturing technologies, from Dornfeld[368].

The motivation for increasingly smaller components parallels the improvements in cutting technology as outlined in Taniguchi's curve. Demand for reduced weight, reduced dimensions, higher surface quality and part accuracy, while at the same time decreasing component costs and reducing batch sizes for components of devices ranging from electro-mechanical instruments to medical devices force us along Taniguchi's curve. These are the forces driving miniaturization. The response of the scientific community has been energetic but mixed. That is, while the development of machine tools for micromachining follows more traditional paths of scaling down conventional components, process research and development is less structured. This is apparent from the literature. As unit removal size decreases, issues of tool edge geometry, grain size and orientation, etc. – effects considered to have little or no influence at larger scales – become dominant factors with strong influences on resulting accuracy, surface quality and integrity of the machined component.

This section aims to put into perspective the earlier work on micromachining and, specially, emphasize some of the "practical" requirements for increased understanding of the fundamental process physics, modeling efforts and experimental validation relative to part quality. In this section, micromachining is strictly defined as mechanical cutting of features with tool engagement less than 1 mm with geometrically defined cutting edges. All chemical, thermal, and abrasive processes are excluded. The machine tool requirements build on those discussed earlier in this book regarding positioning accuracy, thermal and mechanical stability, etc.

We begin first with a discussion of surface and edge considerations, then approaches to modeling of the process, workpiece and design issues and finish with tooling issues and metrology.

11.6.2 Surface and edge finish

Various problems such as surface defects, poor edge finish, and burrs in conventional machining have plagued conventional manufacturing for some time. Some of these problems have been avoided by post processing and process optimization. These problems are also significant in micromachining and require much more attention because, in many cases, inherent material characteristics or limitations in part geometry do not allow some of the solutions used in macromachining. Figure 11.26 shows some typical defects in micromilling.

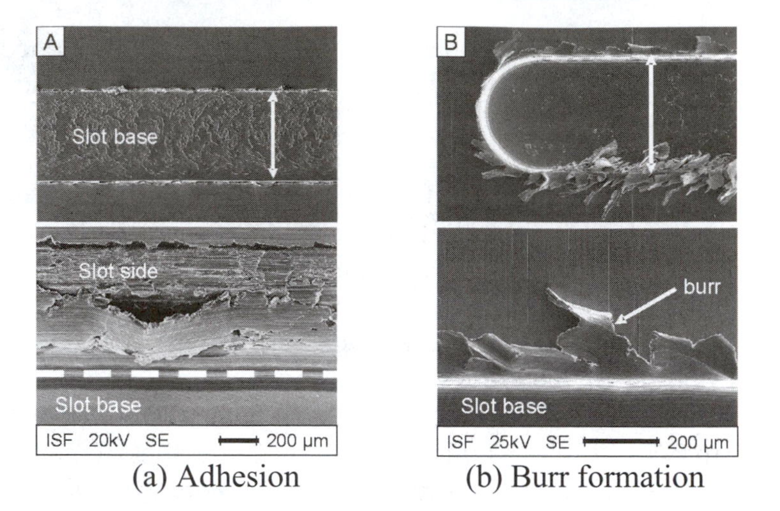

(a) Adhesion (b) Burr formation

Figure 11.26. Typical surface defects in micromilling (work material: NiTi shape memory alloy), from Weinert[369].

In ultra precision diamond machining for optical devices, non-ferrous metals exhibited poorer surface roughness than geometrically predicted due to the steady, small amplitude vibration of the tool which is usually larger for round edge tools than for straight edge tools[370, 371]. This small vibration was investigated by Zhang and Kapoor[372, 373] using a model which can handle deterministic and stochastic excitation and construct a 3-D texture of a machined surface. At the micro level, material properties become non-homogeneous and thus variation in material hardness causes the cutting tool vibration. This effect is significant at low feed and cutting speed and leads to irregular surface roughness.

Vogler et al[374] studied the relationship between surface roughness and microstructure of the ferrous materials and developed a model to predict surface generation in micro-end milling. This is described in the modeling section later in this chapter.

The work material NiTi is used for many medical applications, such as surgical implants, and micromilling is commonly used to fabricate these products. This material is very ductile and easily work hardens during machining causing adhesion and high burr formation. Additionally, high ductility causes adverse chip formation, long and continuously snarled chips. At the micro level, these chips interfere with tool engagement and burrs and contribute to poor surface quality of finished parts[375].

Lee and Dornfeld[375] conducted micro-slot milling experiments on aluminum and copper and found various standard burr types depending on location and work geometry. Interestingly, these burr shapes were similar to those found in macromachining in terms of formation mechanisms and influence of cutting parameters. One major difference found was that the influence of tool run-out on burr formation was significant in micro-slot milling.

Min, et al[376] conducted micro-fly cutting and microdrilling experiments on single crystal and polycrystalline OFHC copper in order to understand the effects of crystal orientation, cutting speed, and grain boundaries on surface roughness, chip formation, and burr formation. Certain crystallographic orientations were found to yield rougher surface finish, as well as significant burrs and breakout at the tool exit edge. The <100> and <110> direction of machining on the workpieces exhibited the greatest amount of variation in formation of burrs and breakout at the exit edge and in chip topology as a function of the angular orientation of the workpiece. This corresponded to a variation in the interaction between the tool and the activate slip systems. They also conducted slot milling experiments on the same material and found a strong dependency of top burr formation on slip systems of each crystal orientation except (100) workpiece[377]. The (100) workpiece did not show a clear correlation due possibly to less anisotropy of the slip systems. Sato et al[378] proposed that the (100) orientation has a relatively smaller anisotropy than the (110) and (111) orientations because it has a greater degree of symmetry, resulting in more equally distributed slip systems than the other orientations.

Bissacco et al [379] found that top burrs are relatively large in micromilling due to the size effect. When the ratio of the depth of cut to the cutting edge radius is small, high biaxial compressive stress pushes material toward the free surface and generates large top burrs. Ahn and Lim[380, 381] proposed a burr formation model in a microgrooving operation based on a side shear plane and an extended deformation area which is caused by the tool edge radius effect. The material near the cutting edge experiences the side shear deformation due to hydrostatic pressure. Aluminum and OFHC generated larger burrs than brass, and thus it was concluded that the thickness of the burr is proportional to the ductility of the material.

Further work by Schaller[382] showed that when fabricating microgrooves in brass, burr formation can be drastically reduced by coating the surface with cyanacrylate. Sugawara and Inagaki[383] investigated the effect of drill diameter and crystal structure on burr formation in microdrilling. They utilized both single crystal and polycrystalline iron with a thickness between 0.06 mm and 2.5 mm and high speed twist drills with diameters from 0.06 mm to 2.5 mm. In general they confirmed that burr size is reduced and cutting ability increased as drill size decreases.

Min[376] found that grain orientation affected burr formation in drilling of polycrystalline copper, Figure 11.27. A single material may produce a ductile-like cutting mode in one grain and brittle-like cutting in another, indicating that favorable and non-favorable cutting orientations for good surface and edge condition exist as a function of crystallographic orientation. These observations demonstrate the importance of further research in this field.

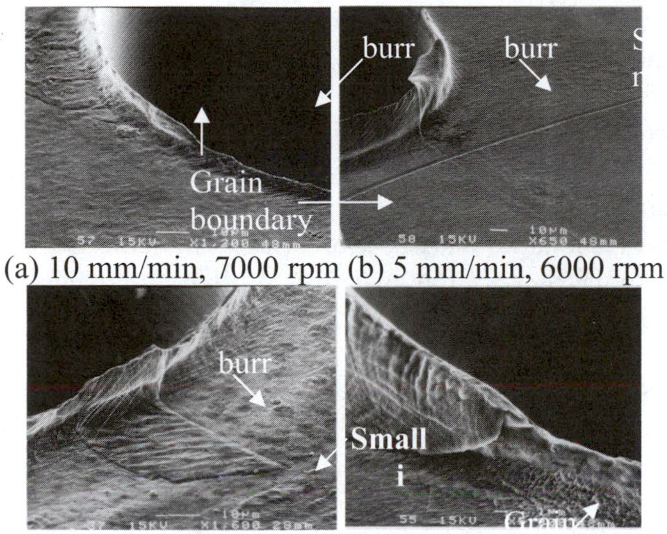

(c) 5 mm/min, 7000 rpm (d) 10 mm/min, 8000 rpm

Figure 11.27. Microdrilling burr formation (250 μm diameter); (a) burr within grain boundary, (b) burr across grain boundary, (c) burr over small grain, (d) grain boundary follows burr topology, from Min[376].

11.6.3 Modeling

One of difficulties in micromachining is the microscopic level of the phenomena. Hence, it is difficult to conduct experiments, and make in-process observations and measure results after the experiment. Moreover, the cutting process itself is very complicated involving elastic/plastic deformation and fracture with high strain rates and temperature and for which material properties vary during the process. Thus, analytical modeling is considered extremely difficult at the current level of understanding of material behavior. Most analytical modeling efforts are based on kinematics from empirical observation combined with classical cutting models at the macro level. The applicability and accuracy of these models are subject to many limitations.

Modeling based on numerical relationships, often accompanied by computer simulation, has become the tool of choice for many researchers. It is a good compliment to experimental approaches as it can overcome some of the limitations, if correctly done. It is not a perfect solution because it also involves many of the same assumptions as in analytical modeling. However, computer based models can offer a reasonable insight into certain verifiable trends or guidance on empirical research to assist further understanding of the process. Finite element modeling (FEM) has been a popular simulation technique but it has one critical limitation for micromachining. FEM is based on principles of continuum mechanics. Hence, material properties are defined as bulk material properties whereas, in reality, the material behaves discontinuously in many cases of micromachining. In most cases of isotropic micromachining, FEM can still be an attractive modeling method because the process can be reasonably treated in continuum space.

Since the molecular dynamic (MD) simulation technique is based on interatomic force calculations, it can accommodate micromaterial characteristics as well as dislocations, crack propagations, specific cutting energy, etc. Hence, many researchers have turned to MD for micromachining studies. However, there are three major obstacles in MD modeling. The core part of MD requires good representation of interatomic forces among various combinations of atoms involved in cutting, referred to as a potential. Formulating a potential requires the equivalent of one good Ph.D. level study and thus very few potentials are available. Second, since MD calculates interatomic forces among all atoms within a certain boundary, intensive computational power is required. Therefore, many MD studies are limited to a very small space, such as at a nanometer or angstrom level. Third, MD analysis lacks a good representation of continuum behavior of material. Therefore, most MD simulations clearly state the boundary of application.

More recently, multi-scale modeling techniques combining FEM and MD have been proposed to overcome the disadvantages of each method and allow the coverage of a wider range of behavior.

Details on these approaches and examples of recent research are given below.

11.6.3.1 Finite element modeling

In the early stages of computer simulation of metal cutting, a FEM approach was not directly used to develop a model. Rather, it was used for intermediate steps to obtain certain values using semi-mechanistic or empirical models. Ueda[384] proposed such an approach in order to analyze the material removal mechanisms in micromachining of ceramics. FEM was used only to calculate the J-integral around a crack in front of the cutting edge. The obtained value was then used to determine likelihood of fracture in micromachining. Using this approach, they were able to model ductile and brittle cutting modes and the results were used to maintain the process in the ductile mode.

Ueda and Manabe[385] modeled chip formation in micromachining of amorphous material using rigid-plastic FEM (RPFEM). Again, they used FEM only for further understanding of localized adiabatic deformation. The model was able to produce a lamellar structure of the chip formation, which was also observed during experimental machining using a SEM. The formation of lamellar structured chip was due to the periodical occurrence of a localized shear band and smooth chip formation, the frequency of which was proportional to the depth of cut.

Moriwaki[386] developed a similar model using RPFEM for micro-orthogonal cutting of copper. They used a two-step FEM approach; first, the model obtains the deformed status of cutting and, second, then obtains specific values such as stress and strain which were used as input values to the thermal FE analysis combined with semi analytical formulations. The roundness of the tool edge was also taken into consideration in the model. The analysis showed that cutting ratio is decreased with an increase in the ratio of the tool edge radius to the depth of cut. It was also found that the temperature gradient in the workpiece increases in front of the cutting edge due to the material flow relative to the cutting tool. Similarly, Fleischer[387] used experimental values to improve a FE model to predict cutting forces in micromachining.

The aforementioned FE modeling is primarily for isotropic micromachining where no crystallographic effects were considered. Chuzhoy[388, 389] developed a FE model for micromachining of heterogeneous material, Figure 11.28. Their model was capable of describing the microstructure of multi phase materials and thus captured the microcutting mechanism in cast iron. Microcutting of multi phase materials exerts larger variations in the resulting chip shape and the cutting force than seen in cutting of a single phase material.

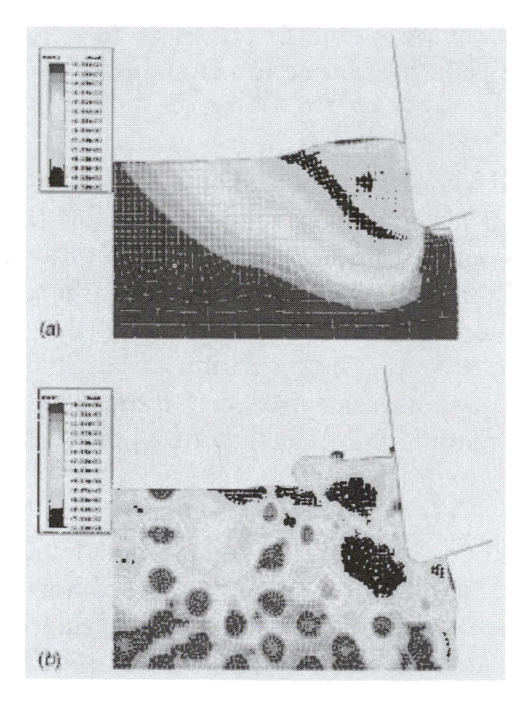

Figure 11.28. Computed equivalent stress for 125 μm depth of cut and 25 μm edge radius at t=0.00012 s with (a) ferritic workpiece, (b) ductile iron workpiece, after Chuzhoy[388].

Park[390] tried to calibrate the mechanistic cutting force through FEM simulation for ferrous materials including ductile and gray irons and carbon steels. Their model is primarily based on analysis of the microstructure of the work materials in their various phases, such as the graphite, ferrite, and pearlite grains seen in ductile iron, gray iron, and carbon steel microstructures. Their model was mainly used to calibrate a cutting force model, Figure 11.29.

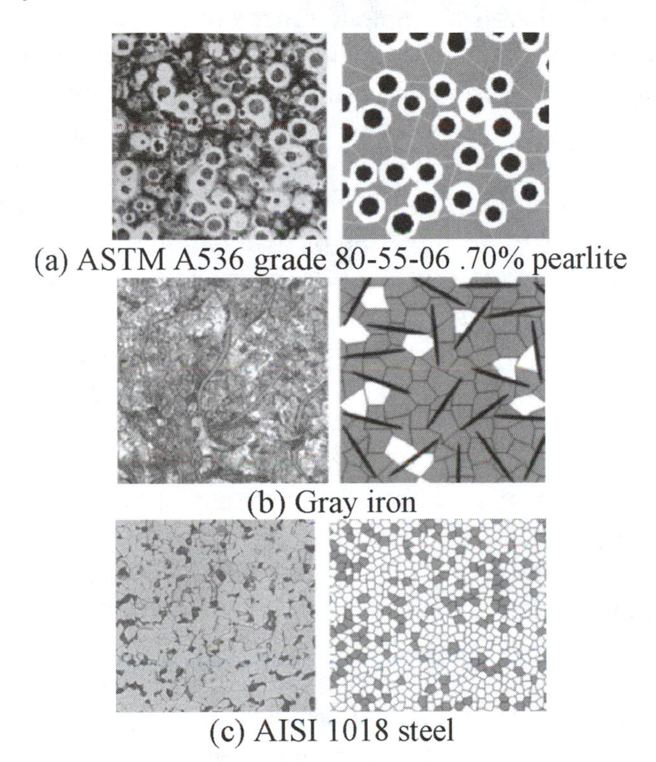

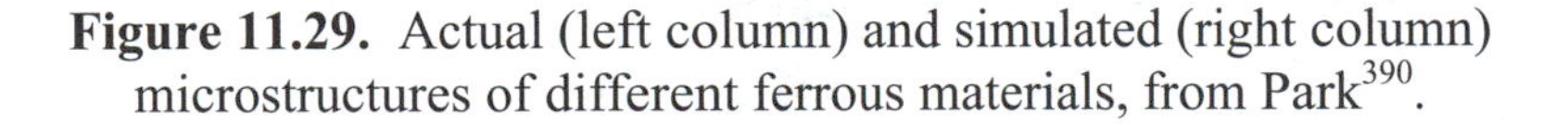
Figure 11.29. Actual (left column) and simulated (right column) microstructures of different ferrous materials, from Park[390].

11.6.3.2 Molecular dynamics

Research work on molecular dynamic simulation of cutting can be traced back to early 1990s. Most of the early researchers used copper as a work material because of its well established structure and potential function. A diamond was used as a cutting tool since it can be reasonably assumed to have a very sharp edge, needed at the MD level[391-395]. The work of Inamura et al [393, 396, 397] focused on a trial of molecular dynamics at an atomic level cutting simulation with a couple of potential functions. This computational study showed that MD is a possible modeling tool for the microcutting process. The simulation was able to correlate the intermittent drop of potential energy accumulated in the workpiece during cutting with the heat generation associated with plastic deformation of the work-piece and impulsive temperature rise on the tool rake face. In a simulation of cutting of polycrystalline copper, the plastic deformation first initiated at the grain boundaries and then propagated into neighboring grains in the direction of dislocation development. They also reported that the rate of energy dissipation in plastic deformation at this scale is larger than in conventional cutting and that a concentrated shear zone did not appear, contrary to what is normally observed in conventional cutting.

Shimada[398] developed a MD model for understanding the chip removal mechanism in micromachining of copper with depths of cut down to 1nm. The model was able to continuously generate dislocations in front of the tool tip and produced chip morphology showing very good agreement with experimental results. In another work[399], they also investigated crystallographic orientation effects on plastic deformation on single crystal copper using MD simulation. Additionally, they found that the ultimate possible surface roughness was about 1 nm for both single crystal and polycrystalline copper materials assuming perfect cutting condition, Figure 11.30.

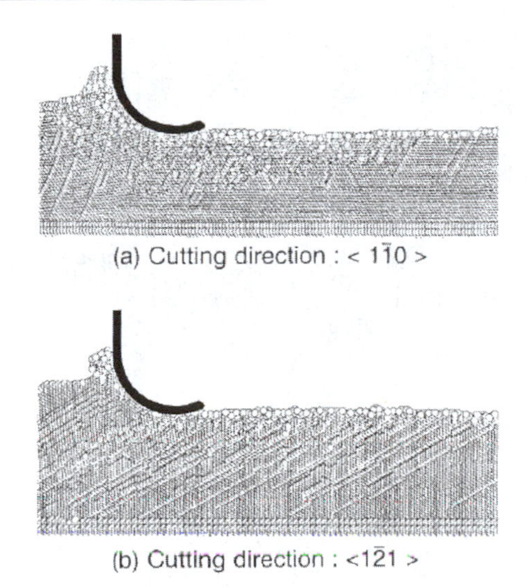

Figure 11.30. Deformation around the cutting edge in microcutting of single crystal copper, from Shimada[399].

In general, MD simulation requires impressive computational power in order to model a cutting process. Hence, many MD models have been applied to two dimensional orthogonal cutting with a very small model size, or unrealistically high cutting speed. Komanduri[400] proposed a new method called a length-restricted molecular dynamics (LRMD) simulation by fixing the length of the work material and shifting atoms along the cutting direction and applied it to nanometric cutting with realistic cutting speeds. They also studied the effect of tool geometry using several ratios of tool edge radius to the depth of cut with various parameters such as cutting force, specific energy, and subsurface damage[401] and further investigated the effect of crystal orientation and direction of cutting on single crystal aluminum and silicon[402-404]. They also applied MD simulation to exit failure and burr formation in ductile and brittle materials with a face centered cubic (FCC) structure. They successfully simulated burr formation on a ductile material and crack propagation in brittle material, Figures 11.31 and 11.32[405].

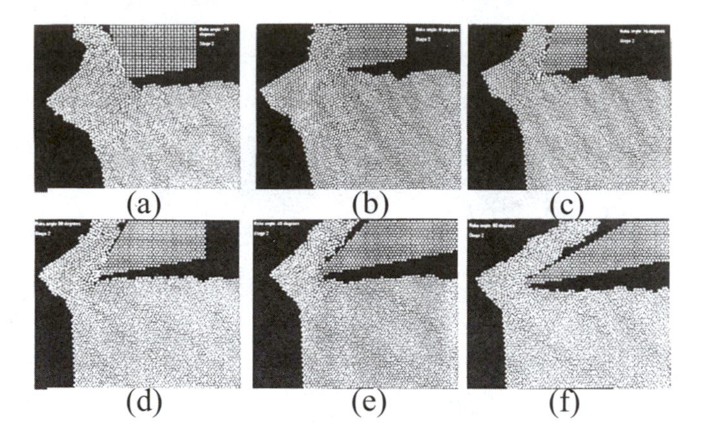

Figure 11.31. MD plots of the nanometric cutting process performed on a ductile work material with no elastic constraint at the exit for various tool rake angles (a) −15° (b) 0° (c) 15° (d) 30° (e) 45° and (f) 60°, from Komanduri[405].

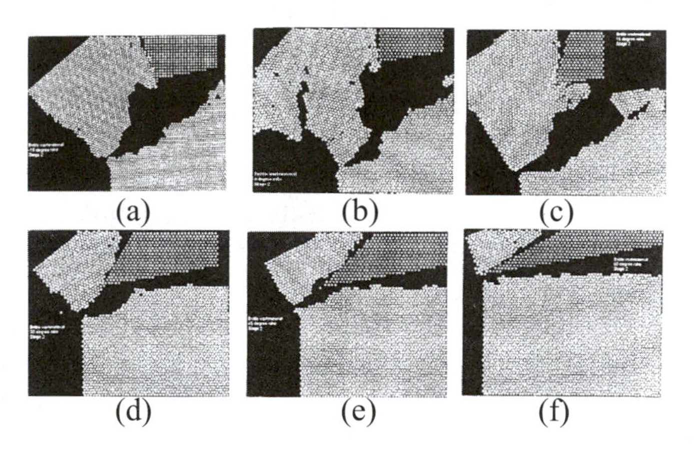

Figure 11.32. MD plots of the nanometric cutting process performed on a somewhat brittle work material with no elastic constraint at the exit for various tool rake angles (a) −15° (b) 0° (c) 15° (d) 30° (e) 45° and (f) 60°, from Komanduri[405].

Komanduri[406] also tried Monte Carlo simulation of nanometric cutting of a single crystal material. This method is only applicable to systems that are neither canonical nor microcanonical. Their model provided a reasonable estimation of the local temperature at the cutting zone. This model was based upon several other models previously developed.

Another typical assumption used in most MD simulations of nano or micro level cutting is that the tool is a rigid body and, thus, dynamic change of tool geometry due to wear during cutting can be ignored. Cheng et al [407, 408] used MD simulation to model the diamond tool as a deformable body. Their model includes heat generation due to cutting and provided analysis of the relationship between the temperature and sublimation energy of the diamond atoms and silicon atoms. The simulation results were compared with experiments using the diamond tip of an atomic force microscope (AFM) on a single crystal silicon plate. The model was able to produce thermo-chemical wear of the diamond tip.

Rentsch and Inasaki[409] and Fang[410] investigated surface integrity of nanomachining and nanoindentation on brittle materials using a conventional MD technique.

11.6.3.3 Multiscale modeling

Inamura[411] used multiscale modeling techniques to cover both atomic and continuum levels of simulation. Using MD, they calculated displacements of interacting atoms in cutting and then transferred into a point in the continuum by weighted mean values of the surrounding atoms to obtain continuum property values such as stress and strain (see Figure 11.33).

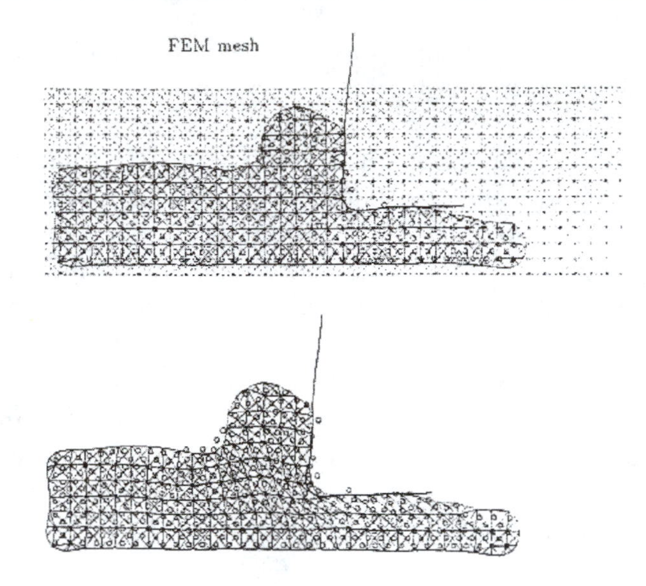

Figure 11.33. Method of transformation from atomic model to finite element model (upper: mesh overlapping, lower: resulting deformation of the finite element model obtained for a tool advancement of 5.1 Å), from Inamura[396].

In other research[411], they found that a very complicated stress state in the workpiece material including a concentrated compressive and shear strain in the primary shear zone, tensile strain along the rake face, and no shear stress inside the workpiece as part of the primary shear zone, exists. They attributed this partially to buckling deformation but left a detailed explanation for future work.

11.6.3.4 Mechanistic modeling

Most of the mechanistic modeling of cutting is based on Merchant's assumption that the tool edge is sharp[412-415]. Kim and Kim[416] proposed an orthogonal cutting model of micromachining with a round edged tool to study sliding along the clearance face of the tool due to elastic recovery and plowing caused by the relatively large edge radius of the tool. They found that the effect of the tool clearance face contributes significantly to the overall cutting force, Figure 11.34.

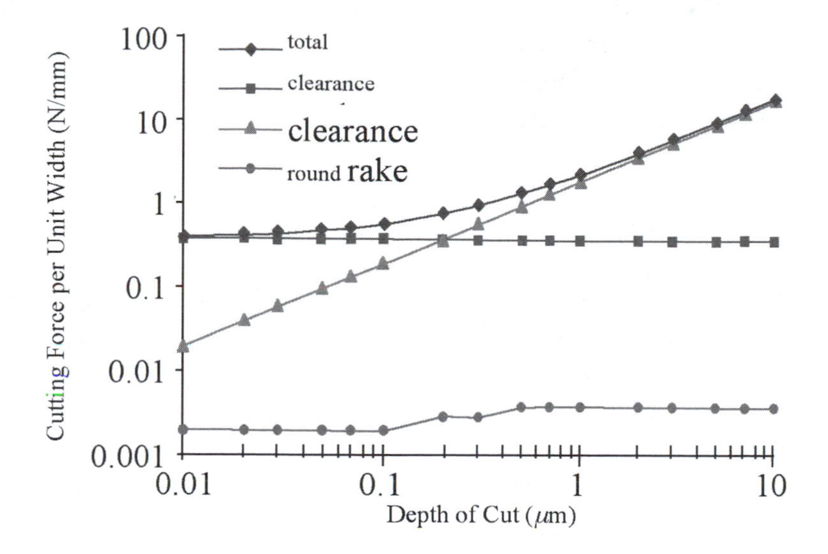

Figure 11.34. Components of cutting force per unit width in round edge cutting model (friction coef.: 0.3, rake angle: 0°, workpiece: copper (CDA110), edge radius of the tool: 0.01 μm), from Kim[416].

In an effort to better understand cutting forces in micro-end milling applications, Bao and Tansel[417-419] developed an analytical model to predict cutting forces. The model is based on geometrical calculation of the chip load by considering the trajectory of the micro-tool tip. The cutting force was calculated by Tlusty and MacNeil's model[420] with previously calculated chip thickness. The model considers many parameters including process conditions, tool geometry, and workpiece material. Their model predicted cutting force within 10% of comparable experimental values and accounted for tool run-out and wear.

Vogler [374, 421, 422] developed a micro-end milling model for various microstructures of workpiece material. The model used tool and surface profile computation and finite element simulation for determination of the minimum chip thickness. With this model, they were able to predict interrupted chip formation over the phase boundaries consistent with experimental results for ductile iron. A slip-line cutting force model for depths of cut larger than minimum chip thickness and an elastic deformation force model for smaller depths of cut were employed assisted by FEM analysis. This model predicted the cutting force for the primary metallurgical phases, ferrite and pearlite, of multiphase ductile iron workpieces and showed good agreement with experimental values.

Kim[423] proposed modeling chip formation and cutting force in a microscale milling process accounting for the coupled minimum chip thickness and edge radius effects. Interestingly, the model predicts that the micromilling tool may rotate without cutting when the feed per tooth is smaller than the minimum chip thickness and that the periodicity of cutting forces is a function of the minimum chip thickness, feed per tooth, and position angle. The model can provide an upper limit of feed per tooth for a given tool diameter, which can be used as a basis for process parameter selection.

Finally, Joshi and Melkote[424] developed a model focused on the material removal mechanism at the micro level. Their model was based on a strain gradient theory whereby the material strength is a function of the strain gradient. The primary deformation zone (PDZ) was modeled and material strength was evaluated as a function of material length scale governing the size effect. The model was able to predict a lower bound value of specific shear energy.

11.6.4 Workpiece and design issues

11.6.4.1 Micromolding

The microturning operation, specially single point diamond turning[425], is commonly used for generating small parts with 3D convex shapes. For 3D concave shapes, microdrilling and micromilling are commonly used. For mass production of microparts, micromachining is not suitable due to low productivity. However, micromachining is key to provide proper tooling for other micro-mass production technology such as micromolding, microforming, and micro-die casting. The quality of products and reliability of these processes is highly dependant on tooling manufactured primarily by micromachining.

As in conventional molding, the precision of the mold is one of the most important factors in micromolding. Mold precision represents the quality of the molded products. As a result, diamond machining is an excellent candidate for micro-mold fabrication. Mold life is another important factor in micromolding since that influences manufacturing costs and part quality. The affinity of diamond to ferrous material causes serious problems in machining the tool steels or hard ferrous materials that are generally preferred for the mold. Hence, various efforts have been made to avoid such problems, for example:

- cutting in a carbon saturated atmosphere[426]
- cooling the cutting process[427]
- modifying the chemical composition of the work material [428]
- superimposing the tool motion with ultrasonic vibration during the cutting process. [429, 430]

These approaches have been partially successful in minimizing wear of the diamond tool but, with the exception of the ultrasonic vibration assisted cutting, the applicability to industry is questionable[431].

Schaller et al [432, 433] have successfully fabricated a micromold of martensitic steel by micromachining with tungsten carbide tools to create features of 200 μm to 330 μm in depth and 220 μm to 420 μm in width. The machining was completed prior to hardening the steel, and then electrochemically polished to remove any remaining burrs. Completion of this type of mold is important because it will allow for long term production of a wide range of plastic as well as metal and ceramic filled polymers.

Freidrich and Vasile[434] were able to produce trench features 8 μm wide and 62 μm deep for an aspect ratio of nearly 8, over a length of 8 mm, Figure 11.35. This aspect ratio is good enough for most micromolding applications including X-ray masks.

Figure 11.35. Portion of micromilled trenches with stepped and straight walls (scale bar = 100 μm), from Friedrich[434].

Micromachining has been successfully employed to fabricate multi-level mold inserts for micromolding of a micro-valve system by Fahrenberg[435]. They combined micromachining with deep etch X-ray lithography to create the micromold with features 60 μm in height and 50 μm wide. This study showed the possibility of stacking several molds for high aspect ratio parts.

Another effort in micro-mold fabrication combines the use of other manufacturing processes with micromachining. Fleischer[436, 437] surveyed potential fabrication processes and listed their characteristics and potential for combining. They found that a combination of micromilling, micro-EDM, and laser processes compensates for inherent individual weaknesses and benefits from the strengths of each process.

Bissacco[438] investigated a wide range of processing methods for mold fabrication, Figure 11.36. illustrates the process chain for micro-injection molding for micro-fluidic systems. MEMS technology provides high precision processing capabilities but is very limited in achievable shapes allowing only molds in 2.5-D and those with a number of stackable layers.

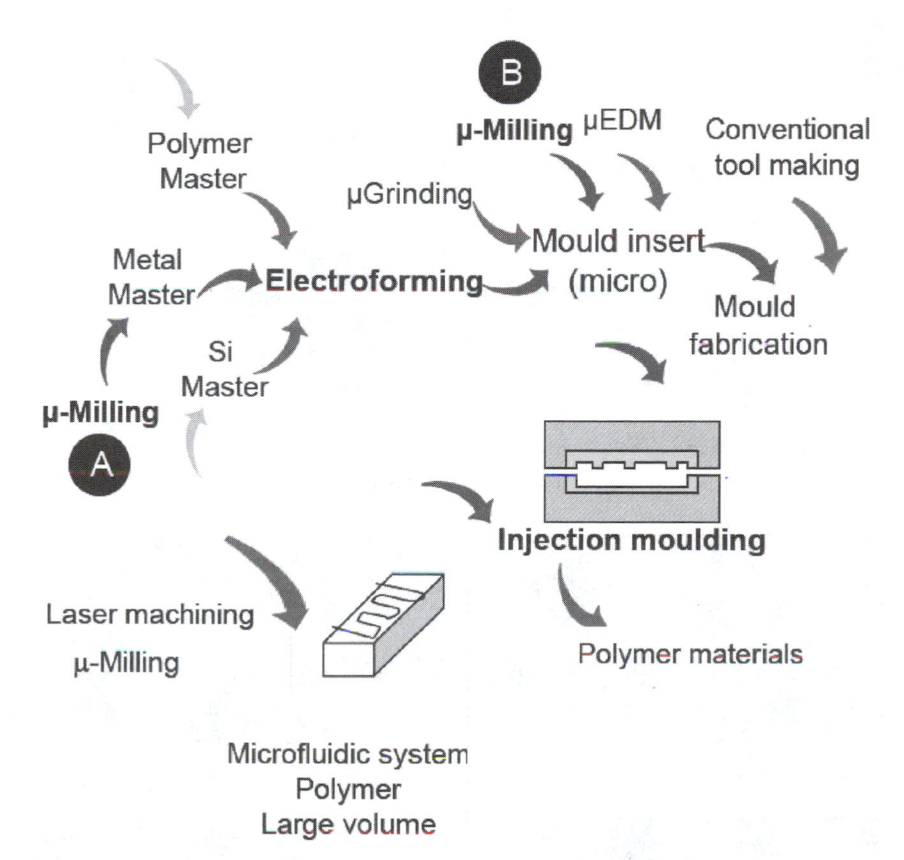

Figure 11.36. Process chain of micro-injection molding, from Bissacco[438].

11.6.4.2 Creation of micropattern and microstructure

One of the more interesting applications of micromachining is in micropattern generation either on flat or curved surfaces. These patterns can work as reflectors, abrasives, and other functions. Traffic signs, Fresnel lenses, and possible CMP (chemical mechanical planarization) pad surfaces are typical examples. Various micro-grooves can be created by feeding a rotating cutting tool along one direction relative to the workpiece, as illustrated in Figure 11.37[439]. We have discussed aspects of this earlier in the book under diamond machining.

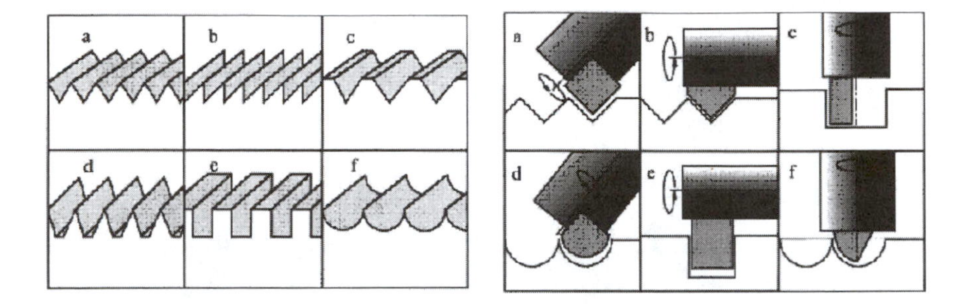

Figure 11.37a. Various microgrooves and their fabrication by rotational cutting tools, from Sawada[439].

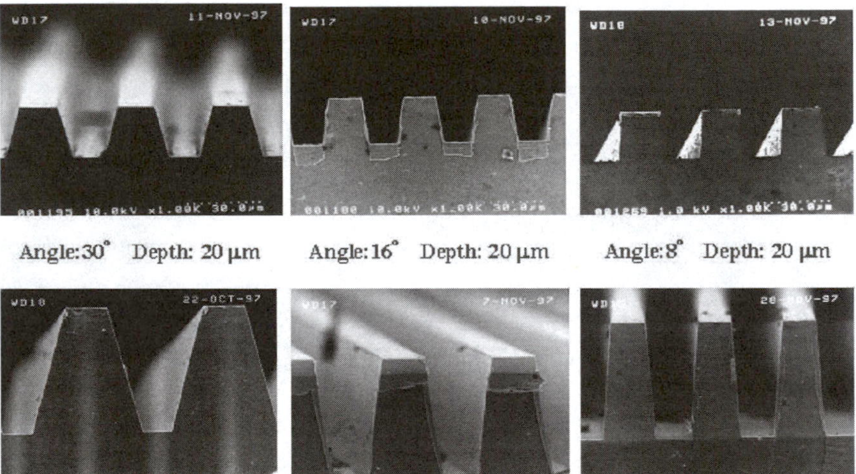

Figure 11.37b. Various kinds of trapezoid microgrooves (first row: 20 µm, second row: 50 µm depth), from Sawada[439].

Changing the angle and depth of groove, six different trapezoid microgrooves were created on a brass surface, as shown in Figure 11.37b. The grooving was done with cutting conditions of 50 mm/min feed rate, 13 m/sec cutting speed, and 1 μm depth of cut for finishing and 5 μm for rough cutting. A diamond tool was used to create the trapezoidal microgrooves. With respect to these grooves, it was found that the machined surfaces are very smooth with sharp edges similar to more conventional V-shaped microgrooves. Trapezoidal microgrooves were successfully applied to bond two small parts without adhesives.

Figure 11.38, from Takeuchi[440] shows micro-lenses on a silicon plate fabricated by rotating a circular arc diamond cutting edge and feeding the tool along an axis perpendicular to the silicon plate. The machining method corresponds to plunge cutting by a rotational tool. A suitable choice of cutting conditions allows for cutting of brittle materials.[441]

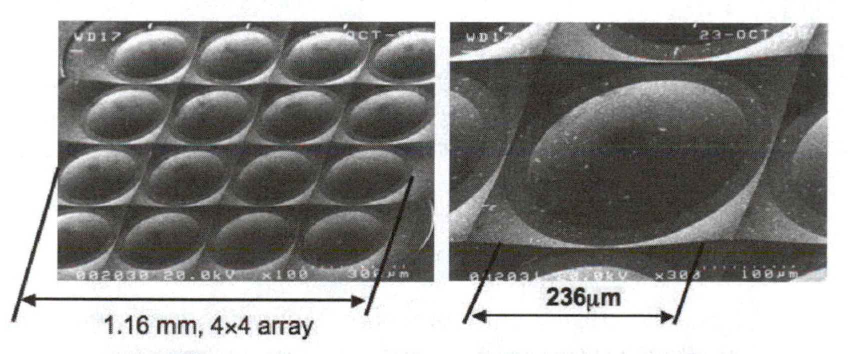

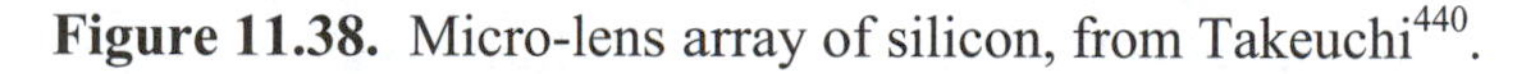

Figure 11.38. Micro-lens array of silicon, from Takeuchi[440].

As for multiple-focus micro-Fresnel lenses, it is difficult to produce those using conventional ultraprecision lathes since the configuration is not rotationally symmetric. The circular grooves are cut at each lens area, and a number of cutting start and end points exist on the same lens plate. Rotational cutting tools are not able to machine these kinds of microgrooves since the rotational tool, due to the rotational radius of the tool, yields long slopes at the cutting start and end points. Thus, two- and three-focus micro-Fresnel lenses have been manufactured utilizing 5-axis ultraprecision machining centers and non-rotational diamond tools. In non-rotational cutting tools, the cutting speed is equal to the feed rate. Although cutting speed is too low compared to rotational tools, a suitable choice of cutting condition allows parts to be cut in this manner.

Figure 11.39, from Takeuchi[442], shows an example of a three-focus micro-Fresnel lens, whose diameter is designed to be 2 mm machined using this technique. From the experimental results, it was found that this method, using a non-rotational cutting tool, is effective for producing optical devices with microgrooves on the order of about 0.1 μm deep[442].

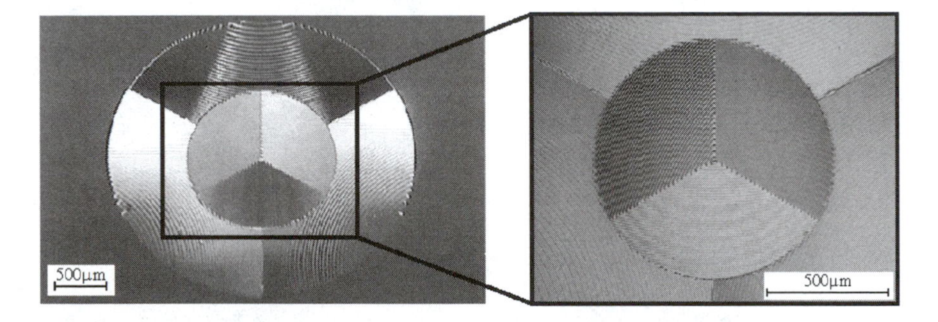

Figure 11.39. Machined 3-focus micro-Fresnel lens, from Takeuchi[442].

Maeda[443] created ultraprecision microgrooves for optical devices such as holographic optical elements using non-rotational cutting tools. V-shaped diamond tools were used to determine an optimal cutting speed by observing burr formation and chip shape. They proposed two cutting methods to create flat end microgrooves, seen in Figure 11.40 and Figure 11.41.

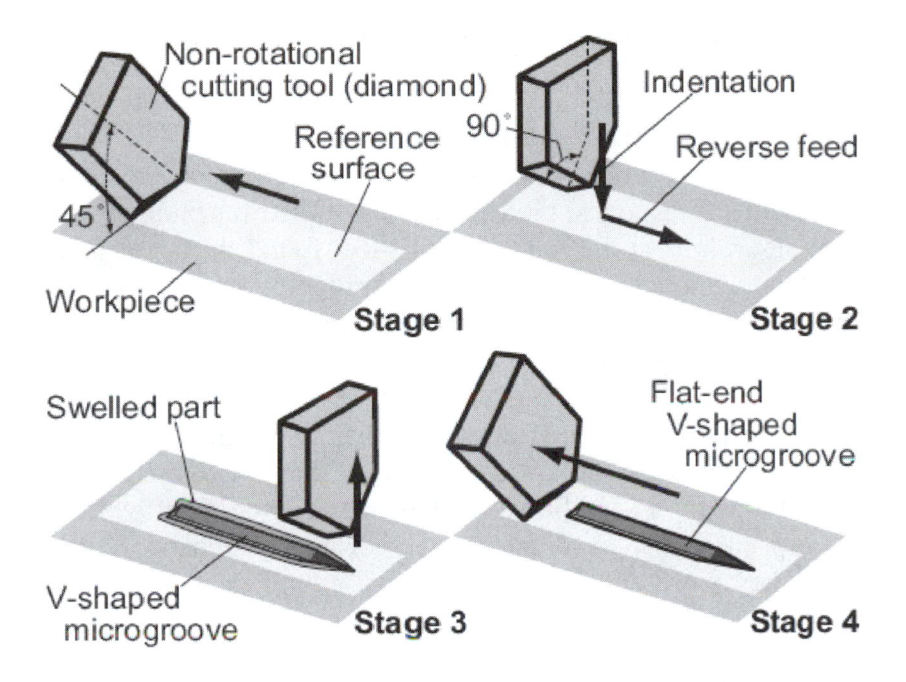

Figure 11.40. Flat-end microgroove machining with reverse feed method, from Maeda[443].

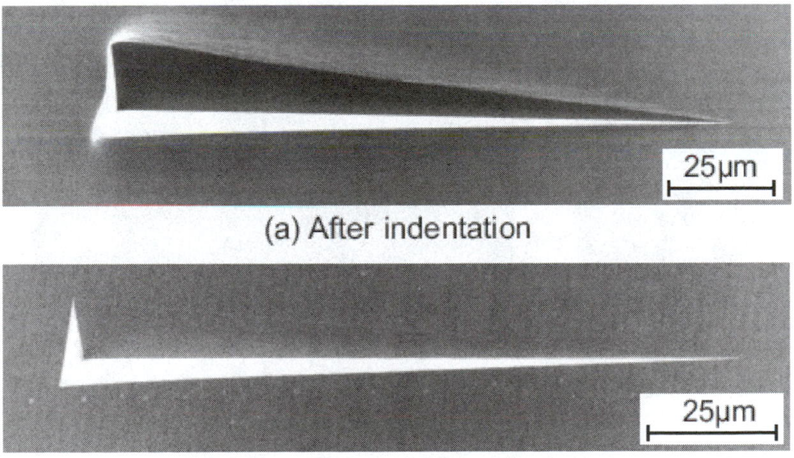

Figure 11.41. Machined flat-end microgroove using reverse feed method, from Maeda[443].

11.6.4.3 Creation of 3-dimensional shapes

One of the advantages of mechanical machining is full 3-D fabrication capability. A number of researchers have tried to create very complicated shapes to test the capabilities of their machine tools and control algorithms at the micro level. Most found that the capabilities of 'conventional' machines are generally applicable with minor modifications. However, the biggest challenge lies in the fabrication of the proper tool geometry for creating various 3-D features.

Takeuchi[445] designed a pseudo diamond ball-end mill. The tool is composed of a half-cut single crystal diamond offset from the rotational axis by a specific amount so that the tool can avoid zero cutting speed at any location on the tool and to enhance chip removal. They were able to create a very complicated small shape from gold using 36 hours of 5-axis control machining, Figure 11.42[446].

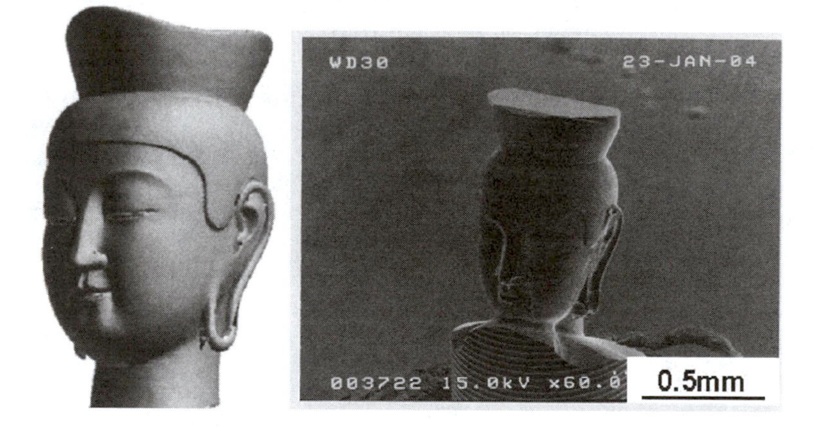

Figure 11.42. Creation of a small statue (Mirokubosatsu) by 5-axis control machining, from Sasaki[446].

11.6.4.4 Ultrasonic vibration assisted micromachining

At the micromachining level, inherent material property characteristics make it difficult to achieve the desired part quality. Hard and brittle materials such as silicon, glass, quartz crystal and ceramic are increasingly used in MEMS devices and reliable methods to process these materials are necessary. One of the efforts directed towards machining these difficult to cut materials is using an ultrasonic vibration cutting method at a frequency range of 20-100 kHz with an amplitude of 25 μm or less. Yu[447] tried microcutting with ultrasonic vibration on some of these materials and they found that tool wear was a significant problem. They applied a uniform wear method developed for micro-electrical discharge machining and experimented with various tool path planning strategies.

In precision or micromachining, diamond tools are usually the preferred cutting tools. However, diamond's affinity for ferrous metals causes severe tool wear. In order to overcome this problem, Kumabe[448] applied an ultrasonic vibration cutting method in diamond micromachining of ferrous material. This method reduced tool wear but sacrificed some degree of surface quality. Later, Moriwaki and Shamoto[429, 430] developed an ultrasonic elliptical vibration cutting method which showed improved cutting performance, Figure 11.43.

Although not vibration based, another approach to reducing wear of the diamond tool on ferrous materials is cryogenic machining. Evan[427] built a special machining setup with constrained liquid nitrogen flows and achieved surface roughness better than 25 nm Ra on 400 series stainless steel work material.

Ohnishi and Onikura[449, 450] applied ordinary ultrasonic vibration to microdrilling on an inclined work surface. The slippage of the drill tip at engagement on an inclined workpiece surface and large partial burrs on the hole exit surface have been difficult problems in industry where inclined surface drilling is commonly required in many applications. In microdrilling, these problems are more serious since tool stiffness is generally much lower and run-out more significant than in conventional drilling. The researchers added a 40 kHz ultrasonic vibration with a 3.5 μm amplitude to the drill during microdrilling of a duralumin workpiece and observed reduced cutting forces and improved cutting accuracy in terms of hole diameter, roundness, and the center location. Also, thinner chips generated by ultrasonic vibration enhanced chip removal during cutting leading to reduced tendency of tool breakage.

Egashira and Mizutani[451] tried ultrasonic vibration drilling (40 kHz and 0.8 μm amplitude) to make a 10 μm diameter hole in glass materials. To avoid either fracture or crack formation around the rim of hole, drilling was conducted in the ductile regime using a depth of cut 0.05 μm.

Egashira and Masuzawa[452] conducted similar microdrilling tests using ultrasonic vibration but the vibration was applied to the workpiece instead of the tool. Quartz glass work material and a sintered diamond tool were used for the tests and they succeeded in drilling holes as small as 5 μm diameter.

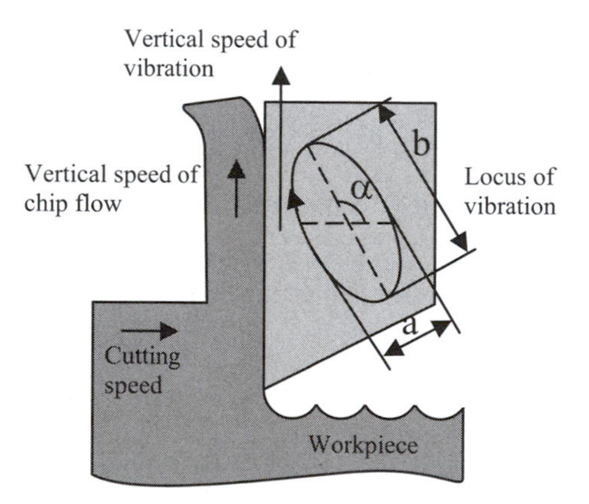

Figure 11.43. Principle of elliptical vibration cutting, from Moriwaki[430].

Machine tool have been developed to employ this approach. An elliptical vibration milling algorithm and machine was developed by Moriwaki and Shamoto[429, 430, 453-457] in order to achieve additional machining precision over that of other ultraprecision machines. The elliptical vibration milling machine used a double spindle mechanism to generate circular vibratory motion of the cutting tool, which resulted in improved surface finish, even with a diamond tool on ferrous materials.

11.6.5 Micro-tools

Commercially-available micro-drills are typically on the order of 50 μm in diameter, and have a similar twist geometry to that of conventional drills. Flat drills with simplified geometries are more common for diameters smaller than 50 μm.

Fabrication of micro-tools is another challenge in micromachining. Imprecise geometry and the irregularity of tools often negate the advantages of ultra precision process control, state of the art machine tools, and ultra fine tuning of process parameters. Also, scaling effects can play a significant role in process physics, which are closely related to the cutting mechanism, caused by a change in tool geometry during cutting.

Due to its hardness, single crystal diamond is the preferred tool material for microcutting. Diamond cutting tools were used in most of the early micromachining research due to their outstanding hardness (for wear resistance) and ease by which a sharp cutting edge could be generated through grinding. However, as diamond has a very high affinity to iron, microcutting is mostly limited to the machining of non-ferrous materials such as brass, aluminum, copper, and nickel. Hence, micromachining tests have been limited to non-ferrous materials[458, 459] with only some exceptions as noted[431, 460-463].

The focused ion beam process has been used to fabricate micro-tools. Using this method, Vasile[434, 464] fabricated 25 μm diameter steel milling tools and Adams[465] fabricated 13 μm diameter micro-grooving and microthreading cutting tools of high speed steel and tungsten carbide. Adams[466] also fabricated micro-end mills having less than 25 μm diameter by sputtering the same materials. They used a focused gallium ion beam to generate a number of cutting edges and tool end clearance and machined trenches with widths nearly the same as the diameter of the tool. They were able to achieve surface finishes of 250 nm (Ra).

Aoki and Takahashi[467] developed a 25 μm diameter thin-wire cut tool of tungsten carbide by cutting an oblique face into the shank. Egashira and Mizutani[468] fabricated micro-ball end mills with a radius of 10 μm using electrical discharge machining (EDM). They also used a wire electrodischarge grinding process (WEDG) to make a micro-drill with a D-shaped cross section and cutting edge radius of 0.5 μm. They used this to study ductile regime drilling and optimal tool geometry for holes with aspect ratios of 4 or higher[469].

Grinding is also used for fabricating micro-tools by many researchers[382, 434, 470, 471]. However, the grinding process has a limitation on tool geometry and is usable only to 65 μm diameter[382]. Onikura[472] proposed ultrasonic vibration grinding to overcome this limitation in fabricating micro-cylindrical tools and flat micro-drills of ultra fine grain cemented carbides. By adding ultrasonic vibration to the grinding process, they were able to produce high aspect ratio tools such as a 11 μm diameter with a length of 160 μm. Ohmori[473] not only focused on fabricating micro-tools, but also investigated the surface quality of the tools since the surface quality is closely related to machining performance, part quality, and tool rupture strength. They developed a machine tool fabrication process utilizing ELID grinding technology to fabricate various cross sectional shapes of the tool with high surface quality, Figure 11.44.

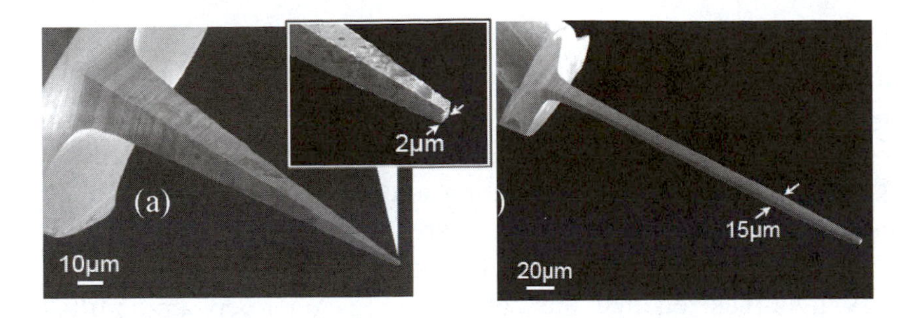

Figure 11.44. Overviews of produced micro-tools under optimum machining conditions: (a) Ultra precise tool (b) Extremely large aspect ratio micro-tool, from Ohmori[473].

Uhlmann and Schauer[474] proposed a new parametric tool design for micro-end mills by dynamic load and strain analysis using FEM analysis. Under micromachining, micro-tools experience a different loading situation from that seen in conventional machining.

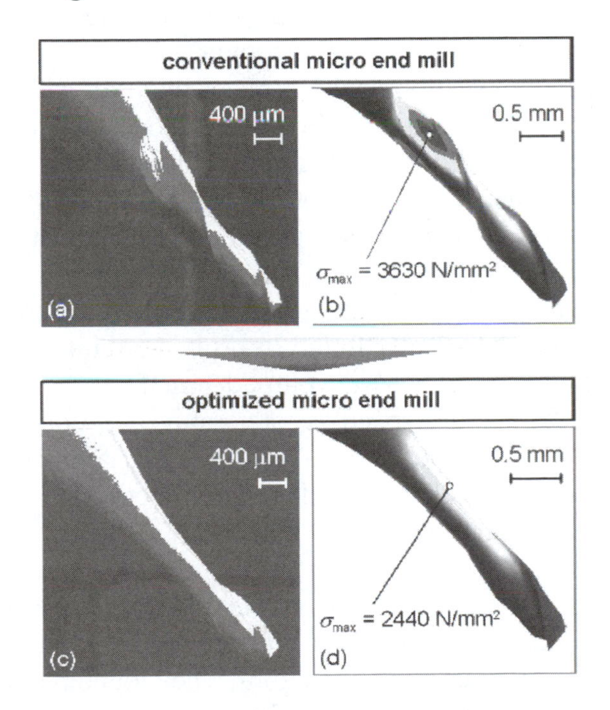

Figure 11.45. Conventional and optimized micro end mills, from Uhlman[474].

The optimized tool has a tapered shape with a reduced diamcter at the tool shank to avoid any contact with workpiece during cutting, Figure 11.45.

Tool life in micromachining is difficult to characterize, especially when feed-per-tooth values are large. Tansel[475, 476] attempted to study the relationship between wear and stress on the tool shank. They hypothesized that the cutting forces on micro-end mills are similar to those of conventional machining, but that the wear mechanisms are different. The first mechanism proposed is chip clogging. When clogging occurs, cutting forces and stresses increase rapidly, and can cause tool breakage in only a few rotations. This mechanism is very unpredictable and happens extremely rapidly. After the tool begins to wear, fatigue related breakage becomes an issue due to increased cutting force and stress on the tool shaft. The last mechanism identified is excessive stress breakage. The concept is that the cutting edge loses its sharpness and becomes dull. The tool is now unable to remove enough material to create satisfactory space for the central section of the shaft of the tool, and the tool begins to deflect. This deflection paired with a constant feed creates an excessive stress that leads to tool failure. Cutting force variations in micromachining increased constantly during machining due to increasing tool wear and can be used as a tool wear indicator.

Chen and Ehmann[477] did wear studies on microdrilling of copper-epoxy stacks typically found in printed circuit board applications. A drill life model was developed based on the independent parameters: feed, spindle speed, aspect ratio, and copper-to-epoxy ratio. It was found that increases in any of these parameters produce a reduction in tool life. The least-dependent factor is the copper-to-epoxy ratio (board density).

Sugano[478] studied wear of single crystal diamond tools and their effects on the microroughness and the residual stress of surface layers under various machining conditions. They found that wear of the diamond tool has less influence on microroughness. Fang[479] investigated relationships between tool geometry and tool life (focusing on tool breakage) in micromilling using FEM analysis and concluded that tapered tool design is optimal for longer tool life and performance as did Uhlmann and Schauer's study[474].

Godlinski[480] tried to minimize the tool wear by optimizing tool material and its microstructure. They used a two-step sintering process on a very fine alumina powder to avoid inter-granular fracture. This tool exhibited stronger wear resistance. Gaebler[481] tried coating spiral micro-drills, two flute end mills and abrasive pencils with CVD diamond to improve tool wear and performance.

In general, micro-tools are used with high rotational speed, which may cause vibration problems. Huang[482] investigated the dynamic characteristics of the microdrilling process and found that the natural frequency of a micro-drill decreases as the thrust force increases. Also, stiffness of the micro-tool is important when fabricating high precision parts. Some studies[483-487] conducted tests on bending stiffness of micro-drills and proposed multi-step shaped drills as a result.

Run-out is another issue which has a larger impact in micromachining. Due to the low strength of micro-tools, machining must be performed on a machine with air bearing stages and spindle and closed loop position and speed control. Furthermore vibration must be minimized and feed rate must be controlled in a smooth and continuous manner. Tool run-out needs to be minimized. Even with a precision tool shank and collets, this interface can lead to undesirable radial run-out. Friedrich[470, 488] were able to minimize this by utilizing a 4-point V-block bearing. Additional run-out was removed by making manual adjustments to the tool in the collet using a video microscope. Adjustments continue until radial run-out can no longer be seen. They found the remaining run-out to be on the order of 3-5 μm.

11.6.6 Cutting fluid

A typical flood supply of lubrication is generally not suitable for micromachining. First, the flow pressure of lubricants may influence cutting tool behavior. Second, even with negligible flow rate or proper control of flow rate, removal of excess working fluid after micromachining is challenging. Hence, special care is needed in lubricating, cooling, and transporting chips and debris during micromachining.

Figure 11.46 shows surface and edge quality under two different lubrication conditions: (a) minimum quantity lubrication (MQL) and (b) dry. Burrs form only at the end of the trench under MQL while a burr is present along the entire trench length under dry conditions. The surface quality of the side walls (second row in figure) is much better and chip adhesion on the tool is much lower under MQL conditions[369].

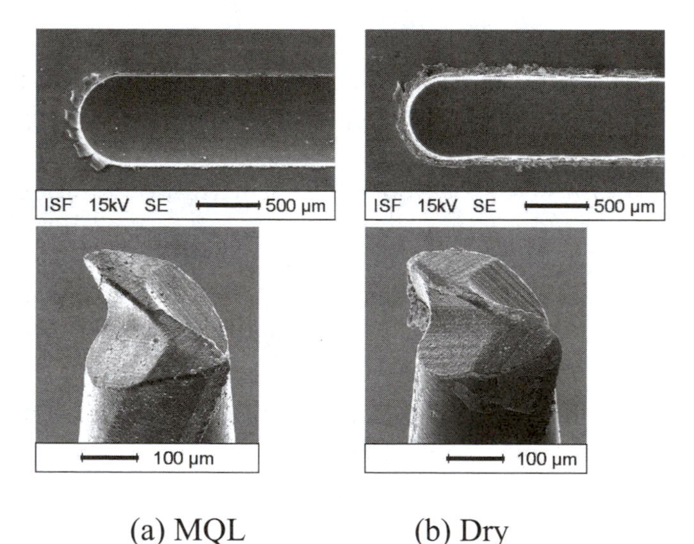

(a) MQL (b) Dry

Figure 11.46. Micro-end milling under MQL and dry (work: NiTi shape memory alloy, cutting speed: $V_c = 33$ m/min, depth of cut: $a_p = 10$ μm, width of cut: $a_e = 40$ μm, feed: $f_z = 12$ μm, micro-end mill: cemented carbide with TiAlN coating, d=400 μm, from Weinert[369].

Several unconventional cutting fluids have been tried in micromachining. Dry ice (CO_2) shows some promise on micro-machining of NiTi materials, Figure 11.47. The proper combination of nozzle distance, supply pressure and supply method (continuous and intermittent) is under investigation for optimal process results.

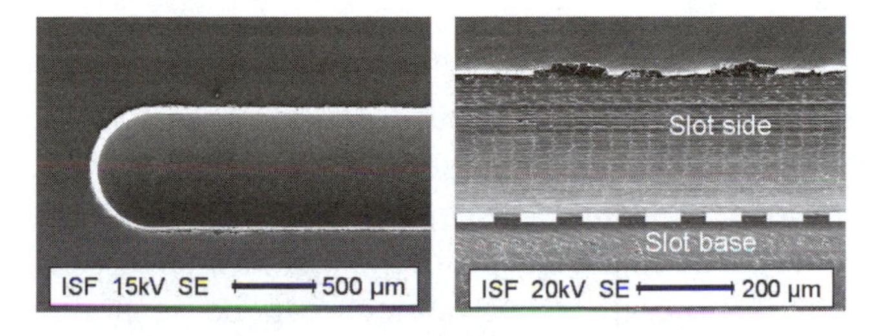

Figure 11.47. Micro-end milling under dry ice supply (work material: NiTi shape memory alloy, cutting speed: $V_c = 19$ m/min, depth of cut: $a_p = 10$ µm, width of cut: $a_e = 10$ µm, feed: $f_z = 20$ µm, micro-end mill: cemented carbide with TiAlN coating, d = 400 µm, nozzle distance: 80 mm, intermittent supply of dry ice), from Weinert[369].

Bissacco[379] investigated the effect of working fluid on the accuracy of micromachining. Commercially available ultraprecision machine tools have a position resolution up to 1 nm with high speed spindles up to 100,000 rpm, but their practical accuracy decreases when thermal deformation is considered. At these levels, a small offset caused by thermal deformation may contribute to significant errors. Axial depth of cut can be significantly offset by thermal expansion of a high speed spindle and/or machine tool column. The temperature difference between the cutting fluid and spindle is also a source of error.

11.6.7 Metrology in micromachining

One of the challenges in micromachining is verifying part geometry or, more generally referred to as metrology. Umeda[489] conducted surveys on measurement technology related to micromachining and found that measurements of material properties, force and displacement dynamics and shape in fabrication at the micro level were the most interesting. As the scale of features and machined parts decreases, the resolution of techniques used to measure and quantify these parts must increase of course. Demands on efficiency and accuracy of measurements are increasing for applications ranging from simple geometry and surface profiles to micro-spheres and micro-holes. Generally, measurement devices can be separated into two categories. The first includes those that measure the distance between edges of a feature, electron microscopes and optical profilometer, for example. These include a sensor (mechanical, magnetic, optical or capacitive), a workpiece holding table, and a transducer to displace the workpiece. The second category of measurement devices record either height or changes in height. This category includes mechanical stylus instruments, optical profile followers, atomic force microscopes, and interference measurement devices such as laser interferometer[490].

However, none of those instruments can really satisfy all the various measurement requirements on micro-devices. Hence, many researchers seek new devices by modifying or combining existing measurement techniques. Howard and Smith[491] modified conventional AFM technology to cover long ranges of surface metrology. They used a precision carriage and slideway mechanism to cover about 20 mm of travel and the AFM force probe, which utilizes the repulsive atomic force, to generate the surface contour.

In many cases, microparts include inside features such as pockets, holes, and channels. No technology exists to measure such features. Hence, Masuzawa[492] developed a vibroscanning method to measure the inside dimensions of micro-holes, Figure 11.48. This method is limited only to conductive materials because it uses a sensitive electrical switch by contacting a vibrating micro-probe onto the workpiece. They added another probe utilizing contact by bending of the probe[493].

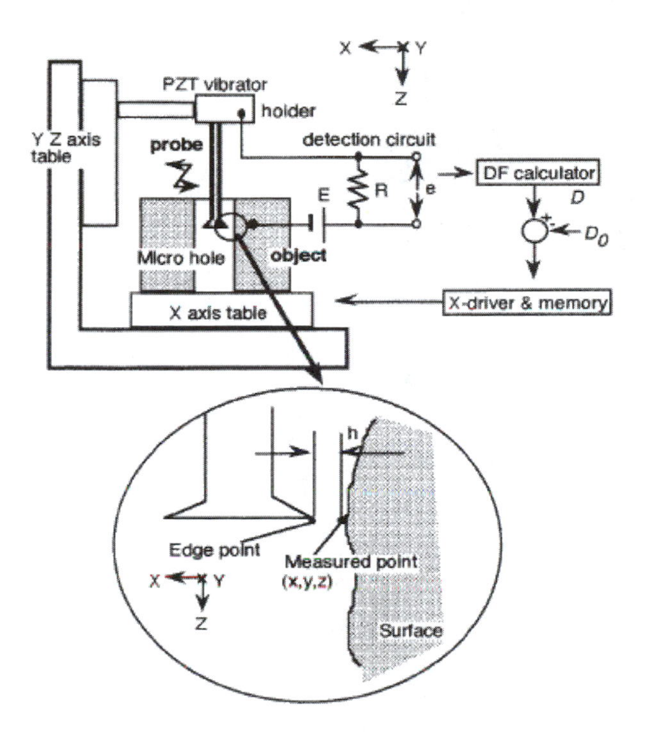

Figure 11.48. The detection principle of the vibroscanning method, from Masuzawa[492].

For a new system that is free of the requirement of electrical conductivity of the material, a new type of VS method, the twin-probe VS method, was developed.[493] This method also solved the problem of measurement instability caused by the microstructure of the basically conductive materials. In order to apply this new VS method, a switch-style probe was substituted for the single cantilever probe. An electrical voltage V is applied between the two elements of the twin probe, which are isolated electrically unless they come directly into contact with each other. These two elements are vibrated in parallel with small amplitude. A low frequency is applied in order to avoid self-resonance and maintain a small gap like an electrical switch. When the twin probe approaches the surface and the front element comes into contact with the surface during vibration, the two elements also make contact because of the bending of the front element, Figure 11.49.

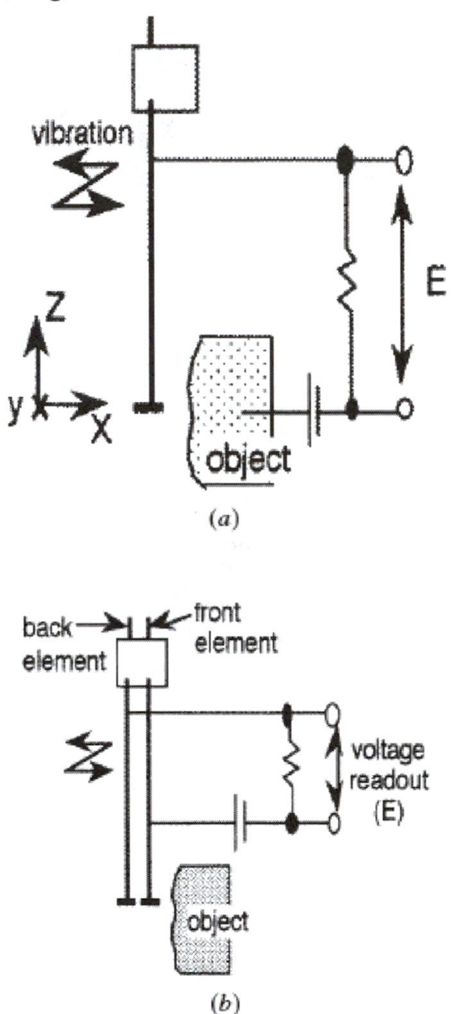

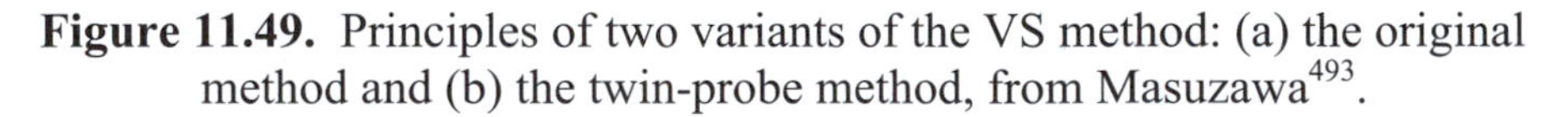
Figure 11.49. Principles of two variants of the VS method: (a) the original method and (b) the twin-probe method, from Masuzawa[493].

Many researchers developed precision CMM (coordinate measuring machine) devices with micron or submicron level resolution[494-497] but they were not generally sufficient for application to present levels of micromachining capability. Further development[498-500] improved the resolution up to few tens of nanometer and Jäger[501] developed a 3D-CMM with a resolution of 1.3 nm using a probe and laser interferometers with angle sensors for guiding deviation, Figure 11.50.

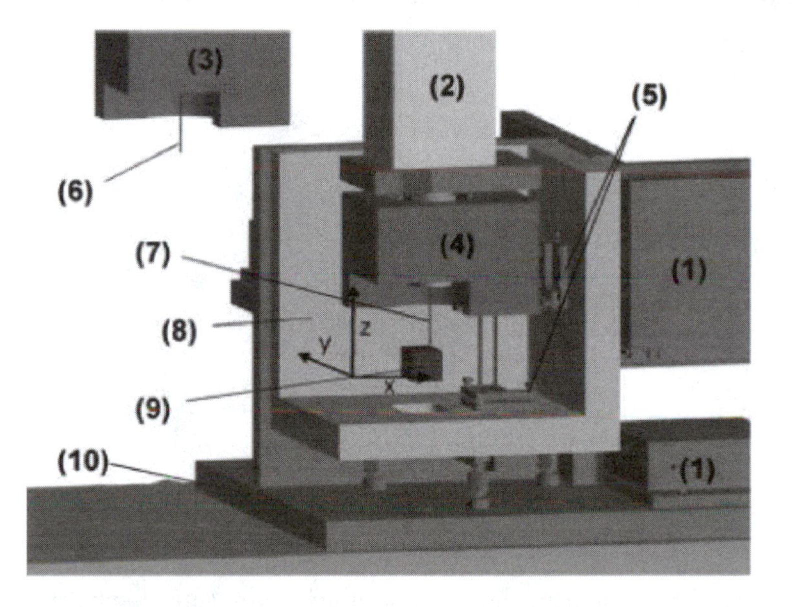

Figure 11.50. Metrology-frame of the μ-CMM (1: compact laser interferometer, 2: second ram, 3: first reference cuboid, 4: second reference cuboid, 5: reference mirrors, 6: 3D-boss-micro-probe, 7: opto-tactile micro-probe, 8: invar frame, 9: measurement volume, 10: aluminum frame), from Jäger[501].

Cao[502, 503] developed a three dimensional micro-CMM for precise three dimensional micro-shape measurements, Figure 11.51. For this, they also developed a 3D opto-tactile sensor for the probe using a silicon boss-membrane with piezo resistive transducers which can simultaneously measure deflections of the probe and force in three dimensions. The system consists of two stage measurements; coarse and fine measurements with a resolution up to 1.22 nm and uncertainty less than 100 nm.

Grigg[504] surveyed the applicability of white light interference microscopy on measurement of micro-systems. They also investigated surface profilometry, integrated profilometry, and lateral metrology for full 3D characterization, defect detection, etc. Figure 11.51 shows an example of an image of a MEMS comb drive structure obtained via white light interference microscopy[504].

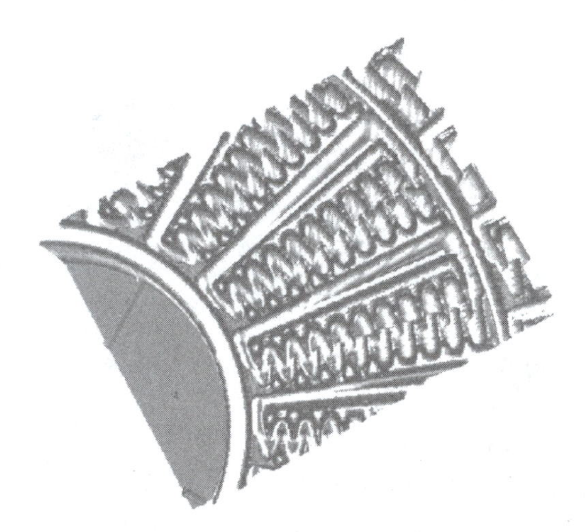

Figure 11.51. Schematic of a section of an electrostatic comb drive rotor used in a micro actuator device. The field of view is 50 μm and the profile range is 20 μm, after Grigg[504].

11.6.8 Conclusion and outlook

This section has attempted to review a field which has been actively researched by precision engineers and manufacturing engineers for some time. The literature, specially that related to modeling and process physics, is impressive. One sees the results of creative researchers responding to natural curiosity and industry demands in ways seldom experienced in the past in manufacturing. And the capability of industrial practice resulting from this is substantial.

But, "grand challenges" still exist in mechanical micromachining. First and foremost is the trade-off between "big" conventional machines doing micromachining and "small" microfactories. Both have their advantages. But a factory you can carry in your pocket may not be able to create microfeatures on a conventionally sized workpiece, for example. Otherwise, machine tool development seems to be progressing nicely, often scaling from conventional machines or conventional instruments with capabilities at these scales.
The other challenges push us to continue our research and development and include the following:

- design and fabrication of cutting tools (which still suffer from inadequate geometry control and edge definition, poor stiffness and excessive tool wear; and coatings have hardly been fully evaluated)
- processing difficult to machine materials (carbon steels are still a challenge for diamond tools, brittle materials frustrate high MRR with reasonable surface quality, and very ductile materials, as often seen in medical devices, are difficult)
- limited productivity (serious challenges remain with respect to fixturing and handling of microparts, increasing material removal rates, and efficient tool path planning; this is specially true as mechanical micromachining is paired up with other microfabrication techniques, MEMS, for example.)
- dry versus wet (what role, and what composition, will lubricants play in helping these challenges?)
- understanding and modeling of process mechanics (this is specially important because of the "non-continuous" nature of the materials at this scale and difficulty of validating model predictions), and
- metrology (although approaches exist, the productive application of the methods is limited).

These are not insurmountable challenges. But they will drive the continued development of mechanical micromachining for some time.

We next look at an aspect of precision manufacturing that is often a cause of quality issues and, more recently, product malfunction in use – edge defects and burrs.

11.7 Burrs – preventing and minimizing burr formation in precision components

The past years have seen an emphasis on increasing the quality of machined workpieces while at the same time reducing the cost per piece. Accompanying this is the decreasing size and increasing complexity of workpieces. This has put continual pressure on improvements in the machining process in terms of new processes, new tooling and tool materials, and new machine tools.[505] Fundamental to this continual improvement is understanding edge finishing of machined components, especially burrs. Deburring, like inspection, is a non-productive operation and, as such, should be eliminated or minimized to the greatest extent possible.

An understanding of the fundamentals of burr formation leads us to procedures for preventing or, at least, minimizing, burr formation. This depends on analytical models of burr formation, studies of tool/workpiece interaction for understanding the creation of burrs and, specially, the material influence, data bases describing cutting conditions for optimal edge quality, and design rules for burr prevention as well as standard terminology for describing edge features and burrs. Ultimately, engineering software tools must be available so that design and manufacturing engineers can use this knowledge interactively in their tasks to yield a mechanical part whose design and production is optimized for burr prevention along with the other critical specifications.

11.7.1 Introduction and background

Burrs in machined workpieces are real "productivity killers." Not only do they require additional finishing operations (deburring) and complicate assembly, but these operations can damage the part. Handling parts with burrs is a challenge for workers. Ideally, we'd like to avoid, or at least minimize, burrs by careful choice of tools, machining parameters and tool path or work material and part design. In fact, most burrs can be prevented or minimized with process control. Recently, more research and interest has been focused on problems associated with burrs from machining. The focus has traditionally been on deburring processes but understanding the burr formation process is critical to burr prevention. However, the level of scientific knowledge in this is just developing, (see Figure 11.52). It is vital to be able to associate details of the part performance and functionality with requirements for edge condition. Standards and specifications are only now being developed for this[506].

To effectively address burr prevention, the entire "process chain" from design to manufacturing must be considered, Figure 11.53. Here we see the importance of integrating all the elements affecting burrs, from the part design, including material selection, to the machining process.

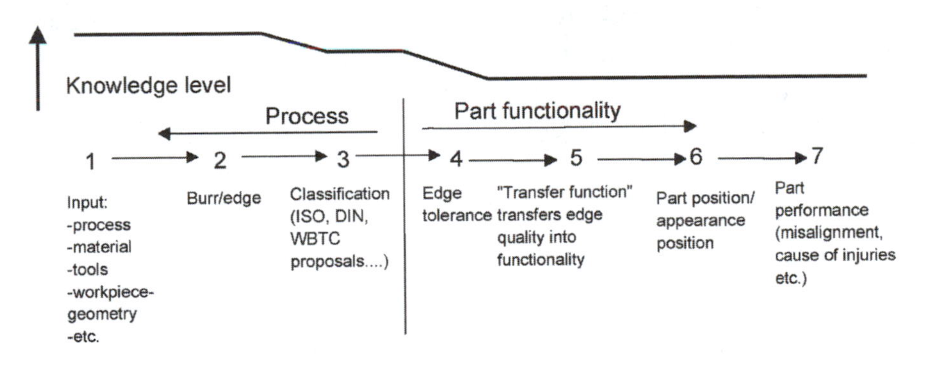

Figure 11.52. State of knowledge in burr formation.

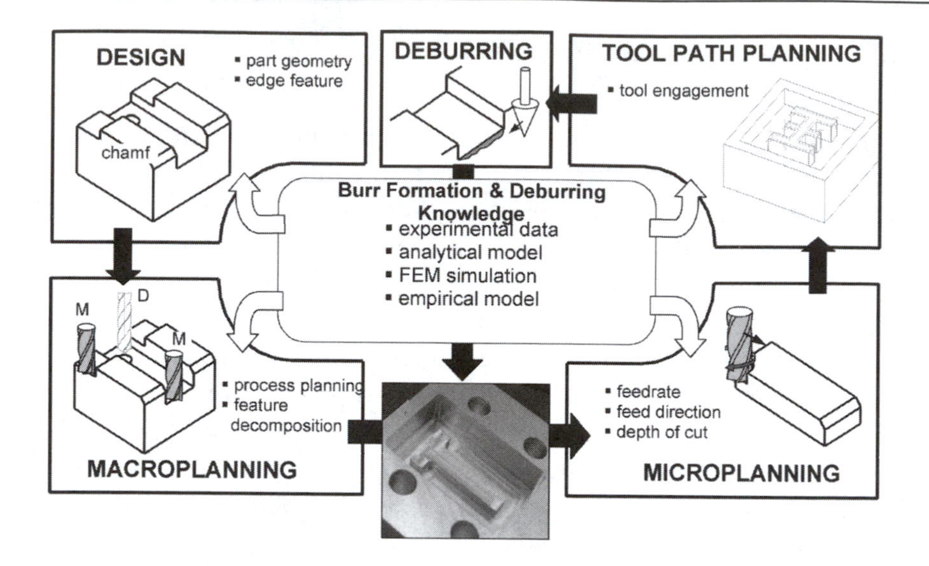

Figure 11.53. Five level integration required for burr minimization.

Burr formation affects workpiece accuracy and quality in several ways; dimensional distortion on part edge, challenges to assembly and handling caused by burrs in sensitive locations on the workpiece and damage done to the work subsurface from the deformation associated with burr formation. A typical burr formed on a metal component due to the exit of a cutting edge is seen in Figure 11.54, after Beier[507]. A number of things are clear from this image– there is substantial subsurface damage and deformation associated with a burr, the shape is quite complex and, hence, the description of a burr can be quite complex, and the presence of a burr can cause problems in manufacturing.

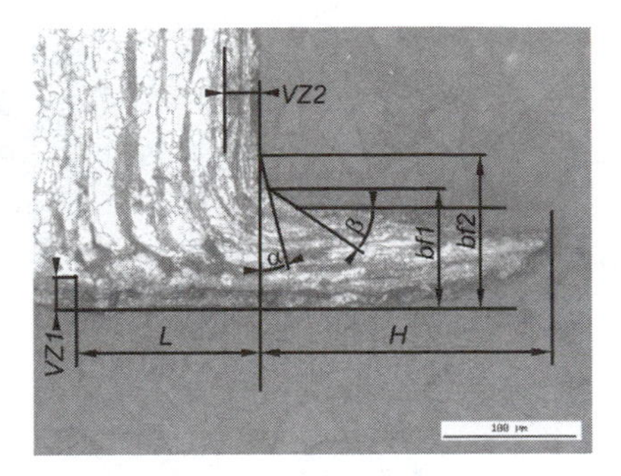

α,β tilting angle
VZ1 deformation zone lateral bore
VZ2 deformation zone main bore
b_{f1}, b_{f2} dimensions of burr root
L,H Measurands of burr

Figure 11.54. Typical burr and proposed measuring Nomenclature, after Beier[507].

In fact, this burr shown in cross-section in Figure 11.54 gives the appearance of a rather simple phenomena. The range of burrs found in machining practice is quite wide, specially when the full range of processes from drilling to grinding is considered. We had previously seen evidence of burr formation in precision machining applications shown in Figures 11.26, 11.27, 11.31 and 11.32 earlier in this chapter related to specific micromachining applications or studies. To emphasize the point, Figure 11.55 shows typical drilling burrs and their classification in stainless steel as an indication of the potential variation[508]. Burrs in milling and turning exhibit wide variation as well.

Classification

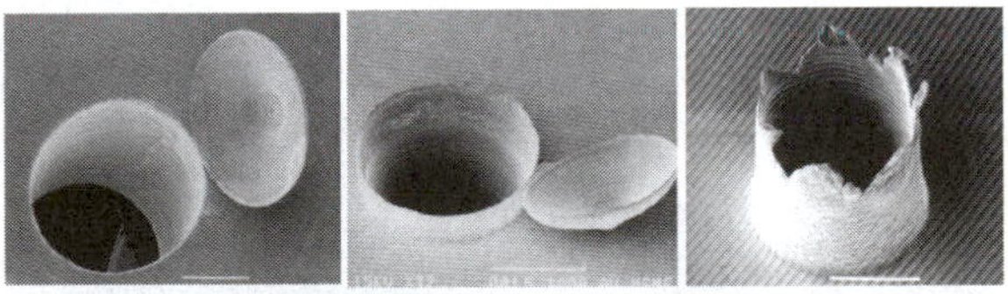

	TYPE I	TYPE II	TYPE III
Burr Shape	Uniform Burr	Uniform Burr	Crown Burr
Burr Height	~0.150 mm	~1.1 mm	(1.1~1.5)(d/2)

Figure 11.55. Three typical burrs in drilling stainless steel.

The costs associated with removing these burrs is substantial. The typical costs as a percentage of manufacturing cost varies up to 30% for high precision components such as aircraft engines, etc. In automotive components, the total amount of deburring cost for a part of medium complexity is approximately 14% of manufacturing expenses.[509] The actual investment in deburring systems increases with part complexity and precision as seen in Figure 11.56 from Berger.[509]

A better strategy is to attempt to prevent of minimize, or prevent, burrs from occurring in the first place. This has two immediate benefits in that, first, it eliminates the additional cost of deburring the component and the likelihood of damage during the deburring process and, second, in the case burrs cannot be eliminated it improves the effectiveness of any deburring strategy due to reduced and more standard burr size and shape. This requires a comprehensive approach to burr prevention and minimization consisting of a number of components.

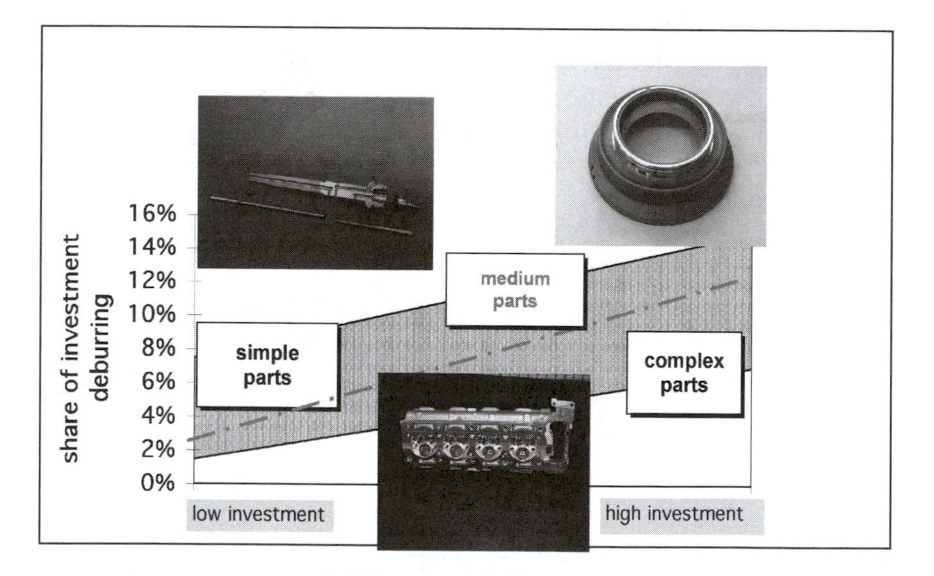

Figure 11.56. Investment in deburring systems as a function of part complexity and total investment in manufacturing system, from Berger[509].

To minimize or prevent burr formation requires that all stages of manufacturing from the design of the component through process planning and production be integrated so that the potential part features and material constraints, tooling and process sequences and process variables be considered from a perspective of the potential for creation of burrs on the workpiece, as seen in Figure 11.53. That is, the inputs (process, material, tools, workpiece geometry, fixturing, etc.) must be considered along with the part functionality (part performance, fit and assembly requirements) as well as any expected or required deburring processes. This is most successful when clear standards and classifications are available, edge tolerances can be specified and the relationship between the edge quality and part functionality is clearly understood. This is not generally the case.

The future development in this regard is seen to depend on the following:

- the development of predictive models with competent databases, including "expert data bases" for process specification
- simulation models of burr formation capable of indicating the interaction and dependencies of key process parameters (finite element models, for example)
- strategies for burr reduction linked to computer aided design (CAD) systems for product design and process planning
- inspection strategies for burr detection and characterization including specialized burr sensors.

One could also add here the development of specialized tooling for deburring and inspection to insure burrs are removed, although that is an area well covered commercially today.

11.7.2 Process-based solutions

The models, databases and strategies mentioned in the previous section must be linked to the process of interest to be most effective.

There are substantial differences between burr formation in drilling and milling for example. In drilling, infeed can play an important role in the development of drilling burrs[510, 511]. In addition, the drill geometry can affect the size and shape of the burr formed as well as prevent burr formation in some cases[512, 513]. Analytical models are increasingly supplemented with finite element method (FEM) models of the drilling process to predict effects of drill geometry, process parameters and workpiece characteristics on size and shape of the burr[514, 515]. Applications to aerospace component manufacturing, specially multi-layer structures, is a primary area of focus for FEM drilling process modeling. This is also applicable to milling but less so to date due to the complexity of the milling process.

11.7.2.1 Milling

Since milling (specially face milling) figures so prominently in the manufacture of so many parts, for example, automotive engines and transmission components, it has been a major focus for burr reduction and prevention for many years. In milling, the kinematics of tool exits from the workpiece are a dominant factor in burr formation and, as a result, substantial success has been realized by adjusting the tool path over the workpiece, Figure 11.57. The principal criteria in tool path determination have been:[516]

- avoiding exits of inserts (or always machining on to the part edge)
- sequencing of process steps to create any burrs on a last, less significant edge
- control of exit order sequence (EOS) by tool geometry and path variation
- maintaining uniform tool chip loads over critical features
- lift and re-contact of milling cutter for some features where maneuverability is limited
- avoiding “push exits” (those with long cutter path/edge contact length

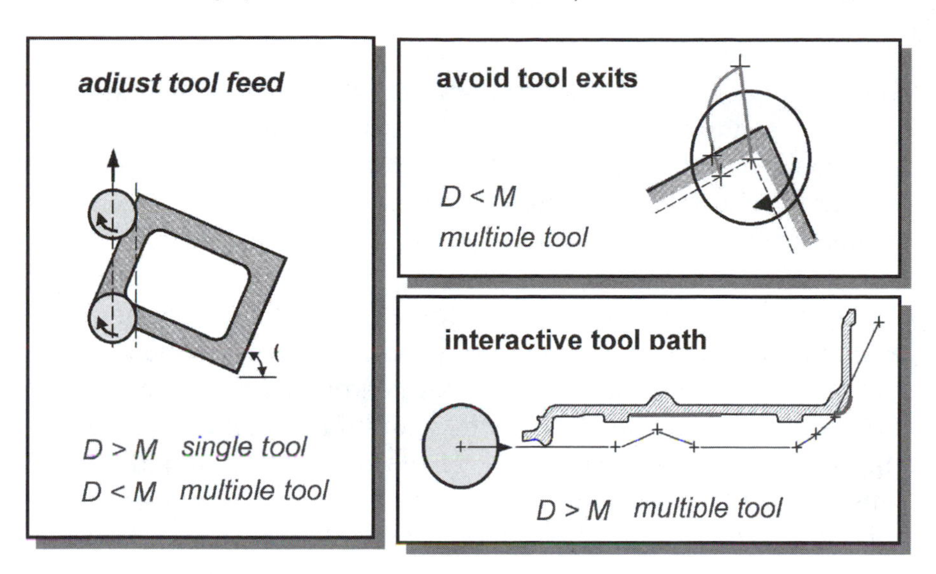

Figure 11.57. Tool path strategies for minimizing and preventing burrs in face milling.

While these criteria are often difficult to apply in all situations they have shown dramatic reductions in burr formation with the corresponding increases in tool life (tools are often changed when burr size reaches a specification limit) and reductions in deburring costs. In all circumstances cycle time constraints must be met with any redesigned tool paths.

With burr expert data bases for different materials and process parameters and the software for tool path planning, the possibility of designers being able to simulate the likely scenario of machining a component and any resulting problems with burrs is becoming a reality, These software systems must also be comprehensive enough to include other process steps and constraints so that other critical specifications (surface roughness, for example) are not compromised.

11.7.2.2 Drilling

Burr formation in drilling is primarily dependent upon the tool geometry and tool/work orientation (that is, whether the hole axis is orthogonal or not to the plane of the exit surface of the hole). The burr types illustrated in Figure 11.55 are created by a sequence of events starting when the drill action first deforms the material on the exit surface of the workpiece through creation of the hole, Figure 11.58[512, 513]. When intersecting holes are drilled, the specific orientation of the axis of the intersecting holes will have a tremendous effect on the location and creation of burrs around the perimeter of the holes. Figure 11.59 shows a schematic of burr formation in intersecting holes. Since the "exit angle" of the drill varies around the circumference of the hole intersection, the potential for burr formation will vary. This means that intersection geometry as well as tool geometries optimized to minimize adverse burr formation conditions can be effective in minimizing burr formation. Burr formation in intersecting holes shows high dependence on angular position under the same cutting conditions. Large exit angles, as seen in Figure 11.59, yield small burrs. There is also a strong dependence on inclination angle (that is the degree of inclination of the intersecting hole from perpendicular.) Research shows that an inclination angle of 45° reduces burr formation.

Further, holes in multilayer materials offer additional challenges. This is specially true in aerospace applications where structures are often composed of "sandwich" configurations of metal, composite and sealant, Figure 11.60. Burr formation here is challenging as interlayer burrs often need to be removed before final assembly. Finite element analysis of these types of specific situations often offers increased understanding of the problems. When drilling multilayer material structures, the fixturing often plays an important role in determining the size and location of burrs. Figure 11.61 illustrates the "gap formation" occurring during drilling of sandwich materials in the absence of proper fixturing. The gap provides space for burr formation at the interface of the two material sheets[517].

Stages	Burr formation mechanism	FEM simulation
Steady-state cutting	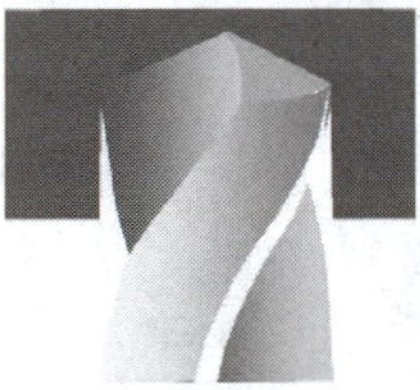	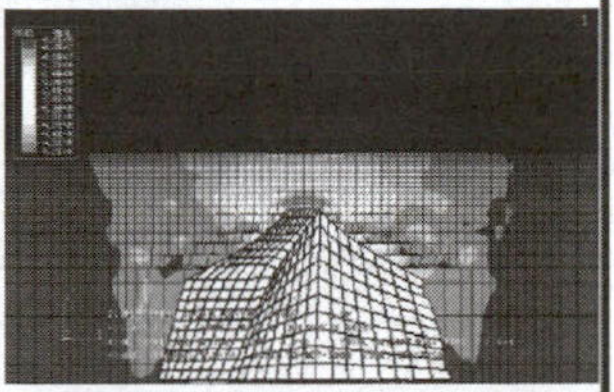
Burr initiation •Plastic deformation at the center (thin)	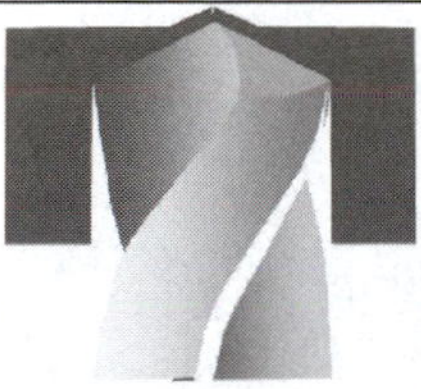	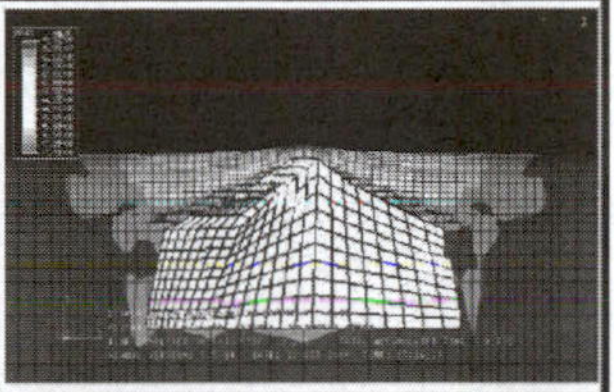
Development Plastic zone expands with little cutting	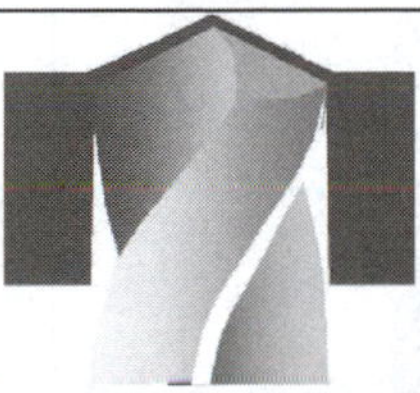	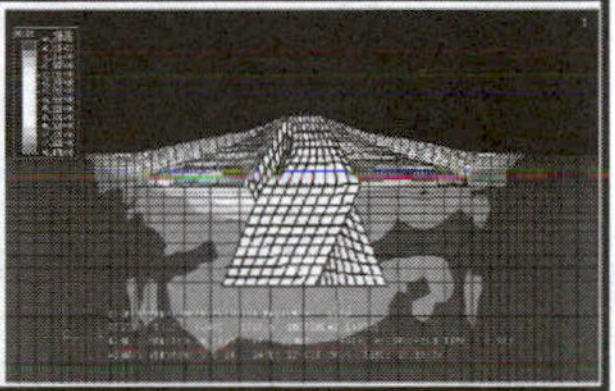
Initial fracture Fracture at the edge of the drill	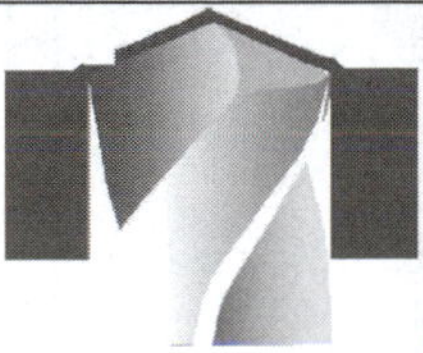	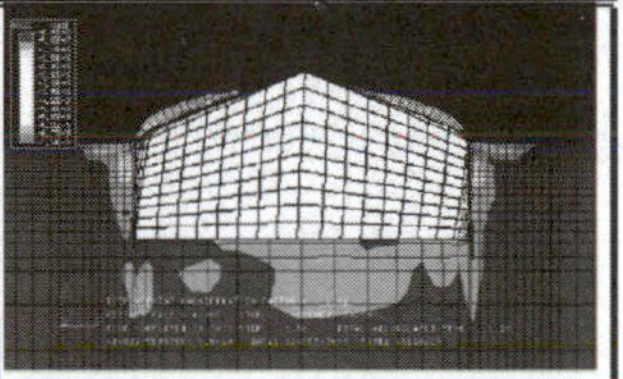
Burr formation Burr and cap formation	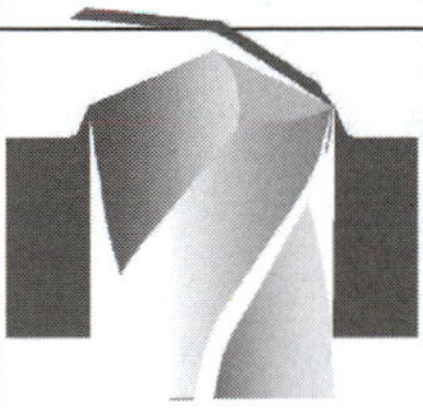	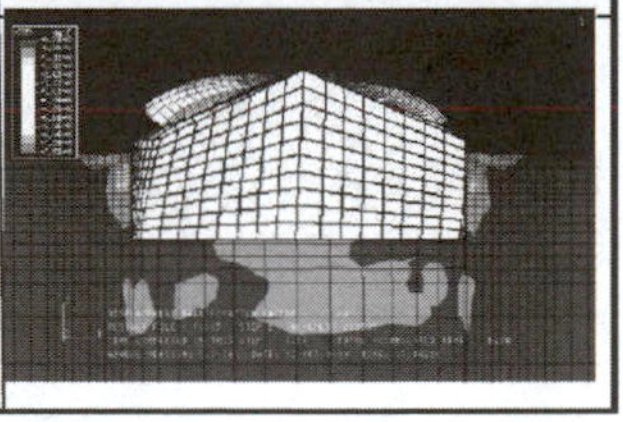

Figure 11.58. Sequence of burr formation in hole drilling for uniform burr with cap.

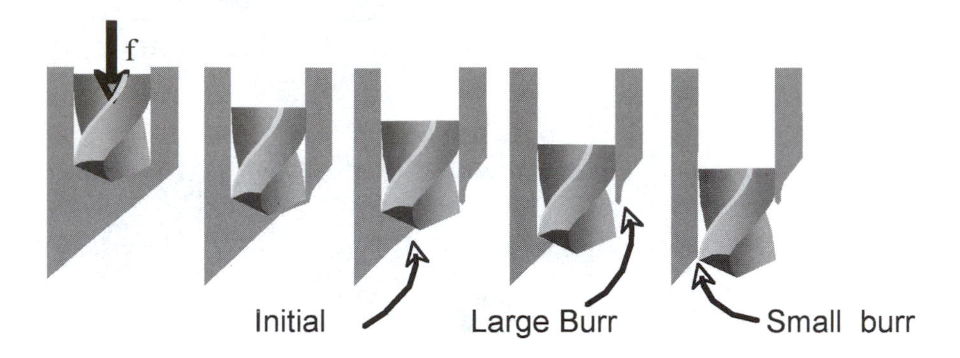

Figure 11.59. Schematic of burr formation in intersecting holes.

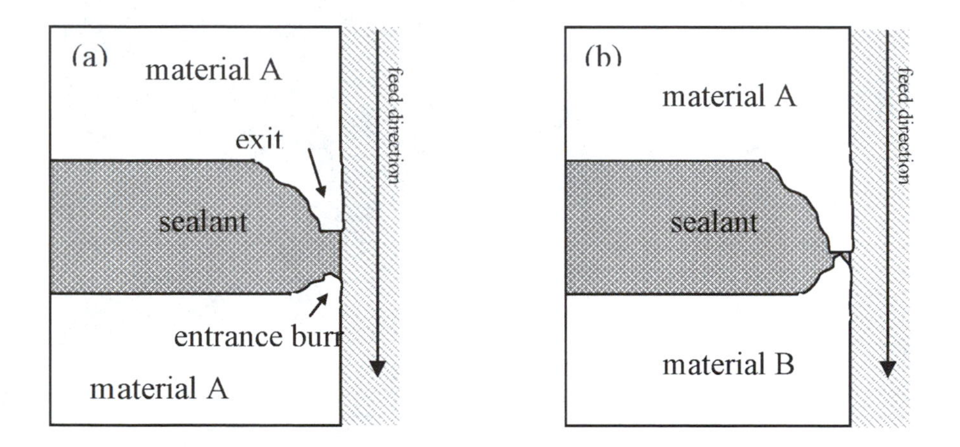

Figure 11.60. Examples of interlayer burrs in sandwich materials (multi-layer)

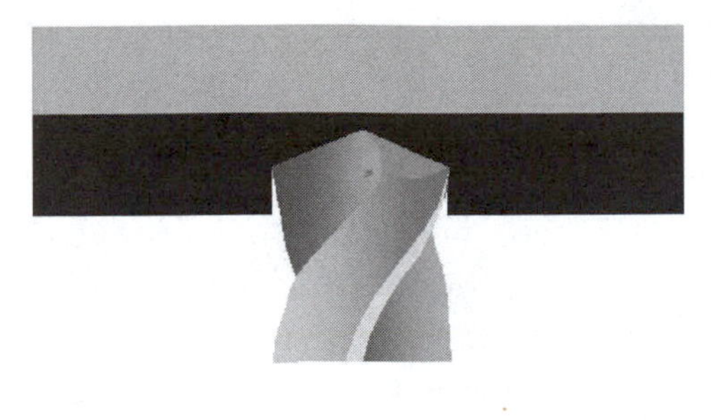

Figure 11.61. Schematic of interlayer burr formation due to gap during drilling.

11.7.3 Examples of application of burr minimization strategies

11.7.3.1 Tool path planning in milling

One of the most successful areas of application of burr minimization strategies is in tool path planning for face milling. To a great extent, burr formation in milling can be prevented by adjusting the path of the milling cutter over the workpiece face. Specific cases have been evaluated in automotive engine manufacturing with major automobile companies. This can be extended to optimization of the process to insure that surface quality, including flatness, specifications are met or exceeded. Figure 11.62 shows a conventional tool path for face milling a surface on a cast AlSi alloy automotive engine block.

The presence of substantial burrs at critical locations required frequent tool changes as well as additional deburring operations. The optimized tool path using the criteria described above is shown in Figure 11.63 and, in Figure 11.64, shows the resulting burr free workpiece. Although the tool path is substantially longer in this example, it was possible to increase the feedrate without loss of surface finish to maintain the required 5 second cycle time for the process. The tool life (as a result of dramatically reduced burr formation) was increased by a factor of 3 and the resulting savings per machine/year were estimated at approximately €40,000[516, 518].

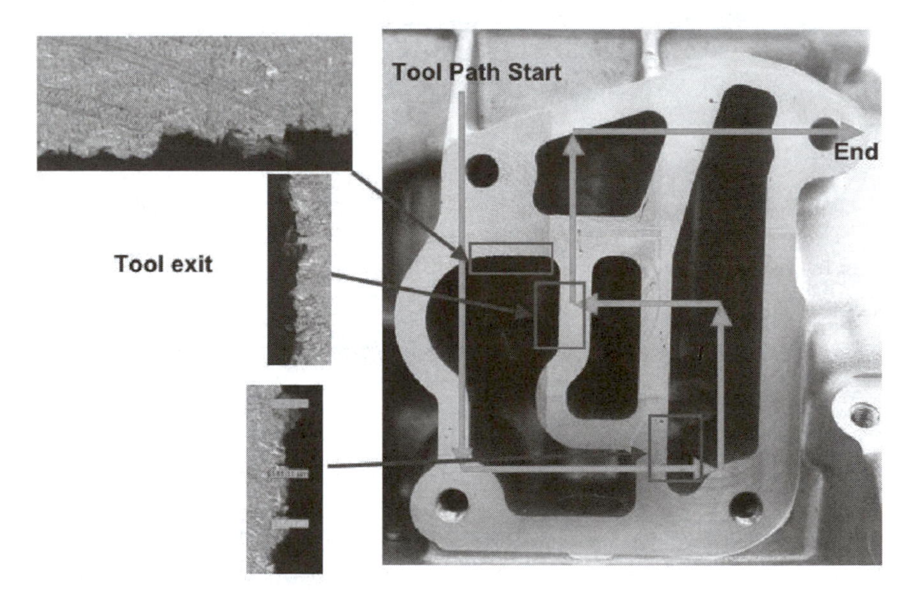

Figure 11.62. Conventional tool path for face milling engine block face and resulting burrs at key locations.

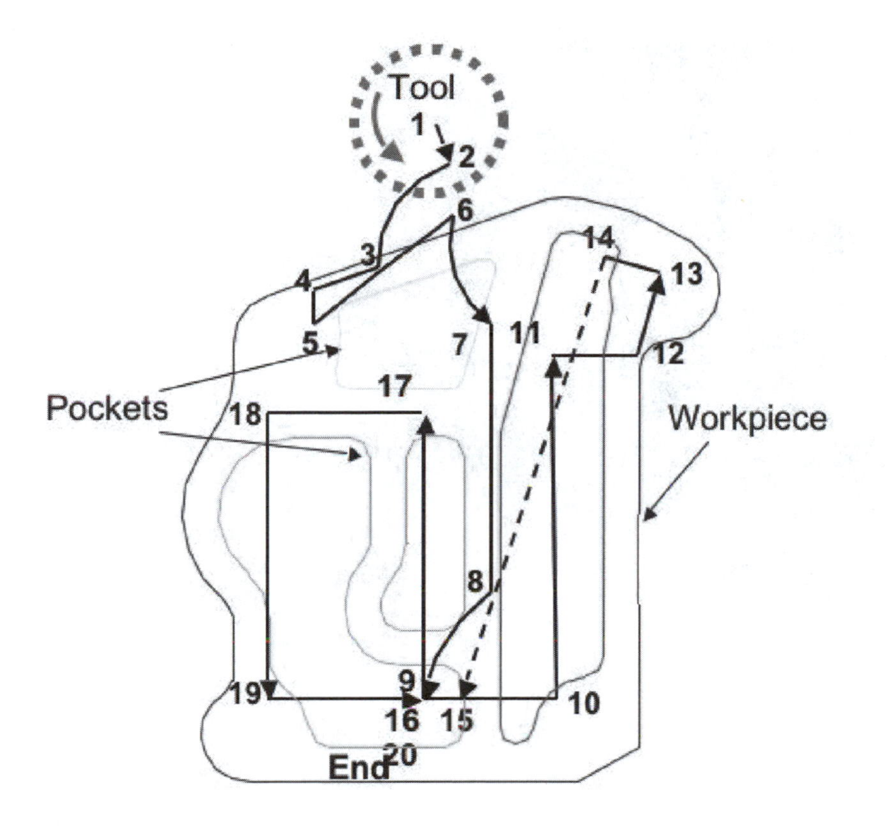

Figure 11.63. Modified tool path for part in Figure 11.62.

Tool Path Length:

Old path – 209 mm
New path – 524 mm

Cycle time (with increased feedrate) stays at 5 seconds

Figure 11.64. Workpiece resulting from optimized tool path and tool path specifics.

11.7.3.2 Burr control chart

Burr minimization and prevention in drilling is strongly related to process conditions (feedrate and speed, for example) and drill geometry. It is possible to represent the reasonable ranges of operating conditions for drilling by use of a "burr control chart" derived from experimental data on burr formation for varying speeds and feeds. This can be normalized to cover a range of drill diameters and, importantly, can be used across similar materials (carbon steels, for example). Data shows the likelihood of creating one of three standard burrs, as shown in Figure 11.55, namely, small uniform (Type I), large uniform (Type II) and crown burr (Type III)[519, 520]. Figure 11.65 below shows a typical burr control chart for 304L stainless steel. Continuous lines delineate different burr types. Type I is preferred. Burr height scales with distance from the origin. This burr control chart can be integrated with an expert system allowing queries of likelihood of burr formation to be shown on the control chart when information on drill diameter, speed, feed, etc. are input. Typical burr sizes expected are shown.

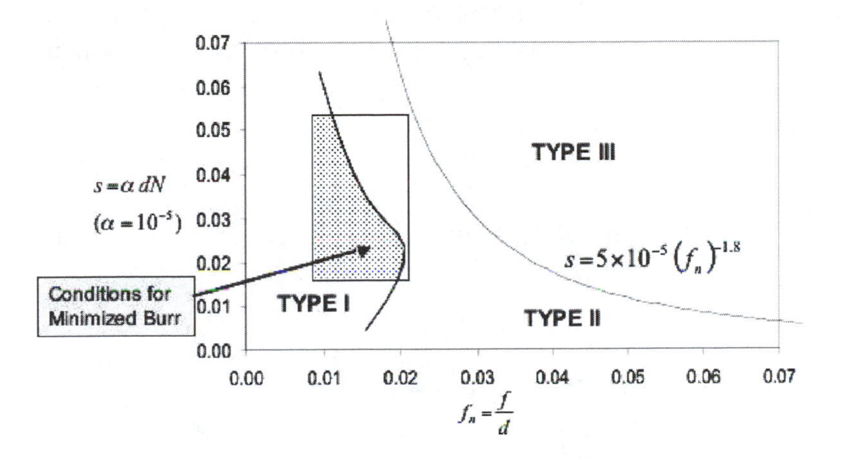

Figure 11.65. Drilling burr control charf for 304L stainless steel material showing normalized speed, s (vertical axis) vs. normalized feed, f (horizontal axis), d is drill diameter. Minimized burr conditions are indicated in crosshatched region.

11.7.3.3 Integrated process planning and burr minimization

It is not sufficient to simply try to adjust process parameters for burr minimization or prevention alone. One should also consider other important constraints in machining, e.g. surface finish and dimensional tolerances. Figure 11.66 shows the process considerations for insuring optimum performance in face milling from the so-called macro planning at a higher level to detailed micro planning selecting machining conditions. The constraints include cycle time, flatness and surface roughness, burr height, surface integrity, etc.[521]

This enhanced process planning can be integrated with the basic design process to insure compliance with design criteria and manufacturing process optimization. This is consistent with recent efforts at implementing the "digital factory" and relying in comprehensive software links between individual elements of design and process planning, with competent process models included, for allowing a view "down the manufacturing pipeline" from any position in the design to manufacturing process. Additional constraints relative to "cleanability" of the component (meaning the degree to which generation of contamination during machining, due to chips and burrs for example, as well as accessibility of internal features for cleaning via brushes or water jets) and minimizing energy and consumable consumption can be added at this stage of process planning also.

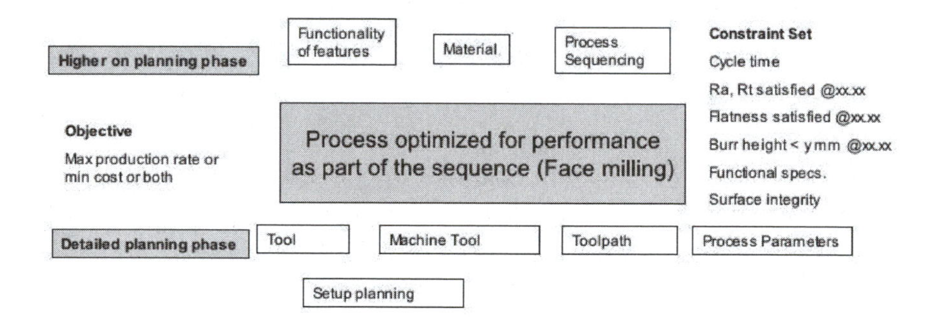

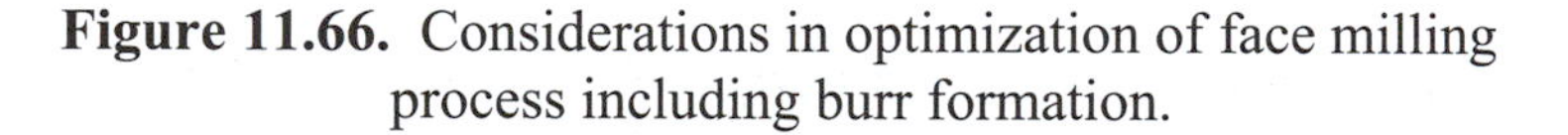

Figure 11.66. Considerations in optimization of face milling process including burr formation.

11.7.4 Summary and conclusions

Although edge finishing in machined components is a constant challenge in precision manufacturing of mechanical components, there are a number of strategies, built on competent process models and extensive data bases, that can substantially minimize or eliminate burrs. These strategies, some illustrated above, can be incorporated in the software relied upon by design and manufacturing engineers in their normal activities to insure that the conditions which can lead to burr formation can be avoided while insuring that production efficiency is maintained. This is part of the development of the "digital factory." Recent experience indicates that the basis for this process optimization may also yield increases in throughput due to decreases in cycle time thanks to optimum part orientation on the machine during machining[522]. In situations where burrs cannot definitely be eliminated there is the possibility, using these tools, to at least control their size over a range of conditions so that commercial deburring techniques are more reliably implemented – techniques such as abrasive filament brushes, for example. Finally, the inclusion of design rules for burr minimization will allow the design engineers to reduce the likelihood of edge defects at the most effective stage – during product design. Future work on burr prevention must focus more on tool design. The potential for substantial improvement, especially in drilling, will depend on analysis of drilling burr formation with the objective of optimization of tool drill design.

It may be some time before we can declare that it is now possible to prevent all burr formation during the machining of mechanical components. But, in the meantime, there is much that can be accomplished towards that goal using the techniques and systems discussed in this section to produce parts with higher edge precision.

XII FUTURE OF PRECISION MANUFACTURING

12.1 Introduction

We have chosen a very specific path in the discussion of precision manufacturing in this book with an emphasis on fundamentals (or roots) of precision engineering driving manufacturing. We've focused on development of basic ways to represent machine performance, evaluate error sources and construct budgets to predict machine and process capability. We then moved to more detailed review of material removal techniques and processes for creating precision features, including process planning. Finally, we reviewed applications of precision manufacturing, including nanoscale device fabrication, and some challenges to creating precision features. What's next?

In this chapter we try to "look over the horizon" a bit to the technologies and needs driving precision manufacturing in the future. As time goes on manufacturing in general and precision manufacturing in particular will play an increasingly important role in new product and process development. Particular areas of impact will be seen in development of new energy sources (based on photovoltaic, hybrid motors, wind, artificial photosynthesis, fuel cells, etc.) The science is just now being developed for many of these technologies and the need for efficient processes and systems to manufacture them will become more apparent.

The ability to predict the outcome of our designs from conception to production, the "manufacturing pipeline," is increasingly important in competitive industry. This so-call "design for manufacturing" (or manufacturing for design) view allows us to preview potential show stoppers down stream and correct or avoid them at the design stage when the flexibility to do this is greatest and the cost the least. The availability of sophisticated software tools makes this within reach if we have capable models of the processes with sufficient detail to reflect the subtle problems we've been discussing.

In addition, we are living in a world with arguably dwindling resources (or at least harder to access and process) and, each day, our consumption further outpaces the ability of our planet and ecosystem to replenish our use. That is, we are not living sustainably. Precision manufacturing will have an impact on this as well both from the point of view of consumption of energy and resources in our clever machines and processes as well as in the production of those machines and processes. Can we design our machines and processes to be sustainable? That is a key question and, in this chapter, we hope you are convinced that there is potential.

12.2 The manufacturing pipeline

In Chapter 9 we discussed process planning and, in Table 9.1, reviewed four levels of integration for the design to fabrication cycle. Level 1 was the highest at the design stage and allowed a lot of flexibility for adjustment of specifications, materials and process requirements. As we got lower in the table, moving towards the factory floor at level 3, our flexibility diminished; that is, our ability to make adjustments to tune the process was limited and the design was fixed. With the assistance of capable process models linked to computer aided design tools it should be possible for the designer to "look down the manufacturing pipeline" to see if there are any problems with the design, from a manufacturing perspective, that will prevent the efficient manufacturing of the product. In addition, the reverse can be true. That is, the manufacturing engineer or can look back up the pipeline to suggest changes to the designer that would make manufacturing easier but, hopefully, not cause any problems with the design. This would also provide early feedback to designers on manufacturing issues during the design process. The goal is end-to-end, science-based simulation of whole manufacturing process. A view of this "pipeline" is shown in Figure 12.1.

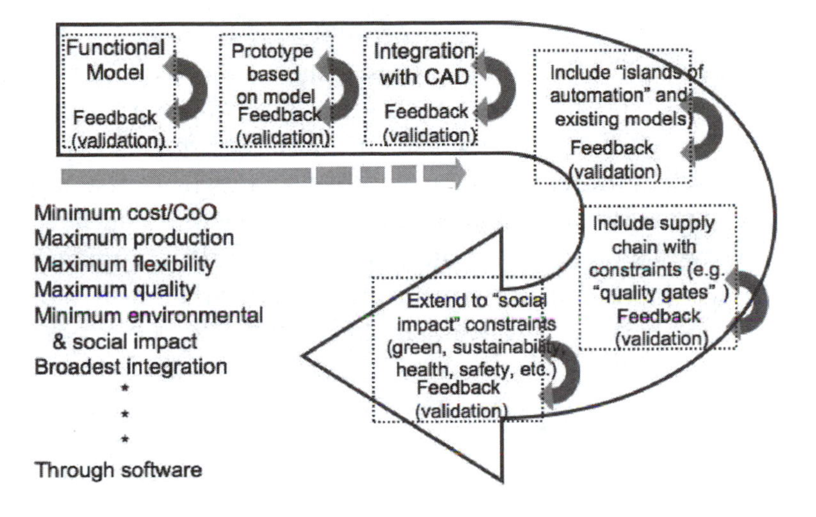

Figure 12.1. Schematic of the manufacturing pipeline based on process models.

This pipeline view, if successful, would connect a number of the existing islands in process modeling and process planning that exist today. These islands include: cutting force prediction tools, tool design, tool path planning, process parameter optimization, feature sequencing and knowledge based expert systems. These are elements that are ultimately important for the successful implementation of precision manufacturing (as discussed in Chapter 9) but very dependent of a detailed level of manufacturing to be successful.

What is really needed is a "Google-earth" view of the design to manufacturing process where each level of view has sufficient detail to understand the features being represented but not too much detail to make it computationally inefficient. That is, we don't need to have the detail to be able to see our office building in a earth view that is focused on western United States. But, when we zoom in on Berkeley, we'd like to see the details at a building by building level.

Similarly with precision manufacturing, we don't need to have the detail on subsurface damage in our view of the process at the machine motion capability level but would need that level of detail when we zoom into the tool-work interface.

The details necessary to accomplish this pipeline view are shown in Figure 12.2 which shows the interconnected chain of elements from design to production and the cooperative use of modeling along the way. Also connected in this view are the supply and distribution chain which input to and distribute from the process. This will require a substantial integration of existing and new software to allow the pipeline to function. For example, computer aided design tools should be able to draw from information at the tool-workpiece interface level to feedback data on the success of creating certain features in a machining process. Recall our discussion of edge defects, burr formation in particular, in Chapter 11 and how design features as affected in the manufacturing environment by tool path can have a dramatic impact of part quality.

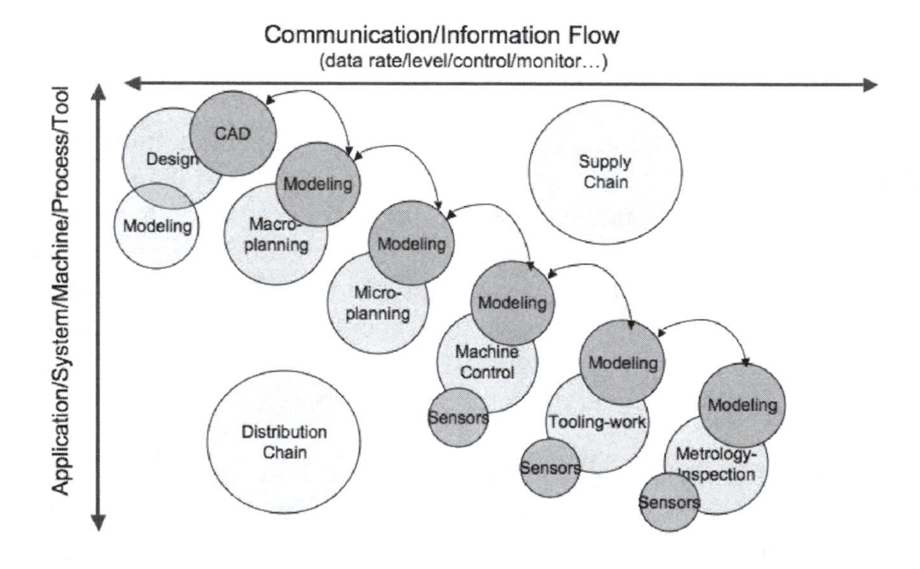

Figure 12.2. Software connectivity driving the manufacturing Pipeline.

Finally, one of the areas to benefit the most from this connected view of the design to manufacturing pipeline is the environmental impact of the production and use of products. This is discussed in the next section. Keep the connected view of the manufacturing pipeline as presented here in mind as you read this next section. This is how the world will operate in the future!

12.3 Sustainable design/environmentally conscious design and manufacturing

We have already discussed some aspects of nanotechnology from the point of view of precision device construction. This will surely play an increasing role in the future developments of precision devices – from medical to energy applications.

It is this second area, energy, and its close compliment, the environment, that we will spend some time discussing with respect to precision manufacturing technology. These two – energy and the environment – will drive the technology and practice of precision manufacturing in the future. For sure, the more incremental developments, following Moore's Law – higher density data storage, advanced optics, more complex medical devices, nano-scale devices, sophisticated telecommunications and IT advances, etc. – will be present and continually push us. But, it is the double header of energy and environmental impacts of product design, manufacturing, use and reuse or disposal that will dominate our future. This is expected to compel us towards:

- technology for advanced energy sources (e.g. batteries, photovoltaic/solar, hydrogen fueled motors, photosynthesis, hybrid vehicles, high efficiency engines, hydropower, wind power, advance bio-fuel manufacture), and
- sustainable manufacturing (e.g. green machines, green factories, green processes and systems (including supply chains), minimizing or eliminating waste, consumables, energy, green products, re-use or re-manufacturing.)

Most of the technologies discussed in this book in earlier chapters are applicable to manufacture of advanced energy products. For sustainable manufacturing, we need to re-think our approach to product design and manufacturing. For example, how can we provide incremental solutions to the large problems of global warming, excessive energy and resource consumption, manufacturing-related atmospheric contamination, etc.?

12.3.1 Technologies for sustainable manufacturing

Increasingly, concerns about the availability and cost of energy and raw materials, the impacts of industrial activity, and the efficient reuse of consumer products at the end of their life are driving efforts towards "green manufacturing." This is part of a much larger international concern over the overall accomplishment of sustainability in design and manufacturing. Sustainability, as defined here, implies a level of resource utilization that is very comprehensive in its scope. It must be in accord with a level necessary to insure that, over time, the resource will a) not only be available; but b) demand is reduced to approach a level of what is naturally sustainable. That is, the "gap" between current use and sustainable use must be understood. This is a hard measure to meet but it is, ultimately, what is necessary.

The topic of "green business" is very popular today in the business and general press. Companies from GE (with its "Ecomagination" campaign) to Wal-Mart (with its sustainable supply chain for some products) are finding it both profitable and responsible to promote green products. Some, like Toyota, even mention the need for the entire manufacturing process, including the supply chain, to be green. Others, like Cummins Engine, are engaged in "remanufacturing" for engine components that have life after first use. Companies like Interface Carpet have aggressively tackled the problem relative to their production and made impressive progress that, at the same time, is on the road to facilities and production that meet the full definition of sustainable manufacturing. Not surprisingly, the basis of claims of "green-ness" and level of commitment vary widely over a wide range of products[523].

It is not the goal of this section to rate the seriousness, or the "sensitivity-to-public-needs," of any particular manufacturing company or product that is sold. Rather, the goal is to help define a methodology by which manufacturing engineers, when tasked to manufacture products in a green manner, can accomplish this within the constraints of the particular product or process being manufactured.

A short introduction to green manufacturing is presented followed by a review of the elements and scope of the system that should logically be included in any discussion of green manufacturing. The concept of "technology wedges" and how they might be applied to identifying opportunities for environmentally benign manufacturing and integrating solutions with existing processes or supply chains is discussed.

12.3.2 Green manufacturing pipeline

It is not possible here to offer a complete review of the literature on green manufacturing and related topics. A number of very comprehensive studies have been made detailing the work ongoing in the world relative to the first attempts at green production. The WTEC report, for example, discussed the status of environmentally benign manufacturing in the US and abroad[524,525]. Integration of green components and analysis into manufacturing systems was considered early on by Sheng[526]. Anastas and Zimmerman[527] address link in design considerations. Some researchers have addressed specific parts of the manufacturing process in more detail with success, e.g. minimum quantity lubrication/dry machining[528,529], which is now being implemented in a number of manufacturing operations internationally. These specific accomplishments are often referred to as "technology wedges" that form part of a comprehensive solution. The wedges are explained in more detail in the next section. Ishi and others[530], have looked at incorporating end-of-life considerations when products are designed.

A schematic of the green elements of design and manufacturing is shown in Figure 12.3. The "life cycle" of the product from design to recycling or reuse at end of life must be considered. The figure refers to life cycle analysis (LCA) tools, see for example the work at Carnegie Mellon University[531], design for assembly (DFA), and design for environment (DFE). The figure distinguishes between the design of a product and the processes or systems that make the product. Of importance is that both the design of the process and how the process operates to manufacture the product are considered. Substantial opportunities exist in the design of machines that make products[532], as well as in their productive operation (macro and microplanning, for example, see Srinivasan, and Sheng[533,534]. Although not specifically shown in Figure 12.1 these elements include the supporting supply chain companies as well.

As we seek green manufacturing practices it is necessary to consider the environmental impact of all the elements in a product's life cycle as shown in Figure 12.4. The figure covers the broadest scope possible from the original source of raw materials (including mining costs) to eventual disposal of the used product. Each element includes waste (or non-utilized resources and by-products). The energy and resources consumed in each of these stages of the process, and the waste output, will "count against" any benefits of the new product or machine resulting from the process. In fact, to be complete, we could consider manufacturing to include elements of most of the stages of the product life cycle since mining, conversion and transport are arguably part of manufacturing. At a more detailed level, sustainable manufacturing can be addressed by improving the individual components to manufacturing, Figure 12.5. Here we see contributors to sustainable manufacturing including power sources and processes.

To be effective, there must be a net energy, resource or environmental impact savings over the life of any product or process produced in an "environmentally benign" manner to meet the sustainability definition. Colloquially, it can be said: "If you consume as much resource in manufacturing a product as you save in use of the product you have not succeeded!" This requires careful accounting for the impact of all the process chain elements and that is often not easy to do.

The individual processes represented by the boxes in Figure 12.4 can be exploded as seen in Figure 12.6. Each process is comprised of many sub-processes and sub-systems each having, on its own, sub processes and systems. Thus is the nature of manufacturing and the ultimate bill of materials (BOM) reflects the diversity of the product and the source of the materials.

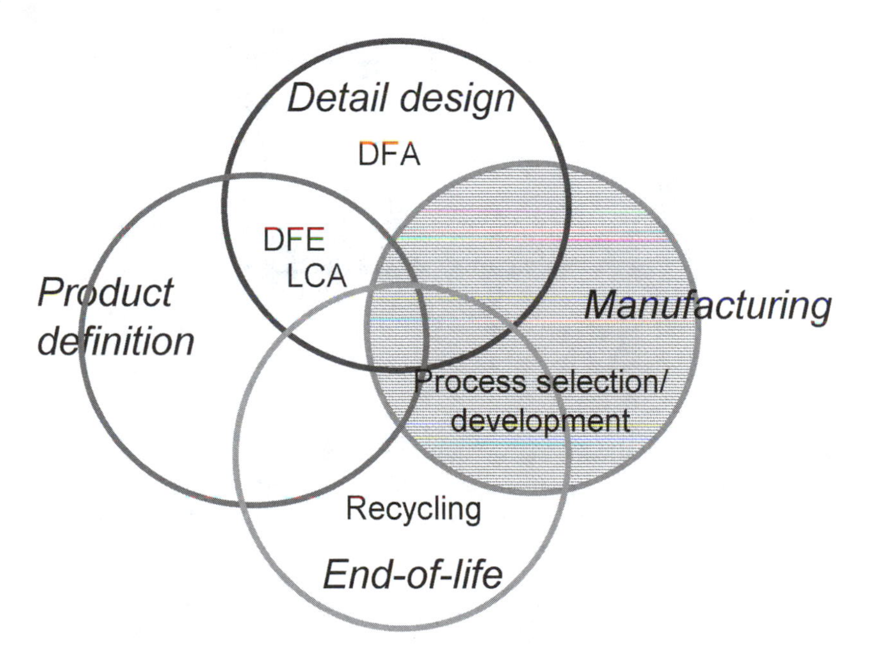

Figure 12.3. Schematic showing relationship of green elements of design, and manufacturing from product definition to end-of-life.

The question is, then: "Where is the most efficient point in this complex broad scope to introduce improvements or enhancements relative to our goal of implementing environmentally benign manufacturing?" Tools exist, for example, life cycle engineering analysis, for example, Hendersen[535] who gives a detailed description of the state of the art that allows the determination of the "environmental cost" of each element. Sometimes these costs are not very accurately computed due to the sources of data, but they provide a valuable start. A good example, besides those in Hendersen cited above, is Williams[536] describing the energy and materials that go into a modern semiconductor device. Many analyses do not consider the "imbedded" energy and materials in manufactured products – only during the use phase.

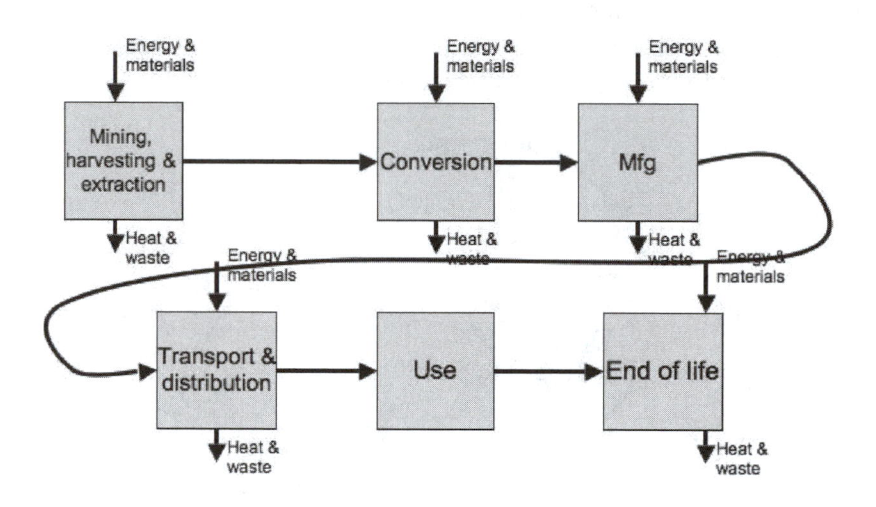

Figure 12.4. Schematic showing relationship of green elements of design, and manufacturing from product definition to end-of-life.

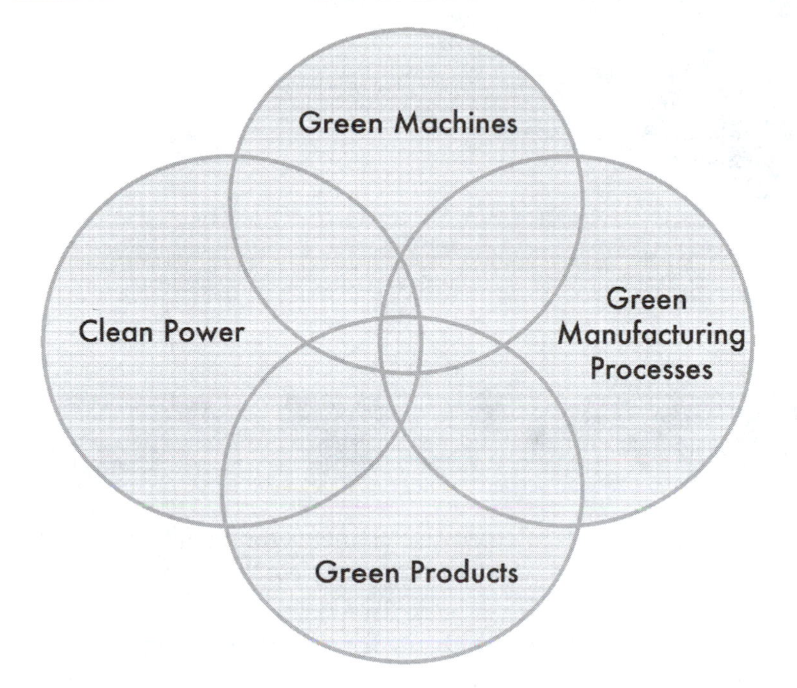

Figure 12.5. Components of green and sustainable manufacturing including machines and processes and power sources.

Other tools include the environmental value system analysis, EnV-S. This has been successfully applied to semiconductor processes, for example chemical mechanical planarization (CMP). It analyzes trade-offs between system components for waste treatment (a sub-system of the CMP process) to achieve the highest environmental enhancement while minimizing cost per piece (another wedge), see for example Krishnan[537].

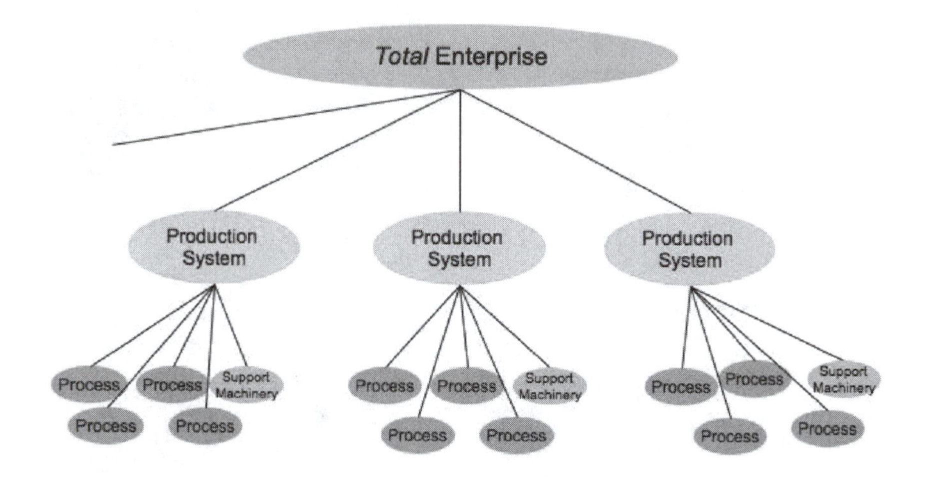

Figure 12.6. Levels of enterprise elements for manufacturing from systems to processes.

With this level of analysis for the entire system of production, a reasonable determination can be made of the cost/benefit of proposed changes. The goal of the changes or enhancements is to provide some improvement in the overall operation of the process or system. This can be implemented as part of agent-based process planning where an "environmental agent" provides feedback on issues that affect the environmental performance of the process or system[538, 539]. The concept of wedges and their role in the improvement of the system is detailed next.

12.3.3 Sustainable manufacturing or "does green = sustainable?"

If one looks at consumption of resources (any resource) or the impact of that consumption over time we can identify a trend, generally increasing over time. The sustainable rate of that consumption, or impact, is often much less. Here we can discuss sustainability in the context of a well cited of the Brundtland Commission, formally the World Commission on Environment and Development (WCED), as part of a study to look at conditions for fostering sustainable development in the world[540]. They defined sustainable development as

> "Sustainable development is development that meets the needs of the present without compromising the ability of future generations to meet their own needs."

This is a high bar to pass when designing products and systems of production. If we define a level of sustainability so that we don't "compromise the ... future", paraphrasing here from Brundtland, it means operating our factories and processes at or below the level of sustainable impact or consumption. Figure 12.7 illustrates the challenge. The "wedge" between our rate of consumption or its impact and the sustainable rate of consumption or impact is often large and growing. In order to return to a sustainable level of consumption, specially from the perspective of manufacturing, we will need to reduce the consumption, or impact, at a rate larger than its natural growth (think population increase times per capita consumption or impact). Just reducing consumption or impact, although preferable than not, is insufficient. That is a challenge! One way to look at this is from the standpoint of "technology wedges."

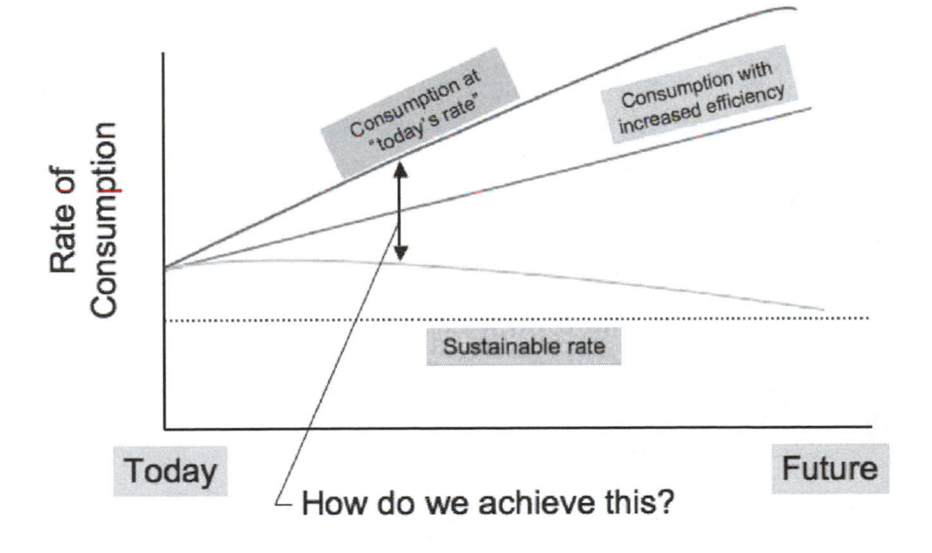

Figure 12.7. Sustainability frame of reference.

12.3.4 Manufacturing technology wedges

A paper in Science in 2004 by Pacala and Socolow describes the concept of "stabilization wedges" with respect to technology that could solve the climate problem (with respect to fossil fuels) for the next 50 years using current technology[541]. The idea is to close the gap between current rate of impact and the sustainable level with "wedges" of technology – each making a small but measurable improvement. The paper introducing the idea focuses on the "gap" between the current trends in fossil fuel emissions (the impact or consumption) relative to the atmosphere's capability to accommodate emissions (the sustainable level). This is a similar gap to that defined earlier relative to sustainable design and manufacturing (normal consumption vs. sustainable consumption) in Figure 12.7. Pacala and Socolow propose a set of "wedges" to solve this problem. Each individual wedge represents the ability of some existing current technology to reduce, on its own, some portion of fossil fuel emission. Then, summed together, these wedges provide the necessary reduction in emissions to achieve an overall "sustainable" situation. Their goal is achieving re CO_2 stabilization at 500ppm by 2125, a sustainable level.

The question is whether or not this strategy could be employed in manufacturing to accomplish environmentally benign manufacturing processes. There is, as noted in the introduction, an increasing interest in developing process enhancements that contribute to reducing an environmental impact. As yet, there is no strategy to coordinate a set of enhancements and new capabilities that will, combined, render a process "sustainable" as in the case of the fossil fuel emission example.

There are some interesting possibilities. With the development of new manufacturing technologies for micro-scale and nanoscale manufacturing, as well as the various alternate energy sources being pursued (from fuel cells to photo-voltaics) attention will need to be paid to ensure that any new processes will have a positive effect in an environmental sense. That is; wedges must be designed to be "net-positive." An improvement in one element of a process or system of manufacture cannot be at the expense of another segment of the cycle, see Hawken[523] for example, for a great discussion on this constraint. This is specially complicated with the complex supply chains employed today. Continuous improvement is as valid here as in other areas of manufacturing. We strive to remove "wasted time and effort." Why not also try to do this for energy/ consumables/waste? In this chapter it is proposed that "technology wedges" – analogous to "stabilization wedges" – offer a framework and potential metric for addressing these energy challenges. The specifics of that metric need a lot of development based on discussion in the community.

There are a number of fundamental rules which govern how wedge technology can be employed in manufacturing. In no particular order, they are:[542]

Rule 1. the cost of materials and manufacturing (in terms of energy consumption and Green House Gas (GHG) emissions, etc.) associated with the wedge cannot exceed the savings generated by the implementation of the wedge (or wedges) over the life of the process or system in which it is employed.

Rule 2. the technology must be able to be applied at the lowest level in the process chain. For example, in Figure 12.6 this would be at one of the root processes in a subsystem.

Rule 3. the cost and impact of the technology must be calculable in terms of the basic metrics of the manufacturing system and the environment. That is, cost and impact must be expressible in units of dollars (or euros, yen, yuan, etc), carbon equivalent, global warming gas creation or reduction, joules, cycle time and production rate, quality measures, lead time, working capital and so on relative to present levels of consumption, use, time, etc.

Rule 4. the technology must take into consideration societal concerns along with business and economy, see Hawken[523] for example, and

Rule 5. there must an accompanying analytical means or design tool so that it can be evaluated at the design stage of the process or system. It must be an integrated approach.

These rules, when applied, will insure a balanced and honest appraisal of the impact of the technology from an environmental perspective and also insure the rules are feasible from a business perspective. This will avoid creation of anomalies in the supply chain caused by a local gain which yields a global net loss. Many examples of a net loss currently exist in society, such as with so-called "high tech/low cost" products, some of which operate efficiently, but which have short useful lives. The consumer "learns" to expect this and, thus, expects to replace the product in a short time. This trend destroys a tremendous amount of product value and resources and, worse, it encourages increased environmental damage. The life cycle of such "throw-away" products is often not considered in the design, fabrication, or use.

12.3.5 Examples of wedge technology application areas for manufacturing

It is instructive to review two short examples and two more detailed examples that could be help define the application areas for "wedge technology" in manufacturing. As mentioned earlier, there are no complete sets of wedges as proposed in the fossil fuel example. The closest so far, perhaps, is the experience of Interface Carpets, see Interface Carpet website[543]. There, thanks to a dedicated and aggressive effort, the evidence of reduction, re-use, and increased efficiency is seen in all three of areas of economy (reduced cost per unit of manufacture, here a square yard of carpet), environment (reduced generation of greenhouse gases per unit of production) and society (where a measure used is an impressive re-direction of material from landfills to re-use; this is in addition to the reduction in global warming gases already noted). Another interesting general example of using wedge technology comes from Boeing and the development of their new 787 "Dreamliner" aircraft[544]. Boeing notes that an aircraft's (or any products for that matter) “environmental footprint” takes form well before its first flight. They describe how they are focusing on the entire product lifecycle from design through manufacturing and eventual retirement and reuse (closer to what is indicated in Figure 12.4.) They are aggressively engaged in comprehensive recycling of retired aircraft. They estimate that some 7,200 aircraft will be removed form active use in the next two decades and Boeing is looking at how to minimize the environmental impact of recycling and insure maximum re-use of materials and components. They also list a number of significant technological advances relative to operating characteristics (fuel consumption and CO_2 emissions, for example, and on plane energy use.) Boeing describes a concept called “tailored arrivals” to better manage arrival and departure schedules for aircraft into and out of airports – they claim this can save between 400-800 pounds of fuel per flight.

We now give two more detailed examples related to more conventional manufacturing and then nanoscale manufacturing.

12.3.5.1 Consumable use in machining

The first example deals with manufacturing in the automotive sector to analyze means to minimize the environmental cost (global warming gas generation, electricity use) and operating cost of a plant compressed air supply and distribution system for a transmission manufacturing line in a domestic auto factory. The study utilizes the environmental values systems analysis (EnV-S) developed at Berkeley, see Krishnan[537], for assessment of manufacturing process alternatives for reducing energy and environmental impacts of manufacturing processes.

Compressed air is one of the most expensive energy sources in manufacturing. This study is an evaluation of the use and supply of compressed air at an automotive transmission plant, aiming towards making recommendations to improve environmental and economic efficiency in future facilities. This study conducted a quantitative analysis of three compressed air supply patterns, namely, plant air, point of use (POU) and local generation, as alternatives for future compressed air usage. Environmental Value systems analysis (EnV-S) was employed to determine the economic and environmental performance of the three alternative supply patterns, by using cost of ownership and environmental impact matrices. Data from shop and facility operation as well as alternate compressed air device suppliers was used.

Compressed air is regarded as the fourth utility, after electricity, natural gas and water, in facilitating production activities. In manufacturing plants, compressed air is widely used for actuating, cleaning, cooling, drying parts, and removing metal chips such operations. However, the cost of compressed air production is one of the most expensive and least understood processes in a manufacturing facility The cost of electric power used to operate an air compressor continuously for a year (about 8200 h) is usually greater than the initial price of the equipment[545]. Per million British Thermal Unit (BTU) of energy delivered, compressed air is more expensive than the other three utilities, as shown in figure 12.8.

Besides cost issues, compressed air production consumes huge amount of energies. It is estimated that about 3% to 9% of total energy consumed in US in 1997 is for air compression in manufacturing[546]. The consumed energy, directly or indirectly, contribute to large amounts of facility CO_2 emissions per vehicle built from automotive manufacturing facilities.

Compressed air is used extensively in automotive manufacturing due to its ease of setup. There is no need for additional maintenance or special machines; usually, the task can be accomplished by adding piping. In addition, as a form of energy, compressed air represents no fire or explosion hazard; as the most natural substance, it is clean and safe and regarded as totally "green"[547].

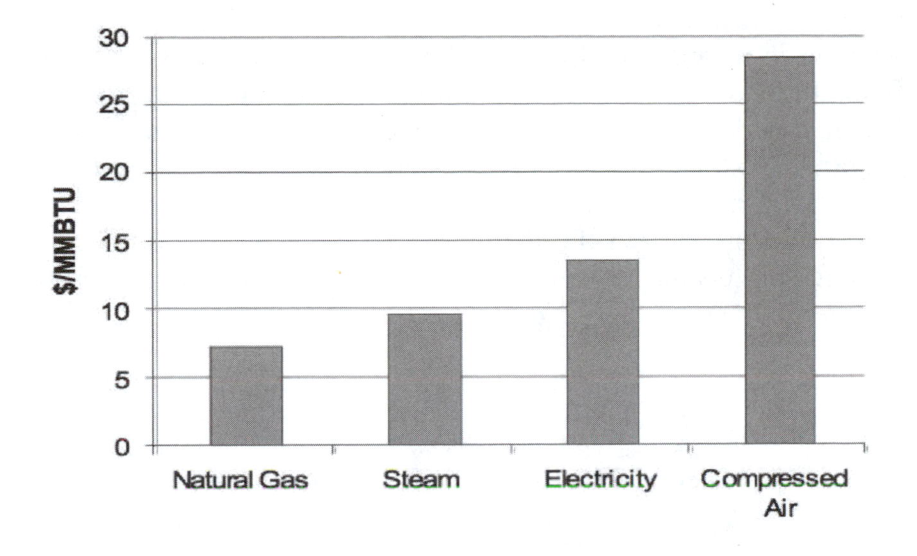

Figure 12.8. Cost of energy delivery modes.

At a facility producing transmissions, compressed air system was identified as a source of potential cost and environmental impact savings. Figure 12.9 illustrates the largest five compressed air consuming processes during the transmission manufacturing processes, among which case and valve body machining are two processes that make particularly heavy use of compressed air. Together, these two processes consume 56% of all compressed air used.

Twenty-four (24) computer numerical control (CNC) milling machines are used for case and valve body machining. A quantitative analysis of the compressed air usage patterns for all these 24 CNC machines was conducted as part of the study. This study compared cost of ownership (COO) and energy use for the machines and facility using scenarios of: existing plant (that is, central) air, compressed air source at each machine (local), and compressors serving a small set of CNC machines.

The COO and energy use analysis favored local generation for cost consideration and energy efficiency. Employment of local generation instead of plant air could potentially save $2,000-3,200 and 95,000 KWH each year on the CNC milling machines at this plant. Meanwhile, local generation offers numerous advantages over plant air in regards to reliability, simplicity, leakage prevention and flexibility. Local generation is supplied by relatively short pipelines, which may lead to a significant reduction of losses due leaks in operation. Extra local compressors may be connected to CNC machines in parallel, which automatically builds a great deal of redundancy into the system. Furthermore, the scale of local generation compressors enables greater flexibility as machines and processes change[548].

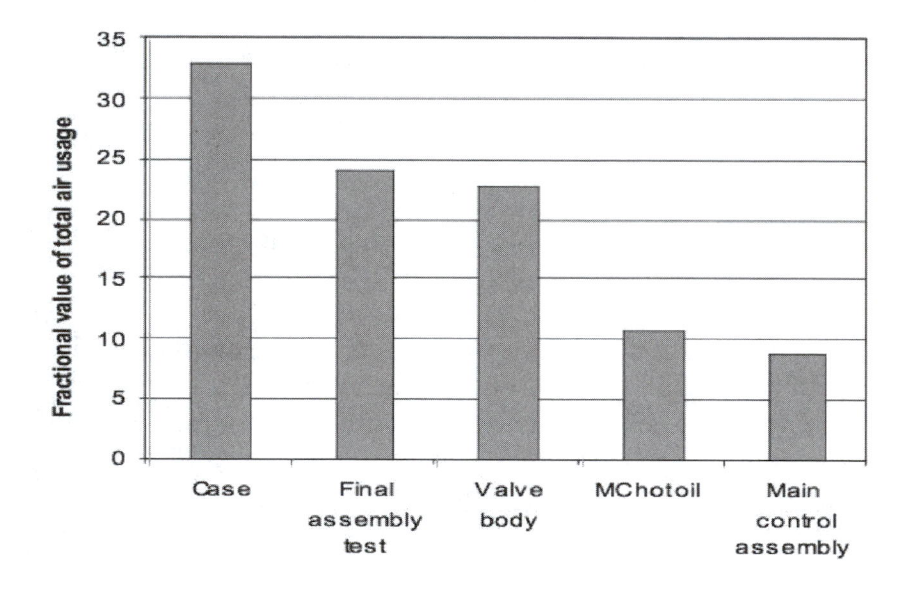

Figure 12.9. Compressed air usage Pareto chart.

This first example gives some sense of the potential for looking for opportunities to utilize wedge technologies in manufacturing for conventional machining processes. This shows how organizational efforts can often yield substantial energy and material savings – even before the process itself is modified or the machine redesigned. This is similar to the "tailored arrivals" concept mentioned as part of Boeing's efforts to improve aircraft environmental performance.

The second example concerns energy use analysis for nanoscale manufacturing.

12.3.5.2 Energy use in nanoscale manufacturing

Nanoscale manufacturing can be defined as the control and manipulation of materials to a precision of one to one hundred nanometers in at least one dimension[549]. As materials exhibit fundamentally different behavior in this scale, new products may be developed with improved performance characteristics. Efficient and scalable nanoscale manufacturing methods are required to harness the benefits of nanotechnology for broader use. As the use of nanoscale manufacturing increases, it becomes necessary to measure and manage the energy efficiency of these processes[550]. Energy usage dictates both the environmental and the economic efficiency of manufacturing processes.

Nanoscale devices have demonstrated the potential to provide energy savings in their use phases[551, 552]. However, increased precision requirements drive up manufacturing costs, particularly those from energy use[553]. Hence, in order for nanoscale devices to be energy efficient across their entire life cycles, care should be taken in the selection of the manufacturing technology used. Moreover, monitoring energy use also leads to the identification of opportunities to improve the efficiency and productivity of the manufacturing method.

This section reviews the major classes of nanoscale manufacturing methods and identifies the direct and indirect process requirements needed for their operation. With this background, it discusses a qualitative comparison of the process requirements and associated energy demands of the various methods. To promote the development of this emerging field, we suggest a roadmap for the comprehensive study of energy use in nanoscale manufacturing. The usage data helps to define the appropriate wedge technologies that can be employed.

Nanoscale manufacturing can be seen as an evolution of semiconductor manufacturing and faces many of the same energy issues, such as large direct process energy requirements, extreme part complexity, highly processed and purified material inputs, and vulnerability to contaminants. Additional considerations specific to certain nano-processing technologies include: low throughput, built-in metrology, and extreme temperature and vacuum requirements. In addition, the usual requirements for manufacturing efficiency and quality are present, Figure 12.10.

We will start with a discussion of various nanoscale manufacturing methods. Past work in this field includes Tseng[554] and Chen[555], which discuss conventional versus next generation lithography NGL) methods. Here, we categorize nanoscale manufacturing technologies by their basic constituent mechanisms, as these mechanisms drive energy use. Nanoscale fabrication methods can be broadly characterized as top-down and bottom-up approaches. Top-down refers to subtractive methods, involving etching, machining or molding larger parts to a desired size, while bottom-up describes the formation of materials or devices additively from individual molecules or atoms. Most of these technologies were discussed in Chapter 11.

Each of the manufacturing technologies discussed have specific requirements for operation. These requirements are broken down into those that can be measured per unit product versus those that are allocated over a production facility. We call these direct and indirect process requirements, respectively. The major process requirements along with their energy demands are discussed in this section.

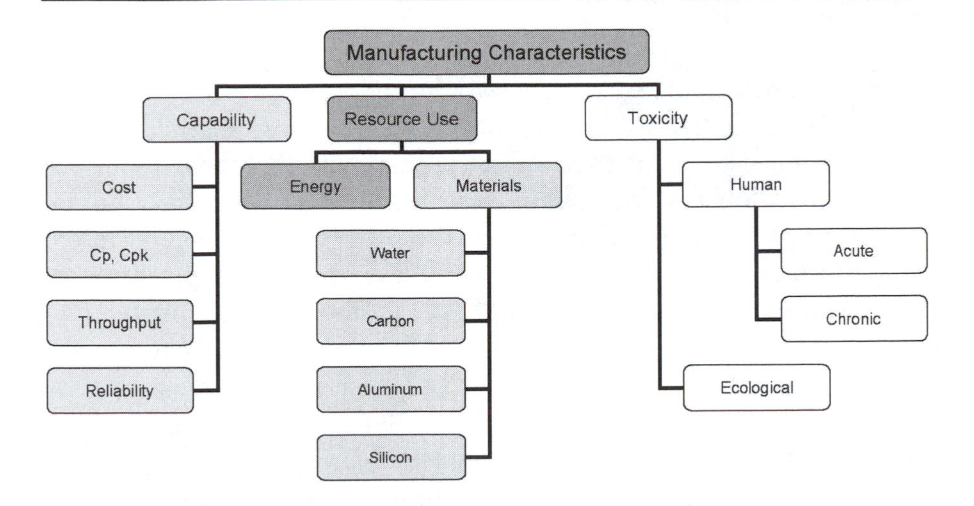

Figure 12.10. Components of manufacturing characterization, including resource use and hazards.

Since many of the manufacturing technologies are still in their developmental stages, it is difficult to obtain data on energy consumption. There are, however, well documented studies of mature semiconductor manufacturing technologies[556-558]. Due to similarities between semiconductor and nanoscale manufacturing technologies, it is appropriate to use some of this data as lower-bound estimates for energy use in nanoscale manufacturing. However, there are numerous gaps in the available knowledge, which are also identified below. These areas should be the focus of the nanotechnology life cycle analysis community.

Direct Requirements Direct requirements are those applied directly at the process or point-of-use (POU). Energy use can be measured directly from the experimental equipment, but must be viewed in light of potentially increased production-scale efficiency. It should also be noted that we are interested in the energy supplied to the process equipment rather than the energy used in the process or the primary energy generated.

1. *Pressure Control:* Pumps are used to maintain the vacuum conditions required for many fabrication methods. Vacuum conditions are crucial for minimizing contaminants in the

parts being produced, as well as to extend the mean free path, or anisotropy, of materials being deposited. Pressures down to 10−3 torr can be attained with roughing, or mechanical, pumps alone. Pressures lower than this threshold require a diffusion or turbo pump in series with the roughing pump. Energy use in pressure control is therefore dependent on the pressures required for each process. The basic machine requirements differ very little from those of semiconductor manufacturing. However, as the throughput of many nanoscale processes is relatively low, the machine must operate over a longer working cycle. Hence, when compared to semiconductor manufacturing, the energy use from pressure control will scale inversely with the throughput of nanoscale operations.

2. *Temperature Control:* Specific high or low chamber temperatures are required for many nanoscale processes. Furnace temperatures, ranging from 400°C to 1400°C are necessary for oxidation, thin film deposition, diffusion, annealing, and sintering processes, while oven temperatures around 100°C are used to dry and harden photoresists and other masking substrates. Cold temperatures are required for processes that require great purity, including scanning probe microscopy, plasma processes, etching, and ion implantation. These processes can be successful only in the absence of uncontrolled kinetic energy sources; hence both vacuum and low temperature conditions are required. While hot and cold temperatures are attained through the same mechanisms as in semiconductor manufacturing, the energy expenditures will again scale with throughput.
3. *Photon Generation:* Ultraviolet light is used to soften or harden photoresist, the resolution of which is inversely dependent on the wavelength of the light source. However, the shorter the wavelength, the more energy is transmitted in the photon. An average arc lamp bulb may dissipate 500-1000W. Since EUV wavelengths are an order of magnitude shorter than that of optical light sources, they are roughly an order of magnitude more energy intensive.

4. *Plasma/Ion Generation:* Plasma sources are used to etch and deposit material, and to implant ions into a substrate. Plasma generation and control is energy intensive and can consume from 50W to 50kW depending on the application. Because lower beam energies are used for single or small scale ion implantation, the energy requirements of nanoscal processes associated with this requirement will be lower with respect to those used in semiconductor manufacturing.
5. *Precision Metrology and Mechanics:* While throughput and process capability are still relatively low, part by part inspection is often necessary. This may rely on the same scanning probe microscopy tools and repeated trials used in feature creation. More work is required in characterizing both the reliability and energy use of high precision equipment used in metrology and manufacturing itself.

Indirect Requirements Facilities provide a layer of "protection" for the manufacturing processes. In experimental environments, appropriate facilities may not yet be available. However, at production scales, it becomes critical to maintain tightly controlled conditions outside of process chambers. The processes inherit the purity and ambient conditions of the facilities and hence it is essential to maintain the facilities at these baseline conditions. This not only takes the load off of direct requirements, but reduces the risk of defects due to contamination. Numerous materials must be continuously supplied to the process chamber, including clean room air, ultra pure water, and process cooling water. All of these must be recaptured to be conditioned and re-circulated or treated and released. In a modern-day semiconductor fabrication facility, a large fraction of energy use goes toward facilities scale support processes.

1. *Purification:* Ultrapure air, water and chemicals are required for all nanoscale manufacturing processes. Clean room recirculating fans are necessary for removing contaminants from the work environment air. Makeup air (MUA) is also used to pressurize the work area, keeping contaminants out. Ultra pure or deionized (DI) water is used in numerous cleaning steps. A ever-widening range of highly purified chemicals are used in both semiconductor and nanoscale manufacturing. The

highest available classes of clean room air are necessary components of both nanoscale manufacturing and semiconductor manufacturing, and the energy use associated with both are comparable, though again dependent on throughput. One issue particularly pertinent to nanoscale manufacturing is how to clean without damaging features. Deionized water may be used differently in nanoscale manufacturing than in semiconductor manufacturing. This is an area in need of further study. While some consumables span the gap from semiconductor manufacturing to nanoscale manufacturing, many others are being developed solely for the latter. On-site chemical purification is a major factor in energy use that will require extensive and continuous study as the field progresses. This is distinct from the embedded energy in process consumables, which will be another priority for the research community.

2. *Temperature Control:* Process cooling water (PCW) must be continually circulated to maintain desired temperatures. Process cooling water consumption for nanoscale manufacturing is similar to that of semiconductor manufacturing, scaling with process heat generation and throughput.
3. *Abatement:* Harmful or regulated chemicals are treated in two steps: once at the point of use (POU) and again at the facilities scale. Because both steps occur outside of the actual processes, they are considered within facilities requirements. POU abatement focuses on the removal or separation of perfluorocompounds (PFCs) by burn and scrub, plasma treatment and scrub, or filtration respectively[558]. Facilities abatement involves a wider range of emissions than POU abatement. Acid waste neutralization, volatile organic compound (VOC) combustion, ammonia neutralization, fluorinated waste water treatment all occur in secondary abatement. The species of chemicals and degree to which they are used and subsequently abated in nanoscale manufacturing is a major area of future work. The qualities and associated energy consumption of process outputs will have to be a priority consideration for any comprehensive study of nanoscale manufacturing.

Given these energy intensive process requirements, we now take a look at the relative demands for them across the manufacturing technology classes. The total energy consumption of a method is driven by the extent to which its process requirements are used. A qualitative assessment of process requirements is presented for each manufacturing technology class in Table 12.1. In some cases, the process requirements are dependent on the material or process parameters used in manufacturing. For example, imprint lithography employs either heat or UV curing. This distinction affects relative demands of the process for photon generation, and direct and indirect temperature control. Abatement for ion beam lithography is chemistry dependent and may emit heavy metals. CVD abatement is process dependent and its emissions may contain perfluorocarbons or hazardous air pollutants. VLS processes occur under a wide range of temperatures (from 1200°C for laser ablation to 3700°C for solar furnace CNT formation), so the degree of temperature control needed is process dependent.

From Table 12.1, we can see that top-down and bottom-up processes with similar mechanisms have similar process requirements, and are therefore likely to have similar energy demands. If the process dependent scores are averaged for a manufacturing class, we find that imprint and plasma/ion processes have the least requirements while single point operations, CVD, VLS and MBE have the most requirements. If we likewise average all the scores for top-down versus bottom-up manufacturing technologies, assigning values from zero for "no/weak requirement" to three for "very strong requirement", we find that top-down processes score an average of 1.3 while bottom-up processes score an average of 1.4. This difference is marginal but suggests a hypothesis, that bottomup manufacturing processes are more energy intensive than top-down processes, to explore in future work.

Table 12.1. Process requirements for manufacturing technology classes.

Manufacturing Technology		Direct					Indirect		
		Pressure Control	Temperature Control	Plasma/Ion Generation	Photon Generation	Mechanics & Metrology	Purification	Temperature Control	Abatement
Top-Down	Photolithography	◉	○	○	●	◍	●	○	◍
	X-Ray Lithography	◉	○	○	●	◍	●	◍	◍
	E-Beam Lithography	◉	○	○	●	◉	●	○	◍
	Ion Beam Lithography	●	○	●	○	◉	◍	○	◍●
	Imprint	○	○◉	○	○◉	◍	●	○◉	◍
	Single Point Operations	●	●	○	○	●	◍	●	○
Bottom-Up	CVD Processes	●	◍●	●	○	◍	◍	○●	◍●
	Vapor-Liquid-Solid Processes	◉●	◉●	○	○	●	◍	◉●	◍●
	Molecular Beam Epitaxy	●	◉	○	○	●	◍	◉	◍●
	Liquid Only Processes	◍	○◍	○	○	●	◍	○◍	◍●
	Plasma/Ion Processes	●	○	●	○	◍	◍	○	○
	Imprint	○	○◉	○	○◉	◍	●	○◉	◍
	Single Point Operations	●	●	○	○	●	◍	●	○

○ No/Weak Requirement ◉ Strong Requirement

◍ Moderate Requirement ● Very Strong Requirement

Multiple circles indicate range of requirements dependent on process parameters

The discussion above presents the basic elements of a systematic energy analysis for nanoscale manufacturing methods. Because the energy use per area depends on the precision and throughput of each process, process requirements serve as a proxy for actual energy use. A framework for a detailed, quantitative study of energy requirements for nanoscale manufacturing is proposed as follows:

- Develop a comprehensive set of requirements for all manufacturing technology class including major consumables and process equipment
- Assess the range of precision attainable through each manufacturing method
- Estimate throughput for each technology class at various levels of precision based on industry projections and trends

- Quantify energy use associated with each process requirement, as a function of precision for each of the process classes
- Quantify energy embedded in materials and equipment using a combination of processed based and economic input-output life cycle assessment (EIO-LCA) energy data
- Identify scaling factors in energy demand as functions of resolution for each method, and compare extrapolated results with currently available data from the semiconductor industry.

Using this model, a rigorous understanding of the energy consumption in nanoscale fabrication methods can be achieved. This framework can also be integrated into a broader lifecycle analysis of nanoscale manufacturing.

12.4 Environmentally conscious design of precision machines

The previous section illustrated examples of identifying operating conditions for manufacturing machines or processes that could benefit form the application of wedge technologies for reducing energy or resource consumption. We have seen in Figure 12.5 earlier in this chapter that "green machines" play a part of the sustainable manufacturing environment. We will introduce some ideas in this section about the design of green machines, specifically green machine tools.

In Chapters 2, 5, and 7 we discussed the constraints associated with precision machines. We're concerned with minimizing, by design, sources of error in the machine motion and performance. These sources included vibrations (forced and self-excited), thermal distortions, compliance (from the machining forces and machine component weight and motion) as well as inaccuracies in basic component causing non-straightness of motion or roll, pitch and yaw effects. In each case, we identified a set of material, component or process metrics needed to minimize or prevent an undesired behavior.

We need a formalism for addressing sustainable design of precision machines just as we do for the basic machine design.

12.4.1 Sustainability budgets

The basis for this formalism was introduced in Chapter 6. In that chapter the ideas of error mapping and developing an error budget were discussed as a methodology for accounting for the magnitude and eventual impact of the numerous potential sources of error in a machine's performance – relative to dimensional accuracy, form error, or surface finish. Recall that we identified a "sensitive direction" (the direction in which an error impacts the part quality: dimension, form, roughness). The error budget allowed us to include all sources of error and an estimate of their relative magnitudes and then determine which of these sources actually impacted a sensitive direction resulting in a part error.

Now consider if we would add constraints on the environmental performance of a machine while insisting that the other quality metrics are met as well as the manufacturing performance (throughput, lead time, cost/piece to operate, etc.). This could be included in our budget analysis but, in this case, we'd call it a *sustainability budget*. A sustainability budget would operate similarly to an error budget except we would be looking for the impact, from environmental metric point of view, of the design and operation of the precision machine, process or system.

Using the idea of sensitive directions (and the complementary concept of non-sensitive directions – that is, those directions for which any error from a specific source has no effect, for example, orthogonal effects) we can imagine an analysis which measures the impact of materials, designs, or operating conditions on the overall environmental behavior. Then we look for instances of materials, design features or operating conditions that give the largest range of variability, form the point of view of design, with the least environmental impact. That is, those instances for which little or no sensitivity is displayed. Following a methodology based on this would allow us to design the machine, or system of machines, in such a way that the basic performance, precision and accuracy, would meet the core error budget constraints but, in addition, we could do so in a way that was more sustainable.

Great idea but how do we do this?! The ultimate goal is to derive an error budget. You will recall from Chapter 6 that the error budget relies on two sets of rules – connectivity and combinational. Connectivity rules define the behavior of machine components and interfaces in the presence of errors. The combinational rules define how the errors are to be combined to determine the impact on the accuracy of the workpiece. The procedure comprised the following three steps:

Step 1 – determination of a kinematic model of the machine and its principal components in the form of a series of homogeneous transformation matrices (HTM).

Step 2 – analyze systematically each type of error in the system and use the HTM to determine the relative tool - work errors.

Step 3 – combine the errors to yield upper and lower bound estimates of the total error of the machine.

Step 1 was referred to as "error mapping".

If we revise this approach for a sustainability budget, we'd follow the same three basic steps but with some different objectives. For example, we would add some elements to the three steps, or, actually, develop a parallel set of "models" and analysis.

Parallel to Step 1 would be:

- determination of an energy, material and resources (consumables, etc.) model of the machine and its principal components in the form of a series of relationships defining the energy consumption, materials use as a function of machine design or operation. (This might be referred to as energy or materials mapping.)

Parallel to Step 2 would be:

- analyze systematically each type of energy and material use in the system and determine the relative performance-energy/material impact.

Parallel to Step 3 would be:

- combine the energy/materials impacts to yield upper and lower bound estimates of the total energy/ material impact of the machine.

Importantly, in this parallel analysis, the embedded energy and materials must be counted. That is, we cannot only look at the energy to move an axis of the machine (for example in a precision diamond turning machine) but we'd need to consider also the energy associated with the earlier material processing and conversion, any subcomponents or subsystems, etc. Also, some measure of global warming gas generation and any other environmental impact effects must be included. We are, essentially, estimating the footprint of this device we are designing. This makes the analysis rather complex and, unfortunately, not as analytical as the construction of the conventional error budget seen in Chapter 6.

12.4.2 Constructing the sustainability budget

The critical part of building these budgets (error or sustainability) is accumulating the data needed to populate the budget. Material data sources are very helpful in determining the basic material-performance characteristics (like modulus of elasticity, thermal properties, density) that are of use in machine design as we have seen. But, these need to be "connected" to embedded energy and operating energy consumption for use in a sustainability budget. Although there are many materials texts available, one excellent source of such "connections" is the text book "Materials: Engineering, Science, Processing and Design" by Ashby, Shercliff and Cebon.[559]

Ashby uses an approach (or strategy) for materials selection which is comprised of four steps:

- translation of design requirements in terms of function, objectives, etc.
- screening to select most usable materials meeting the requirements
- ranking with respect to some set of criteria, and
- documentation on background and history of the material in this or related uses.

This strategy attempts to get the best match between the characteristics of materials (or processes if the four steps are used for process selection) and those required by the design (functionality and constraints). We would add, as one of the screening elements, the need to assess environmental compatibility, energy use and embedded energy, global warming gas emission impacts, etc. Embedded energy is that energy that has gone into the mining, conversion, processing, and transportation of the material up to the point it enters our control or manufacturing facility for use in our product.

This would work as follows. The machine designer first determines the specifications required for the precision device as usual as inputs to the left side of Figure 12.11. A parallel discussion could be had for a process design or system of devices of processes but to simplify the discussion we stay with one device.

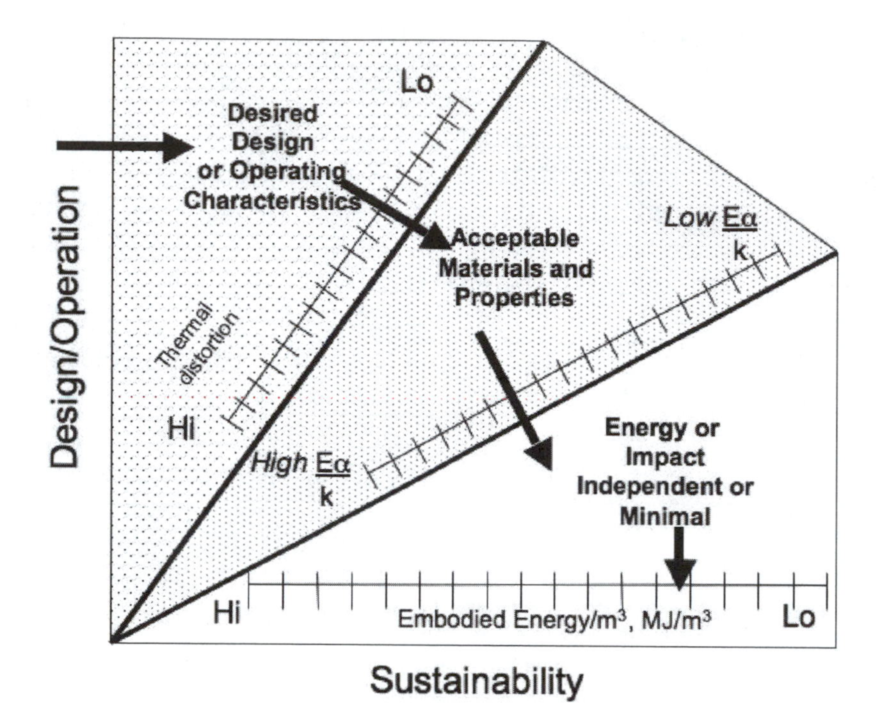

Figure 12.11. Design to sustainability selection chart, thermal stability example.

The designer determines that a critical requirement is that the device should be insensitive to variations in temperature where it operates and, since the device is heavily constrained (recall the discussion in Chapter 5 summarized in Table 5.1) the variable set of interest includes the modulus of elasticity, E, the coefficient of thermal expansion, α, and the conduction coefficient, k. For this situation, the combination of $E\alpha/k$ should be as low as possible to minimize thermal distortion and that will define a set of suitable materials. We saw in Figure 5.5 a wide range of materials with differing expansion and conduction coefficients. Depending on which of these materials we choose, we can determine the energy or sustainability impact by noting the embedded energy (for example) as a function of the weight or volume of the material (or materials) chosen and the amount needed for the design. For example, Figure 12.12 below, from Ashby[559] shows the relationship between embedded energy for a range of materials and the embodied energy/m^3. We see that more "exotic" materials, often used for thermal stability have higher embedded energy due to production requirements. This gets us through Step 2 of the sustainability budget creation.

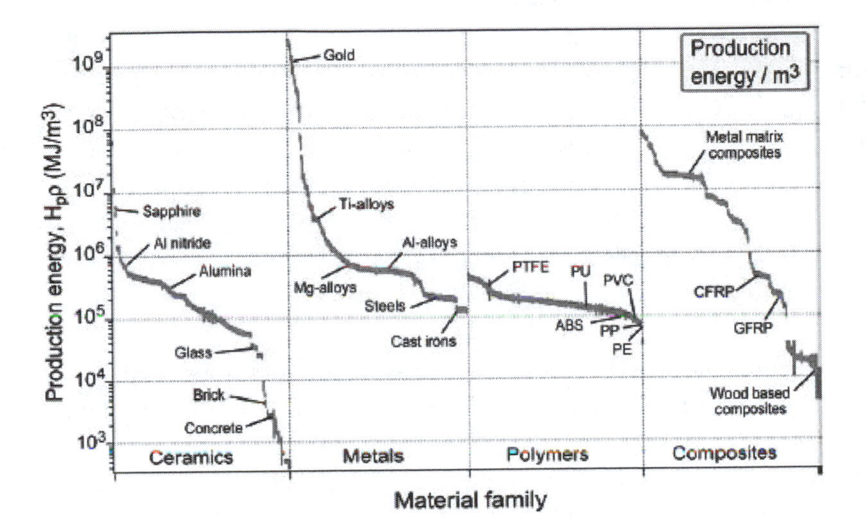

Figure 12.12. Production energy per volume for a range of materials, from Ashby[559].

However, it must be noted that there are usually other issues that need to be considered besides embedded energy (such as societal impacts if the material is toxic or hazardous or comes from a region where damage is done in mining or extraction) for a complete assessment of sustainability. Also, it is clear that we could create a set of charts as in Figure 12.11 for other constraints in machine design (chatter for example, in a milling machine, where the key parameter might include stiffness of the component and the tradeoff could be between cross-sectional area/geometry and stiffness; alternate material choices could be conventional carbon steel, a composite material with high stiffness to weight ratio, or a ceramic (which would also have beneficial thermal properties).

Step 3 of the sustainability budget construction requires combining the energy/materials impacts to yield upper and lower bound estimates of the total energy/material impact of the machine. Summing these for a series of machines in a system would give us a system budget. The most challenging part of this step is determining the "sensitivity" of sustainability to device specifications. If you recall the discussion about directional orientation in Chapter 7 when we were looking at the sensitivity of tool/work deflection to the cutting conditions and induced forces on the structure, you will remember that the key to this analysis was the determination of the array of force vectors and corresponding machine tool and work planes of minimal stiffness. Any component of a cutting force with a trajectory in a plane containing a mode of compliance, especially when that mode of compliance had a component in a sensitive direction relative to work surface, would contribute to an error of form.

We need to make the same type of analysis relative to the sensitive directions that we are designing our device to protect for error sources and the materials in their configurations we are using to accomplish that. Ideally, following our procedure in Figure 12.11, we could determine a range of material properties that can be varied to affect the design requirement of concern, for example thermal stability in the above example, but which would have no or minimal effect on embedded energy. This would be a sort of sensitivity analysis to energy or environmental impact similar to that seen in machine stiffness evaluation. That is, a design/material which allows us to meet design requirements with the maximum of flexibility while having minimal impact on environmental damage would allow the application of the conventional error budget without much additional constraint. It would, in effect, decouple the design and material choice from the sustainability impact for a defined range of conditions.

Let's look at an example. In Figure 12.12 we can see that, at an embodied energy of about 10^5 Mj/m^3, a wide range of materials exist spanning cast irons and some carbon steels to metal foams. Depending on the density of metal foams their modulus of elasticity can be as far from or as close to their parent material. This is not the case for carbon steels and cast irons. Similarly, thermal properties will vary tremendously between metal foams and cast iron, as will damping characteristics (important in machine tool structures). But, from an embodied energy perspective they are all quite similar. So there is an insensitivity we can take advantage of.

Tradeoffs in energy/materials sustainability (depending on what part of the life cycle it is used in) also need to be considered. Some "static" structures such as heavy machine tool bases which support but do not move with the machine axes can be made of heavier materials as their impact on energy of the machine during the "use" phase will not be large. Components making up the moving portions of the machine will logically expend more energy during their life with than stationary components as with each motion, energy will be expended in moving the component proportional to mass (among other things.)

The next major challenge is incorporating these very specific analyses into a broader analysis of sustainability. This can be done using the environmental values system analysis tool, EnV-S, among other tools but will not go into more detail on this here.

12.5 Summary comments/conclusion

This chapter attempts to illustrate where the future of precision manufacturing is heading. As we can see, this future depends a lot more on software interconnected "art-to-part" or design for manufacturing capability. We called that the manufacturing pipeline so that we can look both down the pipe from design to manufacturing as well as back up the pipe from the shop floor into the design studio. These ideas work for any product, precision or not, but as time goes on the demands made for reliable small lot highly efficient production of anything, from consumer products to energy sources to machine tools, demand we consider all as precision products.

This chapter also tried to weave in the need to consider at the same time the sustainability of production and products. Increasingly, our success in precision manufacturing will be measured not only the quality of the product or machine in the traditional precision engineering sense but also in our assurance that the product or machine was made sustainably and will operate sustainably. That is a big challenge and will change the way we work (and live hopefully) as precision manufacturing engineers and practitioners. If we follow some of these ideas, we can assure industry that this can be accomplished without sacrificing the business goals or adversely impacting the public. Then we will have indeed accomplished try sustainable manufacturing.

There have been many shifts in the way manufacturing has evolved over the years Figure 12.13, after Jovane[560], illustrates this well. Here we see the major advances in manufacturing, and the keys to these advances (for example F. W. Taylor's scientific management moving manufacturing from craft production to mass production, with the assistance of Henry Ford). Each transition was

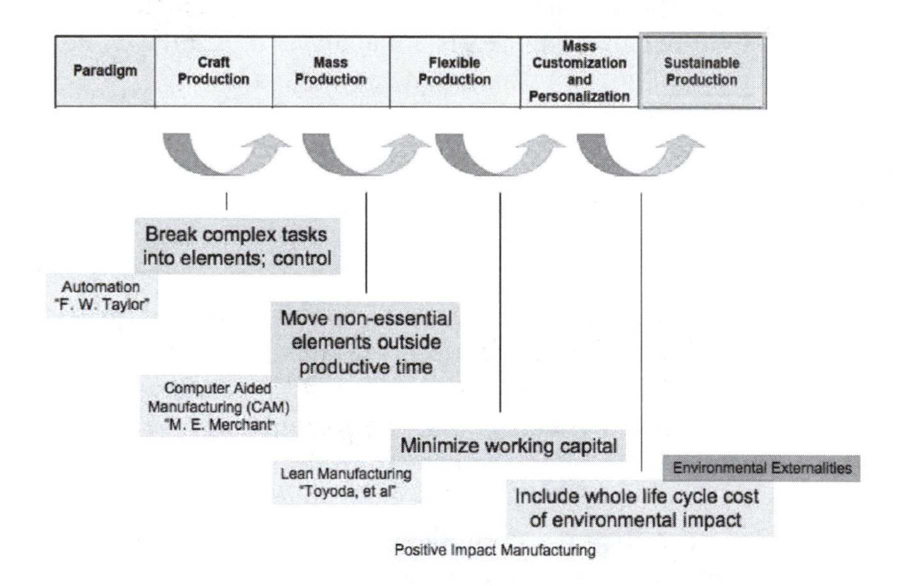

Figure 12.13. Key transitions in manufacturing development, after Jovanne[666].

accomplished by some enabling technology moving some of the elements of production, not contributing to the goal of efficient, cost-effective production, out of the system (that is, externalizing them). This allowed more effective use of production resources for productive use but, at the same time, recognized the cost to production of the external aspects (like inventory, in the case of lean manufacturing.)

The last transition in the diagram is to sustainable production. The key to this transition is externalizing the built in costs of "non-sustainable" manufacturing (for example, embedded energy, recycling and reuse). Until the value of these elements is accounted for, and treated as a cost by the consumer, it will be difficult to drive sustainable manufacturing. At present, many of the non-sustainable aspects of manufacturing are covered by other sources – think who pays to haul away your trash each week. This will be a challenging task to integrate these costs and the efficiencies they will drive into manufacturing. But, this will be the most important change to manufacturing and the kinds of tools we have discussed here, such as the sustainability budget, will be key to making it happen. Let's go to work!

REFERENCES

1 Evans, C., *Precision Engineering: an Evolutionary View*, Cranfield Press, Bedford UK, 1989, pp. 1.

2 ASPE website, http://www.aspe.net

3 McKeown, P. A., "High precision manufacturing in an advanced industrial economy," James Clayton Memorial Lecture, Inst. Mech. Eng., London, UK, April 23, 1986.

4 Taniguchi, N. "Current Status in and future of ultra-precision machining and ultrafine materials processing," *Annals CIRP*, 32, 2, 1983.

5 Evans, C., *ibid.*, p. 22 and Bryan, J. B., "The power of deterministic thinking in machine tool accuracy," *First International Machine Tool Engineers Conference*, JMTBA, Tokyo, Nov. 1984.

6 Donaldson, R. R., "The deterministic approach to machining accuracy," SME Fabrication Technology Symposium, Golden CO, Nov. 1972.

7 Ayres, R. U., "Complexity, reliability and design: manufacturing implications," *Manufacturing Review*, 1, 1, 1988, pp. 26-35.

8 Meister, D., "Reduction of human error" in *Handbook of Industrial Engineering*, G. Salvendy, ed., Wiley, New York, 1982.

9 Kalpakjian, S., *Manufacturing Processes for Engineering Materials*, Addison-Wesley, Reading MA, 1984, pp. 4-5.

10 Boothroyd, G., *Fundamentals of Metal Machining and Machine Tools*, McGraw-Hill, New York, 1975.

11 Taylor, F. W., 1901, "On the Art of Metal Cutting," Trans. ASME, 28, 1901.

12 Burke, J. *Connections*, Little Brown, Boston, 1978.

13 Betts, J., *John Harrison*, Royal Observatory, Greenwich, 1993.

14 Diderot, and D'Alembert, *L'Encyclopédie Diderot et D'Alembert*, Paris, 1759.

15 Roe, J. W., *English and American Tool Builders*, Yale University Press, New Haven, CT, 1916 (republished by Lindsay Publications, Bradley IL, 1987).

16 Shirley, J. V. and Jaikumar, R., "Turing machines and Guttenberg technologies: the postindustrial marriage," *Manufacturing Review*, 1, 1, 1988, p. 37.

17 Moriwaki, T., "Intelligent Machine Tool: Perspective and Themes for Future Development", PED-Vol. 68-2, Manufacturing Science and Engineering, Vol. 2, ASME (1994), pp. 841-849.

18 Slocum, A. H., *Precision Machine Design*, Prentice-Hall, Englewood Cliffs, NJ, 1992.

19 Nakazawa, H., *Principles of Precision Engineering*, Oxford University Press, New York, 1994, original in Japanese translated by R. Takeguichi.

20 Suh, N. P., Bell, A. C., and Gossard, D., "An axiomatic approach to manufacturing systems, *Trans. ASME, J. Eng. Ind.*, 100, 1978.

21 Moore, W. R., *Foundations of Mechanical Accuracy*, Moore Special Tool Company, Bridgeport CT, 1970.

22 Wysk, R. and Chang, T., *An Introduction to Automated Process Planning Systems*, Prentice-Hall, Inc., 1985.

23 Scarr, A. J., *Metrology and Precision Engineering*, McGraw-Hill, 1967.

24 DeVor, R., et al, EM-SIM, End Milling Simulation Software, ARPA/NSF Machine Tool - Agile Manufacturing Institute, University of Ilinois, Champaign-Urbana, 1996.

25 Technology of Machine Tools, The Machine Tool Task Force Study, Lawrence Livermore National Laboratory, October, 1980.

26 Wada, R., "Ultra-precision Machine and Elements Therefor," Proc. 1st IMEC, JMTBA, 1984, pp. 36-60.

27 McKeown, P., *Lectures on Precision Engineering*, University of California, Berkeley, 1994.

28 Bryan, J., "The Abbé principle revisited - an updated interpretation," *Precision Engineering*, Vol. 1, No. 3, 1989, pp. 129-132.

29 *Surface Texture (Roughness, Waviness, Lay)*, ANSI B46. 1-1985, ASME, New York, 1985, pp. 30-37.

30 Ohuchi, K., "Ultra-Precision Machining from the Viewpoint of Work Materials," *Proc. 2nd IMEC, JMTBA*, 1986, pp. 76-87.

31 Gronsky, R., University of California, Berkeley, personal notes, 2007.

32 Fortier, M. F., Parallel Robotic Systems Corporation (PRSCO), Hampton, NH, 2007.

33 Furukawa, Y., in *Frontier Technology in Machine Tools*, Japan Society of Mechanical Engineers, Kogyo Chosakai, 1988 (in Japanese).

34 Bryan, J. B., "International status of thermal error research," *CIRP Annals*, 39, 2, 1990, pp. 645-656.

35 Ashby, M., Shercliff, H. and Cebon, D., Materials: Engineering, Science, Processing and Design," Butterworth-Heinemann/ Elsevier, 2007.

36 ASME/ANSI Standard B89.6.2, "Temperature and humidity environment for dimensional measurement," ASME, 345 E. 47th St., New York, NY, 1973.

37 Bryan, J. B., Brewer, W., McClure, E., and Pearson, J., "Thermal Effects in Dimensional Metrology," Proc. ASME Metals Engineering Conference, Berkeley CA, 1965.

38 Bryan, J., Donaldson, R., Clauser, R. W. and Blewett, W. H., "Reduction of Machine Tool Spindle Growth" UCRL 74672, Proc. First Annual NAMRC conference in Hamilton, Canada, May 14-15, 1973.

39 Krulewich, D., "Error Compensation for Thermally Induced Errors on a Machine Tool," Report UCRL-ID-125496, LLNL, November 8, 1996.

40 Donmez, M.A., "A General Methodology for Machine Tool Accuracy Enhancement - Theory, Application and Implementation," Ph.D. Dissertation, Purdue University, August 1985.

41 Roblee, J., "Precision temperature control for optics manufacturing," Proc. 2nd Int'l Technical Symposium on Optical and Electro-Optical Applied Science and Engineering, Cannes, France, 1985.

42 Bryan, J. and Carter, D., Clouser, R., and Hamilton, "An order of magnitude improvement in thermal stability with the use of liquid shower on a general purpose measuring machine," SME Precision Machining Workshop, St. Paul MN, 1982.

43 Benjamin Thompson (Count Rumford), "Heat is a Form of Motion: An Experiment in Boring Cannon," *Philosophical Transactions*, Vol. 88, 1798.

44 Davies, M., Ueda, T., M'Saoubi, R., Mullany, B., Cooke, A., "On the measurement of temperature in material removal processes," Annals CIRP, 56, 2, 2007.

45 Donaldson, R. R., "Error budgets" in Technology of Machine Tools, vol. 5 of the *Machine Tool Task Force Study*, LLNL, October, 1980, section 9.14, pp. 1-14.

46 Tonshoff, H., *Lectures on Machine Tools Dynamic Behavior*, UC-Berkeley, March, 2001.

47 Tlusty, J., "Criteria for Static and Dynamic Stiffness of Structures," Volume 3, Machine Tool Mechanics of the Technology of Machine Tools - Machine Tool Task Force Study, Lawrence Livermore National Laboratory, October 1980.

48 van den Brink, W., Katz, B., and Wittekeok, S., "New 0.54 aperture i-line wafer stepper with field by field leveling combined with global alignment," SPIE, Vol. 1463, 1991.

49 Ehmann, K. F., S. G. Kapoor, R. E. De Vor, and I. Lazoglu. "Machining Process Modeling; A Review," Journal of Manufacturing Science and Engineering, 119, 1-9, 1997.

50 Altintas, Y., *Manufacturing Automation: Metal Cutting Mechanics, Machine Tool Vibrations, and CNC Design*, Cambridge University Press, Cambridge, 2000.

51 Tlusty, J., *Manufacturing Processes and Equipment*, Prentice-Hall, 2000.

52 Taylor, J. S., Piscotti, M. A., Blaedel, K. L. and Weaver, L. F. (1995), "Investigation of Acoustic Emission as Wheel-to-Workpiece Proximity Sensor in Fixed Abrasive Grinding," *Proc. 1995 ASPE Meeting*, ASPE, Austin, TX., pp. 159-162.

53 Moriwaki, T., 1992, "Intelligent Machining," *Proc. Workshop on Tool Condition Monitoring, Vol. I*, Dornfeld, D. A., Byrne, G., eds. CIRP, Paris.

54 Lee, D. E., Hwang, I., Valente, C. M., Oliveira, J. F., Dornfeld, D. A., 2006, "Precision manufacturing process monitoring with acoustic emission," Int. J. Machine Tools & Manufacture, Vol. 46, No. 2, pp. 176-188.

55 Kegg, R., 1994, "Sensor History - Machine Tool Applications Table," *Proc. Workshop on Tool Condition Monitoring, Vol. III*, Dornfeld, D. A., Byrne, G., eds. CIRP, Paris, p. 51.

56 Byrne, G., D. Dornfeld, I. Inasaki, W. König, and R. Teti, 1995, "Tool Condition Monitoring (TCM) - The Status of Research and Industrial Application," *CIRP Annals*, Vol. 44, No. 2, pp. 541-567.

57 Dornfeld, D. A., König, W., and Kettler, G., 1993, "Present State of Tool and Process Monitoring in Cutting," *Proc. New Developments in Cutting*, VDI Berichte NR 988, Düsseldorf, pp. 363-376.

58 Tönshoff, H. K., Wulfsberg, J. P., Kals, H. J. J., König, W., and van Luttervelt, C. A., 1988, "Development and Trends in Monitoring

and Control of Machining Processes," *Annals CIRP,* 37, 2, pp. 611-622.

59 Tlusty, J. and Andrews, G. C., 1983, "A Critical Review of Sensors for Unmanned Machining," *Annals CIRP,* 32, 2, pp. 563-572.

60 Birla, S., 1980, "Sensors for Adaptive Control and Machine Diagnostics," *Technology of Machine Tools - Machine Tool Task Force Report*, Vol. 4, *Machine Tool Controls*, Miskell R. V., ed., LLNL, Report UCRL-52960, pp. 7.12-1 - 7.12-70.

61 Shiraishi, M., 1988, "Scope of In-Process Measurement, Monitoring and Control Techniques in Machining Processes- Part 1: In Process Techniques for Tools," *Precision Engineering*, Vol. 10, No. 4, pp. 179-189.

62 Shiraishi, M., 1989, "Scope of In-Process Measurement, Monitoring and Control Techniques in Machining Processes- Part 2: In Process Techniques for Workpieces," *Precision Engineering*, Vol. 11, No. 1, pp. 27-37.

63 Shiraishi, M., 1989, "Scope of In-Process Measurement, Monitoring and Control Techniques in Machining Processes - Part 3: In Process Techniques for Cutting Processes and Machine Tools," *Precision Engineering,* Vol. 11, 1, pp. 39-47.

64 Iwata, K., 1988, "Sensing Technologies for Improving the Machine Tool Function," *Proc. 3rd Int'l Machine Tool Engineer's Conference*, JMTBA, Tokyo, pp. 87-109.

65 Hoshi, T., 1990, "Automatic Tool Failure Monitoring in Drilling and Thread Tapping," *Proc. III Int'l Conf. on Automatic Supervision, Monitoring and Adaptive Control in Manuf.*, CIRP, Rydzyna, Poland, pp. 41-58.

66 Teti, R., 1995, "A Review of Tool Condition Monitoring Literature Database," *CIRP Annals*, 44/2.

67 Szafarczyk, M., ed., 1994, *Automatic Supervision in Manufacturing*, Springer-Verlag, London.

68 Dornfeld, D. A., 1995, "Monitoring Technologies for Intelligent Machining," *Proc. CIRP/VDI Conference on Monitoring of Machining and Forming Processes*, VDI Berichte Nr. 1179, Düsseldorf Germany, pp. 71-90.

69 Micheletti, D. F., Koenig, W., and Victor, H. R., 1976, "In Process Tool Wear Sensors for Cutting Operations," *Annals CIRP*, 25, 2, pp. 483-496.

70 Bellmann, B., 1978, Mebverfahren zur kontinuierlichen Erfassung des Freiflächenverschleibes beim Drehen, doctoral thesis, TH Darmstadt.

71 Pfeiffer, T., Fürst, A., and Vollard, W., 1982, Prozebintermittierende Werkzeug- und Werkstückvermessung auf NC-Werkzeugmaschinen, Industrie Anzeiger 104, 27.

72 Hammer, H., 1982, Automatisierung von Fertigungseinrichtungen, *Industrie Anzeiger*, 104, 27.

73 Lierath, F., 1984, Gerätesystem zur Erfassung des Verschleibes an spanenden Werkzeugen bei bedienarmer Fertigung, Feingerätetechnik, Berlin 33, 2.

74 Ernst, P., 1978, Verschleiberfassung beim Bohren mit Wendelbohrern, doctoral thesis, TH Darmstadt.

75 Balakrishnan, P. and Macbain, J. C., 1985, "Monitoring Systems for Unmanned Machining," Technical SME Paper.

76 Dumbs, A. and Laun, H., 1983, Vermessung von Werkzeugen für Metallbearbeitungsmaschinen, VDW-Forschungsbericht 2301,

Fraunhofer-Institut für Physikalische Mebtechnik, Freiburg.

77 Kamm, H., Müller, M., 1975, Kapazitiver Verschleibsensor für das Messerkopf-Fräsen, *Industrie Anzeiger*, 97, 73, pp. 1602-1603.

78 Stöferle, Th. and Bellmann, B., 1975, Kontinuierlich messende Verschleibsensoren für die Drehbearbeitung, VDW-Forschungsbericht A 2551.

79 Essel, K., 1972, Entwicklung einer Optimierregelung für das Drehen, doctoral thesis, RWTH Aachen.

80 Otto, F., 1976, Entwicklung eines gekoppelten AC-Systems für die Drehbearbeitung, doctoral thesis, RWTH Aachen.

81 Leonards, F., Müller, W., and Otto, F., 1976, Prozeblenkungssysteme für die Drehbearbeitung, PDV-Bericht, KfK-PDV 82.

82 Takeyama, H., et al., Sept., 1967, "Sensors of Tool Life for Optimization of Machining", *Proc. 8th Int. MTDR-Conference*, Manchester.

83 Dormehl, E., 1984, Wirtschaftliche Drehbearbeitung mit aufgabenangepabten Automatisierungsbausteinen, VDI-Z 126, 5.

84 Essel, K., Otto, F., and Kirchner, W., 1974, Sensor zum Erfassen des Freiflächenverschleibes an Drehwerkzeugen, VDI-Z 116, 17.

85 Schwan, W., 1979, Verfahren zur Überwachung der Werkzeugschneide sowie NC-gesteuerte Drehmaschine zur Durchführung des Verfahrens, Deutsches Patentamt, OS-Nr. 27 47 487.

86 Heinze, W., Sept., 1985, Mikrorechnergesteuerter Mebplatz zur automatischen Verschleiberfassung bei der Schneidstoffentwicklung für Wendeschneidplatten, 3. Fachtagung "Gestaltung von Fertigungsprozessen im Maschinenbau", 18, Magdeburg, DDR.

87 Lierath, F., Scharfenort, U., and Pieper, H.-J., 1986, Sensorkontrollierte Werkzeugbruch- und Werkzeugverschleibidentifikation an CNC-Drehmaschinen, Fertigungstechnik und Betrieb 36, 5.

88 Der BK. Mikro 2Bohrekontoller schaltet Folgeschäden bei Werkzeugbruch aus, Leukhardt company, Tutlingen, 1981.

89 Sensoren beugen Fehlern vor - Fertigungstechnik setzt auf automatisches Fühlen, Hören und Sehen, VDI-Nachrichten 37, 48, 1983.

90 Schneider, A., 1981, Vorrichtung zum Abtasten von Gegenständen, Deutsches Patentamt, OS-Nr. 30 30 431.

91 Tooltronic - beruhrungslose Bohrbruchkontrolle, Leuze Electronic Company, Owen, 1982.

92 Die rationelle Fertigung für morgen, brochure Fr. Deckel Company, München, 1983.

93 Ductrohet, J.-M., 1975, Vorrichtung zur Feststellung eines umlaufenden Werkzeuges, Deutsches Patentamt, OS-Nr. 24 47 245.

94 Homburg, D., 1983, Methoden der Werkzeugbruchkontrolle, TR Technische Rundschau 75, 44.

95 Thierfelder, A., 1988, Sensoren und Mebsysteme zur Prozebüberwachung, tz für Metallbearbeitung 82, 3.

96 Tüns, J., 1989, Weiter aufwärts - Geräte, Einrichtungen und Systeme zum Überwachen von spanenden Werkzeugen auf der EMO 1989 in Hannover, Maschinenmarkt Würzburg 95, 44.

97 Fertigungssicherheit durch Werkzeugüberwachung, Flexible Fertigung, No. 2., 1989.

98 Vollmer, H. J., Franke, N., and Radtke, U., 1974, Optisch-elektronisches Verfahren zur prozebinternen Verschleibmessung, Fertigungstechnik und Bertrieb 24, 12.

99 Leonards, F., 1978, Ein Beitrag zur mebtechnischen Erfassung von Prozebkenngröben bei der Drehbearbeitung, doctoral thesis, TU Berlin.

100 Uehara, K., 1973, "New Attempts for Short Time Tool-Live Testing," *Annals CIRP*, 22, 1.

101 Aoyama, H., Kishinami, T., and Saito, K., July, 1989, "A Method of Tool Management Based on an Intelligent Cutting Tool," *Adv. Manuf. Eng.*, Vol. 1.

102 Leonards, F., 1974, Die Messung des Schneidkantenversatzes bei der Drehbearbeitung nach dem Ultraschall-Laufzeitverfahren, Industrie Anzeiger 96, 107/108.

103 Werkzeug-korrektursystem, Schneidenkontrolle VT 921, Samsomatic company, Frankfurt, 1980.

104 Brandstätter, A., 1981, Ein Adaptive-Control-Optimization-System für das Fräsen, doctoral thesis, TH Karlsruhe.

105 Kamm, H., 1977, Beitrag zur Optimierung des Messerkopfstirnfräsens, doctoral thesis, TH Karlsruhe.

106 Stöferle, Th., Hartmann, V., 1976, Beitrag zur Lagemessung aller Schneidkanten an Fräswerkzeugen, Werkstatt und Betrieb 109, 3.

107 Krause, W., 1976, Messung des Freiflächenverschleibes beim Umfangsstirnfräsen, Fertigungstechnik und Betrieb 26, 10.

108 Weis, W., Lengeling, A., and Hüntrup, V., 1994, Automatisierte Werkzeugüberwachung und - vermessung beim Fräsen mit Hilfe bildverarbeitender Systeme, tm - Technisches Messen.

109 Crostack, H.-A., Cronjäger, L., Hillmann, K., Müller, P., and Strunck, T., 1988, Rauhtiefe berührungslos gemessen - Erfassung des Oberflächenzustandes spanend bearbeiteter Bauteile mit Ultraschall, tz für Metallbearbeitung 82, 10.

110 Schehl, U., et al., April, 1990, Sicherung des spanabhebenden Bearbeitungsprozesses, KfK Bericht, KfK-PFT 154.

111 Hänsel, W., 1974, Beitrag zur Technologie des Drehprozesses im Hinblick auf Adaptive Control, doctoral thesis, RWTH Aachen.

112 König, W. and Meyen, H. P., 1988, Rechnergestüztes System zur Werkstückkontrolle, Qualitätskontrolle im Prozebtakt beim Schleifen machbar!, Industrie Anzeiger 110, 61/62, pp. 26-29.

113 König, W. and Wünsche, U., 1991, Oberflächen mit Infrarotlicht beurteilen, Industrie Anzeiger 113, 72, pp. 60-61.

114 König, W., Meyen, H. P., Klumpen, T., and Memis, F., 1992, Standzeitüberwachung und integrierte Qualitätskontrolle für den automatischen Schleifprozeb, VDI-Z 134, 5, pp. 71-76.

115 Kluft, W., 1983, Werkzeugüber-wachungssysteme für die Drehbearbeitung, doctoral thesis, RWTH Aachen.

116 König, W., 1990, Fertigungsverfahren, Band 1, Drehen, Bohren, Fräsen, VDI-Verlag, Düsseldorf.

117 Colwell, L. V. and Mazur, J. C., 1979, "Real Time Computer Diagnostics - A Research Tool for Metal Cutting," *Annals CIRP*, 28, 1.

118 Wiele, H. and Menz, P., 1983, Automatische Erfassung des Verschleibes spanender Werkzeuge - Bestandteil der Prozebüberwachung beim bedienarmen Betrieb, Fertigungstechnik und Betrieb 33, 6.

119 Gomoll, P., 1978, Verfahren und Vorrichtung zur Überwachung der Schneidplatten in Werkzeugmaschinen, Deutsches Patentamt, OS-Nr. 30 42 211.

120 Menz, P., Richter, H., 1984, Verschleibmessung an spanenden Werkzeugen durch rechnerbewertete Temperatur-messung, Wiss. Zeitg. TH Magdeburg 28, 5.

121 Uehara, K., Takeshita, H., 1989, "Prognostication of the Chipping of Cutting Tools," *Annals CIRP*, 38, 1.

122 Takata, S., Ahn, J. H., Miki, M., Miyao, Y., and Sata, T., 1986, "A Sound Monitoring System for Fault Detection of Machine and Machining States," *Annals CIRP*, 35, 1, pp. 289-292.

123 Moriwaki, T., 1994, "Intelligent Machine Tool: Perspective and Themes for Future Development," *Manufacturing Science and Engineering*, Vol. 2, ASME, PED-Vol. 68-2, New York, pp. 841-849.

124 Liang, S. and Dornfeld, D. A., 1989, "Tool Wear Detection using Time Series Analysis of Acoustic Emission," *Trans. ASME, J. Eng. Ind.*, Vol. 111, No. 2.

125 Dornfeld, D. A., 1992, "Application of Acoustic Emission Techniques in Manufacturing," *NDT&E Int'l,* 25, 6, 259-269.

126 Hoshi, T., 1990, "Automatic Tool Failure Monitoring in Drilling and Thread Tapping," *Proc. III Intern'l Conf. on Automatic Supervision, Monitoring and Adaptive Control in Manuf.*, CIRP, Rydzyna, Poland, pp. 41-58.

127 Dornfeld, D. A., 1980, "Acoustic Emission and Metalworking-Survey of Potential and Examples of Applications," *Proc. 8th North American Manufacturing Research Conference,* SME, University of Missouri, Rolla, MO., pp. 207-213.

128 Dornfeld, D. A. and Diei, E. N., 1982, "Acoustic Emission from Simple Upsetting of Solid Cylinders," *Trans. ASME, J. Eng. Mat.Tach.*, 104, 2, pp. 145-152.

129 Liang, S. Y. and Dornfeld, D. A., 1987, "Punch Stretching Process Monitoring Using Acoustic Emission Signal Analysis - Part 1: Basic Characteristics," *J. Acoustic Emission,* 6, 1, pp. 37-42.

130 Park, J. S., Han, E. K., Mori, Y., Suzuki, E., Kishigami, T., Teraoka, N., and Ohkoshi, T., 1991, "Acoustic Emission Produced during Wire Drawing Process," *Proc. 4th World Meeting on Acoustic Emission,* ASNT, Boston, MA, pp. 94-101.

131 Miller, R. K. and McIntire, P., 1987, *Nondestructive Testing Handbook, Vol. 5 Acoustic Emission Testing*, 2nd ed., ASNT.

132 Moriwaki, T., 1983, "Application of Acoustic Emission Measurements to Sensing Wear and Breakage of Cutting Tool," *Bull. Japan Soc. of Precision Engin'g*, 17, pp. 154-160.

133 Dornfeld, D. A., 1979, "An Investigation of Orthogonal Cutting via Acoustic Emission Signal Analysis," *Proc. 7th North American Manufacturing Research Conference*, Univ. of Michigan, Ann Arbor.

134 Dornfeld, D. A. and Kannatey-Asibu, E. (1980), "Acoustic Emission during Orthogonal Metal Cutting," *Int. J. Mech. Science,* 22, pp. 285-296.

135 Kannatey-Asibu, E. and Dornfeld, D. A. (1981), "Quantitative Relationships for Acoustic Emission from Orthogonal Metal Cutting," *Trans. ASME, J. Eng. Ind.*, 103, pp. 330-340.

136 Rangwala, S. and Dornfeld, D. A., 1991, "A Study of AE Generated During Orthogonal Metal Cutting - 1: Energy Analysis," *Int. J. Mech. Sci.*, 33, 6, pp. 471-487.

137 Rangwala, S. and Dornfeld, D. A., 1991, "A Study of AE Generated During Orthogonal Metal Cutting - 2: Spectral Analysis," *Int. J. Mech. Sci.*, 33, 6, pp. 489-499.

138 Heiple, C. R., Carpenter, S. H. and Armentrout, D. L. (1991), "Origin of Acoustic Emission Produced during Single Point Machining," *Proc. 4th World Meeting on Acoustic Emission*, ASNT, Boston, MA, pp. 463-470.

139 Teti, R. and Dornfeld, D. A. (1989), "Modeling and Experimental Analysis of Acoustic Emission from Metal Cutting," *Trans. ASME, J. Eng. Ind.*, 111, 3, pp. 229-237.

140 Pan, C. S. and Dornfeld, D. A. (1986), "Modeling the Diamond Turning Process with Acoustic Emission for Monitoring Purposes," *Proc. 14th North American Manufacturing Research Conference*, SME, University of Minnesota, Minneapolis, pp. 257-265.

141 Liu, J. J. and Dornfeld, D. A., 1992, "Monitoring of Micro-machining Process using Acoustic Emission," *Trans. North American Manufacturing Research Institute*, SME, Vol. 20.

142 Lan, M. S. and Dornfeld, D. A., 1986, "Acoustic Emission and Machining- Process Analysis and Control," *Advanced Manufacturing Processes*, 1, 1, 1-21.

143 Lan, M. S. and Naerheim, Y., 1985, "Application of Acoustic Emission Monitoring in Machining," *Proc. 13th North Amrerican Manufacturing Research Conference*, SME, University of California, Berkeley, pp. 310-313.

144 Naerheim, Y. and Lan, M. S., 1988, "Acoustic Emission Reveals Information about Metal Cutting," *Proc. 136h North Amrerican Manufacturing Research Conference*, SME, University of Illinois, Champaign-Urbana, pp. 240-244.

145 Schmenk, M., 1984, "Acoustic Emission and the Mechanics of Metal Cutting," *Acoustic Emission Monitoring and Analysis in Manufacuring*, ed. D. A. Dornfeld, ASME, New York.

146 Diei, E. N. and Dornfeld, D. A., 1987, "Acoustic Emission from the Face Milling Process- The Effects of Process Variables," *Trans. ASME, J. Engineering for Industry*, 109, 2, pp. 92-99.

147 James, D. and Carpenter, S., 1971, "Relationships between Acoustic Emission and Dislocation Kinetics in Crystalline Solids," *J. Appl. Physics*, 42, 4685.

148 Agarwal, A., Frederick, J., and Felbeck, D., 1970, "Detection of Plastic Microstrain in Aluminum by Acoustic Emission," *Met. Trans.*, 1, 1069.

149 Wright, P. and Robinson, J., 1977, "Material Behaviour in Deformation Zones of Machining Operation, *Metals Technology*, 5, 240.

150 Dharan, H. and Hauser, F., 1973, "High Velocity Dislocation Damping in Aluminum, "*J. Appl. Physics*, 44, 1468.

151 Rouby, D., Fleischmann, P. and Gobin, P., 1977, "An Acoustic Emission Source Model Based on Dislocation Movement," in *Internal Friction and Attenuation in Solids*, University of Tokyo Press, Tokyo, pp. 811-815.

152 Rangwala, S., 1988, "Machining Process Characterization and Intelligent Tool Condition Monitoring Using Acoustic Emission," Ph. D. Thesis, Department of Mechanical Engineering, University of California, Berkeley, CA.

153 Dornfeld, D. A., 1987, "Intelligent Sensors for Monitoring Untended Manufacturing Processes," *Proc. 2nd International Machine Tool Research Forum*, NMTBA, Chicago, IL.

154 Chiu, S. L., Morley, D. J. and Martin, J. F., 1987, "Sensor Data Fusion on a Parallel Processor," Proc. 1987 IEEE International Conference on Robotics and Automation, IEEE, Raleigh, NC, pp. 1629-1633.

155 Dornfeld, D. A., 1990, "Neural Network Sensor Fusion for Tool Condition Monitoring," *Annals of CIRP*, 39, 1.

156 Ahmed, N. and Rao, K. K., 1975, Orthogonal Transforms for Digital Signal Processing, Springer-Verlag, New York.

157 Sata, T., K. Matsushima, T. Nagakura and E. Kono, 1973, "Learning and Recognition of the Cutting States by Spectrum Analysis," *Annals of the CIRP*, 22, 1, pp. 41-42.

158 Matsushima, K., and T. Sata, 1980, "Development of the Intelligent Machine Tool," *J. Faculty of Engineering*, Univ. of Tokyo (B), Vol. 35, No. 3, pp. 395-405.

159 Dornfeld, D. A. and C. S. Pan, 1985, "Determination of Chip Forming States Using a Linear Discriminant Function Technique

with Acoustic Emission," Proc. 13th North American Manufacturing Research Conference, SME, University of California, Berkeley, May, pp. 285-303.

160 Emel, E. and E. Kannatey-Asibu, Jr., 1986, "Characterization of Tool Wear and Breakage by Pattern Recognition Analysis of Acoustic Emission Signals," *Proc. 14th North American Manufacturing Research Conference*, SME, University of Minnesota, Minneapolis, pp. 266-272.

161 Balakrishnan, P., Trabelsi, H., Kannatey-Asibu, Jr., E., and Emel, E., 1989, "A Sensor Fusion Approach to Cutting Tool Monitoring," Advances in Manufacturing Systems Integration and Processes, Proc. 15th NSF Conference on Production Research and Technology, SME, University of California, Berkeley, CA, pp. 101-108.

162 Hinton, G. and Fahlman, S. (1987), "Connectionist Architectures for Artificial Intelligence," *IEEE Computer*, pp. 100-109.

163 Hopfield, J., 1982, "Neural Networks and Physical Systems with Emergent Collective Computational Abilities," *Proceedings of the National Academy of Sciences*, Vol. 79, April, pp. 2544-2588.

164 Hopfield, J. J., 1984, "Neurons with Graded Response Have Collective Computation Properties Like Those of Two-State Neurons," *Proc. National Academy of Sciences*, Vol. 81, May, pp. 3088-3092.

165 Rumelhart, D. and McClelland, J., 1986, *Parallel Distributed Processing, Vol. 1*, MIT Press, Cambridge, MA.

166 Le Cun, Y., 1985, "A Learning Procedure for Assymetric Threshold Networks," *Proc. Cognitiva*, Paris.

167 Wu, S. M. and Pandit, S. M., 1983, *Time Series and Systems Analysis with Applications*, Wiley, New York.

168 Liang, S. and Dornfeld, D. A. (1989), "Tool Wear Analysis using Time Series Analysis of Acoustic Emission," *Trans. ASME, J. Eng. Ind.*, 111, 3, pp. 199-205.

169 Rangwala, S., and Dornfeld, D., 1987, "Integration of Sensors via Neural Networks for Detection of Tool Wear States," *Proceedings of the Winter Annual Meeting of the ASME*, PED 25, 109-120.

170 Chryssolouris, G., and Domroese, M., 1988, "Sensor Integration for Tool Wear Estimation in Machining," Proceedings of the Winter Annual Meeting of the ASME, Symposium on Sensors and Controls for Manufacturing, 115-123.

171 Kobayashi, S., and Thomsen, E., 1960, "The Role of Friction in Metal Cutting," *Trans. ASME, J. Engineering Industry*, 82, 324-332.

172 Lan, M., 1983, "Investigation of Tool Wear, Fracture and Chip Formation in Metal Cutting Using Acoustic Emission," Ph.D. Dissertation, University of California at Berkeley, Department of Mechanical Engineering.

173 Wright, P. K., 1983, "Physical Models of Tool Wear for Adaptive Control in Flexible Machining Cells," in *Computer Integrated Manufacturing*, PED-Vol. 8, ASME (M. R. Martinez and M. C. Leu, eds.), New York, pp. 19-31.

174 Cook, N., 1959, "Self Excited Vibrations in Metal Cutting," *Trans. ASME, J. Engineering for Industry*, 183-186.

175 Martin, P., Mutels, B., and Drapier, J., 1974, "Influence of Lathe Tool Wear on theVibrations Sustained in Cutting," *Proc. 15th MTDR Conference*, pp. 251-257, Birmingham, USA.

176 Sze, S. M., editor, 1994, *Semiconductor Sensors*, John Wiley and Sons, New York.

177 Allocca, J. A. and Stuart, A., 1984, *Transducers: Theory and Applications*, Reston Publishing Company, Reston VA.

178 Bray, D. E. and McBride, D., editors, 1992, *Nondestructive Testing Techniques*, John Wiley ands Sons, New York.

179 Webster's Third New International Dictionary, 1971, G. C. Merriam and Co., Springfield MA.

180 Usher, M. J., 1985, *Sensors and Transducers*, McMillan, New Hampshire.

181 Middlehoek, S. and Audet, S. A. 1989, *Silicon Sensors*, Academic Press, New York.

182 White, R. M., 1987, "A Sensor Classification Scheme," *IEEE Transactions Ultrason. Feroelec. Freq. Contr.*, UFFC-34, 124.

183 Goch, G., Schmitz, B., Karpuschewski, B., Geerkins, J, Reigel, M., Sprongl, P., and Ritter R., 1999, "Review of non-destructive measuring methods for the assessment of surface integrity: a survey of new measuring methods for coatings, layered structures and processed surfaces," *Precision Engineering*, 23, pp. 9-33.

184 Oliveira, J.F.G., Dornfeld, D. A., 2001, Application of AE Contact Sensing in Reliable Grinding Monitoring, CIRP Annals, 50(1), 2001.

185 Diei, E. N., Dornfeld, D. A., "A Model of Tool Fracture Generated Acoustic Emission During Machining," ASME Trans., J. Eng. Ind., Vol. 109, No. 3, 1987, pp. 227-234; also appears in Sensors and Controls for Manufacturing, Kannatey-Asibu, E. and Ulsoy, A.G., eds., ASME, New York, 1985.

186 Diei, E.N., Dornfeld, D.A., "Acoustic Emission from the Face Milling Process - the Effects of Process Variables," ASME Trans., J. Eng. Ind., Vol. 109, No. 2, 1987, pp. 92-99.

187 Ansenk, J., Broens, N., "Investigation on Acoustic Emission Mapping Applications in Milling," OPF Report, University of Sao Paulo, Sao Carlos, Brazil, 2004.

188 Dennison, C., "Analysis & Metrology - Post-CMP: Developing effective inspection systems and strategies for monitoring CMP processes," MICRO, February 1998, p 31.

189 Caroli, C., Nozieres, P., Hysteresis and Elastic Interactions of Microasperities in Dry Friction. European Physics Journal B, 4(2), 1998, pp. 233-246.

190 Baumberger, T., Caroli, C., Perrin, B., Nonlinear-analysis of the Stick-slip Bifurcation in the Creep-controlled Regime of Dry Friction. Physics Review E, 51(5), 1995, pp. 4005-4010.

191 Baumberger T, Caroli C., (1998), Multicontact Solid Friction: A Macroscopic Probe of Pinning and Dissipation on the Mesoscopic Scale, MRS Bulletin, 23(6), 1998, pp. 41-46.

192 Chang, Y.P., Hashimura, M., Dornfeld, D.A., "An Investigation of the AE Signals in the Lapping Process," CIRP Annals, 45(1), 1996, pp. 331-334.

193 Tang, J., Dornfeld, D.A., Pangrele, S., Dangca, A., "In-process detection of Micro-scratching During CMP Using Acoustic Emission Sensing Technology", Proceedings TMS Annual Meeting, San Antonio, TX, 1998.

194 Hwang, I., "In-Situ Process Monitoring and Orientation Effects-Based Pattern Design for Chemical Mechanical Planarization", Ph.D. Dissertation, Mech. Eng. Department, University of California at Berkeley, 2004.

195 Choi, J., Lee, D.E., Dornfeld, D.A., "In-Situ Monitoring of a Copper CMP Operation with Acoustic Emission," 10th International Conference on CMP Planarization for ULSI Multilevel Interconnection, Fremont, CA, February 2005.

196 Hernandez, J. Wrschka, P., Oehrlein, G.S., "Surface Chemistry Studies of Copper Chemical Mechanical Planarization", Journal of the Electrochemical Society, 148 (7) G389-G397, 2001.

197 Liang, H, Martin, J., Lee, R. "Influence of Oxides on Friction During Cu CMP", J. of Electronic Materials, Vol. 30, No.4, 2001.

198 Hiratsuka, K., Bohno, A., Kurosawa, M., "Ultra-Low Friction Between Water Droplet and Hydrophobic Surface," Fundamentals of Tribology and Bridging the Gap between the Maco- and Micro/Nanoscales, B. Bhushan (ed.), Kluwer Academic Publishers, 2001, pp. 345.

199 Suzuki, K., Uyeda, Y., "Load-Carrying Capacity and Friction Characteristics of a Water Droplet on Hydrophobic Surfaces," Tribology Letters, Vol. 15 (2), 2003, pp. 77.

200 Dornfeld, D. A., Oliveira, J.F.G., Lee, D., Valente, C.M.O., "Analysis of Tool and Workpiece Interaction in Diamond Turning Using Graphical Analysis of Acoustic Emission," CIRP Annals, 52(1), 2003.

201 Donmez, M. A., Liu, C. R. and Barash, M. M. (1987), "A Generalized Mathematical Model for Machine Tool Errors," *Modeling, Sensing, and Control of Manufacturing Processes*, K. Srinivasan, *et al, eds.*, ASME, Book No. H00370, New York.

202 Donmez, M. A., Blomquist, D. S., Hocken, R. J., Liu, C. R. and Barash, M. M. (1986), "A Generalized Methodology for Machine Tool Accuracy Enhancement by Error Compensation," *Precision Engineering*, Vol. 4.

203 Gilsinn and Donmez, M. A. (1994), "Prediction of Geometric-Thermal Machine Tool Errors by Artificial Neural Networks," NISTIR 5367, NIST, Gaithersburg MD.

204 Ashby, M. F., "Materials Selection in Mechanical Design," Butterworth-Heinemann (Elsevier), Oxford, 1991, p. 190.

205 Bryan, J. B., "Closer Tolerances - Economic Sense," *Annals CIRP*, 19, 1971, pp. 115-120.

206 Nakayama, K., 1997, *Machining Science and Technology*, 1, 2, pg. 251-262.

207 Dornfeld, D, A, and Wright, P. K., "Process Planning for Agent-Based Precision Manufacturing," *Trans. North American Manufacturing Research Institute*, SME, Vol. 25, May 1997, pp. 359-364.

208 Stein, J. and Dornfeld, D. A., "Integrated Design and Manufacturing for Precision Mechanical Components," *Proc. PRIMECA*, Nantes, France, April, 1996.

209 Stein, J. and Dornfeld, D. A., "An Architecture for Integrated Design and Manufacturing for Precision Mechanical Components," *Trans. North American Manufacturing Research Institute*, Vol. 25, May 1997, pp. 249-254.

210 DeVor, R. E., Chang, T. H, and Sutherland, J. W., *Statistical Quality Design and Control,* Macmillan, New York, 1992, pp. 256-267.

211 Dornfeld, D. A., and Wright, P. K., "Intelligent Machining: Global Models, Local Scripts, and Validations" *Trans. North American Manufacturing Research Institute*, SME, Vol. 23, May 1995, pp. 351-356.

212 Wright, P. K. and Dornfeld, D. A. , "Agent-Based Manufacturing Systems," *Trans. North American Manufacturing Research Institute*, SME, Vol. 24, May 1996, pp. 241-246.

213 Shi, J., 2007, *Stream of Variation Modeling and Analysis for Multistage Manufacturing Processes*, CRC Press.

214 Westkamper, E., 1994, "Zero Defect Manufacturing by Means of a Learning Supervision of Process Chains," *CIRP Annals*, 43, 1, pp. 404-407.

215 Wright, P., (1971) "Metallurgical aspects of metal cutting with high speed steel tools, " Ph. D. Thesis, University of Birmingham.

216 Java is a trademark of Sun Microsystems Inc.

217 Betz, M. OMG's CORBA, *Dr. Dobb's Special Report*, Winter 1994/5.

218 Microsoft Corporation, OLE 2 Programmer's Reference: Creating programmable applications with OLE automation, Vol. 2. Microsoft Press, Redmond, Washington 1994.

219 Finin, T., Fritzson, R. and McKay, D., "A Language and Protocol to Support Intelligent Agent Interoperability" *Proc. CE&CALS Washington 1992 Conference*, June.

220 Frost, R., and Cutkosky, M. R. "An Agent Based Approach to Making Rapid Prototyping Processes Manifest to Designers" *Proc. ASME Design Technical Conference on Virtual Design and Manufacturing*, Irvine CA August 1996 On Compact Disc.

221 Sarma, S. E., Schofield, S., Stori, J. A., MacFarlane, J. and Wright, P. K., "Rapid Product Realization from Detail Design," *J. Computer Aided Design*, 23, 5, 1966, pp. 383-392.

222 Smith, C., and Wright, P. K. "CyberCut: A World Wide Web Based Design-to-Fabrication Tool," *The Journal of Manufacturing Systems*, December 1996.

223 Merchant, M. E., Dornfeld, D. A., and Wright, P., K., "Manufacturing - Its Evolution and Future," Trans. North American Manufacturing Research Institute, 2005, Vol. 33, pp. 211-218.

224 Newton, I., 1952, *Optiks*, Dover Publications, Inc., New York, based on the 4th Edition, London, 1730.

225 Nogowa, H., 1988, "Ceramic Processing- State of the Art of R&D in Japan", ASM Int'l, Metals Park OH.

226 Boothroyd, G., *Fundamentals of Metal Machining and Machine Tools*, McGraw-Hill, New York, 1975, p. 57.

227 Whitehouse, D. J., *Handbook of Surface Metrology*, Int. of Physics, London, 1994.

228 Tönshoff, H., Peters, J., Inasaki, I. and Paul, T., "Modelling and Simulation of Grinding Processes," *CIRP Annals*, 41, 2, 1992, p. 677.

229 Shaw, M. C., 1952, "Relationship between Fine Cutting and Shear Stress," Trans. ASME, 74, 1.

230 Lawn, B., and Wilshaw, R., "Indentation fracture: principles and applications," *J. Mat. Sci.*, 10, 1975, 1049-1081.

231 Brinksmeier, E. and Schmütz, J., 1997, "Ultraprecision machining- key to advanced products," *Proc. 8th IMEC*, JMTBA, Osaka, Japan, pp. 285-295.

232 Bifano, T., Dow, T. A., and Scattergood, R. O., 1991, "Ductile-regime grinding: a new technology for machining brittle materials," *Trans. ASME, J. Eng. Ind.*, 113, pp. 184-189.

233 Lawn, B. and Evans, A., "A model for crack initiation in elastic/plastic indentation fields," *J. Mat. Sci.*, 12, 1977, 2195-2199.

234 Lawn, B., Evans, A. and Marshall, D. B., "Elastic-plastic Indentation damage in ceramics: the median/radial crack system," *J. Am. Cer., Soc.*, 63, 1980, 574-581.

235 Hellmold, T. *et al.*, "Ductile mode cutting of silicon micro-parts," *Proc. ASPE Spring Topical Meeting*, Vol. 17, 1998, pp. 117-121.

236 Yoshikawa, H., "Brittle-ductile behavior of crystal surface in finishing," *J. JSPE*, 35, 1967, pp. 662-667 (in Japanese).

237 Moriwaki, T., Experimental Analysis of Ultraprecision Machining," *Int. J. Japan Society of Precision Engineering*, 29, 4, 1995, p. 287.

238 Brinksmeier, E., and Schmütz, J., 1998, "Generation and texture of surfaces in ultraprecision cutting of copper," *Mach. Sci. and Tech.*, 1, 2, pp. 185-193.

239 Courtesy of FANUC, Ltd., Japan.

240 Courtesy of Sodick Company, Japan.

241 Sriyotha, P., Nakamoto, K., Sugai, M., and Yamazaki, K., "A Design Study on, and the Development of, a 5-Axis Linear Motor Driven Super-Precision Machine," *Annals CIRP*, 55, 1, 2006, pp. 381-384.

242 Furakawa, Y., Moronuki, N., Effect of Material Properties on Ultra Precise Cutting Properties, CIRP Annals, 37, 1, pp. 113-116.

243 Ikawa, N., Shimada, S., Donaldson, R. R., Syn, C. K., Taylor, J. S., Ohmori, G., Tanaka H., and Yoshinaga, H., 1993, Chip morphology and minimum thickness of cut in micromachining," *J. Japan Soc. Prec. Eng'g.*, 59, 4, pp. 673-679 (in Japanese).

244 Blake, P. N. and Scattergood, R. O., 1990, "Ductile regime machining of germanium and silicon," *J. Amer. Ceram. Soc.*, 73, 4, p. 949.

245 Yan, J., Syoji, K., Suzuki, H., and Kuriyagawa, T., 1998, "Ductile regime turning of single crystal silicon with a straight-nosed diamond tool," *J. JSPE*, 64, 9, pp. 1345-1349 (in Japanese).

246 Nakajima, I., Uno, Y., Fujiwara, T., 1989, "Cutting mechanism of fine ceramics with a single point diamond," *Prec. Eng.*, 11, 1, pp.19-26.

247 Malkin, S., 1989, *Grinding Technology: Theory and Applications of Machining with Abrasives*, John Wiley, New York.

248 Subramanian, K., Ramanaath, S., and Tricard, M., 1997, "Mechanisms of material removal in the precision production grinding of ceramics," *Trans. ASME, J. Mfg. Sci. and Eng'g.*, 119, pp. 509-519.

249 Inasaki, I., 1998, "Challenges to advanced machining technology," Springer Professor Lecture Notes, University of California-Berkeley, Mechanical Engineering Department.

250 Kato, K., Hokkirigawa, K., Kayaba, T., and Endo, Y., 1986, "Three dimensional shape effect on abrasive wear," *J. Trib.*, 108, pp. 346-351.

251 Williams, J. A., 1999, "Wear modeling: analytical, computational and mapping - a continuum mechanics approach," *Wear*, 225-229, pp. 1-17.

252 Childs, T. H. C., 1988, "The mapping of metallic sliding wear," *Proc. Instn. Mech. Engrs.*, C202, pp. 379-95.

253 Gatzen, H. H., and Maetzig, J. C., 1997, "Nanogrinding," *Precision Engineering*, 21, pp. 134-139.

254 Komanduri, R., Lucca, D. A. and Tani, Y., 1998 , "Technological Advances in Fine Abrasive Processes," *Annals CIRP*, 47, 2, pp. 545-596.

255 Klocke, F., Gerent, O., and Wagemann, A., 1997, "Polishing of advanced ceramics," *Machining Science and Technology*, 1, 2, pp. 263-273.

256 Y. P. Chang, M. Hashimura, and D. A. Dornfeld, 1996, "An Investigation of the AE Signals in the Lapping Process," *CIRP Annals*, Vol. 45, No. 1, pp. 331-334.

257 Xie, Y., and Bhushan, B., 1996, "Effects of particle size, polishing pad and contact pressure in free abrasive polishing," *Wear 200*, pp. 281-295.

258 Kasai, T., Horio, K., Karak-Doy, T., and Kobayashi, A., 1990, "Improvement of conventional polishing conditions for obtaining super smooth surfaces of glass and metal works," *CIRP Annals*, 39, 1, pp. 321-324.

259 Watanabe, J., and Suzuki, J., 1981, "High precision polishing of semiconductor materials using hydrodynamic principle," *CIRP Annals*, 30, 1, pp. 91-95.

260 Namba, Y., and Tsuwa, H., 1977, "Ultra-fine finishing of sapphire single crystals," *CIRP Annals*, 26, 1, pp. 325-329.

261 Namba, Y., Tsuwa, H., and Wada, R., 1987, "Ultra-precision float polishing machine," *CIRP Annals,* 36, 1, pp. 211-214.

262 Stowers, I., Komanduri, R., and Baird, E. D., 1988, "Review of precision surface generating processes and their potential applications to the fabrication of large optical components," *SPIE 32nd Annual Int'l Tech. Symposium on Optical and Optoelectrical Applied Science and Engineering*.

263 Preston, F. W., 1927, "The theory and design of plate polishing machines," *J. Soc. Glass Tech.*, 11, pp. 214-256.

264 Steigerwald, J. M., Murarka, S. P., and R. Gutmann, 1997, *Chemical Mechanical Planarization of Microelectronic Materials*, Wiley and Sons, New York.

265 Evans, J., Paul, E., Dornfeld, D., Lucca, D., Byrne, G., Tricard, M., Klocke, F., Dambon, O., and Mullany, B., "Material Removal Mechanisms in Lapping and Polishing," STC "G" Keynote, CIRP Annals, 52, 2, pp. 611-633, 2003.

266 Runnels, S. R. and L. M. Eyman, 1994, "Tribology analysis of chemical mechanical polishing," *J. Electrochem. Soc.*, vol. 141, pp. 1698-1701.

267 Runnels, S. R., 1994, "Feature scale fluid based erosion modeling for chemical mechanical polishing," *J. Electrochem. Soc.*, Vol. 141, pp. 1900-1904.

268 Cook, L. M., 1990, "Chemical Processes in Glass Processing," *J. Non-Crystalline Solids*, Vol. 120, 152-171.

269 *2005 National Technology Roadmap*, Semiconductor Industry Association, San Jose, CA, 2007.

270 Bibby, T., Presentation on "Endpoint Detection for CMP" at the TMS Meeting, San Antonio, TX, February, 1998.

271 Stell, M., et al, 1994, "Planarization Ability of Chemical Mechanical Planarization (CMP) Process," *Material Research Society Symposium Proceeding*, Vol. 337, pp. 151-156.

272 Davari, B., et al., 1989, "A New Planarization Technique, Using a Combination of RIE and Chemical Mechanical Polish (CMP)," IEEE, *IEDM 89*, pp. 341-344.

273 Ali, I., et al., 1995, "Physical Characterization of Chemical Mechanical Planarized Surface for Trench Isolation," *J. Electrochem. Soc.*, Vol. 142, No. 9, pp. 3088-3092.

274 Kaufman, F. B., et al., 1991, "Chemical Mechanical Polishing for Fabricating Patterned W Metal Features as Chip Interconnects," *J. Electrochem. Soc.*, Vol. 138, No. 11, pp. 3460-3465

275 Phillips, K., Crimes, G. M., and T. R. Wilshaw, 1977, "On the Mechanism of Material Removal by Free Abrasive Grinding of Glass and Fused Silica," *Wear*, Vol 41, pp. 327-350.

276 Marshall, D. B., Lawn, B. R. and A. G. Evans, 1982, "Elastic/ Plastic Indentation Damage in Ceramics: The Lateral Crack System," *J. American Ceramic Society*, Vol 65, No 11, pp. 561-567.

277 Buijs, M. and K. Korpel-Van Houten, 1993, "A Model for Lapping of Glass," *Journal of Materials Science*, 28, pp. 3014-3020.

278 Buijs, M. and K. Korpel-Van Houten, 1993, "Three-Body Abrasion of Brittle Materials as Studied by Lapping," *Wear*, 166, pp. 237-245.

279 Heyboer, W., et al., 1991, "Chemomechanical Silicon Polishing," *J. Electrochem. Soc.*, Vol. 138, No. 8, pp. 774-77.

280 Burke, P., 1991, "Semi-empirical Modeling of SiO_2 Chemical Mechanical Polishing Planarization," *Proc. VMIC Conf.*, pp. 379-384.

281 Warnock, J., 1991, "A Two-dimensional Process Model for Chemimechanical Polish Planarization," *J. Electrochem. Soc.*, Vol. 138, No. 8, pp. 2398-2402.

282 Luo, J. and Dornfeld, D. A., Integrated Modeling of Chemical Mechanical Planarization for sub-micron IC Fabrication: from Particle Scale to Feature, Die and Wafer Scales, Springer-Verlag, Berlin Germany, 2004.

283 Wang, C., Sherman, P., Chandra, A., and Dornfeld. D., "Pad Surface Roughness and Slurry Particle Size Distribution Effects on Material Removal Rate in Chemical Mechanical Planarization, CIRP Annals, 54, 1, 2005, pp. 309-312.

284 N. Taniguchi, "Future trends of nanotechnology," *International Journal of the Japan Society for Precision Engineering*, Vol. 26, No. 1, pp. 1-7, March 1992.

285 Mori, Y., Ikawa, N., Okuda, T. and Yamagata, K., 1976, "Numerically controlled elastic emission machining," *Technology Reports of the Osaka University*, 26, pp. 283-294.

286 Tsuwa, H., Ikawa, N., Mori, Y. and Sugiyama, K., 1979, "Numerically controlled elastic machining," *Annals CIRP*, 28, 1, pp. 193-197.

287 Courtesty of IC Knowledge, http://www.icknowledge.com/economics/productcosts3.html

288 Cassidy, J., "The Force of an Idea," *New Yorker* ; Jan. 12, 1998, pp. 32-37

289 www.sia-online.org/iss_technology.cfm

290 P. Landler, International Semitech presentation, 1999.

291 ITRS Roadmap, Emerging Research Devices, 2003.

292 Franssila, S., *Introduction to Microfabrication*, Wiley, West Sussex, England, 2004.

293 Bhushan, B., ed., *Handbook of Nanotechnology*, 2nd ed., Springer, 2007.

294 Madou, M., *Fundamentals of Microfabrication*, *The Science of Miniaturization*, 2nd ed., CRC Press, 2002.

295 S. A. Campbell: The Science and Engineering of Microelectronic Fabrication, Oxford Univ. Press, New York, 2001.

296 C. J. Jaeger: Introduction to Microelectronic Fabrication, Prentice Hall, New Jersey, 2002.

297 J. D. Plummer, M. D. Deal, P. B. Griffin: Silicon VLSI Technology, Prentice Hall, New Jersey, 2000.

298 M. Gad-el-Hak (Ed.): The MEMS Handbook, CRC Press, Boca Raton, 2002.

299 T.-R. Hsu: MEMS and Microsystems Design and Manufacture, McGraw-Hill, New York, 2002.

300 Courtesy of IBM; www.ibm.com/chips/photolibrary

301 T. W. Zhang, S. W. Bates, and D. A. Dornfeld, "Operational Energy Use of Plasmonic Imaging Lithography," Proceedings of the IEEE International Symposium on Electronics and Environment (ISEE), May 2007.

302 http://www.lithoguru.com/scientist/lithobasics.html

303 Jaeger, Richard C., "Lithography", Introduction to Microelectronic Fabrication. Upper Saddle River: Prentice Hall, 2002.

304 britneyspears.ac/physics/fabrication/photolithography.htm

305 IC Knowledge, 2003, onversity.net/load/immersion _lithography.pdf

306 Courtesy of ASML Corporation.

307 http://www.memsnet.org/mems/what-is.html

308 www-bsac.eecs.berkeley.edu

309 George Harrison in the Introduction to Moore, W. R., Foundations of Mechanical Accuracy, Moore Special Tool Company, Bridgeport Connecticut, 1970.

310 McGraw-Hill Dictionary of Scientific and Technical Terms, Sixth ed., 2003.

311 R. P. Feynmann, "There's plenty of room at the bottom," Eng. Sci., 23, 1960, 22-36.

312 Jannik C. Meyer, A. K. Geim, M. I. Katsnelson, K. S. Novoselov, T. J. Booth and S. Roth, "The structure of suspended graphene sheets, Nature 446, 60-63 (1 March 2007).

313 M. S. Dresselhaus, G. Dresselhaus, P. C. Eklund: Science of Fullerenes and Carbon Nanotubes, Academic, San Diego, 1995.

314 http://www.applied nanotech.com/cntproperties.htm #Carbon%20Nanotube

315 L. Ming, C. Bao-qin, Y. Tian-Chun, Q. He, X. Qiuxia: The sub-micron fabrication technology, Proc. 6th Int. Conf. Solid-State and Integrated-Circuit Technol. (IEEE, 2001) 452-55.

316 S. Y. Chou: Nano-imprint lithography and lithographically induced self-assembly, MRS Bulletin 26 (2001) 512-17.

317 S. Y. Chou, C. Keimel, J. Gu: Ultrafast and direct imprint of nanostructures in silicon, Nature 417 (2002) 835-37.

318 M. A. Herman: Molecular Beam Epitaxy: Fundamentals and Current Status (Springer, New York 1996).

319 J. S. Frood, G. J. Davis, W. T. Tsang: Chemical Beam Epitaxy and Related Techniques (Wiley, New York 1997).

320 S. Mahajan, K. S. Sree Harsha: Principles of Growth and Processing of Semiconductors (McGraw-Hill, New York 1999).

321 S. Kim, M. Razegi: Advances in quantum dot structures. In: Processing and Properties of Compound Semiconductors, ed. by R. Willardson, H. S. Navawa (Academic Press, New York 2001).

322 D. Bimberg, M. Grundmann, N. N. Ledentsov: Quantum Dot Heterostructures (Wiley, New York 1999).

323 G. Seebohm, H. G. Craighead: Lithography and patterning for nanostructure fabrication. In: Quantum Semiconductor Devices and Technologies, ed. By T. P. Pearsall (Kluwer, Boston 2000).

324 E. Kapon: Lateral patterning of quantum well heterostructures by growth on nonplanar substrates. In: Epitaxial Microstructures, ed. by A. C. Gossard (Academic Press, New York 1994).

325 F. Guffarth, R. Heitz, A. Schliwa, O. Stier, N. N. Ledentsov, A. R. Kovsh, V. M. Ustinov, D. Bimberg: Strain engineering of self-organized InAs quantum dots, Phys. Rev. B 64 (2001) 085305(1)–085305(7).

326 M. Sugawara: Self-Assembled InGaAs/GaAs Quantum Dots (Academic Press, New York 1999).

327 B. C. Lee, S. D. Lin, C. P. Lee, H. M. Lee, J. C. Wu, K. W. Sun: Selective growth of single InAs quantum dots using strain engineering, Appl. Phys. Lett. 80 (2002) 326-328.

328 K. Brunner: Si/Ge nanostructures, Rep. Prog. Phys. 65 (2002) 27-72.

329 M. Calleja, J. Anguita, R. Garcia, K. Birkelund, F. Perez-Murano, J. A. Dagata: Nanometer-scale oxidation of silicon surfaces by dynamic force microscopy: reproducibility, kinetics and nanofabrication, Nanotechnology 10 (1999) 34-38.

330 E. S. Snow, P. M. Campbell, F. K. Perkins: Nanofabrication with proximal probes, Proc. IEEE 85 (1997) 601-611.

331 H. Sugimura, T. Uchida, N. Kitamura, H. Masuhara: Tip-induced anodization of titanium surfaces by scanning tunneling microscopy: a humidity effect on nanolithography, Appl. Phys. Lett. 63 (1993) 1288-1290.

332 N. Kramer, J. Jorritsma, H. Birk, C. Schonenberger: Nanometer lithography on silicon and hydrogenated amorphous silicon with low energy electrons, J. Vac. Sci. & Technol. B 13 (1995) 805-811.

333 H. T. Soh, K. W. Guarini, C. F. Quate: Scanning Probe Lithography (Kluwer, Boston 2001).

334 C. A. Mirkin: Dip-pen nanolithography: automated fabrication of custom multicomponent, sub-100 nanometer surface architectures, MRS Bulletin 26 (2001) 535-538.

335 L. L. Sohn, R. L. Willett: Fabrication of nanostructures using atomic-force-microscope-based lithography, Appl. Phys. Lett. 67 (1995) 1552-1554.

336 H. J. Mamin, B. D. Terris, L. S. Fan, S. Hoen, R. C. Barrett, D. Rugar: High-density data storage using proximal probe techniques, IBM J. Res. & Dev. 39 (1995) 681-699.

337 K. Bessho, S. Hashimoto: Fabricating nanoscale structures on Au surface with scanning tunneling microscope, Appl. Phys. Lett. 65 (1994) 2142-2144.

338 I. W. Lyo, P. Avouris: Field-induced nanometer-to atomic-scale manipulation of silicon surfaces with the STM, Science 253 (1991) 173-176.

339 M. F. Crommie, C. P. Lutz, D. M. Eigler: Confinement of electrons to quantum corrals on a metal surface, Science 262 (1993) 218-220.

340 A. de Lozanne: Pattern generation below 0.1 micron by localized chemical vapor deposition with the scanning tunneling microscope, Japan. J. Appl. Physic 33 (1994) 7090-7093.

341 L. A. Nagahara, T. Thundat, S. M. Lindsay: Nanolithography on semiconductor surfaces under an etching solution, Appl. Phys. Lett. 57 (1990) 270-272.

342 T. Thundat, L. A. Nagahara, S. M. Lindsay: Scanning tunneling microscopy studies of semiconductor Part A 5 electrochemistry, J. Vac. Sci. & Technol. A 8 (1990) 539-543.

343 S. C. Minne, S. R. Manalis, A. Atalar, C. F. Quate: Independent parallel lithography using the atomic force microscope, J. Vac. Sci. & Technol. B 14 (1996) 2456-2461.

344 M. Lutwyche, C. Andreoli, G. Binnig, J. Brugger, U. Drechsler, W. Haeberle, H. Rohrer, H. Rothuizen, P. Vettiger: Microfabrication and parallel operation of 5×5 2D AFM cantilever arrays for data storage and imaging, Proc. MEMS '98 (1998), 8-11.

345 G. M. Whitesides, B. Grzybowski: Self-assembly at all scales, Science 295 (2002) 2418-2421.

346 P. Kazmaier, N. Chopra: Bridging size scales with self-assembling supramolecular materials, MRS Bulletin 25 (2000) 30-35.

347 R. Plass, J. A. Last, N. C. Bartelt, G. L. Kellogg: Selfassembled domain patterns, Nature 412 (2001) 875.

348 Y. A. Vlasov, X. -Z. Bo, J. G. Sturm, D. J. Norris: On-chip natural self-assembly of silicon photonic bandgap crystals, Nature 414 (2001) 289-293.

349 C. Gigault, D.-K. Veress, J. R. Dutcher: Changes in the morphology of self-assembled polystyrene microsphere monolayers

produced by annealing, J. Colloid Interface Sci. 243 (2001) 143-155.

350 J. C. Hulteen, P. Van Duyne: Nanosphere lithography: a materials general fabrication process for periodic particle array surfaces, J. Vac. Sci. Technol. A 13 (1995) 1553-1558.

351 J. D. Joannopoulos, P. R. Villeneuve, S. Fan: Photonic crystals: putting a new twist on light, Nature 386 (1997) 143-149.

352 T. D. Clark, R. Ferrigno, J. Tien, K. E. Paul, G. M. Whitesides: Template-directed self-assemblyof 10-μm-sized hexagonal plates, J. Am. Chem. Soc. 124 (2002) 5419-5426.

353 S. A. Sapp, D. T. Mitchell, C. R. Martin: Using template-synthesized micro-and nanowires as building blocks for self-assembly of supramolecular architectures, Chem. Mater. 11 (1999) 1183-1185.

354 Y. Yin, Y. Lu, B. Gates, Y. Xia: Template assisted self-assembly: a practical route to complex aggregates of monodispersed colloids with well-defined sizes, shapes and structures, J. Am. Chem. Soc. 123 (2001) 8718-8729.

355 J. L. Wilbur, G. M. Whitesides: Self-assembly and self-assembles monolayers in micro and nanofabrication. In: Nanotechnology, ed. by G. Timp (Springer, New York 1999).

356 S. R. Wasserman, Y. T. Tao, G.M. Whitesides: Structure and reactivity of alkylsiloxane monolayers formed by reaction of alkyltrichlorosilanes on silicon substrates, Langmuir 5 (1989) 1074-1087.

357 C. P. Tripp, M. L. Hair: An infrared study of the reaction of octadecyltrichlorosilane with silica, Langmuir 8 (1992) 1120-1126.

358 D. R. Walt: Nanomaterials: top-to-bottom functional design, Nature 1 (2002) 17-18.

359 J. Noh, T. Murase, K. Nakajima, H. Lee, M. Hara: Nanoscopic investigation of the self-assembly processes of dialkyl disulfides and dialkyl sulfides on Au(111), J. Phys. Chem. B 104 (2000) 7411-7416.

360 M. Himmelhaus, F. Eisert, M. Buck, M. Grunze: Self-assembly of n-alkanethiol monolayers: a study by IR-visible sum frequency spectroscopy (SFG), J. Phys. Chem. 104 (2000) 576-584.

361 Carbon Nanotechnologies, Inc. www.azonano.com/details.asp?ArticleID=980#_Other_Applications.

362 Hansen, H, Carneiro, K. Haitjema, H., De Chiffre, L., Dimensional Micro and Nano Metrology," Annals CIRP, 55, 2, 2006, pp. 721-743.

363 Masuzawa, T., Toenshoff, H. K., 1997, Three-Dimensional Micromachining by Machine Tools, CIRP Annals, 46/2:621-628.

364 Masuzawa, T., 2000, State of the Art of Micromachining, CIRP Annals, 49/2:473-488.

365 Alting, L., Kimura, F., Hansen, H. N., Bissacco, G., 2003, Micro Engineering, CIRP Annals, 52/2:635-657.

366 Liu, X., DeVor, R. E., Kapoor, S. G., Ehmann, K. F., 2004, The Mechanics of Machining at the Microscale: Assessment of the Current State of the Science, Journal of Manufacturing Science and Engineering, Transactions of the ASME, 126/4:666-678.

367 Ehmann, K. F., Bourell, D., Culpepper, M. L., Hodgson, T. J., Kurfess, T. R., Madou, M., Rajurkar, K., DeVor, R. E., 2005, International Assessment of Research and Development in Micromanufacturing, WTEC Panel Report, World Technology Evaluation Center, Inc., Baltimore, MA.

368 Dornfeld, D., Min, S., and Takeuchi, Y., "Recent Advances in Mechanical Micromachining," CIRP Annals, 55, 2, 2006, pp. 745-768.

369 Weinert, K., Kahnis, P., Petzoldt, V., Peters, C., 2005, Micro-Milling of Steel and Niti SMA, 55th CIRP General Assembly, STC-C section meeting presentation file, Antalya, Turkey.

370 Kobayashi, A., Hoshina, N., Tsukada, T., Ueda, K., 1978, High Precision Cutting with a New Ultra Precision Spindle, CIRP Annals, 27/1:283-287.

371 Takasu, S., Masuda, M., Nishiguchi, T., 1985, Influence of Study Vibration with Small Amplitude Upon Surface Roughness in Diamond Machining, CIRP Annals, 34/1:463-467.

372 Zhang, G. M., Kapoor, S. G., 1991, Dynamic Generation of Machined Surfaces, Part 2. Construction of Surface Topography, Journal of Engineering for Industry, Transactions of the ASME, 113/2:145-153.

373 Zhang, G. M., Kapoor, S. G., 1991, Dynamic Generation of Machined Surfaces, Part 1. Description of a Random Excitation System, Journal of Engineering for Industry, Transactions of the ASME, 113/2:137-144.

374 Vogler, M. P., DeVor, R. E., Kapoor, S. G., 2004, On the Modeling and Analysis of Machining Performance in Micro-Endmilling, Part I: Surface Generation, J. Manufacturing Science and Engineering, Trans. ASME, 126/4:685-694.

375 Lee, K., Dornfeld, D. A., 2002, An Experimental Study on Burr Formation in Micro Milling Aluminum and Copper, Trans. NAMRI/SME, 30:1-8.

376 Min, S., Lee, D., De Grave, A., De Oliveira Valente, C. M., Lin, J., Dornfeld, D., 2004, Surface and Edge Quality Variation in Precision Machining of Single Crystal and Polycrystalline Materials,

Proceedings of the 7th Int'l Conference on Deburring and Surface Finishing, Berkeley, CA, USA, 341-350.

377 Lee, D., Deichmuller, M., Min, S., Dornfeld, D. A., 2005, Variation in Machinability of Single Crystal Material in Micromechanical Machining, Proceedings of the Nanoengineering Symposium 2005, Daejeon, Korea.

378 Sato, M., Kato, Y., Tsutiya, K., 1979, Effects of Crystal Orientation on the Flow Mechanism in Cutting Aluminum Single Crystal, Transactions of the Japan Institute of Metals, 20/8:414-422.

379 Bissacco, G., Hansen, H.N., De Chiffre, L., 2005, Micromilling of Hardened Tool Steel for Mould Making Applications, J. Materials Processing Technology, 167/2-3:201-207.

380 Ahn, J., Lim, H., 1997, Side Burr Generation Model in Micro-Grooving, Proceedings of ASPE, 16:215-219.

381 Ahn, J., Lim, H., Son, S., 2000, Burr and Shape Distortion in Micro-Grooving of Non-Ferrous Metals Using a Diamond Tool, KSME International Journal, 14/11:1244-1249.

382 Schaller, T., Bohn, L., Mayer, J., Schubert, K., 1999, Microstructure Grooves with a Width of Less Than 50 Micro Cut with Ground Hard Metal Micro End Mills, Precision Engineering, 23/4:229-235.

383 Sugawara, A., Inagaki, K., 1982, Effect of Workpiece Structure on Burr Formation in Micro-Drilling, Precision Engineering, 4/1:9-14.

384 Ueda, K., Sugita, T., Hiraga, H., 1991, J-Integral Approach to Material Removal Mechanisms in Microcutting of Ceramics, CIRP Annals, 40/1:61-64.

385 Ueda, K., Manabe, K., 1992, Chip Formation Mechanism in Microcutting of an Amorphous Metal, CIRP Annals, 41/1:129-132.

386 Moriwaki, T., Sugimura, N., Luan, S., 1993, Combined Stress, Material Flow and Heat Analysis of Orthogonal Micromachining of Copper, CIRP Annals, 42/1:75-78.

387 Fleischer, J., Kotschenreuther, J., Loehe, D., Gumbsch, P., Schulze, V., Delonnoy, L., Hochrainer, T., 2005, An Integrated Approach to the Modeling of Size-Effects in Machining with Geometrically Defined Cutting Edges, Proceedings of the CIRP Modeling Workshop, Chemnitz, Germany, 123-129.

388 Chuzhoy, L., DeVor, R. E., Kapoor, S. G., 2003, Machining Simulation of Ductile Iron and Its Constituents, Part 2: Numerical Simulation and Experimental Validation of Machining, J. Manufacturing Science and Engineering, Trans. ASME, 125/2:192-201.

389 Chuzhoy, L., DeVor, R. E., Kapoor, S. G., Beaudoin, A. J., Bammann, D. J., 2003, Machining Simulation of Ductile Iron and Its Constituents, Part 1: Estimation of Material Model Parameters and Their Validation, J. Manufacturing Science and Engineering, Trans. ASME, 125/2:181-191.

390 Park, S., Kapoor, S. G., DeVor, R. E., 2004, Mechanistic Cutting Process Calibration Via Microstructure-Level Finite Element Simulation Model, J. Manufacturing Science and Engineering, Trans. ASME, 126/4:706-709.

391 Stowers, I. F., Belak, J. F., Lucca, D. A., Komanduri, R., Moriwaki, T., Okuda, K., Ikawa, N., Shimada, S., Tanaka, H., Dow, T. A., Drescher, J. D., 1991, Molecular - Dynamics Simulation of the Chip Forming Process in Single Crystal Copper and Comparison with Experimental Data, Proceedings of the ASPE Annual Meeting, 100-104.

392 Ikawa, N., Shimada, S., Tanaka, H., Ohmori, G., 1991, Atomistic Analysis of Nanometric Chip Removal as Affected by Tool-Work Interaction in Diamond Turning, CIRP Annals, 40/1:551-554.

393 Inamura, T., Suzuki, H., Takazawa, N., 1990, Cutting Experiments in a Computer Using Atomic Model of a Copper Crystal and a Diamond Tool, Journal of the JSPE, 56:1480-1486.

394 Shimada, S., Ikawa, N., Ohmori, G., Tanaka, H., Uchikoshi, U., 1992, Molecular Dynamics Analysis as Compared with Experimental Results of Micromachining, CIRP Annals, 41/1:117-120.

395 Stowers, I. F., 1990, Molecular Dynamics Modeling Applied to Indentation and Metal Cutting, Informal Video Presentation at CIRP Meeting, Berlin, Germany.

396 Inamura, T., Takezawa, N., Kumaki, Y., 1993, Mechanics and Energy Dissipation in Nanoscale Cutting, CIRP Annals, 42/1:79-82.

397 Inamura, T., Takezawa, N., Taniguchi, N., 1992, Atomic-Scale Cutting in a Computer Using Crystal Models of Copper and Diamond, CIRP Annals, 41/1:121-124.

398 Shimada, S., Ikawa, N., Tanaka, H., Ohmori, G., Uchikoshi, J., 1993, Molecular Dynamics Analysis of Cutting Force and Chip Formation Process in Microcutting, Seimitsu Kogaku Kaishi/Journal of the Japan Society for Precision Engineering, 59/12:2015-2021.

399 Shimada, S., Ikawa, N., Tanaka, H., Uchikoshi, J., 1994, Structure of Micromachined Surface Simulated by Molecular Dynamics Analysis, CIRP Annals, 43/1:51-54.

400 Chandrasekaran, N., Khajavi, A. N., Raff, L. M., Komanduri, R., 1998, New Method for Molecular Dynamics Simulation of Nanometric Cutting, Philosophical Magazine B: Physics of Condensed Matter; Statistical Mechanics, Electronic, Optical and Magnetic Properties, 77/1:7-26.

401 Komanduri, R., Chandrasekaran, N., Raff, L. M., 1998, Effect of Tool Geometry in Nanometric Cutting: A Molecular Dynamics Simulation Approach, Wear, 219/1:84-97.

402 Komanduri, R., Chandrasekaran, N., Raff, L. M., 1999, Orientation Effects in Nanometric Cutting of Single Crystal Materials: An MD Simulation Approach, CIRP Annals, 48/1:67-72.

403 Komanduri, R., Chandrasekaran, N., Raff, L. M., 2000, M. D. Simulation of Nanometric Cutting of Single Crystal Aluminum-Effect of Crystal Orientation and Direction of Cutting, Wear, 242/1: 60-88.

404 Komanduri, R., Chandrasekaran, N., Raff, L. M., 2001, Molecular Dynamics Simulation of the Nanometric Cutting of Silicon, Philosophical Magazine B: Physics of Condensed Matter; Statistical Mechanics, Electronic, Optical and Magnetic Properties, 81/12: 1989-2019.

405 Komanduri, R., Chandrasekaran, N., Raff, L. M., 2001, MD Simulation of Exit Failure in Nanometric Cutting, Materials Science and Engineering A, 311/1-2:1-12.

406 Komanduri, R., Narulkar, R., Raff, L. M., 2004, Monte Carlo Simulation of Nanometric Cutting, Philosophical Magazine, 84/11: 1155-1183.

407 Cheng, K., Luo, X., Ward, R., Holt, R., 2003, Modeling and Simulation of the Tool Wear in Nanometric Cutting, Wear, 255/ 7-12:1427-1432.

408 Luo, X., Cheng, K., Guo, X., Holt, R., 2003, An Investigation on the Mechanics of Nanometric Cutting and the Development of Its Test-Bed, International Journal of Production Research, 41/7:1449-1465.

409 Rentsch, R., Inasaki, I., 1995, Investigation of Surface Integrity by Molecular Dynamics Simulation, CIRP Annals, 44/1:295-298.

410 Fang, F. Z., Wu, H., Liu, Y. C., 2005, Modelling and Experimental Investigation on Nanometric Cutting of Monocrystalline Silicon, International Journal of Machine Tools and Manufacture, 45/15: 1681-1686.

411 Inamura, T., Takezawa, N., Kumaki, Y., Sata, T., 1994, On a Possible Mechanism of Shear Deformation in Nanoscale Cutting, CIRP Annals, 43/1:47-50.

412 Dautzenberg, J. H., Kals, J. A. G., Van der Wolf, A. C. H., 1983, Forces and Plastic Work in Cutting, CIRP Annals, 32/1:223-227.

413 Dautzenberg, J. H., Veenstra, P. C., van der Wolf, A. C. H., 1981, Minimum Energy Principle for the Cutting Process in Theory and Experiment, CIRP Annals, 30/1:1-4.

414 Merchant, M. E., 1945, Mechanics of Metal Cutting Process, and Type 2 Chip, Journal of Applied Physics, 16/5:267-275.

415 Merchant, M. E., 1945, Mechanics of Metal Cutting Process, and Type 2 Chip, Journal of Applied Physics, 16/6:318-324.

416 Kim, J. D., Kim, D. S., 1996, On the Size Effect of Micro-Cutting Force in Ultraprecision Machining, JSME International Journal, Series C, 39/1:164-169.

417 Bao, W. Y., Tansel, I. N., 2000, Modeling Micro-End-Milling Operations. Part II: Tool Run-Out, International Journal of Machine Tools and Manufacture, 40/15:2175-2192.

418 Bao, W. Y., Tansel, I. N., 2000, Modeling Micro-End-Milling Operations. Part III: Influence of Tool Wear, International Journal of Machine Tools and Manufacture, 40/15:2193-2211.

419 Bao, W. Y., Tansel, I. N., 2000, Modeling Micro-End-Milling Operations. Part I: Analytical Cutting Force Model, International Journal of Machine Tools and Manufacture, 40/15:2155-2173.

420 Tlusty, J., MacNeil, P., 1975, Dynamics of Cutting Forces in End Milling, CIRP Annals, 24/1:21-25.

421 Vogler, M. P., Kapoor, S. G., DeVor, R. E., 2004, On the Modeling and Analysis of Machining Performance in Micro-Endmilling, Part II: Cutting Force Prediction, J. Manufacturing Science and Engineering, Trans. ASME, 126/4:695-705.

422 Vogler, M. P., DeVor, R. E., Kapoor, S. G., 2003, Microstructure-Level Force Prediction Model for Micro-Milling of Multi-Phase Materials, J. Manufacturing Science and Engineering, Trans. ASME, 125/2:202-209.

423 Kim, C. -J., Mayor, J. R., Ni, J., 2004, A Static Model of Chip Formation in Microscale Milling, J. Manufacturing Science and Engineering, Trans. ASME, 126/4:710-718.

424 Joshi, S. S., Melkote, S. N., 2004, An Explanation for the Size-Effect in Machining Using Strain Gradient Plasticity, J. Manufacturing Science and Engineering, Trans. ASME, 126/4:679-684.

425 Klocke, F., Weck, M., Fischer, S., Özmeral, H., Schroeter, R. B., Zamel, S., 1996, Ultrapräzisionsbearbeitung Und Fertigung Von Mikrokomponenten, Verfahren, IDR3:172-177.

426 Casstevens, J. M., 1983, Diamond Turning of Steel in Carbon-Saturated Atmospheres, Precision Engineering, 5/1:9-15.

427 Evans, C., 1991, Cryogenic Diamond Turning of Stainless Steel, CIRP Annals, 40/1:571-575.

428 Brinksmeier, E., Glaebe, R., Osmer, J., 2006, Ultra-Precision Diamond Cutting of Steel Molds, CIRP Annals, 55/1.

429 Moriwaki, T., Shamoto, E., 1991, Ultraprecision Diamond Turning of Stainless Steel by Applying Ultrasonic Vibration, CIRP Annals, 40/1:559-562.

430 Moriwaki, T., Shamoto, E., 1995, Ultrasonic Elliptical Vibration Cutting, CIRP Annals, 44/1:31-34.

431 Weule, H., Huntrup, V., Tritschler, H., 2001, Micro-Cutting of Steel to Meet New Requirements in Miniaturization, CIRP Annals, 50/1:61-64.

432 Schaller, T., Heckele, M., Ruprecht, R., Schubert, K., 1999, Microfabrication of a Mold Insert Made of Hardened Steel and First Molding Results, Proceedings of the ASPE, 20:224-227.

433 Schaller, T., Mayer, J., Schubert, K., 1999, Approach to a Microstructured Mold Made of Steel, Proceedings of the EUSPEN, Bremen, Germany, 1:238-241.

434 Friedrich, C. R., Vasile, M. J., 1996, Development of the Micromilling Process for High-Aspect-Ratio Microstructures, Journal of Microelectromechanical Systems, 5/1:33-38.

435 Fahrenberg, J., Schaller, T., Bacher, W., El-Kholi, A., Schomburg, W.K., 1996, High Aspect Ratio Multi-Level Mold Inserts Fabricated by Mechanical Micro Machining and Deep Etch X-Ray Lithography, Microsystem Technologies 2/4:174-177.

436 Fleischer, J., Haupt, S., 2005, Microstructuring of Hardened Steels by Combining Laser Ablation and EDM - a Comparison with Micromilling, 55th CIRP General Assembly, STC-C section meeting presentation file, Antalya, Turkey.

437 Fleischer, J., Kotschenreuther, J., 2005, Manufacturing of Micro Molds by Conventional and Energy Assisted Processes, Proceedings of the 4M Conference, Research Center Karlsruhe, Karlsruhe.

438 Bissacco, G., Hansen, H. N., Tang, P. T., Fugl, J., 2005, Precision Manufacturing Methods of Inserts for Injection Molding of Microfluidic Systems, Micro Nano Workshop.

439 Sawada, K., Kawai, T., Takeuchi, Y., Sata, T., 2000, Development of Ultraprecision Micro Grooving (Manufacture of V-Shaped Groove), JSME International Journal. Series C: Mechanical Systems, Machine Elements and Manufacturing, 43/1:170-176.

440 Unpublished data from Y. Takeuchi, personal communication.

441 Takeuchi, Y., "Ultraprecision micromachining of complicated shapes by rotational and non-rotational diamond cutting tool," Presentation at STC C, CIRP General Assembly, Antalya, Turkey, August, 2005.

442 Takeuchi, Y., Maeda, S., Kawai, T., Sawada, K., 2002, Manufacture of Multiple-Focus Micro Fresnel Lenses by Means of Nonrotational Diamond Grooving, CIRP Annals, 51/1:343-346.

443 Maeda, S., Takeuchi, Y., Sawada, K., Kawai, T., Sata, T., 2000, Creation of Ultraprecision Microgrooves Using Non-Rotational Cutting Tools, Journal of the JSPE, 66/9:1456-1460.

444 Takeuchi, Y., Murota, M., Kawai, T., Sawada, K., 2003, Creation of Flat-End V-Shaped Microgrooves by Non-Rotational Cutting Tools, CIRP Annals, 52/1:41-44.

445 Takeuchi, Y., Sawada, K., Kawai, T., 1997, Three-Dimensional Micromachining by Means of Ultraprecision Milling, Proceedings of the 9th Int. Prec. Eng. Seminar and 4th Int. Conf. on Ultraprecision in Manuf. Eng., Braunschweig, 596-599.

446 Sasaki, T., Takeuchi, Y., Kawai, T., Sakaida, Y., 2004, 5-Axis Control Ultraprecision Micromachining of Micro 3-D Body, Proceedings of the Spring Annual Meeting of JSPE, 1075-1076.

447 Yu, Z. Y., Rajurkar, K. P., Tandon, A., 2004, Study of 3D Micro-Ultrasonic Machining, J. Manufacturing Science and Engineering, Trans. ASME, 126/4:727-732.

448 Kumabe, J., 1979, Vibration Cutting, Jikkyou Publishing Co., Tokyo, Japan.

449 Ohnishi, O., Onikura, H., 2003, Effects of Ultrasonic Vibration on Microdrilling into Inclined Surface, Journal of the JSPE, 69/9:1337-1341.

450 Ohnishi, O., Onikura, H., Feng, J., Kanda, T., Morita, T., 1996, Effects of Ultrasonic Vibration on Machining Accuracy in Microdrilling, Journal of the JSPE, 62/5:676-680.

451 Egashira, K., Mizutani, K., Nagao, T., 2002, Ultrasonic Vibration Drilling of Microholes in Glass, CIRP Annals, 51/1:339-342.

452 Egashira, K., Masuzawa, T., 1999, Microultrasonic Machining by the Application of Workpiece Vibration, CIRP Annals, 48/1:131-134.

453 Moriwaki, T., Shamoto, E., Song, Y., Kohda, S., 2004, Development of a Elliptical Vibration Milling Machine, CIRP Annals, 53/1:341-344.

454 Shamoto, E., Moriwaki, T., 1994, Study on Elliptical Vibration Cutting, CIRP Annals, 43/1:35-38.

455 Shamoto, E., Moriwaki, T., 1999, Ultraprecision Diamond Cutting of Hardened Steel by Applying Elliptical Vibration Cutting, CIRP Annals, 48/1:441-444.

456 Shamoto, E., Song, Y., Yoshida, H., Suzuki, N., Moriwaki, T., Kohda, S., Yamanishi, T., 2003, Development of Elliptical Vibration Cutting Machine by Utilizing Mechanical Vibrator, Journal of the JSPE, 69/4:542-543.

457 Shamoto, E., Suzuki, N., Moriwaki, T., Naoi, Y., 2002, Development of Ultrasonic Elliptical Vibration Controller for Elliptical Vibration Cutting, CIRP Annals, 51/1:327-330.

458 Brinksmeier, E., Malz, R., Riemer, O., 1996, Mikrozerspanung Duktiler Und Sproeder Werkstoffe in Optischer Qualitaet, VDI Berichte, 1276:229.

459 Weck, M., Fischer, S., Vos, M., 1997, Fabrication of Microcomponents Using Ultraprecision Machine Tools, Nanotechnology, 8/3:145-148.

460 Schulze, V., Löhe, D., Wick, A., 1999, Influence of Heat Treatment State on Material Changes Close to Surface of Micromachined Steel SAE 1045, Proceedings of the EUSPEN, Bremen, Germany.

461 Spath, D., Hüntrup, V., 1999, Micro-Milling of Steel for Mold Manufacturing - Influences of Material, Tools and Process Parameters, Proceedings of the EUSPEN, Bremen, Germany.

462 Weinert, K., Guntermann, G., Schwietering, C., 1998, Mirofräsbearbeitung Schwerzerspanbarer Werkstoffe, Werkstattstechnik 88, Heft 11/12:503-506.

463 Weule, H., Schmidt, J., Huntrup, V., Tritschler, H., 1999, Micromilling of Ferrous Materials, Production Engineering, 6/2:17-20.

464 Vasile, M. J., Friedrich, C. R., Kikkeri, B., McElhannon, R., 1996, Micrometer-Scale Machining: Tool Fabrication and Initial Results, Precision Engineering, 19/2-3:180-186.

465 Adams, D. P., Vasile, M. J., Krishnan, A. S. M., 2000, Microgrooving and Microthreading Tools for Fabricating Curvilinear Features, Precision Engineering, 24/4:347-356.

466 Adams, D. P., Vasile, M. J., Benavides, G., Campbell, A. N., 2001, Micromilling of Metal Alloys with Focused Ion Beam-Fabricated Tools, Precision Engineering, 25/2:107-113.

467 Aoki, I., Takahashi, T., 1999, Micropattern Fabrication by Specially Designed Micro Tool, Proceedings of SPIE - The International Society for Optical Engineering, 3874:365-372.

468 Egashira, K., Mizutani, K., 2003, Milling Using Ultra-Small Diameter Ball End Mills Fabricated by Electrical Discharge Machining, Journal of the JSPE, 69/1:1449-1453.

469 Egashira, K., Mizutani, K., 2002, Micro-Drilling of Monocrystalline Silicon Using a Cutting Tool, Precision Engineering, 26/3:263-268.

470 Friedrich, C., Coane, P., Goettert, J., Gopinathin, N., 1998, Direct Fabrication of Deep X-Ray Lithography Masks by Micromechanical Milling, Precision Engineering, 22/3:164-173.

471 Masuzawa, T., Fujino, M., 1990, Process for Manufacturing Very Fine Pin Tools, SME Technical Papers, MS90-307.

472 Onikura, H., Ohnishi, O., Take, Y., 2000, Fabrication of Micro Carbide Tools by Ultrasonic Vibration Grinding, CIRP Annals, 49/1: 257-260.

473 Ohmori, H., Katahira, K., Uehara, Y., Watanabe, Y., Lin, W., 2003, Improvement of Mechanical Strength of Micro Tools by Controlling Surface Characteristics, CIRP Annals, 52/1:467-470.

474 Uhlmann, E., Schauer, K., 2005, Dynamic Load and Strain Analysis for the Optimization of Micro End Mills, CIRP Annals, 54/1:75-78.

475 Tansel, I., Rodriguez, O., Trujillo, M., Paz, E., Li, W., 1998, Micro-End-Milling - I. Wear and Breakage, International Journal of Machine Tools & Manufacture, 38/12:1419-1436.

476 Tansel, I. N., Arkan, T. T., Bao, W. Y., Mahendrakar, N., Shisler, B., Smith, D., McCool, M., 2000, Tool Wear Estimation in

Micro-Machining. Part I: Tool Usage-Cutting Force Relationship, Intl Journal of Machine Tools and Manufacture, 40/4:599-608.

477 Chen, W. S., Ehmann, K. F., 1994, Experimental Investigation on the Wear and Performance of Micro-Drills, Proceedings of the 1994 International Mechanical Engineering Congress and Exposition, Tribology in Manufacturing Processes, Chicago, IL, USA, 30:145-157.

478 Sugano, T., Takeuchi, K., Goto, T., Yoshida, Y., 1987, Diamond Turning of an Aluminum Alloy for Mirror, CIRP Annals, 36/1:17-20.

479 Fang, F. Z., Wu, H., Liu, X.D., Liu, Y. C., Ng, S. T., 2003, Tool Geometry Study in Micromachining, Journal of Micromechanics and Microengineering, 13/5:726-731.

480 Godlinski, D., Grathwohl, G., Kuntz, M., 1999, Development of Ceramic Micro-Tools for Precision Machining, Proceedings of the EUSPEN, Bremen, Germany, 298-301.

481 Gaebler, J., Schaefer, L., Westermann, H., 2000, Chemical Vapour Deposition Diamond Coated Microtools for Grinding, Milling and Drilling, Diamond and Related Materials, 9/3:921-924.

482 Huang, B.-W., 2004, The Drilling Vibration Behavior of a Twisted Microdrill, Journal of Manufacturing Science and Engineering, Transactions of the ASME, 126/4:719-726.

483 Chyan, H. C., Ehmann, K. F., 1998, Development of Curved Helical Micro-Drill Point Technology for Micro-Hole Drilling, Mechatronics, 8/4:337-358.

484 Chyan, H. C., Ehmann, K. F., 2002, Curved Helical Drill Points for Microhole Drilling, Proceedings of the Institution of Mechanical Engineers, Part B: Journal of Engineering Manufacture, 216/1:61-75.

485 Hinds, B. K., Treanor, G. M., 2000, Analysis of Stresses in Micro-Drills Using the Finite Element Method, International Journal of Machine Tools and Manufacture, 40/10:1443-1456.

486 Yang, Z., Li, W., Chen, Y., Wang, L., 1998, Study for Increasing Micro-Drill Reliability by Vibrating Drilling, Reliability Engineering & System Safety, 61/3:229-233.

487 Yang, Z., Tan, Q., Wang, L., 2002, Principle of Precision Micro-Drilling with Axial Vibration of Low Frequency, International Journal of Production Research, 40/6:1421-1427.

488 Friedrich, C., Coane, P., Goettert, J., Gopinathin, N., 1997, Precision of Micromilled X-Ray Masks and Exposures, Microsystem Technologies, 4/1:21-24.

489 Umeda, A., 1996, Review on the Importance of Measurement Technique in Micromachine Technology, Proceedings of SPIE - The International Society for Optical Engineering, 2880:26-38.

490 McGeough, J. A., 2002, Micromachining of Engineering Materials, Marcel Dekker, New York.

491 Howard, L. P., Smith, S. T., 1994, Metrological Constant Force Stylus Profiler, Review of Scientific Instruments, 65/4 pt 1:892-901.

492 Masuzawa, T., Hamasaki, Y., Fujino, M., 1993, Vibroscanning Method for Nondestructive Measurement of Small Holes, CIRP Annals, 42/1:589-592.

493 Kim, B., Masuzawa, T., Bourouina, T., 1999, Vibroscanning Method for the Measurement of Micro-Hole Profiles, Measurement Science and Technology, 10/8:697-705.

494 Miyoshi, T., Takaya, Y., Saito, K., 1996, Micromachined Profile Measurement by Means of Optical Inverse Scattering Phase Method, CIRP Annals, 45/1:497-500.

495 Kim, S.-W., Rhee, H.-G., Chu, J.-Y., 2003, Volumetric Phase-Measuring Interferometer for Three-Dimensional Coordinate Metrology, Precision Engineering, 27/2:205-215.

496 Pril, W. O., Struik, K. G., Schellekens, P.H.J., 1997, Development of a 2D Probing System with Nanometer Resolution, Proceedings of ASPE, 16:438-442.

497 Vermeulen, M.M.P.A., Rosielle, P.C.J.N., Schellekens, P.H.J., 1998, Design of a High-Precision 3D-Coordinate Measuring Machine, CIRP Annals, 47/1:447-450.

498 Kramar, J. A., Jun, J. S., Penzes, W. B., Scire, F. E., Teague, E.C., Villarrubia, J.S., 1999, Grating Pitch Measurements with the Molecular Measuring Machine, Proceedings of SPIE - The International Society for Optical Engineering, 3806:46-53.

499 Peggs, G. N., Lewis, A. J., Oldfield, S., 1999, Design for a Compact High-Accuracy CMM, CIRP Annals, 48/1:417-420.

500 Shiozawa, H., Fukutomi, Y., Ushioda, T., Yoshimura, S., 1998, Development of Ultra-Precision 3D-CMM Based on 3-D Metrology Frame, Proceedings of ASPE, 18:15-18.

501 Jäger, G., Manske, E., Hausotte, T., Büchner, H.-J., 2000, Laserinterferometrische Nanomessmaschinen, VDI Berichte, 1530 (Sensoren und Messsysteme 2000), VDI Verlag GmbH, Düsseldorf: 271-278.

502 Brand, U., Cao, S., Kleine-Besten, T., Hoffmann, W., Schwenke, H., Butefisch, S., Buttgenbach, S., 2002, Recent Developments in Dimensional Metrology for Microsystem Components, Microsystem Technologies, 8/1:3-6.

503 Cao, S., Hassler-Grohne, W., Brand, U., Gao, S., Wilke, R., Buttgenbach, S., 2002, A Three Dimensional Measurement System

with Micro Tactile Sensor, Proceedings of SPIE - The International Society for Optical Engineering, 4902:52-59.

504 Grigg, D., Felkel, E., Roth, J., De Lega, X. C., Deck, L., De Groot, P., 2004, Static and Dynamic Characterization of MEMS and MOEMS Devices Using Optical Interference Microscopy, Proceedings of SPIE - The International Society for Optical Engineering, 5455:429-435.

505 Byrne, G., Dornfeld, D., and Denkena, B., "Advancing Cutting Technology," STC "C" Keynote, CIRP Annals, 52, 2, 2003.

506 Berger, K., "An Overview of Status and Trends in the Automotive Industry," 7th Int'l Conf. on Deburring and Surface Finishing, University of California, Berkeley, June, 2004.

507 Beier, H., "Handbuch Entgrattechnik," Hanser, München, 1999.

508 Kim, J., and Dornfeld, D., Kim, J. and Dornfeld, D., "Cost Estimation of Drilling Operation by a Drilling Burr Control Chart and Bayesian Statistics," SME J. Manufacturing Systems, 2001, Vol. 20, No. 2, pp. 89-97.

509 Berger, K., in "Burr Reduction Investment - Production Costs - Burr Reduction - Prediction of Burrs," Presentation at HPC Workshop, CIRP, Paris, January 23, 2002.

510 Gillespie, L., 1975, "Burrs Produced by Drilling," Bendix Corporation, Unclassified Topical Report, BDX-613-1248.

511 Pande, S. and Relakar, H., "Investigation on Reducing Burr Formation in Drilling," Int. J. Mach. Tool Des. Res., 26, 3, 1986, pp. 339-348.

512 Dornfeld, D., Kim, J., Dechow, H., Hewson, J., and Chen, L., "Drilling Burr Formation in Titanium Alloy, Ti-6AL-4V," Annals of the CIRP, 48, 1, 1999, pp. 73-76.

513 Kim, J, and Dornfeld, D., "Development of an Analytical Model for Drilling Burr Formation Ductile Materials," Trans. ASME, J. Eng. Mats Tech., 124, 2, 2002, pp. 192-198.

514 Min, S. Dornfeld, D., Kim, J., and Shyu, B., "Finite element modeling of burr formation in metal cutting," Proc. 4th CIRP Int'l Workshop on Modeling of Machining Operations, Delft, August, 2001, pp. 97-104.

515 Min, S., Dornfeld, D., Kim, J., and Shyu, B., "Finite element modeling of burr formation in metal cutting," Machining Science and Technology, 5(3), 2001, pp. 307-322.

516 Dornfeld, D., "Burr minimization strategies as part of process planning," Presentation at HPC Workshop, CIRP, San Sebastian, August 22, 2002.

517 Choi, J., "Formation of burrs when drilling multi-layer materials," Presented at CODEF Annual Meeting, UC-Berkeley, 2003.

518 Rangarajan, A., Chu, C. H., and Dornfeld, D. A., "Avoiding Tool Exit in Planar Milling by Adjusting Width of Cut," Proc. ASME IMECE, ASME, MED-Vol. 11, 2000, pp. 1017-1025.

519 Kim, J. Min, S. and Dornfeld, D., "Optimization and Control of Drilling Burr Formation of AISI 304L and AISI 4118 Based on Drilling Burr Control Charts," Int. J. of Machine Tools and Manufacture, 2001, 41, 7, pp. 923-936.

520 Kim, J. and Dornfeld, D., "Cost Minimization of Drilling Operation by a Drilling Burr Control Chart and Bayesian Statistics," SME J. Manufacturing Systems, 2001, Vol. 20, No. 2, pp. 89-97.

521 Rangarajan, A., Dornfeld, D., and Wright, P., "Probabilistic Precision Process Planning - P4," Trans. NAMRC/SME, 31, 2003, 539-546.

522 Rangarajan, A., and Dornfeld, D., "Efficient Tool Paths and Part Orientation for Face Milling," CIRP Annals, 53, 1, 2004, pp. 73-76.

523 Hawken, P. (1993), *The Ecology of Commerce*, Harper Collins, New York.

524 Allen, D., Bauer, D., Bras. B., Gutkowski, T., Murphy, C., Piwonka, T., Sheng, P., Sutherland, J., Thurston, D., and Wolff, E. (2002), "Environmentally Benign Manufacturing: Trends in Europe, Japan and USA., Trans. ASME, J. Manufacturing Science and Engineering, 124, pp. 908-920.

525 WTEC (2000), *Environmentally Benign Manufacturing*, WTEC Panel Report, Baltimore, MD, Loyola College.

526 Sheng, P., Dornfeld, D. and Worhach, P. (1995), "Integration Issues in Green Design and Manufacturing," *Manufacturing Review*, Vol. 8, No. 2, pp. 95-105.

527 Anastas, P., and Zimmerman, J. (2003), "Design through the 12 principles of green engineering," Environ. Sci. Technol. A-Pages; 37(5); 94A-101A.

528 Filipovic, A., and Sutherland, J. (2005), "Development of Magnetostrictive Tool-Holder for Dry Deep Hole Drilling," *Trans. North American Manufacturing Research Institute*, Vol. 33.

529 Zimmerman, J., Hayes, K., and Skerlos, S. (2003), 2003, "Influence of Ion Type and Concentration on the Emulsion Stability and Machining Performance of Two Semi-Synthetic Metalworking Fluids", Environmental Science and Technology, Vol. 38, pp. 2482-2490.

530 Ishii, K. (1999), "Incorporating End-of-Life Strategy in Product Definition," EcoDesign '99: First Int'l Symp. on Environmentally Conscious Design and Inverse Manufacturing, Tokyo, Japan.

531 http://www.life-cycle.org/, accessed June, 2007.

532 Taniguchi, M., Kakinuma, Y., Aoyama, T. and Inasaki, I. (2006), "Influences of Downsized Machine Tools on the Environment," Proc. MTTRF 2006 Meeting, San Francisco.

533 Srinivasan, M. and Sheng, P. (1999a), "Feature-based process planning for environmentally conscious machining - Part 1: microplanning," *Robotics and Computer Integrated Manufacturing*, 15, pp. 257-270.

534 Srinivasan, M. and Sheng, P. (1999b), "Feature-based process planning for environmentally conscious machining - Part 2: macroplanning," *Robotics and Computer Integrated Manufacturing*, 15, pp. 271-281.

535 Hendersen, C., Lave, S., and Matthews, S. (2006), *Environmental Life Cycle Assessment of Goods and Services*, Resources for the Future, Washington, DC.

536 Williams E. D., Ayres R. U., Heller M. (2002). "The 1.7 Kilogram Microchip: Energy and Material Use in the Production of Semiconductor Devices." *Environmental Science and Technology*, 36 (24), pp. 5504-5510.

537 Krishnan, N., Raoux, S., and Dornfeld, D. (2004), Quantifying the Environmental Footprint of Semiconductor Process Equipment using the Environmental Value Systems (EnV-S) Analysis, *IEEE Trans. Semiconductor Manufacturing*, 17, 4, pp. 554-561.

538 Dornfeld, D., and Wright, P. (1997), "Process planning for agent-based precision systems," *Trans. NAMRC*, SME, Vol. 25. pp. 359-364.

539 Dornfeld, D., Wright, P., Wang, F., Sheng, P. Stori, J., Sundarajaran, V., Krishnan, N, and Chu, C. (1999), "Multi-agent

Process Planning for a Networked Machining Service," *Trans. NAMRC*, pp. 191-196.

540 Our Common Future (1987), Oxford: Oxford University Press. ISBN 0-19-282080-X.

541 Pacala, S., and Socolow, R. (2004), "Stabilization Wedges: Solving the Climate Problem for the Next 50 Years with Current Technologies," *Science*, 305: pp. 968-972.

542 Dornfeld, D. and Wright, P., "Technology Wedges for Implementing Green Manufacturing," Trans. North American Manufacturing Research Institute, 2007, Vol. 35, pp. 193-200.

543 Interface Carpet sustainability website: www.interfacesustainability.com/, accessed January, 2007.

544 http://boeing.com/commercial/787/family/, accessed June, 26, 2007.

545 Kaya, A. D., Phelan, P., Chau, D., Sarac, H. I., 2002, "Energy conservation in compressed air systems," International Journal of Energy Research, Vol. 26, pp. 837-849.

546 Curtner, K. L., O'Neill, P. J., Winter, D., Bursch, P., 1997 "Simulation-Based Features of the Compressed Air System Description Tool, XCEEDTM," Proceedings of International Building Performance Simulation Association conference, Prague, Czech Republic, Sept. 8-10.

547 Cox, R., 1996 "Compressed air - clean energy in a green world," Glass International. Vol. 19, no. 2, pp. 2.

548 Yuan, C., "A Decision-based Analysis of Compressed Air Usage Patterns in Automotive Manufacturing," Ford Motor Company,

Seminar, Dearborn MI, April 27, 2005 and SME J. Manufacturing Systems.

549 NNI, "What is nano," http://www.nano.gov/ html/facts/ whatIsNano.html, accessed Oct. 17, 2005, 2005.

550 Zhang, T., Boyd, S., Vijayaraghavan, A., and Dornfeld, D., "Energy Use in Nanoscale Manufacturing," Proc. 2006 IEEE International Symposium on Electronics and Environment (ISEE), IEEE, May, 2006.

551 S. Lloyd, L. B. Lave, and H. S. Matthews, "Life cycle benefits of using nanotechnology to stabilize platinum-group metal particles in automotive catalysts," *Environmental Science & Technology*, Vol. 39, pp. 1384-1392, 2005.

552 S. Lloyd and L. Lave, "Life cycle economic and environmental implications of using nanocomposites in automobiles," *Environmental Science & Technology*, Vol. 37, No. 15, pp. 3458-3466, 2003.

553 A. Tseng and A. Notargiacomo, "Nanoscale fabrication by nonconventional approaches," *Journal of Nanoscience and Nanotechnology*, Vol. 5, pp. 683-702, 2005.

554 Y. Chen and A. Pépin, "Nanofabrication: Conventional and nonconventional methods," *Electrophoresis*, Vol. 22, pp. 187-207, 2001.

555 C. F. Murphy, G. Kenig, D. T. Allen, J. Laurent, and D. Dyer, "Development of parametric material, energy, and emission inventories for wafer fabrication in the semiconductor industry," Environmental Science & Technology, Vol. 37, No. 23, pp. 5373-5382, 2003.

556 E. Williams, R. Ayres, and M. Heller, "The 1.7 kilogram microchip: Energy and material use in the production of semiconductor devices," Environmental Science & Technology, Vol. 36, pp. 5504-5510, 2002.

557 E. Williams, "Energy intensity of computer manufacturing: Hybrid assessment combining process and economic input-output methods," Environmental Science & Technology, Vol. 38, pp. 6166-6174, 2004.

558 N. Krishnan, S. Raoux, and D. Dornfeld, "Quantifying the environmental footprint of semiconductor equipment using the environmental value systems analysis (env-s)," IEEE Transactions on Semiconductor Manufacturing, Vol. 17, pp. 554-561, 2004.

559 Ashby, M., Shercliff, H. and Cebon, D., "*Materials: Engineering, Science, Processing and Design*," Butterworth-Heinemann/ Elsevier, 2007.

560 F. Jovane, et al., "Present and Future of Flexible Automation: Towards New Paradigms," CIRP Annals, 52, 2, 2003, 543.

INDEX